Duncan Buell (Ed.)

Algorithmic Number Theory

6th International Symposium, ANTS-VI
Burlington, VT, USA, June 13-18, 2004
Proceedings

Springer

Volume Editor

Duncan Buell
University of South Carolina
Department of Computer Science and Engineering
Columbia, SC 29208, USA
E-mail: buell@cse.sc.edu

Library of Congress Control Number: 2004106274

CR Subject Classification (1998): F.2, G.2, E.3, I.1

ISSN 0302-9743
ISBN 3-540-22156-5 Springer-Verlag Berlin Heidelberg New York

Springer-Verlag is a part of Springer Science+Business Media

springeronline.com

Printed in Germany

Typesetting: Camera-ready by author, data conversion by PTP-Berlin, Protago-TeX-Production GmbH
Printed on acid-free paper SPIN: 11012283 06/3142 5 4 3 2 1 0

Lecture Notes in Computer Science 3076

Springer
Berlin
Heidelberg
New York
Hong Kong
London
Milan
Paris
Tokyo

Preface

The sixth Algorithmic Number Theory Symposium was held at the University of Vermont, in Burlington, from 13–18 June 2004. The organization was a joint effort of number theorists from around the world.

There were four invited talks at ANTS VI, by Dan Bernstein of the University of Illinois at Chicago, Kiran Kedlaya of MIT, Alice Silverberg of Ohio State University, and Mark Watkins of Pennsylvania State University. Thirty contributed talks were presented, and a poster session was held. This volume contains the written versions of the contributed talks and three of the four invited talks. (Not included is the talk by Dan Bernstein.)

ANTS in Burlington is the sixth in a series that began with ANTS I in 1994 at Cornell University, Ithaca, New York, USA and continued at Université Bordeaux I, Bordeaux, France (1996), Reed College, Portland, Oregon, USA (1998), the University of Leiden, Leiden, The Netherlands (2000), and the University of Sydney, Sydney, Australia (2002). The proceedings have been published as volumes 877, 1122, 1423, 1838, and 2369 of Springer-Verlag's *Lecture Notes in Computer Science* series.

The organizers of the 2004 ANTS conference express their special gratitude and thanks to John Cannon and Joe Buhler for invaluable behind-the-scenes advice.

June 2004 Duncan Buell

ANTS VI Organizing Committee

Duncan Buell (University of South Carolina)
David S. Dummit (University of Vermont)
David Joyner (United States Naval Academy)
Blair Kelly III (United States Department of Defense)
Jonathan W. Sands (University of Vermont)

Program Committee

Duncan Buell (University of South Carolina, USA, Chair)
Eric Bach (University of Wisconsin, USA)
Johannes Buchmann (Darmstadt University, Germany)
Richard Brent (Oxford University, UK)
Henri Cohen (Université de Bordeaux I, France)
Everett Howe (Center for Communications Research, La Jolla, California, USA)
David Joyner (United States Naval Academy, USA)
Bjorn Poonen (University of California at Berkeley, USA)
Renate Scheidler (University of Calgary, Alberta, Canada)
Igor Shparlinski (Macquarie University, Sydney, Australia)
Jonathan Sorenson (Butler University, Indianapolis, Indiana, USA)

Table of Contents

Invited Talks

Contributed Papers

Computing Zeta Functions via p-Adic Cohomology

Kiran S. Kedlaya*

Department of Mathematics
Massachusetts Institute of Technology
77 Massachusetts Avenue
Cambridge, MA 02139
kedlaya@math.mit.edu
http://math.mit.edu/~kedlaya/

Abstract. We survey some recent applications of p-adic cohomology to machine computation of zeta functions of algebraic varieties over finite fields of small characteristic, and suggest some new avenues for further exploration.

1 Introduction

1.1 The Zeta Function Problem

For X an algebraic variety over $\mathbb{F}_q$ (where we write $q = p^n$ for p prime), the zeta function

$$Z(X,t) = \exp\left(\sum_{i=1}^{\infty} \frac{t^i}{i} \#X(\mathbb{F}_{q^i})\right)$$

is a rational function of t. This fact, the first of the celebrated Weil Conjectures, follows from Dwork's proof using p-adic analysis [12], or from the properties of étale (ℓ-adic) cohomology (see [14] for an introduction).

In recent years, the algorithmic problem of determining $Z(X,t)$ from defining equations of X has come into prominence, primarily due to its relevance in cryptography. Namely, to perform cryptographic functions using the Jacobian group of a curve over $\mathbb{F}_q$, one must first compute the order of said group, and this is easily retrieved from the zeta function of the curve (as $Q(1)$, where $Q(t)$ is as defined below). However, the problem is also connected with other applications of algebraic curves (e.g., coding theory) and with other computational problems in number theory (e.g., determining Fourier coefficients of modular forms).

Even if one restricts X to being a curve of genus g, in which case

$$Z(X,t) = \frac{Q(t)}{(1-t)(1-qt)}$$

* Thanks to Michael Harrison, Joe Suzuki, and Fré Vercauteren for helpful comments, and to David Savitt for carefully reading an early version of this paper.

D.A. Buell (Ed.): ANTS 2004, LNCS 3076, pp. 1–17, 2004.

with $Q(t)$ a polynomial over $\mathbb{Z}$ of degree $2g$, there is no algorithm known[1] for computing $Z(X,t)$ which is polynomial in the full input size, i.e., in g, n, and $\log(p)$. However, if one allows polynomial dependence in p rather than its logarithm, then one can obtain a polynomial time algorithm using Dwork's techniques, as shown by Lauder and Wan [30]. The purpose of this paper is to illustrate how these ideas can be converted into more practical algorithms in many cases.

This paper has a different purpose in mind than most prior and current work on computing zeta functions, which has been oriented towards low-genus curves over large fields (e.g., elliptic curves of "cryptographic size"). This problem is well under control now; however, we are much less adept at handling curves of high genus or higher dimensional varieties over small fields. It is in this arena that p-adic methods shoud prove especially valuable; our hope is for this paper, which mostly surves known algorithmic results on curves, to serve as a springboard for higher-genus and higher-dimensional investigations.

1.2 The Approach via p-Adic Cohomology

Historically, although Dwork's proof predated the advent of ℓ-adic cohomology, it was soon overtaken as a theoretical tool[2] by the approach favored by the Grothendieck school, in which context the Weil conjectures were ultimately resolved by Deligne [8]. The purpose of this paper is to show that by contrast, from an algorithmic point of view, "Dworkian" p-adic methods prove to be much more useful.

A useful analogy is the relationship between topological and algebraic de Rham cohomology of varieties over $\mathbb{C}$. While the topological cohomology is more convenient for proving basic structural results, computations are often more convenient in the de Rham setting, since it is so closely linked to defining equations. The analogy is more than just suggestive: the p-adic constructions we have in mind are variants of and closely related to algebraic de Rham cohomology, from which they inherit some computability.

1.3 Other Computational Approaches

There are several other widely used approaches for computing zeta functions; for completeness, we briefly review these and compare them with the cohomological point of view.

The method of Schoof [45] (studied later by Pila [40] and Adleman-Huang [1]) is to compute the zeta function modulo ℓ for various small primes ℓ, then apply bounds on the coefficients of the zeta function plus the Chinese remainder theorem. This loosely corresponds to computing in ℓ-adic and not p-adic cohomology.

[1] That is, unless one resorts to quantum computation: one can imitate Shor's quantum factoring algorithm to compute the order of the Jacobian over $\mathbb{F}_{q^n}$ for n up to about $2g$, and then recover $Z(X,t)$. See [26].

[2] The gap has been narrowed recently by the work of Berthelot and others; for instance, in [25], one recovers the Weil conjectures by imitating Deligne's work using p-adic tools.

This has the benefit of working well even in large characteristic; on the downside, one can only treat curves, where ℓ-adic cohomology can be reinterpreted in terms of Jacobian varieties, and moreover, one must work with the Jacobians rather concretely (to extract division polynomials), which is algorithmically unwieldy. In practice, Schoof's method has only been deployed in genus 1 (by Schoof's original work, using improvements by Atkin, Elkies, Couveignes-Morain, etc.) and genus 2 (by work of Gaudry and Harley [17], with improvements by Gaudry and Schost [18]).

A more p-adic approach was given by Satoh [43], based on iteratively computing the Serre-Tate canonical lift [46] of an ordinary abelian variety, where one can read off the zeta function from the action of Frobenius on the tangent space at the origin. A related idea, due to Mestre, is to compute "p-adic periods" using a variant of the classical AGM iteration for computing elliptic integrals. This method has been used to set records for zeta function computations in characteristic 2 (e.g., [33]). The method extends in principle to higher characteristic [27] and genus (see [41], [42] for the genus 3 nonhyperelliptic case), but it seems difficult to avoid exponential dependence on genus and practical hangups in handling not-so-small characteristics.

We summarize the comparison between these approaches in the following table. (The informal comparison in the n column is based on the case of elliptic curves of a fixed small characteristic.)

Table 1. Comparison of strategies for computing zeta functions

Algorithm class	Applicability	Dependence on: p	n	g
Schoof	curves	polylog	big polynomial	at least exponential
Canonical lift/AGM	curves	polynomial	small polynomial	at least exponential
p-adic cohomology	general	nearly linear	medium polynomial	polynomial

2 Some p-Adic Cohomology

In this section, we briefly describe some constructions of p-adic cohomology, amplifying the earlier remark that it strongly resembles algebraic de Rham cohomology.

2.1 Algebraic de Rham Cohomology

We start by recalling how algebraic de Rham cohomology is constructed. First suppose $X = \operatorname{Spec} A$ is a smooth affine variety[3] over a field K of characteristic zero. Let $\Omega^1_{A/K}$ be the module of Kähler differentials, and put $\Omega^i_{A/K} = \wedge^i \Omega^1_{A/K}$;

[3] By "variety over K" we always mean a separated, finite type K-scheme.

these are finitely generated locally free A-modules since X is smooth. By a theorem of Grothendieck [21], the cohomology of the complex $\Omega^i_{A/K}$ is finite dimensional.

If X is smooth but not necessarily affine, one has similar results on the sheaf level. That is, the hypercohomology of the complex formed by the sheaves of differentials is finite dimensional. In fact, Grothendieck proves his theorem first when X is smooth and proper, where the result follows by a comparison theorem to topological cohomology (via Serre's GAGA theorem), then uses resolution of singularities to deduce the general case.

For general X, one can no longer use the modules of differentials, as they fail to be coherent. Instead, following Hartshorne [22], one (locally) embeds X into a smooth scheme Y, and computes de Rham cohomology on the formal completion of Y along X.

As one might expect from the above discussion, it is easiest to compute algebraic de Rham cohomology on a variety X if one is given a good compactification $\overline{X}$, i.e., a smooth proper variety such that $\overline{X} \setminus X$ is a normal crossings divisor. Even absent that, one can still make some headway by computing with $\mathcal{D}$-modules (where $\mathcal{D}$ is a suitable ring of differential operators), as shown by Oaku, Takayama, Walther, et al. (see for instance [52]).

2.2 Monsky-Washnitzer Cohomology

We cannot sensibly work with de Rham cohomology directly in characteristic p, because any derivation will kill p-th powers and so the cohomology will not typically be finite dimensional. Monsky and Washnitzer [38], [36], [37] (see also [49]) introduced a p-adic cohomology which imitates algebraic de Rham cohomology by lifting the varieties in question to characteristic zero in a careful way.

Let $X = \operatorname{Spec} A$ be a smooth affine variety over a finite field $\mathbb{F}_q$ with $q = p^n$, and let W be the ring of Witt vectors over $\mathbb{F}_q$, i.e., the unramified extension of $\mathbb{Z}_p$ with residue field $\mathbb{F}_q$. By a theorem of Elkik [13], we can find a smooth affine scheme $\tilde{X}$ over W such that $\tilde{X} \times_W \mathbb{F}_q \cong X$. While $\tilde{X}$ is not determined by X, we can "complete along the special fibre" to get something more closely bound to X.

Write $\tilde{X} = \operatorname{Spec} \tilde{A}$ and let $A^\dagger$ be the *weak completion* of $\tilde{A}$, which is the smallest subring containing $\tilde{A}$ of the p-adic completion of $\tilde{A}$ which is p-adically saturated (i.e., if $px \in A^\dagger$, then $x \in A^\dagger$) and closed under the formation of series of the form

$$\sum_{i_1,\dots,i_m \geq 0} c_{i_1,\dots,i_m} x_1^{i_1} \cdots x_m^{i_m}$$

with $c_{i_1,\dots,i_m} \in W$ and $x_1, \dots, x_m \in pA^\dagger$. We call $A^\dagger$ the *(integral) dagger algebra* associated to X; it is determined by X, but only up to *noncanonical* isomorphism.

In practice, one can describe the weak completion a bit more concretely, as in the following example.

Lemma 1. *The weak completion of* $W[t_1, \dots, t_n]$ *is the ring* $W\langle t_1, \dots, t_n\rangle^\dagger$ *of power series over* W *which converge for* $t_1, \dots, t_n$ *within the disc (in the integral closure of* W*) around* 0 *of some radius greater than* 1.

In general, $A^\dagger$ is always a quotient of $W\langle t_1, \dots, t_n\rangle^\dagger$ for some n.

We quickly sketch a proof of this lemma. On one hand, $W\langle t_1, \dots, t_n\rangle^\dagger$ (which is clearly p-adically saturated) is weakly complete: if $x_1, \dots, x_m \in pW\langle t_1, \dots, t_n\rangle^\dagger$, then for $t_1, \dots, t_n$ in some disc of radius strictly greater than 1, the series defining $x_1, \dots, x_m$ converge to limits of norm less than 1, and so $\sum c_{i_1,\dots,i_m} x_1^{i_1} \cdots x_m^{i_m}$ converges on the same disc. On the other hand, any element of $W\langle t_1, \dots, t_n\rangle^\dagger$ has the form

$$\sum_{j_1,\dots,j_n \geq 0} e_{j_1,\dots,j_n} t_1^{j_1} \cdots t_n^{j_n}$$

where $v_p(e_{j_1,\dots,j_n}) + a(j_1 + \cdots + j_n) > -b$ for some a, b with $a > 0$ (but no uniform choice of a, b is possible). We may as well assume that $1/a$ is an integer, and that $b > 0$ (since the weak completion is saturated). Then it is possible to write this series as

$$\sum_{i_1,\dots,i_m \geq 0} c_{i_1,\dots,i_m} x_1^{i_1} \cdots x_m^{i_m}$$

where the x's run over $p^j t_k$ for $j = 1, \dots, 1/a$ and $k = 1, \dots, n$; hence it lies in the weak completion.

The module of continuous differentials over $A^\dagger$ can be constructed as follows: given a surjection $W\langle t_1, \dots, t_n\rangle^\dagger \to A^\dagger$, $\Omega^1_{A^\dagger}$ is the $A^\dagger$-module generated by $dt_1, \dots, dt_n$ modulo enough relations to obtain a well-defined derivation $d : A^\dagger \to \Omega^1_{A^\dagger}$ satisfying the rule

$$d\left(\sum_{j_1,\dots,j_n \geq 0} c_{j_1,\dots,j_n} t_1^{j_1} \cdots t_n^{j_n}\right) = \sum_{i=1}^n \sum_{j_1,\dots,j_n \geq 0} j_i e_{j_1,\dots,j_n} (t_1^{j_1} \cdots t_i^{j_i - 1} \cdots t_n^{j_n}) dt_i.$$

Then the Monsky-Washnitzer cohomology (or MW-cohomology) $H^i_{\mathrm{MW}}(X)$ of X is the cohomology of the “de Rham complex”

$$\cdots \xrightarrow{d} \Omega^i_{A^\dagger} \otimes_W W[\frac{1}{p}] \xrightarrow{d} \cdots,$$

where $\Omega^i_{A^\dagger} = \wedge^i_{A^\dagger} \Omega^i_{A^\dagger}$. Implicit in this definition is the highly nontrivial fact that this cohomology is independent of all of the choices made. Moreover, if $X \to Y$ is a morphism of $\mathbb{F}_q$-varieties, and $A^\dagger$ and $B^\dagger$ are corresponding dagger algebras, then the morphism lifts to a ring map $B^\dagger \to A^\dagger$, and the induced maps $H^i_{\mathrm{MW}}(Y) \to H^i_{\mathrm{MW}}(X)$ do not depend on the choice of the ring map. The way this works (see [38] for the calculation) is that there are canonical homotopies (in the homological algebra sense) between any two such maps, on the level of the de Rham complexes.

MW-cohomology is always finite dimensional over $W[\frac{1}{p}]$; this follows from the analogous statement in rigid cohomology (see [5]). Moreover, it admits an analogue of the Lefschetz trace formula for Frobenius: if X is purely of dimension d, and $F : A^\dagger \to A^\dagger$ is a ring map lifting the q-power Frobenius map, then for all $m > 0$,

$$\#X(\mathbb{F}_{q^m}) = \sum_{i=0}^{d} (-1)^i \operatorname{Trace}(q^{dm} F^{-m}, H^i_{\mathrm{MW}}(X)).$$

This makes it possible in principle, and ultimately in practice, to compute zeta functions by computing the action of Frobenius on MW-cohomology.

2.3 Rigid Cohomology

As in the algebraic de Rham setting, it is best to view Monsky-Washnitzer cohomology in the context of a theory not limited to affine varieties. This context is provided by Berthelot's rigid cohomology; since we won't compute directly on this theory, we only describe it briefly. See [4] or [19, Chapter 4] for a somewhat more detailed introduction.[4]

Suppose X is an $\mathbb{F}_q$-variety which is the complement of a divisor in a smooth proper Y which lifts to a smooth proper formal W-scheme. Then this lift gives rise to a rigid analytic space Y^{an} via Raynaud's "generic fibre" construction (its points are the subschemes of the lift which are integral and finite flat over W). This space comes with a specialization map to Y, and the inverse image of X is denoted $]X[$ and called the *tube* of X. The rigid cohomology of X is the (coherent) cohomology of the direct limit of the de Rham complexes over all "strict neighborhoods" of $]X[$ in Y^{an}. (Within Y^{an}, $]X[$ is the locus where certain functions take p-adic absolute values less than or equal to 1; to get a strict neighborhood, allow their absolute values to be less than or equal to $1+\epsilon$ for some $\epsilon > 0$.)

For general X, we can do the above locally (e.g., on affines) and compute hypercohomology via the usual spectral sequence; while the construction above does not sheafify, the complexes involves can be glued "up to homotopy", which is enough to assemble the hypercohomology spectral sequence.

For our purposes, the relevance of rigid cohomology is twofold. On one hand, it coincides with Monsky-Washnitzer cohomology for X affine. On the other hand, it is related to algebraic de Rham cohomology via the following theorem. (This follows, for instance, from the comparison theorems of [5] plus the comparison theorem between crystalline and de Rham cohomology from [3].)

Theorem 1. *Let $\tilde{Y}$ be a smooth proper W-scheme, let $\tilde{Z} \subset \tilde{Y}$ be a relative normal crossings divisor, and set $\tilde{X} = \tilde{Y} \setminus \tilde{Z}$. Then there is a canonical isomorphism*

$$H^i_{\mathrm{dR}}(\tilde{X} \times_W (\operatorname{Frac} W)) \to H^i_{\mathrm{rig}}(\tilde{X} \times_W \mathbb{F}_q).$$

[4] We confess that a presentation at the level of detail we would like does not appear in print anywhere. Alas, these proceedings are not the appropriate venue to correct this!

In particular, if X is affine in this situation, its Monsky-Washnitzer cohomology is finite dimensional and all of the relations are explained by relations among algebraic forms, i.e., relations of finite length. This makes it much easier to construct "reduction algorithms", such as those described in the next section.

One also has a comparison theorem between rigid cohomology and crystalline cohomology, a p-adic cohomology built in a more "Grothendieckian" manner. While crystalline cohomology only behaves well for smooth proper varieties, it has the virtue of being an *integral* theory. Thus the comparison to rigid cohomology equips the latter with a canonical integral structure. By repeating this argument in the context of log-geometry, one also obtains a canonical integral structure in the setting of Theorem 1; this is sometimes useful in computations.

3 Hyperelliptic Curves in Odd Characteristic

The first[5] class of varieties where p-adic cohomology was demonstrated to be useful for numerical computations is the class of hyperelliptic curves in odd characteristic, which we considered in [24]. In this section, we summarize the key features of the computation, which should serve as a prototype for more general considerations.

3.1 Overview

An overview of the computation may prove helpful to start with. The idea is to compute the action of Frobenius on the MW-cohomology of an affine hyperelliptic curve, and use the Lefschetz trace formula to recover the zeta function. Of course we cannot compute exactly with infinite series of p-adic numbers, so the computation will be truncated in both the series and p-adic directions, but we arrange to keep enough precision at the end to uniquely determine the zeta function.

Besides worrying about precision, carrying out this program requires making algorithmic two features of the Monsky-Washnitzer construction.

- We must be able to compute a Frobenius lift on a dagger algebra.
- We must be able to identify differentials forming a basis of the relevant cohomology space, and to "reduce" an arbitrary differential to a linear combination of the basis differentials plus an exact differential.

3.2 The Dagger Algebra and the Frobenius Lift

Suppose that $p \neq 2$, and let $\overline{X}$ be the hyperelliptic curve of genus g given by the affine equation

$$y^2 = P(x)$$

[5] Although this seems to be the first overt use of MW-cohomology for numerically computing zeta functions in the literature, it is prefigured by work of Kato and Lubkin [23]. Also, similar computations appear in more theoretical settings, such as Gross's work on companion forms [20].

with $P(x)$ monic of degree $2g+1$ over $\mathbb{F}_q$ with no repeated roots; in particular, $\overline{X}$ has a rational Weierstrass point[6] at infinity. Let X be the affine curve obtained from $\overline{X}$ by removing all of the Weierstrass points, i.e., the point at infinity and the zeroes of y.

Choose a lift $\tilde{P}(x)$ of $P(x)$ to a monic polynomial of degree $2g+1$ over W. Then the dagger algebra corresponding to X is given by

$$W\langle x,y,z\rangle^\dagger/(y^2-\tilde{P}(x), yz-1),$$

whose elements can be expressed as $\sum_{i\in\mathbb{Z}} A_i(x)y^i$ with $A_i(x)\in W[x]$, $\deg(A_i)\leq 2g$, and $v_p(A_i)+c|i|>d$ for some constants c,d with $c>0$.

The dagger algebra admits a p-power Frobenius lift σ given by

$$\begin{aligned} x &\mapsto x^p \\ y &\mapsto y^p\left(1+\frac{\tilde{P}(x)^\sigma-\tilde{P}(x)^p}{\tilde{P}(x)^p}\right)^{1/2}, \end{aligned}$$

which can be computed by a Newton iteration. Here is where the removal of the Weierstrass points come in handy; the simple definition of σ above clearly requires inverting $\tilde{P}(x)$, or equivalently y. It is possible to compute a Frobenius lift on the dagger algebra of the full affine curve (namely $W\langle x,y\rangle^\dagger/(y^2-\tilde{P}(x))$), but this requires solving for the images of both x and y, using a cumbersome two-variable Newton iteration.

3.3 Reduction in Cohomology

The hyperelliptic curve defined by $y^2=\tilde{P}(x)$, minus its Weierstrass points, forms a lift $\tilde{X}$ of X of the type described in Theorem 1, so its algebraic de Rham cohomology coincides with the MW-cohomology $H^1_{\mathrm{MW}}(X)$. That is, the latter is generated by

$$\frac{x^i dx}{y} \quad (i=0,\ldots,2g-1), \qquad \frac{x^i dx}{y^2} \quad (i=0,\ldots,2g)$$

and it is enough to consider "algebraic" relations. Moreover, the cohomology splits into plus and minus eigenspaces for the hyperelliptic involution $y\mapsto -y$; the former is essentially the cohomology of $\mathbb{P}^1$ minus the images of the Weierstrass points (since one can eliminate y entirely), so to compute the zeta function of $\overline{X}$ we need only worry about the latter. In other words, we need only consider forms $f(x)dx/y^s$ with s odd.

The key reduction formula is the following: if $A(x)=\tilde{P}(x)B(x)+\tilde{P}'(x)C(x)$, then

$$\frac{A(x)\,dx}{y^s}\equiv\left(B(x)+\frac{2C'(x)}{s-2}\right)\frac{dx}{y^{s-2}}$$

[6] The case of no rational Weierstrass point is not considered in [24]; it has been worked out by Michael Harrison, and has the same asymptotics.

as elements of $H^1_{\mathrm{MW}}(X)$. This is an easy consequence of the evident relation

$$d\left(\frac{C(x)}{y^{s-2}}\right) \equiv 0$$

in cohomology.

We use this reduction formula as follows. Compute the image under Frobenius of $\frac{x^i dx}{y}$ (truncating large powers of y or y^{-1}, and p-adically approximating coefficients). If the result is

$$\sum_{j=-M}^{N} \frac{A_j(x)\,dx}{y^{2j+1}},$$

use the reduction formula to eliminate the $j = N$ term in cohomology, then the $j = N-1$ term, and so on, until no terms with $j > 0$ remain. Do likewise with the $j = -M$ term, the $j = -M+1$ term, and so on (using a similar reduction formula, which we omit; note that there are relatively few terms on that side anyway). Repeat for $i = 0, \ldots, 2g-1$, and construct the "matrix of the p-power Frobenius" Φ. Of course the p-power Frobenius is not linear, but the matrix of the q-power Frobenius is easily obtained as $\Phi^{\sigma^{n-1}} \cdots \Phi^{\sigma}\Phi$, where σ here is the Witt vector Frobenius and $q = p^n$.

3.4 Precision

We complete the calculation described above with a p-adic approximation of a matrix whose characteristic polynomial would exactly compute the numerator $Q(t)$ of the zeta function. However, we can bound the coefficients of that numerator using the Weil conjectures: if $Q(t) = 1 + a_1 t + \cdots + a_{2g}t^{2g}$, then for $1 \leq i \leq g$, $a_{g+i} = q^i a_{g-i}$ and

$$|a_i| \leq \binom{2g}{i} q^{i/2}.$$

In particular, computing a_i modulo a power of p greater than twice the right side determines it uniquely.

As noted at the end of the previous section, it is critical to know how much p-adic precision is lost in various steps of the calculation, in order to know how much initial precision is needed for the final calculation to uniquely determine the zeta function. Rather than repeat the whole analysis here, we simply point out the key estimate [24, Lemmas 2 and 3] and indicate where it comes from.

Lemma 2. *For $A_k(x)$ a polynomial over W of degree at most $2g$ and $k \geq 0$ (resp. $k < 0$), the reduction of $A_k(x)y^{2k+1}\,dx$ (i.e., the linear combination of $x^i\,dx/y$ over $i = 0, \ldots, 2g-1$ cohomologous to it) becomes integral upon multiplication by p^d for $d \geq \log_p((2g+1)(k+1)-2)$ (resp. $d \geq \log_p(-2k-1)$).*

This is seen by considering the polar part of $A_k(x)y^{2k+1}\,dx$ around the point at infinity if $k < 0$, or the other Weierstrass points if $k > 0$. Multiplying by p^d ensures that the antiderivatives of the polar parts have integral coefficients, which forces the reductions to do likewise.

It is also worth pointing out that one can manage precision rather simply by working in p-adic fixed point arithmetic. That is, approximate all numbers modulo some fixed power of p, regardless of their valuation (in contrast to p-adic floating point, where each number is approximated by a power of p times a mantissa of fixed precision). When a calculation produces undetermined high-order digits, fill them in arbitrarily once, but do not change them later. (That is, if x is computed with some invented high-order digits, each invocation of x later must use the *same* invented digits.) The analysis in [24], using the above lemma, shows that most of these invented digits cancel themselves out later in the calculation, and the precision loss in the reduction process ends up being negligible compared to the number of digits being retained.

3.5 Integrality

In practice, it makes life slightly[7] easier if one uses a basis in which the matrix of Frobenius is guaranteed to have p-adically *integral* coefficients. The existence of such a basis is predicted by the comparison with crystalline cohomology, but an explicit good basis can be constructed "by hand" by careful use of Lemma 2. For instance, the given basis $x^i dx/y$ $(i = 0, \dots, 2g-1)$ is only good when $p > 2g+1$; on the other hand, the basis $x^i dx/y^3$ $(i = 0, \dots, 2g-1)$ is good for all p and g.

3.6 Asymptotics

As for time and memory requirements, the runtime analysis in [24] together with [15] show that the algorithm requires time $\tilde{O}(pn^3g^4)$ and space $\tilde{O}(pn^3g^3)$, where again g is the genus of the curve and $n = \log_p q$. (Here the "soft O" notation ignores logarithmic factors, arising in part from asymptotically fast integer arithmetic.)

4 Variations

In this section, we summarize some of the work on computing MW-cohomology for other classes of curves. We also mention some experimental results obtained from implementations of these algorithms.

[7] But only slightly: the fact that there is some basis on which Frobenius acts by an integer matrix means that the denominators in the product $\Phi^{\sigma^{n-1}} \cdots \Phi^{\sigma}\Phi$ can be bounded independently of n.

4.1 Hyperelliptic Curves in Characteristic 2

The method described in the previous section does not apply in characteristic 2, because the equation $y^2 = P(x)$ is nonreduced and does not give rise to hyperelliptic curves. Instead, one must view the hyperelliptic curve as an Artin-Schreier cover of $\mathbb{P}^1$ and handle it accordingly; in particular, we must lift somewhat carefully. We outline how to do this following Denef and Vercauteren [9], [10]. (Analogous computations based more on Dwork's work have been described by Lauder and Wan [31], [32], but they seem less usable in practice.)

Let $\overline{X}$ be a hyperelliptic curve of degree g over $\mathbb{F}_q$, with $q = 2^n$; it is defined by some plane equation of the form

$$y^2 + h(x)y = f(x),$$

where f is monic of degree $2g+1$ and $\deg(h) \leq g$. Let H be the monic squarefree polynomial over $\mathbb{F}_q$ with the same roots as h. By an appropriate substitution of the form $y \mapsto y + a(x)$, we can ensure that f vanishes at each root of H.

Let X be the affine curve obtained from $\overline{X}$ by removing the point at infinity and the zero locus of H. Choose lifts $\tilde{H}, \tilde{h}, \tilde{f}$ of H, h, f to polynomials over W of the same degree, such that each root of $\tilde{h}$ is also a root of $\tilde{H}$, and each root of $\tilde{f}$ whose reduction mod p is a root of H is also a root of $\tilde{H}$. The dagger algebra corresponding to X is now given by

$$W\langle x, y, z\rangle^\dagger / (y^2 + \tilde{h}(x)y - \tilde{f}(x), \tilde{H}(x)z - 1),$$

and each element can be written uniquely as

$$\sum_{i\in\mathbb{Z}} A_i(x)\tilde{H}_1(x)^i + \sum_{i\in\mathbb{Z}} B_i(x)y\tilde{H}_1(x)^i$$

with $\tilde{H}_1(x) = x$ if $\tilde{H}$ is constant and $\tilde{H}_1(x) = \tilde{h}(x)$ otherwise, $\deg(A_i) < \deg(\tilde{H}_1)$ and $\deg(B_i) < \deg(\tilde{H}_1)$ for all i, and $v_p(A_i) + c|i| > d$ and $v_p(B_i) + c|i| > d$ for some c, d with $c > 0$. The dagger algebra admits a Frobenius lift sending x to x^2, but this requires some checking, especially to get an explicit convergence bound; see [51, Lemma 4.4.1] for the analysis.

By Theorem 1, the MW-cohomology of X coincides with the cohomology of the hyperelliptic curve $y^2 + \tilde{h}(x)y - \tilde{f}(x)$ minus the point at infinity and the zero locus of $\tilde{H}$. Again, it decomposes into plus and minus eigenspaces for the hyperelliptic involution $y \mapsto -y - \tilde{h}(x)$, and only the minus eigenspace contributes to the zeta function of $\overline{X}$. The minus eigenspace is spanned by $x^i y\,dx$ for $i = 0, \dots, 2g-1$, there are again simple reduction formulae for expressing elements of cohomology in terms of this basis, and one can again bound the precision loss in the reduction; we omit details.

In this case, the time complexity of the algorithm is $\tilde{O}(n^3g^5)$ and the space complexity is $\tilde{O}(n^3g^4)$. If one restricts to ordinary hyperelliptic curves (i.e., those where H has degree g), the time and space complexities drop to $\tilde{O}(n^3g^4)$ and $\tilde{O}(n^3g^3)$, respectively, as in the odd characteristic case. It may be possible to optimize better for the opposite extreme case, where the curve has p-rank close to zero, but we have not tried to do this.

4.2 Other Curves

Several variations on the theme developed above have been pursued. For instance, Gaudry and Gürel [15] have considered superelliptic curves, i.e., those of the form

$$y^m = P(x)$$

where m is not divisible by p. More generally still, Denef and Vercauteren consider the class of $C_{a,b}$-curves, as defined by Miura [35]. For a, b coprime integers, a $C_{a,b}$-*curve* is one of the form

$$y^a + \sum_{i=1}^{a-1} f_i(x) y^i + f_0(x) = 0,$$

where $\deg f_0 = b$ and $a \deg f_i + bi < ab$ for $i = 1, \dots, a-1$, and the above equation has no singularities in the affine plane.

These examples fit into an even broader class of potentially tractable curves, which we describe following Miura [35]. Recall that for a curve C and a point P, the *Weierstrass monoid* is defined to be the set of nonnegative integers which occur as the pole order at P of some meromorphic function with no poles away from P. Let $a_1 < \cdots < a_n$ be a minimal set of generators of the Weierstrass monoid, and put $d_i = \gcd(a_1, \dots, a_i)$. Then the monoid is said to be *Gorenstein* (in the terminology of [39]) if for $i = 2, \dots, n$,

$$\frac{a_i}{d_i} \in \frac{a_1}{d_{i-1}} \mathbb{Z}_{\geq 0} + \cdots + \frac{a_{i-1}}{d_{i-1}} \mathbb{Z}_{\geq 0}.$$

If the Weierstrass monoid of C is Gorenstein for some P, the curve C is said to be *telescopic*; its genus is then equal to

$$\frac{1}{2}\left(1 + \sum_{i=1}^{n} \left(\frac{d_{i-1}}{d_i} - 1\right) a_i\right).$$

The cohomology of telescopic curves is easy to describe, so it seems likely that one can compute Monsky-Washnitzer cohomology on them. The case $n = 2$ is the $C_{a,b}$ case; for larger n, this has been worked out by Suzuki [47] in what he calls the "strongly telescopic" case. This case is where for each i, the map from C to its image under the projective embedding defined by $\mathcal{O}(a_i P)$ is a *cyclic* cover (e.g., if C is superelliptic).

We expect that these can be merged to give an algorithm treating the general case of telescopic curves. One practical complication (already appearing in the $C_{a,b}$ case) is that using a Frobenius lift of the form $x \mapsto x^p$ necessitates inverting an unpleasantly large polynomial in y; it seems better instead to iteratively compute the action on both x and y of a Frobenius lift without inverting anything.

4.3 Implementation

The algorithms described above have proved quite practicable; here we mention some implementations and report on their performance. Note that time and space usage figures are only meant to illustrate feasibility; they are in no way standardized with respect to processor speed, platform, etc. Also, we believe all curves and fields described below are "random", without special properties that make them easier to handle.

The first practical test of the original algorithm from [24] seems to have been that of Gaudry and Gürel [15], who computed the zeta function of a genus 3 hyperelliptic curve over $\mathbb{F}_{3^{37}}$ in 30 hours (apparently not optimized). They also tested their superelliptic variant, treating a genus 3 curve over $\mathbb{F}_{2^{53}}$ in 22 hours.

Gaudry and Gürel [16] have also tested the dependence on p in the hyperelliptic case. They computed the zeta function of a genus 3 hyperelliptic curve over $\mathbb{F}_{251}$ in 42 seconds using 25 MB of memory, and over $\mathbb{F}_{10007}$ in 1.61 hours using 1.4 GB.

In the genus direction, Vercauteren [51, Sections 4.4–4.5] computed the zeta function of a genus 60 hyperelliptic curve over $\mathbb{F}_2$ in 7.64 minutes, and of a genus 350 curve over $\mathbb{F}_2$ in 3.5 days. We are not aware of any high-genus tests in odd characteristic; in particular, we do not know whether the lower exponent in the time complexity will really be reflected in practice.

Vercauteren [51, Section 5.5] has also implemented the $C_{a,b}$-algorithm in characteristic 2. He has computed the zeta function of a $C_{3,4}$ curve over $\mathbb{F}_{2^{288}}$ in 8.4 hours and of a $C_{3,5}$ curve over $\mathbb{F}_{2^{288}}$ in 12.45 hours.

Finally, we mention an implementation "coming to a computer near you": Michael Harrison has implemented the computation of zeta functions of hyperelliptic curves in odd characteristic (with or without a rational Weierstrass point) in a new release of MAGMA. At the time of this writing, we have not seen any performance results.

5 Beyond Hyperelliptic Curves

We conclude by describing some of the rich possibilities for further productive computations of p-adic cohomology, especially in higher dimensions. A more detailed assessment, plus some explicit formulae that may prove helpful, appear in the thesis of Gerkmann [19] (recently completed under G. Frey).

5.1 Simple Covers

The main reason the cohomology of hyperelliptic curves in odd characteristic is easily computable is that they are "simple" (Galois, cyclic, tamely ramified) covers of a "simple" variety (which admits a simple Frobenius lift). As a first step into higher dimensions, one can consider similar examples; for instance, a setting we are currently considering with de Jong (with an eye toward gathering data on the Tate conjecture on algebraic cycles) is the class of double covers of $\mathbb{P}^2$ of fixed small degree.

One might also consider some simple wildly ramified covers, like Artin-Schreier covers, which can be treated following Denef-Vercauteren. (These are also good candidates for Lauder's deformation method; see below.)

5.2 Toric Complete Intersections

Another promising class of varieties to study are smooth complete intersections in projective space or other toric varieties. These are promising because their algebraic de Rham cohomology can be computed by a simple recipe; see [19, Chapter 5].

Moreover, some of these varieties are of current interest thanks to connections to physics. For instance, Candelas et al. [6] have studied the zeta functions of some Calabi-Yau threefolds occurring as toric complete intersections, motivated by considerations of mirror symmetry.

5.3 Deformation

We mention also a promising new technique proposed by Lauder. (A related strategy has been proposed by Nobuo Tsuzuki [48] for computing Kloosterman sums.) Lauder's strategy is to compute the zeta function of a single variety not in isolation, but by placing it into a family and studying, after Dwork, the variation in Frobenius along the family as the solution of a certain differential equation.[8]

A very loose description of the method is as follows. Given an initial X, say smooth and proper, find a family $f : Y \to B$ over a simple one-dimensional base (like projective space) which is smooth away from finitely many points, includes X as one fibre, and has another fibre which is "simple". We also ask for simplicity that the whole situation lifts to characteristic zero. For instance, if X is a smooth hypersurface, Y might be a family which linearly interpolates between the defining equation of X and that of a diagonal hypersurface.

One can now compute (on the algebraic lift to characteristic zero) the Gauss-Manin connection of the family; this will give in particular a module with connection over a dagger algebra corresponding to the part of B where f is smooth. One then shows that there is a Frobenius structure on this differential equation that computes the characteristic polynomial of Frobenius on each smooth fibre. That means the Frobenius structure itself satisfies a differential equation, which one solves iteratively using an initial condition provided by the simple fibre. (In the hypersurface example, one can write down by hand the Frobenius action on the cohomology of a diagonal hypersurface.)

Lauder describes explicitly how to carry out the above recipe for Artin-Schreier covers of projective space [28] and smooth projective hypersurfaces [29]. The technique has not yet been implemented on a computer, so it remains to be seen how it performs in practice. It is expected to prove most advantageous for higher dimensional varieties, as one avoids the need to compute in multidimensional polynomial rings. In particular, Lauder shows that in his examples,

[8] Lest this strategy seem strangely indirect, note the resemblance to Deligne's strategy [8] for proving the Riemann hypothesis component of the Weil conjectures!

the dependence of this technique on $d = \dim X$ is exponential in d, and not d^2. (This is essentially best possible, as the dimensions of the cohomology spaces in question typically grow exponentially in d.)

5.4 Additional Questions

We conclude by throwing out some not very well-posed further questions and suggestions,.

- Can one can collect data about a class of "large" curves (e.g., hyperelliptic curves of high genus) over a fixed field, and predict (or even prove) some behavioral properties of the Frobenius eigenvalues of a typical such curve, in the spirit of Katz-Sarnak?
- With the help of cohomology computations, can one find nontrivial instances of cycles on varieties whose existence is predicted by the Tate conjecture? As noted above, we are looking into this with Johan de Jong.
- The cohomology of Deligne-Lusztig varieties furnish representations of finite groups of Lie type. Does the p-adic cohomology in particular shed any light on the modular representation theory of these varieties (i.e., in characteristic equal to that of the underlying field)?
- There is a close link between p-adic Galois representations and the p-adic differential equations arising here; this is most explicit in the work of Berger [2]. Can one extend this analogy to make explicit computations on p-adic Galois representations, e.g., associated to varieties over $\mathbb{Q}_p$, or modular forms? The work of Coleman and Iovita [7] may provide a basis for this.

References

1. L.M. Adleman and M.-D. Huang, Counting rational points on curves and abelian varieties over finite fields, in H. Cohen (ed.), *ANTS-II, Lecture Notes in Comp. Sci.* **1122**, Springer-Verlag, 1996, 1–16.
2. L. Berger, Représentations p-adiques et équations différentielles, *Invent. Math.* **148** (2002), 219–284.
3. P. Berthelot, Cohomologie cristalline des schémas de caractéristique $p > 0$, *Lecture Notes in Math.* **407**, Springer-Verlag, 1974.
4. P. Berthelot, Géométrie rigide et cohomologie des variétés algébriques de caractéristique p, in Introductions aux cohomologies p-adiques (Luminy, 1984), *Mém. Soc. Math. France* **23** (1986), 7–32.
5. P. Berthelot, Finitude et pureté cohomologique en cohomologie rigide (with an appendix in English by A.J. de Jong), *Invent. Math.* **128** (1997), 329–377.
6. P. Candelas, X. de la Ossa and F. Rodriguez-Villegas, Calabi-Yau manifolds over finite fields, I, preprint (`arXiv: hep-th/0012233`).
7. R. Coleman and A. Iovita, Revealing hidden structures, preprint (URL `http://math.berkeley.edu/~coleman/`).
8. P. Deligne, La conjecture de Weil. I, *Publ. Math. IHES* **43** (1974), 273–307.
9. J. Denef and F. Vercauteren, An extension of Kedlaya's algorithm to Artin-Schreier curves in characteristic 2, in C. Fieker and D.R. Kohel (eds.), *ANTS-V, Lecture Notes in Comp. Sci.* **2369**, Springer-Verlag, 2002, 308–323.

10. J. Denef and F. Vercauteren, An extension of Kedlaya's algorithm to hyperelliptic curves in characteristic 2, to appear in *J. Crypt.*
11. J. Denef and F. Vercauteren, Computing zeta functions of C_{ab} curves using Monsky-Washnitzer cohomology, preprint (2003).
12. B. Dwork, On the rationality of the zeta function of an algebraic variety, *Amer. J. Math.* **82** (1960), 631–648.
13. R. Elkik, Solutions d'équations à coefficients dans un anneau hensélien, *Ann. Sci. Éc. Norm. Sup.* **6** (1973), 553–603.
14. E. Freitag and R. Kiehl, Étale cohomology and the Weil conjectures (translated by B.S. Waterhouse and W.C. Waterhouse), Ergebnisse der Math. 13, Springer-Verlag, 1998.
15. P. Gaudry and N. Gürel, An extension of Kedlaya's point-counting algorithm to superelliptic curves, in *Advances in Cryptology – ASIACRYPT 2001 (Gold Coast), Lecture Notes in Comp. Sci.* **2248**, Springer-Verlag, 2001, 480–494.
16. P. Gaudry and N. Gürel, Counting points in medium characteristic using Kedlaya's algorithm, preprint (URL `http://www.inria.fr/rrrt/rr-4838.html`).
17. P. Gaudry and R. Harley, Counting points on hyperelliptic curves over finite fields, in W. Bosma (ed.), *ANTS-IV, Lecture Notes in Comp. Sci.* **1838**, Springer-Verlag, 2000, 313–332.
18. P. Gaudry and É. Schost, Construction of secure random curves of genus 2 over prime fields, to appear in *Eurocrypt 2004.*
19. R. Gerkmann, The p-adic cohomology of varieties over finite fields and applications to the computation of zeta functions, thesis, Universität Duisberg-Essen, 2003.
20. B.H. Gross, A tameness criterion for Galois representations associated to modular forms (mod p), *Duke Math. J.* **61** (1990), 445–517.
21. A. Grothendieck On the de Rham cohomology of algebraic varieties, *Publ. Math. IHES* **29** (1966), 95–103.
22. R. Hartshorne, On the De Rham cohomology of algebraic varieties, *Publ. Math. IHES* **45** (1975), 5–99.
23. G.C. Kato and S. Lubkin, Zeta matrices of elliptic curves, *J. Number Theory* **15** (1982), 318–330.
24. K.S. Kedlaya, Counting points on hyperelliptic curves using Monsky-Washnitzer cohomology, *J. Ramanujan Math. Soc.* **16** (2001), 323–338; errata, *ibid.* **18** (2003), 417–418.
25. K.S. Kedlaya, Fourier transforms and p-adic "Weil II", preprint (URL `http://math.mit.edu/~kedlaya/papers/`).
26. K.S. Kedlaya, Quantum computation of zeta functions of curves, preprint (URL `http://math.mit.edu/~kedlaya/papers/`).
27. D. Kohel, The AGM-$X_0(N)$ Heegner point lifting algorithm and elliptic curve point counting, in *Asiacrypt '03, Lecture Notes in Comp. Sci.* **2894**, Springer-Verlag, 2003, 124–136.
28. A.G.B. Lauder, Deformation theory and the computation of zeta functions, to appear in *Proc. London Math. Soc.*
29. A.G.B. Lauder, Counting solutions to equations in many variables over finite fields, to appear in *Foundations of Comp. Math.*
30. A.G.B. Lauder and D. Wan, Counting points on varieties over finite fields of small characteristic, to appear in J.P. Buhler and P. Stevenhagen (eds.), *Algorithmic Number Theory: Lattices, Number Fields, Curves and Cryptography*, MSRI Publications, Cambridge Univ. Press.
31. A.G.B. Lauder and D. Wan, Computing zeta functions of Artin-Schreier curves over finite fields, *London Math. Soc. J. Comp. Math.* **5** (2002), 34–55.

32. A.G.B. Lauder and D. Wan, Computing zeta functions of Artin-Schreier curves over finite fields II, *J. Complexity*, to appear.
33. R. Lercier and D. Lubicz, email to the NMBRTHRY mailing list, 5 December 2002 (URL `http://listserv.nodak.edu/archives/nmbrthry.html`).
34. J.-F. Mestre, Algorithmes pour compter des points en petite caractéristique en genre 1 et 2, preprint (URL `http://www.math.univ-rennes1.fr/crypto/2001-02/mestre.ps`).
35. S. Miura, Error correcting codes based on algebraic curves (in Japanese), thesis, University of Tokyo, 1997.
36. P. Monsky, Formal cohomology. II. The cohomology sequence of a pair, *Ann. of Math. (2)* **88** (1968), 218–238.
37. P. Monsky, Formal cohomology. III. Fixed point theorems, *Ann. of Math. (2)* **93** (1971), 315–343.
38. P. Monsky and G. Washnitzer, Formal cohomology. I, *Ann. of Math. (2)* **88** (1968), 181–217.
39. A. Nijenhuis and H.S. Wilf, Representations of integers by linear forms in nonnegative integers, *J. Number Th.* **4** (1972), 98–106.
40. J. Pila, Frobenius maps of abelian varieties and finding roots of unity in finite fields, *Math. Comp.* **55** (1990), 745–763.
41. C. Ritzenthaler, Problèmes arithmétiques relatifs à certaines familles de courbes sur les corps finis, thesis, Université Paris 7, 2003 (URL `http://www.math.jussieu.fr/~ritzenth/`).
42. C. Ritzenthaler, Point counting on genus 3 non hyperelliptic curves, preprint (URL `http://www.math.jussieu.fr/~ritzenth/`).
43. T. Satoh, The canonical lift of an ordinary elliptic curve over a finite field and its point counting, *J. Ramanujan Math. Soc.* **15** (2000), 247–270.
44. T. Satoh, On p-adic point counting algorithms for elliptic curves over finite fields, *ANTS-V*, *Lecture Notes in Comp. Sci.* **2369**, Springer-Verlag, 2002, 43–66.
45. R. Schoof, Elliptic curves over finite fields and the computation of square roots mod p, *Math. Comp.* **44** (1985), 483–494.
46. J.-P. Serre and J. Tate, Good reduction of abelian varieties, *Ann. Math.* (2) **88** (1968), 492–517.
47. J. Suzuki, An extension of Kedlaya's order counting based on Miura theory, preprint.
48. N. Tsuzuki, Bessel F-isocrystals and an algorithm of computing Kloosterman sums, preprint.
49. M. van der Put, The cohomology of Monsky and Washnitzer, in Introductions aux cohomologies p-adiques (Luminy, 1984), *Mém. Soc. Math. France* **23** (1986), 33–59.
50. F. Vercauteren, Extensions of Kedlaya's algorithm, notes from ECC 2002 talk (URL `http://www.cs.bris.ac.uk/~frederik/`).
51. F. Vercauteren, Computing zeta functions of curves over finite fields, thesis, Katholieke Universiteit Leuven, 2003 (URL `http://www.cs.bris.ac.uk/~frederik/`).
52. U. Walther, Algorithmic determination of the rational cohomology of complex varieties via differential forms, in *Symbolic computation: solving equations in algebra, geometry, and engineering (South Hadley, MA, 2000)*, *Contemp. Math.* **286**, Amer. Math. Soc. (Providence), 2001, 185–206.

Using Primitive Subgroups to Do More with Fewer Bits

K. Rubin[1*] and A. Silverberg[2**]

[1] Department of Mathematics
Stanford University
Stanford CA, USA
rubin@math.stanford.edu

[2] Department of Mathematics
Ohio State University
Columbus, OH, USA
silver@math.ohio-state.edu

Abstract. This paper gives a survey of some ways to improve the efficiency of discrete log-based cryptography by using the restriction of scalars and the geometry and arithmetic of algebraic tori and abelian varieties.

1 Introduction

This paper is a survey, intended to be readable by both mathematicians and cryptographers, of some of the results in [24,25,26], along with a new result in §3.6. It can be viewed as a sequel to the Brouwer-Pellikaan-Verheul paper "Doing more with fewer bits" [8].

The overall objective is to provide greater efficiency for the same security. The idea is to shorten transmissions by a factor of $\frac{n}{\varphi(n)}$, by going from a finite field $\mathbb{F}_q$ up to the larger field $\mathbb{F}_{q^n}$, and using "primitive subgroups". Here, $\varphi(n)$ is the Euler φ-function. Note that $n/\varphi(n)$ goes to infinity (very slowly), as n goes to infinity.

The first goal is to obtain the same security as the classical Diffie-Hellman and ElGamal cryptosystems, while sending shorter transmissions. More precisely, the goal is to do discrete log-based cryptography, relying on the security of $\mathbb{F}_{q^n}^\times$, while transmitting only $\varphi(n)$ elements of $\mathbb{F}_q$, instead of n elements of $\mathbb{F}_q$ (i.e., one element of $\mathbb{F}_{q^n}$). We use algebraic tori. The next goal is to improve pairing-based cryptosystems. Here, we use elliptic curves E and primitive subgroups of $E(\mathbb{F}_{q^n})$.

As pointed out by Dan Bernstein, the techniques discussed here can be viewed as "compression" techniques, adding more flexibility for the user, who might

* Rubin was partially supported by NSF grant DMS-0140378.

** Silverberg thanks NSA for support, and the ANTS VI organizers for inviting her to speak.

D.A. Buell (Ed.): ANTS 2004, LNCS 3076, pp. 18–41, 2004.

choose to send compressed information when the network is the bottleneck and uncompressed information when computational power is the bottleneck.

In §2 we discuss some background and past results on compressing the transmissions in discrete log-based cryptography for the multiplicative group. In §3 we give an exposition of torus-based cryptography; we give a new implementation of CEILIDH in §3.6. In §4 we show how to compress the transmissions in pairing-based cryptosystems. In §5 we discuss some of the underlying mathematics, including an elementary introduction to the Weil restriction of scalars; we define "primitive subgroup" in §5.5. In §6 we discuss the mathematics underlying torus-based cryptography, and interpret some earlier systems in terms of quotients of algebraic tori.

For technical details, see the original papers. See also [11] (especially §3.2) for the use of primitive subgroups in cryptography.

2 Some Background

We first recall the classical Diffie-Hellman key agreement scheme [10,21].

2.1 Classical Diffie-Hellman

In classical Diffie-Hellman key agreement, a large finite field $\mathbb{F}_q$ is public ($q \approx 2^{1024}$), as is an element $g \in \mathbb{F}_q^\times$ of large (public) multiplicative order ℓ ($> 2^{160}$). Alice chooses a private integer a, random in the interval between 1 and $\ell - 1$, and Bob similarly chooses a private integer b.

- Alice sends g^a to Bob.
- Bob sends g^b to Alice.
- They share $g^{ab} = (g^a)^b = (g^b)^a$.

Tautologically, the security is based on the difficulty of the Diffie-Hellman Problem in $\mathbb{F}_q^\times$.

Note that when this is performed using $\mathbb{F}_{q^n}$ in place of $\mathbb{F}_q$, then the transmissions are elements of $\mathbb{F}_{q^n}$ (i.e., n elements of $\mathbb{F}_q$). If one can do Diffie-Hellman transmitting only $\varphi(n)$ elements of $\mathbb{F}_q$ while relying on security coming from $\mathbb{F}_{q^n}^\times$, then one would like to have $n\log(q)$ large for high security, and $\varphi(n)\log(q)$ small for high bandwidth efficiency. In particular, for maximal efficiency per unit of security (i.e., to achieve a system that is $\frac{n}{\varphi(n)}$ times as efficient as Diffie-Hellman), one would like $\frac{n}{\varphi(n)}$ to be as large as possible. Thus, the most useful n's to consider are those in the sequence

$$1, \quad 2, \quad 2 \cdot 3 = 6, \quad 2 \cdot 3 \cdot 5 = 30, \quad 2 \cdot 3 \cdot 5 \cdot 7 = 210, \quad \ldots$$

(whose i-th entry is the product of the first $i - 1$ primes). We will discuss some ways to do this, below.

2.2 A Brief Tour of Some History

As noted in [17,8], one can achieve greater efficiency per unit of security by choosing g in the subgroup of $\mathbb{F}_{q^n}^\times$ of order $\Phi_n(q)$, where $\Phi_n(x)$ is the n-th cyclotomic polynomial. (The polynomial $\Phi_n(x)$ has integer coefficients, and its (complex) roots are the primitive n-th roots of unity.)

Diffie-Hellman key agreement is based on the full multiplicative group $\mathbb{F}_q^\times$, which is a group of order $q-1 = \Phi_1(q)$.

In [22,31,32,28,29,25], analogues of the classical Diffie-Hellman key agreement scheme are introduced that rely on the security of $\mathbb{F}_{p^2}^\times$ while transmitting only one element of $\mathbb{F}_p$. One now takes the element g to lie in the subgroup of $\mathbb{F}_{p^2}^\times$ of order $p+1$ $(= \Phi_2(p))$. Since $n = 2$, we have $n/\varphi(n) = 2$, and achieve twice the efficiency of Diffie-Hellman for comparable security. The papers [22,31,32,28,29] use Lucas sequences [20], to give what are known as Lucas-based cryptosystems. See [4] for a critique of [28,29]. In [25] (see §3.4 below) we introduced the $\mathbb{T}_2$-cryptosystem,

which is a torus-based system. It is related to the Lucas-based cryptosystems (see §6.5 below), and has some advantages over them.

The Gong-Harn system [13] uses linear feedback shift register sequences. In this case $n = 3$, so $n/\varphi(n) = 1.5$. This cryptosystem relies on the security of $\mathbb{F}_{p^3}^\times$ while transmitting only two elements of $\mathbb{F}_p$, using the subgroup of $\mathbb{F}_{p^3}^\times$ of order p^2+p+1 $(= \Phi_3(p))$.

The case where $n = 6$ (so $n/\varphi(n) = 3$) is considered in [8], [19] (the XTR system), and [25] (the CEILIDH system). These systems give three times the efficiency of Diffie-Hellman, for the same security. They rely on the security of $\mathbb{F}_{p^6}^\times$ while transmitting only two elements of $\mathbb{F}_p$, using the subgroup of $\mathbb{F}_{p^6}^\times$ of order p^2-p+1 $(= \Phi_6(p))$.

Arjen Lenstra [18] has asked whether one can use $n = 30$ to do better than XTR. Note that $\varphi(30) = 8$ and

$$\Phi_{30}(x) = x^8 + x^7 - x^5 - x^4 - x^3 + x + 1.$$

Building on a conjecture in [8], conjectures for arbitrary n were given in [6]. Those conjectures were disproved in [6,25,26], and it was proposed in [25,26] that a conjecture of Voskresenskii should replace those conjectures.

2.3 Classical ElGamal Encryption

As before, the public information is a large finite field $\mathbb{F}_q$ and an element $g \in \mathbb{F}_q^\times$ of order ℓ, along with q and ℓ.
Alice's private key: an integer a, random in the interval $[1, \ell-1]$
Alice's public key: $P_A = g^a \in \mathbb{F}_q$

- Bob represents the message M in $\langle g \rangle$ and chooses a random integer r between 1 and $\ell-1$. Bob send Alice the ciphertext (c, d) where $c = g^r$ and $d = M \cdot P_A^r$.
- To decrypt a ciphertext (c, d), Alice computes

$$d \cdot c^{-a} = M \cdot (g^a)^r \cdot (g^r)^{-a} = M.$$

2.4 Classical ElGamal Signatures

With public information as before, also fix a public cryptographic hash function $H : \{0,1\}^* \to \mathbb{Z}/\ell\mathbb{Z}$ (i.e., H takes bit strings to integers modulo ℓ, is easy to compute and hard to invert, and its images look "random").
Alice's private key: an integer a, random in the interval $[1, \ell - 1]$
Alice's public key: $P_A = g^a \in \mathbb{F}_q$

- To sign a message $M \in \{0,1\}^*$, Alice chooses a random integer r between 1 and $\ell - 1$ with $\gcd(r, \ell) = 1$. Alice's signature on M is (c, d) where $c = g^r$ and $d = r^{-1}(H(M) - aH(g^r)) \pmod{\ell}$.
- Bob accepts Alice's signature if and only if

$$g^{H(M)} = P_A^{H(c)} \cdot c^d$$

in the field $\mathbb{F}_q$.

Remark 1. Note that Diffie-Hellman key agreement only requires exponentiations (i.e., computing powers of elements in the group generated by g), while the ElGamal encryption and signature schemes require multiplications in the finite field (i.e., $M \cdot P_A^r$, $c^{-a} \cdot d$, and $P_A^{H(c)} \cdot c^d$).

2.5 Using XTR to Illustrate the Idea

We give an illustration, in the case $n = 6$, of the idea behind [8,13,19] and the Lucas-based cryptosystems.

XTR is short for ECSTR, which stands for *Efficient Compact Subgroup Trace Representation.*

The *trace* is the trace map from $\mathbb{F}_{p^6}$ to $\mathbb{F}_{p^2}$, which is defined by

$$\text{Tr}(h) = h + h^{p^2} + h^{p^4} = h + \sigma(h) + \sigma^2(h),$$

where σ generates the Galois group $\text{Gal}(\mathbb{F}_{p^6}/\mathbb{F}_{p^2})$. (Note that $h^{p^6} = h$.)

The *subgroup* is the subgroup of $\mathbb{F}_{p^6}^\times$ of order $p^2 - p + 1 = \Phi_6(p)$. Choose a generator g of this subgroup.

- Alice sends $\text{Tr}(g^a)$ to Bob.
- Bob sends $\text{Tr}(g^b)$ to Alice.
- They share $\text{Tr}(g^{ab})$.

Since the transmissions are elements of $\mathbb{F}_{p^2}$, Alice and Bob are sending 2 $(= \varphi(6))$ elements of $\mathbb{F}_p$, rather than 6 elements of $\mathbb{F}_p$ (i.e., one element of $\mathbb{F}_{p^6}$, as would be the case in classical Diffie-Hellman over the field $\mathbb{F}_{p^6}$). The point is that the trace gives an *efficient compact representation* of elements in the subgroup $\langle g \rangle$.

We claim that Alice and Bob now share $\text{Tr}(g^{ab}) \in \mathbb{F}_{p^2}$. This is proved in [19], where an efficient way to compute $\text{Tr}(g^{ab})$ is given. Let's convince ourselves that

Alice and Bob really do have enough information to compute $\mathrm{Tr}(g^{ab})$. Suppose that h is an element of the subgroup of $\mathbb{F}_{p^6}^{\times}$ of order $p^2 - p + 1$. Let

$$C_h = \{h, \sigma(h), \sigma^2(h)\}.$$

The three elementary symmetric polynomials of the set C_h are:

$$\begin{aligned}
\Pi_1(C_h) &= h + \sigma(h) + \sigma^2(h) = \mathrm{Tr}(h), \\
\Pi_2(C_h) &= h \cdot \sigma(h) + h \cdot \sigma^2(h) + \sigma(h) \cdot \sigma^2(h) = \mathrm{Tr}(h \cdot \sigma(h)), \\
\Pi_3(C_h) &= h \cdot \sigma(h) \cdot \sigma^2(h) = \mathrm{N}(h),
\end{aligned}$$

where $\mathrm{N} : \mathbb{F}_{p^6} \to \mathbb{F}_{p^2}$ is the norm map. It turns out that if h is in the subgroup of order $p^2 - p + 1$, then $\Pi_2(C_h) = \mathrm{Tr}(h)^p$ and $\Pi_3(C_h) = 1$.

Thus, knowing $\mathrm{Tr}(h)$ is equivalent to knowing the values of all the elementary symmetric polynomials of C_h, which is equivalent to knowing the set C_h. However, if you know C_h and you know a, then you know C_{h^a}, just by taking every element of C_h to the power a. But we have already noted that knowing C_{h^a} is equivalent to knowing $\mathrm{Tr}(h^a)$.

To sum up, if h is in the subgroup of $\mathbb{F}_{p^6}^{\times}$ of order $p^2 - p + 1$, then a and $\mathrm{Tr}(h)$ together determine $\mathrm{Tr}(h^a)$. Since Alice knows $\mathrm{Tr}(g^b)$ and a, she has enough information to compute $\mathrm{Tr}((g^b)^a)$, and similarly Bob can compute $\mathrm{Tr}((g^a)^b)$.

Note that knowing C_h is equivalent to knowing the characteristic polynomial of h over $\mathbb{F}_{p^2}$, since that characteristic polynomial is

$$\prod_{c \in C_h} (X - c) = X^3 - \Pi_1(C_h)X^2 + \Pi_2(C_h)X - \Pi_3(C_h).$$

Remark 2. In XTR [19], the Gong-Harn system [13], and the Lucas-based cryptosystems, Alice can compute $f(g^{ab})$ from $f(g^b)$ and a, for a suitable function f (usually a trace). In other words, these cryptosystems can exponentiate, as is needed for doing (analogues of) Diffie-Hellman. However, they cannot multiply in a straightforward way. If you know $\mathrm{Tr}(g)$ and $\mathrm{Tr}(h)$, that does not give you enough information to compute $\mathrm{Tr}(gh)$, since C_g and C_h do not determine the set C_{gh} (knowing only C_g and C_h, you do not have enough information to distinguish C_{gh} from $C_{g \cdot \sigma(h)}$, for example). These are examples of "lossy" compression. If one orders the conjugates of h and transmits a couple of extra bits to specify which conjugate h is, then one can reconstruct h from $\mathrm{Tr}(h)$, and perform multiplications in $\mathbb{F}_{p^6}$.

3 Torus-Based Cryptography

The goal is to find a computable function f satisfying the following properties:

- the number of bits needed to represent $f(h)$ is less than the number of bits needed to represent h (ideally, $f(h)$ is $\frac{\varphi(n)}{n}$ as long as h),

- $f(h)$ and a determine $f(h^a)$ and h^a,
- $f(g)$ and $f(h)$ determine $f(gh)$ and gh,
- f is defined on almost all elements of the subgroup of $\mathbb{F}_{q^n}^\times$ of order $\Phi_n(q)$.

Note that these conditions imply that f has a computable inverse function.

From now on, fix a square-free integer n and a prime power q. (Square-free means that the only square that divides n is 1.)

Definition 3. Let T_n denote the subgroup of $\mathbb{F}_{q^n}^\times$ of order $\Phi_n(q)$.

Example 4. (i) Diffie-Hellman is based on the group $T_1 = \mathbb{F}_q^\times$.
(ii) If q is not a power of 2, one can write $\mathbb{F}_{q^2} = \mathbb{F}_q(\sqrt{d})$. Then

$$\begin{aligned} T_2 &= \{a + b\sqrt{d} : a, b \in \mathbb{F}_q \text{ and } (a + b\sqrt{d})^{q+1} = 1\} \\ &= \{a + b\sqrt{d} : a, b \in \mathbb{F}_q \text{ and } a^2 - db^2 = 1\} \subset \mathbb{F}_{q^2}^\times, \end{aligned}$$

since $(a + b\sqrt{d})^q = a - b\sqrt{d}$.

Choose a prime power q of about $1024/n$ bits, such that $\Phi_n(q)$ is divisible by a large prime. Choose $g \in T_n$ whose order ℓ is divisible by that large prime. Suppose for now that one has efficiently computable maps

$$\mathbb{F}_q^{\varphi(n)} \underset{f}{\overset{j}{\dashrightarrow\!\!\!\!\dashleftarrow}} T_n \tag{1}$$

that are inverses of each other. The dotted arrows signify that these maps need not be defined everywhere; they might be undefined at a "small" number of elements. In §3.4, §3.6, §6.3, and [25] we discuss the maps f and j, and give explicit examples. The following protocols are generalized Diffie-Hellman and ElGamal [21], using the subgroup T_n of $\mathbb{F}_{q^n}^\times$. In §3.7 below we discuss how to represent the message in $\langle g \rangle$. Note that the maps f and j allow one to compress transmissions not only for Diffie-Hellman and ElGamal, but also for any discrete log-based system that can use a general group.

3.1 Torus-Based Diffie-Hellman Key Agreement

Alice chooses an integer a randomly in the interval $[1, \ell - 1]$. Similarly, Bob chooses a random integer b from the same range.

- Alice sends $P_A = f(g^a) \in \mathbb{F}_q^{\varphi(n)}$ to Bob.
- Bob sends $P_B = f(g^b) \in \mathbb{F}_q^{\varphi(n)}$ to Alice.
- They share $(j(P_B))^a = g^{ab} = (j(P_A))^b$, and also $f(g^{ab})$.

3.2 Torus-Based ElGamal Encryption

Alice's private key: an integer a, random in the interval $[1, \ell - 1]$
Alice's public key: $P_A = f(g^a) \in \mathbb{F}_q^{\varphi(n)}$

- Bob represents the message M in $\langle g \rangle$ and picks a random r between 1 and $\ell - 1$. The ciphertext is (c, d) where $c = f(g^r)$ and $d = f(M \cdot j(P_A)^r)$.
- To decrypt a ciphertext (c, d), Alice computes $M = j(d) \cdot j(c)^{-a}$.

3.3 Torus-Based ElGamal Signatures

Fix a cryptographic hash function $H : \{0,1\}^* \to \mathbb{Z}/\ell\mathbb{Z}$.
Alice's private key: an integer a, random in the interval $[1, \ell - 1]$
Alice's public key: $P_A = f(g^a) \in \mathbb{F}_q^{\varphi(n)}$

- To sign a message $M \in \{0,1\}^*$, Alice chooses a random integer r between 1 and $\ell - 1$ with $\gcd(r, \ell) = 1$. Alice's signature on M is (c, d) where $c = f(g^r) \in \mathbb{F}_q^{\varphi(n)}$ and $d = r^{-1}(H(M) - aH(c)) \pmod{\ell}$.
- Bob accepts Alice's signature if and only if

$$g^{H(M)} = j(P_A)^{H(c)} \cdot j(c)^d.$$

The signature length is $\varphi(n)\log_2(q) + \log_2(\ell)$ bits, as opposed to $n\log_2(q) + \log_2(\ell)$ bits in the classical ElGamal signature scheme over $\mathbb{F}_{q^n}$.

3.4 The $\mathbb{T}_2$-Cryptosystem

Here, $n = 2$. Choose a prime power q that has about 512 bits, and such that $\frac{q+1}{2}$ is a prime. One can write $\mathbb{F}_{q^2} = \mathbb{F}_q(\sqrt{d})$ for some non-square $d \in \mathbb{F}_q^\times$. Define

$$j : \mathbb{F}_q \to T_2 \qquad \text{by} \qquad j(a) = \frac{a + \sqrt{d}}{a - \sqrt{d}}.$$

Define an inverse map (defined on $T_2 - \{1, -1\}$):

$$f : T_2 \dashrightarrow \mathbb{F}_q \qquad \text{by} \qquad f(a + b\sqrt{d}) = \frac{1 + a}{b}.$$

It is easy to check that if $a, b \in \mathbb{F}_q$ and $a \neq -b$, then

$$j(a)j(b) = j\left(\frac{ab + d}{a + b}\right).$$

In the $\mathbb{T}_2$-cryptosystem, one does Diffie-Hellman key agreement and ElGamal encryption and signatures, using the group law on the group T_2, while representing the elements in $\mathbb{F}_q$. Here, it is not necessary to go back and forth between $\mathbb{F}_q$ and T_2, since the previous equation translates T_2's multiplication to $\mathbb{F}_q$, i.e., multiplication in T_2 translates into the following operation on $\mathbb{F}_q$:

$$(a, b) \mapsto \frac{ab + d}{a + b},$$

giving a way to compose elements of $\mathbb{F}_q$ without having to pass to T_2 each time.

3.5 The CEILIDH Public Key System

The acronym **CEILIDH** (pronounced "cayley") stands for **C**ompact, **E**fficient, **I**mproves on **L**UC, **I**mproves on **D**iffie-**H**ellman. The CEILIDH key agreement (resp., encryption, resp., signature) scheme is torus-based Diffie-Hellman (resp., ElGamal encryption, resp., ElGamal signatures) in the case $n = 6$.

Examples 11 and 12 of [25] give explicit examples of maps f and j (called ρ and ψ there) when $n = 6$. We give a new example in §3.6 (and use it in §3.7).

3.6 An Explicit Example of Maps f and j

Take an odd prime power q congruent to 2, 6, 7, or 11 (mod 13) and such that $\Phi_6(q)$ is prime. Then $\mathbb{F}_q(\zeta_{13}) \cong \mathbb{F}_{q^{12}}$, where ζ_{13} is a primitive 13-th root of unity, and $\mathbb{F}_q(z) \cong \mathbb{F}_{q^6}$, where $z = \zeta_{13} + \zeta_{13}^{-1}$. Let

$$y = \zeta_{13} + \zeta_{13}^{-1} + \zeta_{13}^{5} + \zeta_{13}^{-5} \in \mathbb{F}_{q^3}.$$

For $u, v \in \mathbb{F}_q$, define

$$j(u,v) = \frac{r - s\sqrt{13}}{r + s\sqrt{13}} \in T_6$$

where

$$r = (3(u^2+v^2) + 7uv + 34u + 18v + 40)y^2 + 26uy \\ - (21u(3+v) + 9(u^2+v^2) + 28v + 42),$$

$$s = 3(u^2+v^2) + 7uv + 21u + 18v + 14.$$

For $t \in T_6$, define

$$f(t) = \left(\frac{u}{w+1}, \frac{v-3}{w+1}\right) \in \mathbb{F}_q^2,$$

with

$$t = a + b\sqrt{13} \quad \text{and} \quad \frac{1+a}{b} = wy^2 + u\left(y + \frac{y^2}{2}\right) + v$$

where t is written with respect to the basis $\{1, \sqrt{13}\}$ for $\mathbb{F}_{q^6}/\mathbb{F}_{q^3}$, with $a, b \in \mathbb{F}_{q^3} = \mathbb{F}_q(y)$, and $\frac{1+a}{b}$ is written with respect to the basis $\{y^2, y + \frac{y^2}{2}, 1\}$ for $\mathbb{F}_{q^3}/\mathbb{F}_q$, with $u, v, w \in \mathbb{F}_q$.

Then f and j are inverses. The map $j : \mathbb{F}_q^2 \to T_6$ is defined on all of $\mathbb{F}_q^2$. The map $f : T_6 \dashrightarrow \mathbb{F}_q^2$ is defined except at 1 and $-2z^5 + 6z^3 - 4z - 1 \in T_6$.

3.7 Representing Elements of $\mathbb{F}_q^{\varphi(n)}$ in $\langle g \rangle$

For torus-based ElGamal encryption, how does one represent a message as an element of $\langle g \rangle$? First, represent the message as an element M in $\mathbb{F}_q^{\varphi(n)}$.

If g is taken to be a generator of T_n, then taking $j(M)$ represents the message in $\langle g \rangle$ (where j is as in (1)). Note that g is a generator of T_n whenever $\Phi_n(q)$ is prime.

If g is taken to be in an index s subgroup of T_n for some small integer s, then by adding a few bits of redundancy to M, after at most a few tries one obtains an M such that $j(M)$ is in $\langle g \rangle$. If g has order ℓ, one can test whether $j(M)$ is in $\langle g \rangle$ by checking whether $j(M)^{\ell} = 1$.

How does one represent the message in $\langle g \rangle$ when $n = 6$?

Take a prime r and an odd prime power q such that the order of $q \pmod r$ is divisible by 6 but is not 6 itself, and such that $\Phi_6(q)$ is prime. (One expects, but cannot prove, that there are infinitely many such q; it is not hard to find some in a suitable range for cryptography, e.g., such that q has about 170 bits, to get 1024-bit security.) These conditions ensure that $\mathbb{F}_q(\zeta_r)$ contains $\mathbb{F}_{q^6}$, where ζ_r is a primitive r-th root of unity. (Note that if the order of $q \pmod r$ is 6, then $\Phi_6(q)$ is divisible by 6, so is not prime. Note also that the condition that the order of $q \pmod r$ is divisible by 6 implies that $r \equiv 1 \pmod 6$.) In the case $r = 13$, one can use the example given in §3.6. Here, one represents the message in $\mathbb{F}_q^2$, and uses the map j to put it in the prime order group $T_6 = \langle g \rangle$.

In Example 11 of [25], we have $q \equiv 2$ or $5 \pmod 9$. Here, $\Phi_6(q)$ is divisible by 3. One can choose the prime power q so that $\Phi_6(q)/3$ is prime. If one takes g to have order $\Phi_6(q)$, then $j(M)$ is in $\langle g \rangle = T_6$.

Similarly for Example 12 of [25], we have $q \equiv 3$ or $5 \pmod 7$. Now $\Phi_6(q)$ is divisible by 7. One can choose q so that $\Phi_6(q)/7$ is prime. If g is taken to have order $\Phi_6(q)$, then $j(M) \in \langle g \rangle = T_6$.

The following sample parameters are all the primes q between $2^{170} - 10^5$ and $2^{170} + 10^5$ such that $q^2 - q + 1$ is prime and q has order 12 modulo 13:

$$
\begin{gathered}
1496577676626844588240573268701473812127674923933621,\\
1496577676626844588240573268701473812127674923946773,\\
1496577676626844588240573268701473812127674923949251,\\
1496577676626844588240573268701473812127674924018047,\\
1496577676626844588240573268701473812127674924027533.
\end{gathered}
$$

3.8 Comparison between CEILIDH and XTR

The security of CEILIDH is exactly the same as that of XTR, with the same security proof; they both rely on the security of the "hardest" subgroup of $\mathbb{F}_{q^6}^{\times}$ (see §3.11). Parameter selection for CEILIDH is exactly the same as for XTR.

The advantage of the $\mathbb{T}_2$-cryptosystem and CEILIDH over LUC and XTR is that $\mathbb{T}_2$ and CEILIDH make full use of the multiplication in the group T_n (for $n = 2$ and 6). This is especially useful for signature schemes. XTR is efficient for key agreement and hybrid encryption (i.e., using a Diffie-Hellman-like protocol to exchange a secret key, and using symmetric key encryption, not public key encryption). CEILIDH can do efficient key agreement, public key (i.e., non-hybrid) encryption, and signatures.

XTR has computational efficiency advantages over CEILIDH (key agreement can be performed with fewer operations).

3.9 Conjectural $\mathbb{T}_n$-Cryptosystems

Whenever f and j exist as in (1), one has a "$\mathbb{T}_n$-cryptosystem", or $\mathbb{T}_n$ compression technique. As in §3.1–§3.3, use f to compactly represent transmissions in $\mathbb{F}_q^{\varphi(n)}$, and use j to send elements of $\mathbb{F}_q^{\varphi(n)}$ to the group T_n, where group operations can be performed.

3.10 Parameter Selection when $n = 30$

For torus-based ElGamal signatures, finding good parameters when $n = 30$ amounts to finding prime powers q of about $1024/30 \approx 35$ bits such that $\Phi_{30}(q)$ has a prime factor ℓ of about 160 bits. Here is a method for doing this:

- choose a 20–30 bit prime $p \equiv 1 \pmod{30}$,
- find the $x_1, \ldots, x_8$ with $1 < x_i < p$ whose orders modulo p are 30,
- find 35-bit primes q congruent to some $x_i \pmod{p}$,
- factor out small ($<$ 90–100 bits) prime divisors from the integer $\Phi_{30}(q)/p$,
- see if what is left is a prime of about 160-bits.

Paul Leyland suggested doing the factorization step by using the Elliptic Curve Method optimized for 90 – 100 bit factors. Using this, he can obtain a few examples per hour on a laptop.

Note that the parameters are like Diffie-Hellman parameters — they do not need to be changed often, and the same q and g can be used for all users.

The table below gives some pairs of primes q and ℓ where q has 35 bits, ℓ has 160 or 161 bits, and ℓ divides $\Phi_{30}(q)$. One expects there to be about

$$717267168(\ln(161) - \ln(160)) \approx 4.47 \times 10^6$$

35-bit primes q such that $\Phi_{30}(q)$ has a 160-bit prime divisor (717267168 is the number of 35-bit primes).

q	ℓ
18849585563	2721829278598645763229135555203875381215025850251
18859507111	1145377552213689334808880803247608425700596690441
18918018433	2191067457957167273280468413326196522745324110911
18937704077	2622917550423816956639040650402145314798081975731
19020912667	2009907944188511109843286107856362388569736938661
19096959863	2670351518767065322212846696686298421468094820481
19123281371	1089731979081189465083403285791765213322453796291
19200181867	1382108007746224782292716444254570494753142184301
19241156549	1292631930593942028414888386684571922308680383411

3.11 Security

The security of all the systems discussed thus far is the discrete log security of the "hardest" subgroup of $\mathbb{F}_{q^n}^\times$, in the following sense. The group $\mathbb{F}_{q^n}^\times$ is "almost the same" as the direct product

$$\prod_{d|n} T_d = T_n \times \prod_{\substack{d|n \\ d \neq n}} T_d$$

(there are homomorphisms between them for which the prime divisors of the orders of the kernel and cokernel all divide n); see pp. 60–61 of [30].

We have $T_d \subset \mathbb{F}_{q^d}^\times$ for all d, so for $d < n$ the elements of these subgroups lie in a strictly smaller field than $\mathbb{F}_{q^n}$. Therefore, these groups T_d are weaker for cryptographic purposes — they are vulnerable to attacks on the discrete logarithm problem in $\mathbb{F}_{q^d}^\times$, where now $d < n$.

Almost none of the elements of T_n lie in a smaller field than $\mathbb{F}_{q^n}$ (see Lemma 1 of [6]). Therefore, T_n can be viewed as the cryptographically strongest subgroup of $\mathbb{F}_{q^n}^\times$.

4 Improving Pairing-Based Cryptography

Inspired by and building on a paper of Galbraith [12], in [24] we use the theory of supersingular abelian varieties to improve the efficiency of pairing-based cryptosystems.

Pairing-based cryptography was conceived of independently by Joux [14] and by Sakai, Ohgishi, and Kasahara [27]. There are numerous applications of pairing-based cryptography, including tripartite Diffie-Hellman, identity-based encryption, and short signatures. See [1] for numerous references and information.

The Boneh-Lynn-Shacham (BLS) short signature scheme [5] uses pairings associated with elliptic curves. The question of whether one can use abelian varieties (which are higher dimensional generalizations of elliptic curves) to obtain shorter signatures was stated as an open problem in [5], and answered in the affirmative in [24]. While we arrived at our method (see §4.2 below) for compressing BLS signatures by studying the arithmetic of abelian varieties, in fact our final algorithm can be performed entirely using elliptic curve arithmetic, without going to higher dimensional abelian varieties.

The Rubin-Silverberg (RS) modification of the BLS signature scheme multiplies the security of BLS signatures by n while multiplying the signature size by $\varphi(n)$. Implementations when $n = 3$ and $n = 5$ are given in [24]. We give an example when $n = 5$ in §4.2 below.

Our methods can be used to improve the bandwidth efficiency of any pairing-based cryptosystem, not just the BLS signature scheme.

4.1 BLS Short Signature Scheme

We give an example of the Boneh-Lynn-Shacham signature scheme, with fixed parameters.

Let $q = 3^{97}$. Consider the elliptic curve $E^+ : y^2 = x^3 - x + 1$ over $\mathbb{F}_q$, and take $P \in E^+(\mathbb{F}_q)$ of (prime) order

$$\ell = 2726865189058261010774960798134976187171462721.$$

Note that $\#E^+(\mathbb{F}_q) = 7\ell$.

Use a pairing

$$e \colon \langle P \rangle \times \langle P \rangle \to \mathbb{F}_{q^6}^\times$$

that satisfies

$$e(aP, bP) = e(P,P)^{ab} \quad \text{for every } a, b \in \mathbb{Z},$$
$$e(P,P) \neq 1.$$

One can use a modified Weil or Tate pairing [15].

The public information is q, E^+, P, ℓ, e, and a cryptographic hash function

$$H \colon \{0,1\}^* \to \langle P \rangle.$$

Alice's private key: an integer a, random in the interval $[1, \ell]$
Alice's public key: $P_A = aP$

- To sign a message $M \in \{0,1\}^*$, Alice computes $P_M = H(M)$ and $aP_M = (s,t) \in \langle P \rangle$.
- Alice's signature is $s \in \mathbb{F}_q$ (and 1 bit to recover the sign of t).
- To verify the signature, Bob computes

  $$t = \pm\sqrt{s^3 - s + 1} \in \mathbb{F}_q,$$

  lets

  $$P' = (s,t) \ (= aP_M),$$

  and checks that

  $$e(P, P') = e(P_A, P_M).$$

4.2 RS Compression of BLS Signatures

We give an example with fixed parameters, with $n = 5$. Let $q' = 3^{19}$ and let $q = (q')^5 = 3^{95}$. Consider the elliptic curve $E^- : y^2 = x^3 - x - 1$, and take $P \in E^-(\mathbb{F}_q)$ of (prime) order

$$\ell = 67332385860403367623388760960599521.$$

Note that

$$\#E^-(\mathbb{F}_q) = 271 \cdot 1162320517 \cdot \ell,$$

$$\#E^-(\mathbb{F}_{3^5}) = 271, \qquad \#E^-(\mathbb{F}_{q'}) = 1162320517.$$

Take a pairing e and a hash function H as before. Let σ be a generator of $\mathrm{Gal}(\mathbb{F}_q/\mathbb{F}_{q'})$. For $Q \in E^-(\mathbb{F}_q)$,

$$\mathrm{Tr}_{\mathbb{F}_q/\mathbb{F}_{q'}}(Q) = Q + \sigma(Q) + \sigma^2(Q) + \sigma^3(Q) + \sigma^4(Q).$$

Let

$$A_0 = \{Q \in E^-(\mathbb{F}_q) : \mathrm{Tr}_{\mathbb{F}_q/\mathbb{F}_{q'}}(Q) = \mathcal{O}_{E^-}\},$$

the "trace-0 subgroup" of $E^-(\mathbb{F}_q)$. Then A_0 has order $271 \cdot \ell$. Since P has order ℓ, we have $P \in A_0$.

Alice's private key: an integer a, random in the interval $[1, \ell]$

Alice's public key: $P_A = aP$

- To sign M, as before, Alice computes $P_M = H(M)$ and $aP_M = (s, t)$.
- Letting $(s_0, s_1, s_2, s_3, s_4)$ be the coordinates of s with respect to a basis for $\mathbb{F}_q$ over $\mathbb{F}_{q'}$, Alice's signature is (s_1, s_2, s_3, s_4) (and 6 bits to recover s_0 and t).
- To verify the signature, Bob first uses that $\mathrm{Tr}_{\mathbb{F}_q/\mathbb{F}_{q'}}(P) = \mathcal{O}_{E^-}$ to reconstruct s_0 (see below).
- Bob then, as before, computes

  $$t = \pm\sqrt{s^3 - s - 1} \in \mathbb{F}_q,$$

  lets

  $$P' = (s, t) \ (= aP_M),$$

  and checks that

  $$e(P, P') = e(P_A, P_M).$$

The process of reconstructing s_0 and t from s_1, s_2, s_3, s_4 is as follows. The input is $(s_1, s_2, s_3, s_4) \in \mathbb{F}_{q'}^4$ and the output will be $s_0, t \in \mathbb{F}_{q'}$. Viewing $\mathbb{F}_q$ as $\mathbb{F}_{q'}(z)$ with $z^5 - z + 1 = 0$, let $c = S + \sum_{i=1}^4 s_i z^i$ and define $a_0, \cdots, a_4 \in \mathbb{F}_{q'}[S]$ by

$$\prod_{i=0}^{4}(Y - \sigma^i(c)) = Y^5 + a_4Y^4 + a_3Y^3 + a_2Y^2 + a_1Y + a_0.$$

The trace-0 condition can (eventually) be reduced to finding simultaneous solutions of $p_1 = 0$ and $p_2 = 0$, where p_1 and p_2 are as follows:

$$\begin{aligned}
p_1 = {} & X^8 - a_4X^7 + (1 + a_4^2 - a_3)X^6 + (a_4 - a_4^3 - a_2)X^5 + (a_4 - a_4^2 + a_4^4 - a_3 - a_4a_2)X^4 \\
& + (1 - a_4 + a_4^2 - a_4^5 - a_3 + a_4^3a_3 + a_2 - a_3a_2 + a_0)X^3 \\
& + (-1 + a_4^2 - a_4^3 + a_4^4 + a_4^6 + a_3 + a_4a_3 - a_3^2 - a_3^3 - a_2 - a_4^3a_2 + a_4a_3a_2 + a_2^2)X^2 \\
& + (-1 - a_4^2 - a_4^3 - a_4^4 - a_4^5 - a_4^7 + a_3 + a_4a_3 - a_4^2a_3 - a_4^3a_3 - a_3^2 \\
& - a_4a_3^2 + a_4a_3^3 - a_2 - a_4^2a_2 - a_4^4a_2 + a_3a_2 - a_4^2a_3a_2 - a_3^2a_2)X \\
& + 1 - a_4^2 - a_4^6 + a_4^8 + a_3 - a_4^6a_3 + a_3^3 - a_4^2a_3^3 + a_3^4,
\end{aligned}$$

$$p_2 = X^6 - X^4 + (-1 - a_4 - a_4^3 + a_2)X^3 + (-1 + a_4^2 - a_3 - a_4a_2 + a_1)X^2 \\ + (-1 - a_4 + a_4^2 + a_4^3 - a_3 - a_4a_3 - a_2 + a_4^2a_2 - a_3a_2)X - 1 + a_4^6 - a_3^3.$$

Taking the resultant of p_1 and p_2 eliminates the variable X, and gives a degree 27 polynomial $h \in \mathbb{F}_{q'}[S]$ that has s_0 as a root. The extra 6 bits allow one to decide which root of h to take for s_0, and to determine t. The polynomial $h(S)$ is of the form $h_1(S^3 - S)$ for a certain degree 9 polynomial $h_1(S) \in \mathbb{F}_{q'}[S]$, and this simplifies finding the roots of h. See §5.1 of [24] for an explanation of this reconstruction step.

RS compression was arrived at by studying the Weil restriction of scalars of elliptic curves (which are abelian varieties), and understanding the theory of abelian varieties. In §5.7 we discuss some of the underlying mathematics.

Remark 5. In elliptic curve point compression and in BLS, an elliptic curve point (x, y) is compressed to its x-coordinate, giving lossy compression. One can transmit an extra bit that determines the y-coordinate, in order to fully reconstruct the point. The signature (s_1, s_2, s_3, s_4) above is similarly an example of lossy compression; the extra 6 bits and the reconstruction step allow one to fully recover the elliptic curve point (s, t).

4.3 Comparison

RS compression (§4.2) produces signatures that are roughly $\frac{4}{5}$ as large as BLS signatures with comparable security. In both cases, the security is based on the difficulty of the Elliptic Curve Diffie-Hellman Problem in $\langle P \rangle$. RS signing is no more work than for BLS. Compared with BLS, RS verification requires an additional reconstruction step to recover s_0. For applications in which the verifier is powerful, this is not a significant problem.

Note that RS compression (like BLS) only uses elliptic curve arithmetic, and does not use any abelian variety arithmetic.

Bernstein and Bleichenbacher have compressed RSA and Rabin signatures ([2,3]). In Table 1 below, BCR stands for Bleichenbacher's Compressed Rabin signatures, DSA is the Digital Signature Algorithm, and ECDSA is the Elliptic Curve Digital Signature Algorithm. In the middle column of Table 1, the signatures are all scaled to 1024-bit RSA security. In the remaining columns the signatures are scaled to the MOV security of the RS scheme. The MOV security refers to attacks on the discrete log problem in $\mathbb{F}_{q^6}^\times$. The DL security refers to generic attacks on the group $\langle P \rangle$; the relevant value for DL security is $\log_2(\ell)$-bits, where ℓ is the order of P. (See [5,24].)

There is an RS scheme similar to the one in §4.2 (see §5.2 of [24]) that uses elliptic curves over binary fields $\mathbb{F}_{2^w}$. Working over binary fields might yield some efficiency advantages. However, due to Coppersmith's attack on the discrete log problem in low characteristic [9], larger parameters should be used.

To achieve the flexibility of higher characteristic, in §6 of [24] we suggest the use of (Jacobian varieties of) certain twists of Fermat curves. In a recent preprint giving an expanded version of [5], Boneh, Lynn, and Shacham suggest using MNT elliptic curves.

Table 1. Signature lengths, in bits, for comparable MOV security

system			
RSA	904	1024	2045
BCR	452	512	1024
DSA		320	
ECDSA		320	
BLS	152	172	342
RS	127	143	279

5 The Underlying Mathematics

5.1 Varieties and Algebraic Groups

Definition 6. Loosely speaking, an *algebraic variety* (over a field k) is the solution set of a system of polynomial equations (whose coefficients are in k). An *algebraic group* (or *group variety*) over a field k is a variety over k such that the group law and the inverse map are quotients of polynomials whose coefficients are in k.

5.2 The Weil Restriction of Scalars

Suppose that V is a variety over a field L. This means that V is the solution set of a system of polynomial equations $f_i(x_1, \ldots, x_r) = 0$, $1 \le i \le s$, where the polynomials f_i have coefficients in the field L. Suppose k is a subfield of L, and n is the degree of L over k. Fix a basis $\{v_1, \ldots, v_n\}$ for L over k. Write $x_i = \sum_{j=1}^{n} y_{ij} v_j$ with variables y_{ij}. Substitute this into the equations $f_i(x_1, \ldots, x_r) = 0$. Multiplying out, writing everything with respect to the basis $\{v_1, \ldots, v_n\}$, and equating coefficients, one obtains a system of polynomials in the variables $\{y_{ij}\}$, with coefficients in the field k. The variety defined by these new equations is denoted $\mathrm{Res}_{L/k}V$, and is called the (Weil) restriction of scalars from L down to k. It is a variety over k with the property that its k-points are the L-points of V:

$$(\mathrm{Res}_{L/k}V)(k) \cong V(L).$$

Its dimension is $n \cdot \dim(V)$. See for example §3.12 in Chapter 1 of [30] for more information.

5.3 The Multiplicative Group $\mathbb{G}_m$

Diffie-Hellman is based on the multiplicative group, denoted $\mathbb{G}_m$. Over any field F, the F-points on $\mathbb{G}_m$ are

$$\mathbb{G}_m(F) = F^\times = F - \{0\},$$

the multiplicative group of invertible elements of the field F. The algebraic variety $\mathbb{G}_m$ is defined by the equation $xy = 1$, i.e., it consists of the elements x such that there exists a y with $xy = 1$. It is an algebraic group over any field k. We will view $\mathbb{G}_m$ as an algebraic group over the field $\mathbb{F}_q$.

5.4 The Restriction of Scalars $\mathrm{Res}_{\mathbb{F}_{q^n}/\mathbb{F}_q}\mathbb{G}_m$

The Weil restriction of scalars $\mathrm{Res}_{\mathbb{F}_{q^n}/\mathbb{F}_q}\mathbb{G}_m$ is an algebraic variety (in fact, an algebraic group) over $\mathbb{F}_q$. We have

$$(\mathrm{Res}_{\mathbb{F}_{q^n}/\mathbb{F}_q}\mathbb{G}_m)(\mathbb{F}_q) \cong \mathbb{F}_{q^n}^\times.$$

Example 7. To find equations defining the two-dimensional algebraic variety $\mathrm{Res}_{\mathbb{F}_9/\mathbb{F}_3}\mathbb{G}_m$, write $\mathbb{F}_9 = \mathbb{F}_3(\sqrt{-1})$, and write $x = x_1 + x_2\sqrt{-1}$ and $y = y_1 + y_2\sqrt{-1}$. Substituting into $xy = 1$ and equating coefficients gives the equations:

$$x_1y_1 - x_2y_2 = 1, \qquad x_1y_2 + x_2y_1 = 0.$$

5.5 The Primitive Subgroup G_0

Suppose that G is a commutative algebraic group over a field k. In the cases of interest to us, V will be the multiplicative group $\mathbb{G}_m$ or an elliptic curve. For now, we write G's group operation as multiplication.

If L is a field that is a finite extension of k, define the *primitive subgroup* G_0 of $\mathrm{Res}_{L/k}G$ to be

$$G_0 = \ker[\mathrm{Res}_{L/k}G \xrightarrow{\oplus \mathrm{N}_{L/F}} \bigoplus_{k\subseteq F\subsetneq L} \mathrm{Res}_{F/k}G],$$

where the norm maps $\mathrm{N}_{L/F}$ induce the usual norm maps

$$\mathrm{N}_{L/F} : G(L) \to G(F), \qquad x \mapsto \prod_{\sigma\in\mathrm{Gal}(L/F)} \sigma(x).$$

Then G_0 is an algebraic group over k, and $G_0(k)$ consists of all elements of $G(L)$ whose norm down $G(F)$ is the identity, for every intermediate field F with $F \neq L$.

The group $\mathrm{Res}_{L/k}G$ is "almost the same" as the product $G \times G_0$ (there are homomorphisms between them with "small" kernel and cokernel).

5.6 The Algebraic Torus $\mathbb{T}_n$

Let $\mathbb{T}_n$ (or $\mathbb{T}_{n,q}$ when it is important to keep track of the ground field) denote the primitive subgroup of $\mathrm{Res}_{\mathbb{F}_{q^n}/\mathbb{F}_q}\mathbb{G}_m$, i.e.,

$$\mathbb{T}_n = \mathbb{T}_{n,q} = \ker\left[\mathrm{Res}_{\mathbb{F}_{q^n}/\mathbb{F}_q}\ \mathbb{G}_m \xrightarrow{\oplus \mathrm{N}_{\mathbb{F}_{q^n}/\mathbb{F}_{q^d}}} \bigoplus_{\substack{d|n \\ d\neq n}} \mathrm{Res}_{\mathbb{F}_{q^d}/\mathbb{F}_q}\ \mathbb{G}_m\right].$$

By definition, $\mathbb{T}_n(\mathbb{F}_q)$ is the group of elements of $\mathbb{F}_{q^n}^\times$ that have norm 1 down to every intermediate field $\mathbb{F}_{q^d}$ (for $d \neq n$). By Lemma 7 of [25],

$$\mathbb{T}_n(\mathbb{F}_q) = T_n. \tag{2}$$

Example 8. Continuing Example 7, where $q = 3$ and $n = 2$, it is easy to write down embeddings:

$$\mathbb{G}_m \hookrightarrow \mathrm{Res}_{\mathbb{F}_9/\mathbb{F}_3}\mathbb{G}_m, \quad x \mapsto (x, 0, x^{-1}, 0),$$

$$\mathbb{T}_2 \hookrightarrow \mathrm{Res}_{\mathbb{F}_9/\mathbb{F}_3}\mathbb{G}_m, \quad x_1 + x_2\sqrt{-1} \mapsto (x_1, x_2, x_1, -x_2).$$

The compositions (in both orders) of the resulting map

$$\mathbb{G}_m \times \mathbb{T}_2 \to \mathrm{Res}_{\mathbb{F}_9/\mathbb{F}_3}\mathbb{G}_m$$

with the map

$$\mathrm{Res}_{\mathbb{F}_9/\mathbb{F}_3}\mathbb{G}_m \to \mathbb{G}_m \times \mathbb{T}_2$$

defined by

$$(x_1, x_2, y_1, y_2) \mapsto (x_1^2 + x_2^2, x_1y_1 + x_2y_2 + 2x_2y_1\sqrt{-1})$$

are the squaring maps. Thus, $\mathrm{Res}_{\mathbb{F}_9/\mathbb{F}_3}\mathbb{G}_m$ is “almost the same” as $\mathbb{G}_m \times \mathbb{T}_2$.

5.7 The Trace-0 Subgroup of $\mathrm{Res}_{\mathbb{F}_q/\mathbb{F}_{q'}}(E^-)$

Abelian varieties are, by definition, projective algebraic groups. Elliptic curves are exactly the one-dimensional abelian varieties.

With E^-, q', q, ℓ, and P as in §4.2, let

$$B = \mathrm{Res}_{\mathbb{F}_q/\mathbb{F}_{q'}}(E^-),$$

and let A be the primitive subgroup of B:

$$A = \ker[B \xrightarrow{\mathrm{N}_{\mathbb{F}_q/\mathbb{F}_{q'}}} E^-].$$

Then A and B are abelian varieties over $\mathbb{F}_{q'}$ of dimensions 4 and 5, respectively, and B is isogenous to $E^- \times A$. (See also §3.2 of [11].) The abelian variety A is simple. Since the group law on an abelian variety is written additively, the norm map now corresponds to the sum of the conjugates, i.e., the trace defined in §4.2. We have

$$\begin{array}{ccccc} \langle P \rangle \subset A_0 = \{Q \in & E^-(\mathbb{F}_q) & : \mathrm{Tr}_{\mathbb{F}_q/\mathbb{F}_{q'}}(Q) = \mathcal{O}_{E^-}\} & \cong & A(\mathbb{F}_{q'}) \\ & \cap & & & \cap \\ & E^-(\mathbb{F}_q) & & \cong & B(\mathbb{F}_{q'}) \end{array}$$

Note that the underlying four-dimensional abelian variety A is invisible in the algorithms in §4.2.

6 Cryptographic Applications of Algebraic Tori and Their Quotients

We give an exposition of some of the mathematics underlying torus-based cryptography (i.e., the $\mathbb{T}_n$-cryptosystems) and the cryptosystems discussed in §2. We discuss how the latter schemes are based on quotients of tori by the actions of symmetric groups.

6.1 Algebraic Tori

Definition 9. An *algebraic torus* is an algebraic group that over some larger field is a product of multiplicative groups. A field over which the torus becomes isomorphic to a product of multiplicative groups is called a *splitting field* for the torus; one says that the torus *splits* over that field. See [23,30] for expositions.

Example 10. (i) For every positive integer r, $\mathbb{G}_m^r$ is an r-dimensional algebraic torus.
(ii) $\mathrm{Res}_{\mathbb{F}_{q^n}/\mathbb{F}_q}\mathbb{G}_m$ is an n-dimensional algebraic torus over $\mathbb{F}_q$ that splits over $\mathbb{F}_{q^n}$.

By Proposition 2.6 of [26], the group $\mathbb{T}_n$ defined in §5.6 is a $\varphi(n)$-dimensional torus.

6.2 Rationality and Birational Isomorphisms

If r is a positive integer, write $\mathbb{A}^r$ for affine r-space. For any field F, we have $\mathbb{A}^r(F) = F^r$, the direct sum of r copies of F.

Definition 11. A *rational* map between algebraic varieties is a function defined by polynomials or quotients of polynomials that is defined almost everywhere. A *birational isomorphism* between algebraic varieties is a rational map that has a rational inverse (the maps are inverses wherever both are defined). A d-dimensional variety is *rational* if it is birationally isomorphic to $\mathbb{A}^d$.

Note that birational isomorphisms are not necessarily group isomorphisms. Note also that rational maps are not necessarily functions — they might fail to be defined on a lower dimensional set.

By (2), if $\mathbb{T}_n$ is rational (i.e., birationally isomorphic to $\mathbb{A}^{\varphi(n)}$), then almost all elements of T_n can be represented by $\varphi(n)$ elements of $\mathbb{F}_q$.

The maps f and j in §3 are only birational. The sets T_n and $\mathbb{F}_q^{\varphi(n)}$ are of size approximately $q^{\varphi(n)}$. The "bad" sets where f and j are not defined correspond to algebraic subvarieties of dimension at most $\varphi(n) - 1$, and therefore have at most $cq^{\varphi(n)-1}$ elements for some constant c. Thus the probability that an element lands in the bad set is at worst c/q, which will be small for large q. In any given case the bad sets might be even smaller. For example, in §3.6 the bad sets have 2 and 0 elements, respectively.

6.3 Obtaining the Rational Maps f and j

How were the maps in Examples 11 and 12 of [25] and in §3.6 above arrived at? The idea is as follows.

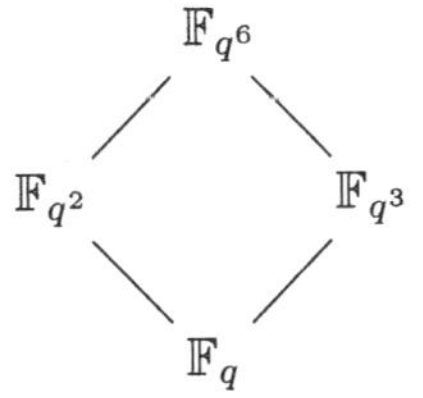

The one-dimensional torus $\mathbb{T}_{2,q^3}$ is, by definition, the kernel of the norm map $N_{\mathbb{F}_{q^6}/\mathbb{F}_{q^3}}$. The torus

$$\mathcal{T} := \mathrm{Res}_{\mathbb{F}_{q^3}/\mathbb{F}_q}(\mathbb{T}_{2,q^3})$$

has dimension 3. As in §3.4, the torus $\mathbb{T}_{2,q^3}$ is rational (i.e., is birationally isomorphic to $\mathbb{A}^1$), and thus the torus $\mathcal{T}$ is rational (i.e., birationally isomorphic to $\mathbb{A}^3$). The two-dimensional torus $\mathbb{T}_6$ is the hypersurface cut out by the equation $\mathrm{N}_{\mathbb{F}_{q^6}/\mathbb{F}_{q^2}} = 1$ inside the torus $\mathcal{T}$. This hypersurface is defined by a quadratic equation that can be used to parametrize the hypersurface. We gave examples of this in Examples 11 and 12 of [25]. Section 3.6 gives an additional example.

6.4 A Group Action on the Torus

Next, we define actions of symmetric groups on the tori $\mathbb{T}_n$. Suppose e is a divisor of n, and let $d = n/e$. Since n is square-free, we have $\gcd(e, d) = 1$, so

$$\mathbb{Z}/n\mathbb{Z} \quad \cong \quad \mathbb{Z}/e\mathbb{Z} \times \mathbb{Z}/d\mathbb{Z}.$$

The symmetric group on e letters, S_e, acts on $\mathbb{Z}/e\mathbb{Z}$. Extend this action to an action of S_e on $\mathbb{Z}/n\mathbb{Z}$, by acting trivially on $\mathbb{Z}/d\mathbb{Z}$. Now define an action of S_e on $\mathbb{A}^n$ $(= \mathbb{A}^{\mathbb{Z}/n\mathbb{Z}})$ as follows. For $\pi \in S_e$,

$$(x_i)_{i\in\mathbb{Z}/n\mathbb{Z}} \quad \mapsto \quad (x_{\pi^{-1}(i)})_{i\in\mathbb{Z}/n\mathbb{Z}}.$$

We have

$$\mathbb{A}^n \underset{\mathbb{F}_{q^n}}{\cong} \mathrm{Res}_{\mathbb{F}_{q^n}/\mathbb{F}_q}\mathbb{A}^1 \supset \mathrm{Res}_{\mathbb{F}_{q^n}/\mathbb{F}_q}\mathbb{G}_m \supset \mathbb{T}_n.$$

The action of S_e on $\mathbb{A}^n$ preserves $\mathrm{Res}_{\mathbb{F}_{q^n}/\mathbb{F}_q}\mathbb{G}_m$. However, it does not necessarily preserve the torus $\mathbb{T}_n$.

Theorem 12 (Lemma 3.5 of [26]) *If p is a prime divisor of n, then the above action of S_p on $\mathbb{A}^n$ preserves the torus $\mathbb{T}_n$.*

6.5 Interpreting the Other Systems in Terms of Quotients of Tori

- The Lucas-based cryptosystems are “based on” the quotient variety $\mathbb{T}_2/S_2$.
- The Gong-Harn system is based on the quotient variety $\mathbb{T}_3/S_3$.
- XTR is based on the quotient variety $\mathbb{T}_6/S_3$.
- Conjectural “Looking beyond XTR” systems would rely on the quotient variety $\mathbb{T}_{30}/(S_3 \times S_5)$ or $\mathbb{T}_{30}/(S_2 \times S_3 \times S_5)$.

These quotient varieties are *not* groups. This is why the Lucas-based systems and XTR do not do straightforward multiplication.

- The $\mathbb{T}_2$-cryptosystem is based on the group (and torus) $\mathbb{T}_2$.
- CEILIDH is based on the group (and torus) $\mathbb{T}_6$.

– The (sometimes conjectural) $\mathbb{T}_n$-cryptosystems are based on the group (and torus) $\mathbb{T}_n$.

We therefore call the $\mathbb{T}_n$-cryptosystems "torus-based cryptosystems".

What do we mean when we say that these systems are "based on" certain algebraic varieties?

XTR works because the variety $\mathbb{T}_6/S_3$ is rational, and the trace map $\mathbb{F}_{p^6} \to \mathbb{F}_{p^2}$ induces a birational isomorphism:

$$\mathbb{T}_6/S_3 \dashrightarrow \mathbb{A}^2 = \mathrm{Res}_{\mathbb{F}_{q^2}/\mathbb{F}_q}\mathbb{A}^1.$$

Similarly for the Lucas-based cryptosystems, the trace map $\mathbb{F}_{p^2} \to \mathbb{F}_p$ induces a birational isomorphism:

$$\mathbb{T}_2/S_2 \dashrightarrow \mathbb{A}^1.$$

More precisely, let $B_{(d,e)}$ denote the image of $\mathbb{T}_n$ in $(\mathrm{Res}_{\mathbb{F}_{q^n}/\mathbb{F}_q}\mathbb{G}_m)/S_e$ (where $n = de$). By Theorem 3.7 of [26], $B_{(d,e)}$ is birationally isomorphic to $\mathbb{T}_n/(S_{p_1} \times \cdots \times S_{p_r})$ where $e = p_1 \cdots p_r$ is the prime factorization of e. Note that the quotient map $\mathbb{T}_n \to \mathbb{T}_n/S_e$ induces a (non-surjective) map on $\mathbb{F}_q$-points:

$$T_n = \mathbb{T}_n(\mathbb{F}_q) \to (\mathbb{T}_n/S_e)(\mathbb{F}_q).$$

Let

$$\mathrm{XTR}(d,e) = \{\mathrm{Tr}_{\mathbb{F}_{q^n}/\mathbb{F}_{q^d}}(\alpha) : \alpha \in T_n\} \subset \mathbb{F}_{q^d}.$$

When $(d,e) = (1,2)$ or $(2,3)$, then $\mathrm{XTR}(d,e)$ is the set of traces that occur in the Lucas-based systems and XTR, respectively. In these two cases, $\mathrm{XTR}(d,e)$ can be naturally identified with the image of $\mathbb{T}_n(\mathbb{F}_q)$ in $(\mathbb{T}_n/S_e)(\mathbb{F}_q)$. More precisely (see Theorem 13 of [25]), when $(d,e) = (1,2)$ or $(2,3)$, the trace map $\mathrm{Tr}_{\mathbb{F}_{q^n}/\mathbb{F}_{q^d}}$ induces a birational embedding

$$\mathbb{T}_n/S_e \hookrightarrow \mathrm{Res}_{\mathbb{F}_{q^d}/\mathbb{F}_q}\mathbb{A}^1$$

such that $\mathrm{XTR}(d,e)$ is the image of the composition

$$T_n = \mathbb{T}_n(\mathbb{F}_q) \longrightarrow (\mathbb{T}_n/S_e)(\mathbb{F}_q) \hookrightarrow (\mathrm{Res}_{\mathbb{F}_{q^d}/\mathbb{F}_q}\mathbb{A}^1)(\mathbb{F}_q) \cong \mathbb{F}_{q^d}.$$

6.6 "Looking beyond XTR"

The paper "Looking beyond XTR" [6], building on a conjecture in [8], asks whether, for $n > 6$, some set of elementary symmetric polynomials can be used in place of the trace. In particular, [6] asks whether, when $d \mid n$ and $d \mid \varphi(n)$, one can recover the values of all the elementary symmetric polynomials (i.e., the entire characteristic polynomial) for $\mathrm{Gal}(\mathbb{F}_{p^n}/\mathbb{F}_{p^d})$ from the first $\varphi(n)/d$ of them (this was already answered in the affirmative in some cases in [8,13]). If this were true, one could use the first $\varphi(n)/d$ elementary symmetric polynomials on the set of $\mathrm{Gal}(\mathbb{F}_{p^n}/\mathbb{F}_{p^d})$-conjugates of an element $h \in T_n$ to represent h by $\varphi(n)$ elements of $\mathbb{F}_q$. More generally, [6] asks whether, for $d \mid n$, one can recover the entire characteristic polynomial over $\mathbb{F}_{p^d}$ from its first $\lceil \varphi(n)/d \rceil$ coefficients.

The answer is no. In particular, in [25] we show that when $n = 30$ and $p = 7$, then:

- for $d = 1$, no 8 ($= \varphi(n)/d$) elementary symmetric polynomials determine *any* of the remaining ones (except those determined by the symmetry of the characteristic polynomial),
- for $d = 1$, no 10 elementary symmetric polynomials determine *all* of them;
- for $d = 2$, no 4 ($= \varphi(n)/d$) elementary symmetric polynomials determine all of them.

Reinterpreted in terms of algebraic tori, the conjectures in [6] imply (see [26]) that the first eight elementary symmetric polynomials induce a birational isomorphism over $\mathbb{F}_p$:

$$\mathbb{T}_{30}/(S_2 \times S_3 \times S_5) \dashrightarrow \mathbb{A}^8,$$

and the first four elementary symmetric polynomials on the $\mathrm{Gal}(\mathbb{F}_{p^{30}}/\mathbb{F}_{p^2})$-conjugates of an element in T_{30} induce a birational isomorphism over $\mathbb{F}_p$:

$$\mathbb{T}_{30}/(S_3 \times S_5) \dashrightarrow \mathrm{Res}_{\mathbb{F}_{p^2}/\mathbb{F}_p} \mathbb{A}^4 \cong \mathbb{A}^8.$$

In [26] we prove that these statements are both false, for all but possibly finitely many primes p.

More generally, we have

$$\mathbb{T}_n \twoheadrightarrow B_{(d,e)} \hookrightarrow (\mathrm{Res}_{\mathbb{F}_{q^d}/\mathbb{F}_q} \mathbb{A}^1)^e \cong \mathbb{A}^n,$$

where the middle map $\oplus_{i=1}^{e} s_i$ is induced by the e elementary symmetric polynomials $s_1, \ldots, s_e$ on $\mathrm{Gal}(\mathbb{F}_{q^n}/\mathbb{F}_{q^d})$-conjugacy classes. (Recall that $B_{(d,e)}$ was defined at the end of §6.5, and $de = n$.)

The conjectures in [6] would imply that, when d divides $\varphi(n)$, then the first $\varphi(n)/d$ functions $s_1, \ldots, s_{\varphi(n)/d}$ induce a birational isomorphism

$$B_{(d,e)} \dashrightarrow (\mathrm{Res}_{\mathbb{F}_{q^d}/\mathbb{F}_q} \mathbb{A}^1)^{\varphi(n)/d} \cong \mathbb{A}^{\varphi(n)}.$$

This is true when the pairs (d, e) are $(1, 1)$ (this is Diffie-Hellman), $(1, 2)$ (Lucas-based systems), $(1, 3)$ (Gong-Harn), and $(2, 3)$ (XTR). It is also true (see [8]) when ℓ is a prime and $(d, e) = (1, \ell)$ or $(2, \ell)$. As noted above, we showed in [25,26] that this is false for $(d, e) = (1, 30)$ and $(2, 15)$ (in all but at most finitely many characteristics).

When $(d, e) = (n, 1)$, the underlying variety $B_{(d,e)}$ is $\mathbb{T}_n$ itself, corresponding to the $\mathbb{T}_n$-cryptosystems.

In summary, elementary symmetric polynomials are not the correct functions to use. In the next section we state a conjecture (of Voskresenskii) that seems to be closer to the truth.

6.7 Voskresenskii's Conjecture

Conjecture 13 (Voskresenskii) $\mathbb{T}_n$ *is rational; i.e., for every* n, *there is a birational isomorphism*

$$\mathbb{T}_n \dashrightarrow \mathbb{A}^{\varphi(n)}.$$

The conjecture is true, and not difficult to prove, if n is a prime power [30]. The conjecture was proved by Klyachko [16] when n is a product of two prime powers. Explicit birational isomorphisms are given in §5 of [25] and §3.6 above (see also §3.4 above), in the cases $n = 2$ and 6. A $\mathbb{T}_n$-cryptosystem arises for every n for which Voskresenskii's Conjecture is true with efficiently computable birational maps.

When n is divisible by more than two distinct primes, Voskresenskii's Conjecture is still an open question. In particular, the conjecture is not known when $n = 30 = 2 \cdot 3 \cdot 5$. We have tried unsuccessfully to construct a birational isomorphism between $\mathbb{T}_{30}$ and $\mathbb{A}^8$. It would be interesting to know whether Voskresenskii's Conjecture is true or false when $n = 30$. We have been able to construct explicit rational maps of low degree in this case, which might be useful if no birational map exists. For example, an s-to-1 map from $\mathbb{T}_{30}$ to $\mathbb{A}^8$ would provide a lossy compression scheme, and would allow one to represent elements of T_{30} in $\mathbb{F}_q^8 \times \{1, \dots, s\}$.

Rationality of the varieties $B(1, n)$ (or more generally the varieties $B(d, e)$) would imply the conjecture in [8].

6.8 Stable Rationality

One reason that Voskresenskii's Conjecture would be difficult to disprove is that the tori $\mathbb{T}_n$ are known to always be stably rational over $\mathbb{F}_q$ (see the Corollary on p. 61 of [30]).

Definition 14. A variety V over k is called *stably rational* over k if for some r and s, $V \times \mathbb{A}^r$ is birationally isomorphic over k to $\mathbb{A}^s$ (i.e., $V \times \mathbb{A}^r$ is rational for some $r \geq 0$).

Although the stable rationality of $\mathbb{T}_n$ does not allow one to represent elements of T_n in $\mathbb{F}_q^{\varphi(n)}$, it does allow one to represent elements of $T_n \times \mathbb{F}_q^r$ in $\mathbb{F}_q^s$ for suitable r and s, and this might be useful.

7 Open Problems

Some goals for the future are:

- Improve the efficiency of CEILIDH.
- Obtain more efficient key agreement, encryption, and signature schemes, by generalizing to $\mathbb{T}_{30}$-cryptosystems:
  - find explicit and efficient birational isomorphisms f and j between $\mathbb{T}_{30}$ and $\mathbb{A}^8$, if such exist,

  - look for special attacks on the discrete log problem in $\mathbb{F}_{q^{30}}^{\times}$.
- Use non-supersingular (i.e., ordinary) abelian varieties to further improve pairing-based cryptography.

Progress has been made on the last point in the case of elliptic curves; see for example [7].

Acknowledgments. The authors thank Dan Bernstein, Steven Galbraith, and Paul Leyland for helpful comments on a draft of the paper.

References

1. P. Barreto, *Pairing-based crypto lounge:* `http://planeta.terra.com.br/informatica/paulobarreto/pblounge.html`
2. D. Bernstein, *A state-of-the-art public-key signature system,* `http://cr.yp.to/sigs.html`
3. D. Bleichenbacher, *Compressing Rabin Signatures,* in Topics in Cryptology — CT-RSA 2004, Lect. Notes in Comp. Sci. **2964**, Springer, Berlin, 2004, 126–128.
4. D. Bleichenbacher, W. Bosma, A. K. Lenstra, *Some remarks on Lucas-based cryptosystems,* in Advances in Cryptology — CRYPTO '95, Lect. Notes in Comp. Sci. **963**, Springer, Berlin, 1995, 386–396.
5. D. Boneh, B. Lynn, H. Shacham, *Short signatures from the Weil pairing,* in Advances in Cryptology — Asiacrypt 2001, Lect. Notes in Comp. Sci. **2248**, Springer, Berlin, 2001, 514–532.
6. W. Bosma, J. Hutton, E. R. Verheul, *Looking beyond XTR,* in Advances in Cryptology — Asiacrypt 2002, Lect. Notes in Comp. Sci. **2501**, Springer, Berlin, 2002, 46–63.
7. F. Brezing, A. Weng, *Elliptic curves suitable for pairing based cryptography,* Cryptology ePrint Archive, Report 2003/143.
8. A. E. Brouwer, R. Pellikaan, E. R. Verheul, *Doing more with fewer bits,* in Advances in Cryptology — Asiacrypt '99, Lect. Notes in Comp. Sci. **1716**, Springer, Berlin, 1999, 321–332.
9. D. Coppersmith, *Fast evaluation of logarithms in fields of characteristic two,* IEEE Trans. Inform. Theory **30** (1984), 587–594.
10. W. Diffie, M. E. Hellman, *New Directions in Cryptography,* IEEE Trans. Inform. Theory **22** (1976), 644–654.
11. G. Frey, *Applications of arithmetical geometry to cryptographic constructions,* in Finite fields and applications (Augsburg, 1999). Springer, Berlin, 2001, 128–161.
12. S. Galbraith, *Supersingular curves in cryptography,* in Advances in Cryptology — Asiacrypt 2001, Lect. Notes in Comp. Sci. **2248**, Springer, Berlin, 2001, 495–513.
13. G. Gong, L. Harn, *Public-key cryptosystems based on cubic finite field extensions,* IEEE Trans. Inform. Theory **45** (1999), 2601–2605.
14. A. Joux, *A one round protocol for tripartite Diffie-Hellman,* in Algorithmic Number Theory Symposium (ANTS-IV), Lect. Notes in Comp. Sci. **1838**, Springer, Berlin, 2000, 385–394.
15. A. Joux, *The Weil and Tate Pairings as Building Blocks for Public Key Cryptosystems,* in Algorithm Number Theory Symposium (ANTS-V), Lect. Notes in Comp. Sci. **2369**, Springer, Berlin, 2002, 20–32.

16. A. A. Klyachko, *On the rationality of tori with cyclic splitting field*, in Arithmetic and geometry of varieties, Kuybyshev Univ. Press, Kuybyshev, 1988, 73–78 (Russian).
17. A. K. Lenstra, *Using Cyclotomic Polynomials to Construct Efficient Discrete Logarithm Cryptosystems Over Finite Fields*, in Information Security and Privacy, Proc. ACISP '97, Lect. Notes in Comp. Sci. **1270**, Springer, Berlin, 1997, 127–138.
18. A. K. Lenstra, *The XTR public key system*, lecture at MSRI Number-Theoretic Cryptography Workshop, October 20, 2000.
19. A. K. Lenstra, E. R. Verheul, *The XTR public key system*, in Advances in Cryptology — CRYPTO 2000, Lect. Notes in Comp. Sci. **1880**, Springer, Berlin, 2000, 1–19.
20. E. Lucas, *Théorie des fonctions numériques simplement périodiques*, Amer. J. Math. **1** (1878), 184–239, 289–321.
21. A. J. Menezes, P. C. van Oorschot, S. A. Vanstone, Handbook of applied cryptography, CRC Press, Boca Raton, FL, 1997.
22. W. B. Müller, W. Nöbauer, *Some remarks on public-key cryptosystems*, Studia Sci. Math. Hungar. **16** (1981), 71–76.
23. T. Ono, *Arithmetic of algebraic tori*, Ann. of Math. **74** (1961), 101–139.
24. K. Rubin, A. Silverberg, *Supersingular abelian varieties in cryptology*, in Advances in Cryptology — CRYPTO 2002, Lect. Notes in Comp. Sci. **2442**, Springer, Berlin, 2002, 336–353.
25. K. Rubin, A. Silverberg, *Torus-based cryptography*, in Advances in Cryptology — CRYPTO 2003, Lect. Notes in Comp. Sci. **2729** (2003), Springer, Berlin, 2003, 349–365.
26. K. Rubin, A. Silverberg, *Algebraic tori in cryptography*, to appear in High Primes and Misdemeanours: lectures in honour of the 60th birthday of Hugh Cowie Williams, Fields Institute Communications Series, AMS, Providence, RI, 2004.
27. R. Sakai, K. Ohgishi, M. Kasahara, *Cryptosystems based on pairing*, SCIS2000 (The 2000 Symposium on Cryptography and Information Security), Okinawa, Japan, January 26–28, 2000, C20.
28. P. J. Smith, M. J. J. Lennon, *LUC: A New Public Key System*, in Proceedings of the IFIP TC11 Ninth International Conference on Information Security IFIP/Sec '93, North-Holland, Amsterdam, 1993, 103–117.
29. P. Smith, C. Skinner, *A public key cryptosystem and a digital signature system based on the Lucas function analogue to discrete logarithms*, in Advances in Cryptology — Asiacrypt 1994, Lect. Notes in Comp. Sci. **917**, Springer, Berlin, 1995, 357–364.
30. V. E. Voskresenskii, Algebraic groups and their birational invariants, Translations of Mathematical Monographs **179**, AMS, Providence, RI, 1998.
31. H. C. Williams, *A* $p+1$ *method of factoring*, Math. Comp. **39** (1982), 225–234.
32. H. C. Williams, *Some public-key crypto-functions as intractable as factorization*, Cryptologia **9** (1985), 223–237.

Elliptic Curves of Large Rank and Small Conductor

Noam D. Elkies[1*] and Mark Watkins[2**]

[1] Department of Mathematics, Harvard University, Cambridge, MA 02138
elkies@math.harvard.edu
[2] The Pennsylvania State University
watkins@math.psu.edu

Abstract. For $r = 6, 7, \ldots, 11$ we find an elliptic curve $E/\mathbf{Q}$ of rank at least r and the smallest conductor known, improving on the previous records by factors ranging from 1.0136 (for $r = 6$) to over 100 (for $r = 10$ and $r = 11$). We describe our search methods, and tabulate, for each $r = 5, 6, \ldots, 11$, the five curves of lowest conductor, and (except for $r = 11$) also the five of lowest absolute discriminant, that we found.

1 Introduction and Motivation

An elliptic curve over the rationals is a curve E of genus 1, defined over $\mathbf{Q}$, together with a $\mathbf{Q}$-rational point. A theorem of Mordell [23] states that the rational points on E form a finitely generated abelian group under a natural group law. The rank of E is the rank of the free part of this group. Currently there is no general unconditional algorithm to compute the rank. Elliptic curves of large rank are hard to find; the current record is a curve of rank at least 24 (see [16]).[1]

We investigate a slightly different question: instead of seeking curves of large rank, we fix a small rank r (here $5 \le r \le 11$) and try to make the conductor N as small as possible, which, due to the functional equation for the L-function of the elliptic curve, is more natural than trying to minimize the absolute discriminant $|\Delta|$. The question of how fast the rank can grow as a function of N has generated renewed interest lately, partially due to the predictions made by random matrix theory about ζ-function analogues [8]. However, there are at present two different conjectures, one that comes from a function field analogue, and another from analytic number theory considerations. We shall return to this in Section 5.

* Supported in part by NSF grant DMS-0200687.

** Supported in part by VIGRE postdoctoral funding from the NSF, and was a Visiting Scholar with the MAGMA Computer Algebra Group at the University of Sydney during part of the time in which this work was done.

[1] Conrey wrote of a curve of rank 26 [7, p. 353], but confirms in e-mail to the authors that "26" was a typographical error for "24" as in [16].

D.A. Buell (Ed.): ANTS 2004, LNCS 3076, pp. 42–56, 2004.

We try to find $E/\mathbf{Q}$ of high rank and low conductor by searching for elliptic curves that have many integral points. As stated, this strategy is ill-posed, as integrality of points is not invariant under change of model (defining equation). However, if we only consider (say) Néron models then the question makes sense, and a conjecture of Lang [15, p. 140] links the number of integral points to the rank, at least in one direction. More explicitly, one might conjecture that there is an absolute constant C such that the number of integral points on an elliptic curve E of rank r is bounded by C^{r+1}. The best result is due to Silverman [27], who shows the conjecture is true when the j-invariant $j(E)$ is integral, and in fact proves that for every number field K there is a constant C_K such that the number of S-integral points over K is bounded by $C_K^{(1+r)(1+\delta)+|S|}$, where δ is the number of primes of K at which $j(E)$ is nonintegral. Explicit constants appear in [13]. Szpiro's conjecture [31], which is equivalent to the Masser-Oesterlé ABC conjecture [24], states that $\Delta \ll N^{6+o(1)}$; Hindry and Silverman [14] show this implies that the number of S-integral points on a quasi-minimal model of E/K is bounded by $C_K^{(1+r)\sigma_{E/K}+|S|}$ where $\sigma_{E/K}$ is the Szpiro ratio, which is the ratio of the logarithms of the norms of the discriminant and the conductor of E/K. Finally, Abramovich [1] has shown that the Lang-Vojta conjecture (which states that the integral points on a quasi-projective variety of log general type are not Zariski dense, see [35, 4.4]) implies the uniform boundedness of the number of integral points on rational semistable elliptic curves, but the lack of control over the Zariski closure of the integral points makes this result ineffective.

Conversely, it is frequently the case that elliptic curves of high rank, and especially those with relatively small conductor, have many integral points, and thus our search method is likely to find these curves. In fact, for each r in our range $5 \leq r \leq 11$ we found a curve E of rank at least r whose conductor N is the smallest known. For $r = 5$ this was a previously known (see [5]) curve with $N = 19047851$. For the other r our curve is new, with N smaller than the previous record by a factor ranging from 1.0136 for $r = 6$ to over 100 for $r = 10$ and $r = 11$. As a byproduct we also find the curves of rank r whose discriminants Δ have the smallest absolute values known. We estimate that finding a similarly good rank 12 curve would take 20–25 times as much work as for rank 11.

Since rational elliptic curves are modular [36,32,9,6,3], the tables of Cremona [10] are complete for $N \leq 20000$. Hence the lowest conductors for ranks 0–3 are respectively 11, 37, 389, and 5077. The rank 4 record was found by McConnell and appears in his Maple™ package APECS [17]; the curve has $[a_1, a_2, a_3, a_4, a_6] = [1, -1, 0, -79, 289]$ (see Section 2.1 for notation) and its conductor of 234446 is more than twice as small as the best example in [5]. Stein, Jorza, and Balakrishnan have verified [28] that there is no rank 4 elliptic curve of prime conductor less than 234446.

The rest of this paper is organized as follows. In the next section we describe the methods we used to search efficiently for curves with many small integral points. We then report on the curves of low conductor and/or absolute discriminant that we found, and compare them with previous records. The next section reports on our computation of further integral points on each of these record curves and on many others found in our search. Finally we compare our numerical results with previous speculations on the growth of the minimal N as a function of r.

2 Algorithms

We describe two algorithms that each find elliptic curves with numerous integral points whose x-coordinates have small absolute value. The input to our algorithms is an ordered triple (h, I, b_2) where h is a height parameter, I is a lower bound on the number of integral points we want, and $b_2 \in \{-4, -3, 0, 1, 4, 5\}$, these being the possible values of $b_2 = a_1^2 + 4a_2$ for an elliptic curve in minimal Weierstrass form (see below). We then try to find elliptic curves E with an equation $y^2 = 4x^3 + b_2x^2 + 2b_4x + b_6$ such that there are at least I integral points on E with $0 \leq y \leq 2h^3$, $|x| \leq h^2$, and $|2b_4| \leq 4h^4$. In modifications of the algorithm, we use variants of these bounds, and in general only have a high probability of finding the desired curves.

2.1 First Algorithm

An elliptic curve $E/\mathbf{Q}$ can be written in its minimal Weierstrass form as $Y^2 + a_1XY + a_3Y = X^3 + a_2X^2 + a_4X + a_6$, where a_1 and a_3 are 0 or 1 and $|a_2| \leq 1$. We can obtain the "2-torsion" equation[2]

$$y^2 = 4x^3 + b_2x^2 + 2b_4x + b_6$$

by completing the square via $y = 2Y + a_1X + a_3$ and $x = X$, so that we get $b_2 = a_1^2 + 4a_2$, $b_4 = a_1a_3 + 2a_4$, and $b_6 = a_3^2 + 4a_6$. Note that this transformation preserves integral points; we use the 2-torsion equation rather than the minimal equation since it is relatively fast to check whether its right-hand side is square. Fixing a choice of $b_2 \in \{-4, -3, 0, 1, 4, 5\}$ and a height-bound h, we search for curves with integral points by looping over the coordinates of such points. In particular, we first fix a b_4-value with $|2b_4| \leq 4h^4$ and then loop over integral values of x and y with $|x| \leq h^2$ and $0 \leq y \leq 2h^3$, and finally calculate the value of b_6 from the above equation, counting how often each b_6-value occurs. Note that the above bounds imply that $|y^2|$, $|4x^3|$, and $|2b_4x|$ are all bounded by $4h^6$.

This algorithm takes on the order of h^9 time, with memory requirements around h^5 for the recording of the b_6-values. There are various methods of speeding this up. We can note that neither positive b_4 nor negative b_6 are likely to give curves with many integral points, due to the shape of the cubic. From Table 1 we see that b_4 cannot be odd when b_6 is even. Also, we know that b_6 is a square modulo 4. We can extend this idea to probabilistic considerations; for instance, a curve with $b_2 = 1$ is not that likely to have numerous integral points unless

[2] The second-named author suggests this term because such a model makes it easy to locate the 2-torsion points on E: $(x, y) \in E[2]$ if and only if $y = 0$.

b_4 is odd and b_6 is 1 mod 8. We ran this algorithm for $h = 20$, and an analysis showed that the congruence restrictions most likely to produce good curves had (b_2, b_4, b_6) mod 8 equal to one of $(1,1,1)$, $(1,3,1)$, $(5,2,4)$, $(5,0,0)$, $(0,2,1)$, $(0,0,0)$, or $(4,0,1)$. Of course, there are curves that have many integral points yet fail such congruence restrictions, but the percentage of such is rather low (only 10–20%), and even those that do have numerous integral points appear less likely to have high rank. However, our table of records does contain some curves that fail these congruence restrictions, so there is some loss in making them. With these congruence restrictions, our computation took 15–20 hours on an Athlon MP 1600 to handle one b_2-value for $h = 20$; with no congruence restrictions, this would be about 5 days. Note that our congruence restrictions imply that the trials for $b_2 = \pm 4$ should only take half as long as the others. With this algorithm, we broke the low-conductor records of Tom Womack (from whose work this sieve search was adapted) for ranks 6, 7, and 8 (see Tables 2 and 3).[3]

Table 1. Congruence relations with a_1 and a_3

a_1	a_3	b_4	b_6	x and y
0	0	even	even	y even
0	1	even	odd	y odd
1	0	even	even	$y \equiv x\ (2)$
1	1	odd	odd	$y \not\equiv x\ (2)$

2.2 Second Algorithm

The number of elliptic curves E with $b_4 \ll h^4$ and $b_6 \ll h^6$ grows as h^{10}. The typical such curve has no small integral points at all: as we have seen, the number of (E, P), with E as above and $P \in E(\mathbf{Q})$ a small integral point, grows only as h^9, as does the time it takes to find all these (E, P). But we expect that even in this smaller set the typical E does not interest us, because it has no integral points other than $\pm P$. We shall see that there are (up to at most a logarithmic factor) only $O(h^8)$ curves E in this range together with a **pair** of integral points P, P' such that $P' \neq \pm P$, and that again we can find all such (E, P, P') with given b_2, b_4 in essentially constant time per curve. We thus gain a factor of almost h compared to our first algorithm.[4] Further improvements might be available by searching for elliptic curves with three or more points, but we do not know how to do this with the same time and space efficiency.

We wish to compute all $b_4, b_6, x_1, y_1, x_2, y_2$ in given ranges that satisfy the pair of equations $y_j^2 = 4x_j^3 + b_2 x_j^2 + 2b_4 x_j + b_6$ $(j = 1, 2)$. Subtracting these two

[3] In our tables the stated value of the "rank" is actually the rank of the subgroup generated by small integral points on the curve, which is very likely to be the actual rank, though in general such results can be quite difficult to prove.

[4] This pair-finding idea is also used in [11] to find curves $x^3 + y^3 = k$ of high rank.

equations, we find that

$$(y_2 - y_1)(y_2 + y_1) = (x_2 - x_1)[2b_4 + b_2(x_2 + x_1) + 4(x_2^2 + x_1x_2 + x_1^2)].$$

We can thus write

$$x_2 - x_1 = rt \quad \text{and} \quad y_2 - y_1 = rs \quad \text{and} \quad y_2 + y_1 = tu$$

for some integers r, s, t, u. From the latter two equations, we see that we need rs and tu to have the same parity in order for the y's to be integral. Our expectation is that generically we shall have $r, t \ll h$ and $s, u \ll h^2$ when the x-values are bounded by h^2 and the y-values by h^3. It is unclear how often this expectation is met. One way of estimating the proportion is to consider pairs of points $(x_1, y_1), (x_2, y_2)$ with $|x_i| \le h^2$ on various curves and see what values of (r, s, t, u) are obtained. This is not quite well-defined from the above; for instance, the quadruple $(x_1, x_2, y_1, y_2) = (7, 3, 6, 2)$ could have (r, s, t, u) as either $(4, 1, 1, 8)$ or $(2, 2, 2, 4)$. However, it becomes well-defined upon imposing the additional condition that $r = \gcd(y_2 - y_1, x_2 - x_1)$. Experiments show that about 18% of the (r, s, t, u) obtained from this process satisfy $1 \le r, t \le h$, though the exact percentage can vary significantly with the curve. Note that swapping r and t or negating either leads either to a switching of (x_1, y_1) and (x_2, y_2) or to a negation of y-values. Thus we can assume that $1 \le r \le t$.

We rewrite the above equation in the form

$$rstu = rt[2b_4 + b_2(x_1 + x_2) + 3(x_1 + x_2)^2 + (x_1 - x_2)^2]$$

and define $z = x_1 + x_2$ so that $su = 2b_4 + b_2 z + 3z^2 + (rt)^2$. Our algorithm is now the following. Given one of the six possible values of b_2, we loop over $2b_4$-values between $-4h^4$ and 0 (implementing our above comment that positive b_4-values are not that likely to give curves with many integral points). For each value of b_4 we loop over pairs of integers (r, t) that satisfy $1 \le r \le t \le h$. We then compute $l = rt$ and loop over values of $2x_2$ (that is, $z + l$) with $-2h^2 \le 2x_2 \le 2h^2$. Next we compute the quantity $W = 2b_4 + b_2 z + (l^2 + 3z^2)$ and factor this in all possible ways as $W = su$. We then take $y_2 = (rs + tu)/2$ (assuming that rs and tu have the same parity) and compute $b_6 = y_2^2 - 4x_2^3 - b_2x_2^2 - 2b_4x_2$. As before, we record the b_6-values and count how many times each occurs. This algorithm takes about $h^8 \log h$ time, where the logarithmic factor comes from solutions of $W = su$, assuming we can find these relatively fast via a lookup table. Already at $h = 20$, a version of this algorithm ran in under an hour and found most of the curves found by the first algorithm. One can view this algorithm as looping over pairs of x-values (both of size h^2), or more precisely the sum (given by z) and difference (given by l) of such a pair, and then reconstructing the y-values by factoring. Thus the inner loop takes time $h^4 \log h$ instead of h^5 as in the first algorithm.

2.3 Implementation Tricks

We now describe the various tricks we used in the implementation. We shall see that our b-congruence restrictions allow us to limit the z and l values in a productive way. First we consider the cases where b_2 is odd. Given a fixed b_4-value we only loop over z's and l's that are both odd, and can note that this makes W odd. Actually we do not loop over l but determine it as $l = rt$; thus we are looping over odd r and t with $1 \leq r \leq t \leq h$. It may seem that this loses a factor of 4 of (r, s, t, u) quadruples (with the z-restriction losing nothing because z must be of the same parity as rt), but we claim that it is actually only a factor of 2 for "interesting" curves. Indeed, though our yield of b_6-values will drop by a factor of 4 because of this parity restriction on both r and t, many of these values of b_6 will correspond to curves on which all integral points have x-coordinates of the same parity. Since l is the difference of two x-coordinates, this implies that l must be even for all pairs of integral points. These curves, which are plainly less likely to have a large number of integral points, are over-represented in the curves we ignore through not considering even l. From this we get our heuristic assertion that restricting to odd l loses only a factor of about 2.

We also consider only the values of z for which $|W|$ is less than a certain bound. This serves a dual purpose in that it speeds up the algorithm and also reduces the size of the tables used for factoring. We see that $W = 2b_4 + b_2z + (l^2 + 3z^2)$ should be of size h^4, and so we restrict the size of W via the inequality $|W| \leq 2h^4/U$, where U is a parameter we can vary (we had $U = 1$ for the experiments with $h = 30$ and $h = 40$). Again it is not immediately clear how many (r, s, t, u) quadruples we miss by making this restriction on W, and again the proportion can depend significantly upon the curve (curves with b_4 near $-2h^4$ lead to more quadruples with large $|W|$ than those with b_4 close to 0). Experimentation showed that with $U = 1$ we catch on average about 83% of the relevant b_6-values under this restriction. Our expectation might be approximate inverse linearity of the catch rate in U, though only in the limit as $U \to \infty$. Experimentation showed that with $U = 8$ our catch rate is down to 27%, while at $U = 32$ it is about 10%. However, there is interdependence between this restriction and that on the size of r and t — when r and t are both small, this corresponds to a small x-difference, which implies a small y-difference, and so $W = su$ should also be diminished in size. In a final accounting of the proportion of (r, s, t, u) quadruples, including the loss of a factor of 2 from the parity restriction on l, we find that with $U = 1$ we catch 7.4% of the quadruples, with $U = 8$ we catch 2.5%, and with $U = 32$ we catch just under 1%. Most of the curves of interest to us have at least 40 integral points within the given bounds $|x| \leq h^2$ and $0 \leq y \leq 2h^3$, and thus have at least 780 pairs of integral points. So missing 93% or more of the (r, s, t, u) quadruples does not trouble us — indeed, our "laziness" in not finding all the possible (r, s, t, u) is quadratically efficient compared to what we would achieve via similar "laziness" in our first algorithm.

So far we assumed that b_2 was odd, but similar ideas apply also for even values of b_2. When $b_2 = \pm 4$, we took l and z to be even but not congruent modulo 4. This ensures that W is 4 mod 8. Similarly, when $b_2 = 0$ and b_6 is odd, we take l and z to be even and congruent modulo 4, again ensuring that W is 4 mod 8. To implement these restrictions, we took $r \equiv 2 \bmod 4$ with no restriction on t other than $t \geq r$ if t itself is also 2 mod 4 (with both variables less than h as before).

Again we required that $|W| \leq 2h^4/U$, and here we have various restrictions on the decomposition $W = su$ depending on t mod 4. Specifically, we can always take s odd and $4|u$, we can take both s and u even if t is odd, and we can take $4|s$ and u odd if t is 2 mod 4, as we need for y to be odd in these cases. When $b_2 = 0$ and b_6 is even, we take z and l both to be odd, which makes W be 4 mod 8, and we need both s and u to be even (hence each 2 mod 4) for y to be even. As above, the loss in the number of interesting b_6-values from these restrictions is not much more than a factor of 2.

2.4 More Tricks

To reduce the memory needed for the counting of b_6 values, we used the following idea. We create a array of 2^L counters (of size 16 bits each); for instance, for $h = 30$ we used $L = 19$. Then for each b_6-value we obtain from above, we reduce $\lfloor b_6/8 \rfloor$ modulo 2^L, and increment the corresponding counter. In other words, we only record b_6 modulo 2^{L+3}. At the end of the loops over r, t, and z, we extract the counters with at least 10 hits. These residue classes are then passed to a secondary test phase. Here we set up counters for the values of b_6 with $0 \leq b_6 \leq 4h^6$ that are in the desired residue class $b_6^\#$ modulo 2^{L+3}. We then run through integral x with $|x| \leq h^2$, and for each x-value determine the corresponding positive y-values such that $y^2 \equiv 4x^3 + b_2x^2 + 2b_4x + b_6^\#$ mod 2^{L+3} via a lookup table of square roots modulo 2^{L+3}. Most of these y-values exceed $2h^3$, and we thus ignore them. If not, we compute b_6 exactly from x and y, and increment the corresponding counter. After running over all the x-values, we then check for large counter values. By taking 2^L somewhere around h^4 (note that this is about how many b_6-values we generate), we can use this method to handle a b_6-congruence-class in essentially h^2 time. This is generically small compared to the h^4 time for the loops over r, t, and z; when $h = 30$ we averaged about 100 congruence classes checked for each b_4-value, but the time for the loops still dominated.

3 Experimental Results

We ran this algorithm with $h = 30$ and $U = 1$ with a few more congruence classes in consideration, taking about a day for each (b_2, b_4, b_6) class. We then proceeded to run it for $h = 40$ and $U = 1$, and then $h = 60$ and $U = 8$, taking a few weeks for each (b_2, b_4, b_6) class. Other runs were done with the "better" congruence restrictions of (b_2, b_4, b_6) with varied parameters up to $h = 90$ and $U = 48$. Though with the $U = 48$ restriction we are catching less than 1% of the (r, s, t, u) quadruples, by this time we expect that most interesting curves have 60 or more integral points with $|x| \leq h^2$ and $0 \leq y \leq 2h^3$; indeed, even with the $h = 60$ search all the record curves we found had at least 70 integral points in this range.

Table 2 lists minimal equations for each of the five curves of smallest conductor N for each rank from 5–11 that were found by the above method. Table

Table 2. Low conductor records for ranks 5–11

$[a_1, a_2, a_3, a_4, a_6]$	N	$\lvert\Delta\rvert/N$	I	r
$[0,0,1,-79,342]$	19047851	1	39	5
$[1,0,0,-22,219]$	20384311	1	29	5
$[0,0,1,-247,1476]$	22966597	1	40	5
$[1,-1,0,-415,3481]$	34672310	10	52	5
$[0,0,0,-532,4420]$	37396136	32	52	5
$[1,1,0,-2582,48720]$	5187563742	6	71	6
$[0,0,1,-7077,235516]$	5258110041	243	67	6
$[1,-1,0,-2326,43456]$	5739520802	2	60	6
$[1,-1,0,-16249,799549]$	6601024978	184	68	6
$[1,-1,1,-63147,6081915]$	6663562874	32768	88	6
$[0,0,0,-10012,346900]$	382623908456	32	101	7
$[1,0,1,-14733,694232]$	536670340706	8	77	7
$[0,0,1,-36673,2704878]$	814434447535	5	84	7
$[1,-1,0,-92656,10865908]$	858426129202	142	92	7
$[1,-1,0,-18664,958204]$	896913586322	26	109	7
$[1,-1,0,-106384,13075804]$	249649566346838	14	124	8
$[1,-1,0,-222751,40537273]$	292246301470558	2	101	8
$[0,0,0,-481663,128212738]$	314214346667560	160	141	8
$[1,-1,0,-71899,5522449]$	314658846776578	34	130	8
$[1,-1,0,-124294,14418784]$	315734078239402	106	131	8
$[1,-1,0,-135004,97151644]$	32107342006814614	122	191	9
$[1,-1,0,-613069,98885089]$	43537345103385386	242	203	9
$[0,0,1,-3835819,2889890730]$	62986816173592807	67	142	9
$[1,0,1,-1493028,701820182]$	72070075910145406	2	139	9
$[1,0,1,-1076185,496031340]$	77211251506212554	344	156	9
$[0,0,1,-16312387,25970162646]$	10189285026863130793	1331	262	10
$[1,-1,0,-10194109,12647638369]$	22006161865320788846	58	241	10
$[0,0,1,-21078967,35688990786]$	22630148490190627609	2173	238	10
$[1,-1,0,-1536664,648294124]$	25440555737235843986	2	207	10
$[1,-1,0,-4513546,3716615296]$	39432942782223365758	2	179	10
$[0,0,1,-16359067,26274178986]$	18031737725935636520843	1	229	11
$[1,-1,0,-38099014,115877816224]$	66484354768372183177742	34	281	11
$[1,-1,0,-41032399,106082399089]$	219576020293485812169274	2	236	11
$[1,-1,0,-34125664,69523358164]$	227946110025657660240686	2	215	11
$[1,-1,0,-56880994,168642718624]$	252948166615918192888894	2	235	11

Table 3. Value of $\log N$ for old and new rank records

	6	7	8	9	10	11
Old	22.383	27.703	33.962	40.721	49.033	55.852
New	22.370	26.670	33.151	38.008	43.768	51.246

4 lists similar data for smallest absolute discriminant $|\Delta|$. The rank 5 data agree with the data from the Elliptic Curve Database [29]. The I column gives how many x-coordinates of integral points we found (see Section 4) for the given

equation. Some of the curves fail our congruence conditions on (b_2, b_4, b_6), but they still can be found via a non-minimal model; indeed, letting c_4 and c_6 be the invariants of the minimal model, the model with invariants $12^4 c_4$ and $12^6 c_6$ has $b_2 = 0$ and $4|b_4$ and $8|b_6$ and is thus in the $(0, 0, 0)$ class. In this way, from $(b_2, b_4, b_6) = (0, -1826496, 2637633024)$ we recover the curve $[1, 0, 0, -22, 219]$.

Table 4. Low absolute discriminant records for ranks 5–10

$[a_1, a_2, a_3, a_4, a_6]$	$\|\Delta\|$	I	r
[0, 0, 1, −79, 342]	19047851	39	5
[1, 0, 0, −22, 219]	20384311	29	5
[0, 0, 1, −247, 1476]	22966597	40	5
[0, 1, 1, −100, 110]	55726757	33	5
[0, 0, 1, −139, 732]	59754491	32	5
[1, 0, 0, −9227, 340354]	6822208199	36	6
[0, 0, 1, −277, 4566]	7647224363	49	6
[0, 0, 1, −379, 5172]	8072781371	51	6
[0, 0, 1, −889, 9150]	8796007189	54	6
[0, 1, 1, −390, 5460]	9694585723	43	6
[0, 0, 1, −1387, 68046]	1829517077483	71	7
[0, 0, 1, −5707, 151416]	1991659717477	68	7
[1, 0, 1, −5983, 164022]	2010552189452	72	7
[1, 0, 1, −14505, 667472]	2132568452204	71	7
[0, 0, 1, −15577, 744876]	2206378706437	71	7
[0, 1, 1, −23846, 1022562]	409086620841461	78	8
[0, 0, 1, −23737, 960366]	457532830151317	96	8
[0, 1, 1, −16440, 1394010]	561715239383323	84	8
[1, −1, 0, −222751, 40537273]	584492602941116	101	8
[1, −1, 0, −201814, 34925104]	643509175703572	109	8
[0, 0, 1, −167419, 30261330]	95276302704064331	135	9
[1, 0, 1, −1493028, 701820182]	144140151820290812	139	9
[0, 0, 1, −514507, 140806716]	151673348057775877	126	9
[0, 0, 1, −402157, 96291336]	157107745029925477	131	9
[0, 0, 1, −826609, 289956150]	172539371946838571	120	9
[1, −1, 0, −1536664, 648294124]	50881111474471687972	207	10
[0, 0, 1, −1788817, 843180666]	59202439687694448757	176	10
[1, −1, 0, −4513546, 3716615296]	78865885564446731516	179	10
[0, 1, 1, −1856500, 1072474760]	87950374485438204043	154	10
[0, 0, 1, −2438527, 1545098346]	103294665688000244363	173	10

How good is this method at finding elliptic curves of low conductor N and relatively high rank? Obviously if such a curve has few integral points then we will not find it. Indeed, it was suggested to us by J. Silverman that for large ranks r the smallest conductor curve might not have r independent integral points. However, for the ranks we consider there are sufficiently many independent integral points; the same is true for Mestre's rank 15 curve [21], but apparently not for later rank records.

Note also that our search operates by increasing b_4 and b_6 corresponding to some height parameter, which is not exactly the same as simply increasing the absolute value $|\Delta|$ of the discriminant, which again is not quite the same as just increasing N. Finally, the probabilistic nature of our algorithm and the necessity of restricting to "likely" congruence classes also cast doubt on the exhaustiveness of our search procedure. However, we are still fairly certain that the curves we found for ranks 5–8 are indeed the actual smallest conductor curves for those ranks. Note that our methods were almost exhaustive in the region of interest (h up to about 30), and were verified with the first algorithm in much of this range. For rank 9 we could be missing some curves with large $|\Delta|/N$ and h around 50 or so; perhaps this range should be rechecked with a smaller U-parameter. Indeed, for a long time the second curve on the $r = 10$ list, which we found with an $h = 60$ search, was our rank 10 record, but then a run with $h = 80$ found the first and third curves which have h-values of about 64 and 68 respectively. We have yet to find many rank 11 curves with large $|\Delta|/N$; note that our current record curve in fact has prime conductor. This suggests that there still could be significant gains here. However, Table 3, which lists values of $\log N$ for the old records and our new ones, indicates that our method has already shown its usefulness. There does not seem to have been any public compilation of such records before Womack [37] did so on his website in the year 2000, soon after he had found the records for ranks 6–8 via a sieve-search.[5] The records for ranks 9–10 were again due to Womack but from a Mestre-style construction [20], with Mestre listing the rank 11 record in [19] (presumably found by the methods of [18]). There was no particular reason to expect the old records for ranks 9–11 to be anywhere near the true minima, as they were constructed without a concentrated attempt to make the conductor or any related quantity as small as possible.

4 Counting Integral Points

Since we have curves that have many integral points of small height, it is natural to ask how many integral points these curves have overall, with no size restriction. For our curves of rank higher than 8, current methods, as described in [30], do not yet make it routine to list all the integral points and to prove that the list is complete. Indeed, even verification that the rank is actually what it seems to be is not necessarily routine.

However, we have at least two ways to search for integral points and thus obtain at least a lower bound for the number of integral points. One method is a simple sieve-assisted search, which can reach x-values up to 10^{12} in just over an hour on an Athlon MP 1600. The other method is to write down a linearly independent set from the points we have, and then take small linear combinations

[5] Prior to Womack, there were records listed in the Edinburgh dissertation of Nigel Suess (2000); it appears that Womack and Suess enumerated these lists in part to help dispense Cremona from answering emails about the records. Indeed, the rank 4 record of McConnell [17] mentioned above was relatively unknown for quite a while.

of these.[6] For this, we took the linearly independent set that maximized the minimal eigenvalue of the height-pairing matrix (as in the "c_1-optimal basis" of [30]), subject to the condition that the set must generate (as a subgroup of $E(\mathbf{Q})$) all the integral points in our list. We then tried all $((2m+1)^r-1)/2$ relevant linear combinations with coefficients bounded in absolute value by m.[7] With $r=11$ and $m=3$ this takes about an hour.

The maximal number of integral points we found on a curve was 281×2 for $[1,-1,0,-38099014,115877816224]$. This can be compared with the rank 14 curve $[0,0,1,-2248232106757,1329472091379662406]$ that is listed by Mestre [19], which we find has 311×2 integral points with $|x|\le 10^{12}$, plus at least 7×2 more that were found with linear combinations as above. Note that amongst the curves of a given rank there is not much correspondence between number of integral points and smallness of conductor. For instance, we have no idea which curve of rank 7 has the maximal number of integral points;[8] trying a few curves found in our search turned up $[1,-1,0,-22221159,40791791609]$ which has at least 165×2 integral points, but conductor 13077343449126, more than 34 times larger than our record. Note that $|\Delta|/N = 2\cdot 3^4 11^4 23^2$ for this curve; in general, large values of $|\Delta|/N$ seem to correlate with large counts of integral points (see Table 2).

5 Growth of Maximal Rank as a Function of Conductor

We review two different heuristics and conjectures for the growth of the maximal rank of an elliptic curve as a function of its conductor, and then indicate which is more likely according to our data. The first conjecture is due to Murty and appears in the appendix to [25]. He first notes that, similar to a heuristic of Montgomery [22, pp. 512–513] regarding the ζ-function, it is plausible that $\arg L_E(1+it) \ll \sqrt{\log(Nt)/\log\log(Nt)}$ as $t\to\infty$. Murty speculatively applies this bound in a small circle of radius $1/\log\log N$ about $s=1$. He then claims that Jensen's Theorem implies that the order of vanishing of $L_E(s)$ at $s=1$ is bounded by $C\sqrt{\log N/\log\log N}$, though we cannot follow the argument. Assuming the Birch–Swinnerton-Dyer conjecture [2], the same upper bound holds for the rank of the elliptic curve. However, the Montgomery heuristic comes from taking the approximation $\log\zeta(s)=\sum_{p\le t} p^{-\sigma-it}+O_\sigma(1)$ for $\sigma>1/2$ and assuming that the p^{-it} act like random variables; upon taking a limit as $\sigma\to 1/2$, this implies the asserted bound of $\sqrt{\log t/\log\log t}$, but only for large t. Indeed, in our elliptic curve case with small t, we should have an approximation (see

[6] The possible size of coefficients in such linear combinations can be bounded via elliptic logarithms (possibly p-adic) as in [30] and later works. Also, as indicated by Zagier [38], one can combine elliptic logarithms with lattice reduction to search for large integral points, but we did not do this.

[7] We do not compute sums of points on elliptic curves directly over the rationals, but instead work modulo a few small primes and use the Chinese Remainder Theorem.

[8] The maximal count of integral points may well be attained by a curve with nontrivial torsion, whose discriminant would then be too large to be found by our search.

[12]) more like $\log L_E(s) \sim \sum_{p\le X} a_p/p^{-s}$; it is unclear whether the variation of the a_p or that of p^{-s} should have the greater impact. Finally, Conrey and Gonek [8] contest that Montgomery's heuristic could be misleading; they suggest that $\log|\zeta(1/2+it)|$ (and maybe analogously the argument) can be as big as $C\log t/\log\log t$ instead of the square root of this. One idea is that the above limit as $\sigma \to 1/2$ disregards a possibly larger error term coming from zeros of $\zeta(s)$; the asymmetry of upper and lower bounds for $\log|\zeta(1/2+it)|$ makes the analysis delicate. A classic paper of Shafarevich and Tate [26] shows that in function fields the rank grows at least as fast as the analogue of $\frac{\log N}{2\log\log N}$. However, the curves used in this construction were isotrivial, and thus fairly suspect for evidence toward a conjecture over number fields. Ulmer recently gave non-isotrivial function field examples with this growth rate, and conjectured [33, Conjecture 10.5] that this should be the proper rate of growth even in the number field case, albeit possibly with a different constant. In a different paper [34, p. 19], Ulmer notes that certain random matrix models suggest that the growth rate is as in the function field case; presumably this is an elliptic curve analogue of the work of [8].

Figure 1 plots the rank r versus $\log N/\log\log N$, where N is the smallest known conductor for an elliptic curve of rank r. A log-regression gives us that the best-fit exponent is 0.975, much closer to the exponent of 1.0 of Ulmer than

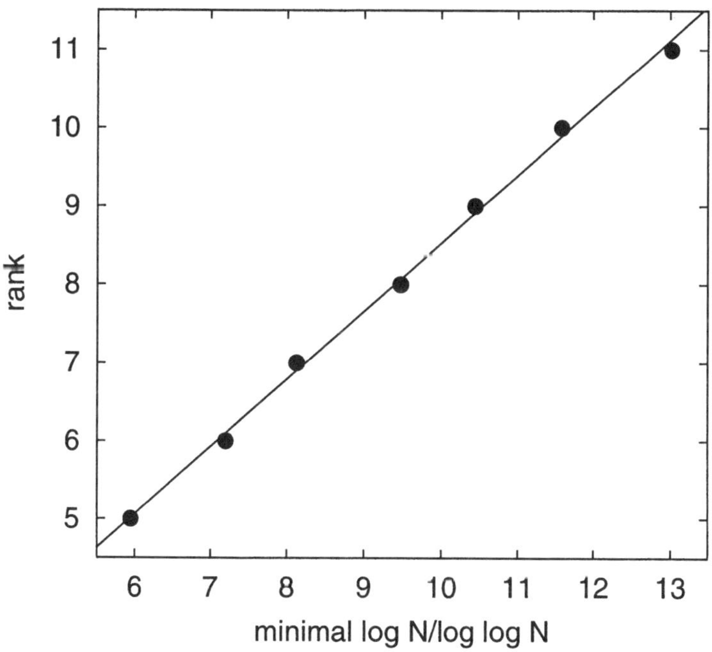

Fig. 1. Plot of rank versus $\log N/\log\log N$

to the 0.5 of Murty. Note that an improvement in the records for ranks 9–11 would most likely increase the best-fit exponent.

Assuming the growth is linear, the line of best fit is approximately $r = 0.865\frac{\log N}{\log\log N} - 0.126$. But this could mislead; GRH plus BSD implies

$$r \le \frac{1}{2}\frac{\log N}{\log\log N}\left(1 + \frac{\log 8e}{\log\log N} + O\left(\frac{1}{(\log\log N)^2}\right)\right).$$

The main term in the above already appears in Corollary 2.11 of [4] (see also Proposition 6.11), and we have simply calculated the next term in the expansion. To get more reliable data, we would need to consider curves with $\log\log N$ rather large, which is of course quite difficult.

Finally we mention a possible heuristic refinement of the above upper bound. The bound comes from a use of the Weil explicit formula (see [4, 2.11]) to obtain the relation $\sum_\gamma h(\gamma \log\log N) \sim \frac{\log N}{2\log\log N}$, where $h(t) = \left(\frac{\sin t}{t}\right)^2$ and the sum is over imaginary parts of nontrivial zeros of $L_E(s)$, counted with multiplicity. When only the high-order zero at $\gamma = 0$ contributes, we get the stated upper bound. In the function field case, the other zeros contribute little because they are all near the minima of h. This is unlikely to occur in the number field case. Also unlikely is the idea that the other zeros have negligible contribution due to the $1/t^2$ decay of h. Thus the other zeros are likely to have some impact; however, it is not clear how large this impact will be.

References

1. D. Abramovich, *Uniformity of stably integral points on elliptic curves.* Invent. Math. **127** (1997), no. 2, 307–317.
2. B. J. Birch, H. P. F. Swinnerton-Dyer, *Notes on elliptic curves. I.* J. reine angew. Math. **212** (1963), 7–25.
3. C. Breuil, B. Conrad, F. Diamond, R. Taylor *On the modularity of elliptic curves over* $\mathbf{Q}$*: wild 3-adic exercises.* J. Amer. Math. Soc. **14** (2001), no. 4, 843–939.
4. A. Brumer, *The average rank of elliptic curves. I.* Invent. Math. **109** (1992), no. 3, 445–472.
5. A. Brumer, O. McGuinness, *The behavior of the Mordell-Weil group of elliptic curves.* Bull. Amer. Math. Soc. (N.S.) **23** (1990), no. 2, 375–382.
6. B. Conrad, F. Diamond, R. Taylor, *Modularity of certain potentially Barsotti-Tate Galois representations.* J. Amer. Math. Soc. **12** (1999), no. 2, 521–567.
7. J. B. Conrey, *The Riemann hypothesis.* Notices Amer. Math. Soc. **50** (2003), no. 3, 341–353.
8. J. B. Conrey, S. M. Gonek, *High moments of the Riemann zeta-function.* Duke Math. J. **107** (2001), no. 3, 577–604.
9. F. Diamond, *On deformation rings and Hecke rings.* Ann. of Math. (2) **144** (1996), no. 1, 137–166.
10. J. E. Cremona, *Algorithms for modular elliptic curves.* Cambridge University Press, Cambridge, 1992. Second edition 1997. Book and data electronically available at `www.maths.nottingham.ac.uk/personal/jec/`
11. N. D. Elkies, N. F. Rogers, *Elliptic curves* $x^3 + y^3 = k$ *of high rank.* To appear in the ANTS-VI proceedings, `www.arxiv.org/abs/math.NT/0403116`

12. D. Goldfeld, *Sur les produits partiels eulériens attachés aux courbes elliptiques.* (French. English summary) [On the partial Euler products associated with elliptic curves] C. R. Acad. Sci. Paris Sér. I Math. **294** (1982), no. 14, 471–474.
13. R. Gross, J. Silverman, *S-integer points on elliptic curves.* Pacific J. Math. **167** (1995), no. 2, 263–288.
14. M. Hindry, J. H. Silverman, *The canonical height and integral points on elliptic curves.* Invent. Math. **93** (1988), no. 2, 419–450.
15. S. Lang, *Elliptic curves: Diophantine analysis.* Grundlehren der mathematischen Wissenschaften [Fundamental Principles of Mathematical Sciences], 231. Springer-Verlag, Berlin-New York, 1978.
16. R. Martin, W. McMillen, *An elliptic curve over* **Q** *with rank at least* 24. Posted to the NMBRTHRY mailing list on May 2, 2000 by V. Miller. Electronically available from `listserv.nodak.edu/archives/nmbrthry.html` and directly from `www.math.harvard.edu/~elkies/rk24_1.html`
17. I. McConnell, *Arithmetic of Plane Elliptic Curves (APECS).* Computer program written in Maple^{TM}, available from `ftp.math.mcgill.ca/pub/apecs`
18. J.-F. Mestre, *Construction d'une courbe elliptique de rang* $\geq$ 12. (French. English summary) [Construction of an elliptic curve of rank $\geq$ 12] C. R. Acad. Sci. Paris Sér. I Math. **295** (1982), no. 12, 643–644.
19. J.-F. Mestre, *Formules explicites et minorations de conducteurs de variétés algébriques.* (French) [Explicit formulas and lower bounds for the conductors of algebraic varieties] Compositio Math. **58** (1986), no. 2, 209–232.
20. J.-F. Mestre, *Courbes elliptiques de rang* $\geq$ 11 *sur* $\mathbf{Q}(t)$. (French) [Elliptic curves with rank $\geq$ 11 over $\mathbf{Q}(t)$] C. R. Acad. Sci. Paris Sér. I Math. **313** (1991), no. 3, 139–142.
21. J.-F. Mestre, *Un exemple de courbe elliptique sur Q de rang* $\geq$ 15. (French) [An example of an elliptic curve over **Q** with rank $\geq$ 15] C. R. Acad. Sci. Paris Sér. I Math. **314** (1992), 453–455.
22. H. L. Montgomery, *Extreme values of the Riemann zeta function.* Comment. Math. Helv. **52** (1977), no. 4, 511–518.
23. L. J. Mordell, *On the Rational Solutions of the Indeterminate Equations of the Third and Fourth Degrees.* Proc. Cambridge Philos. Soc. **21** (1922–1923), 179–192.
24. J. Oesterlé, *Nouvelles approches du "théorème" de Fermat.* (French) [New approaches to Fermat's "theorem"] Séminaire Bourbaki, Vol. 1987/88, Astérisque No. 161–162 (1988), Exp. No. 694, 4, 165–186 (1989).
25. C. S. Rajan, *On the size of the Shafarevich-Tate group of elliptic curves over function fields.* Compositio Math. **105** (1997), no. 1, 29–41.
26. I. R. Šafarevič, J. T. Tate, *The rank of elliptic curves.* Soviet Math. Dokl. **8** (1967), 917–920. [trans. J. W. S. Cassels from *Dokl. Akad. Nauk SSSR* **175** (1967), 770–773.]
27. J. H. Silverman, *A quantitative version of Siegel's theorem: integral points on elliptic curves and Catalan curves.* J. reine angew. Math. **378** (1987), 60–100.
28. W. Stein, A. Jorza, J. Balakrishnan, *The Smallest Conductor for an Elliptic Curve of Rank Four is Composite.* Available at `modular.fas.harvard.edu/papers/rank4`
29. W. A. Stein, M. Watkins, *A Database of Elliptic Curves—First Report.* In *Algorithmic number theory* (Sydney 2002), 267–275, edited by C. Fieker and D. Kohel, Lecture Notes in Comput. Sci., 2369, Springer, Berlin, 2002.
30. R. J. Stroeker, N. Tzanakis, *On the elliptic logarithm method for elliptic Diophantine equations: reflections and an improvement.* Experiment. Math. **8** (1999), no. 2, 135–149.

31. L. Szpiro, *Discriminant et conducteur des courbes elliptiques.* (French) [Discriminant and conductor of elliptic curves] Séminaire sur les Pinceaux de Courbes Elliptiques (Paris, 1988), Astérisque No. 183, (1990), 7–18.
32. R. Taylor, A. Wiles, *Ring-theoretic properties of certain Hecke algebras.* Ann. of Math. (2) **141** (1995), no. 3, 553–572.
33. D. Ulmer, *Elliptic curves with large rank over function fields.* Ann. of Math. (2) **155** (2002), no. 1, 295–315.
34. D. Ulmer, *Elliptic curves and analogies between number fields and function fields.* To appear in *Heegner points and Rankin L-series*, MSRI Publications 48, Cambridge University Press, New York, 2004. Preprint available at `www.arxiv.org/abs/math.NT/0305320`
35. P. Vojta, *A higher-dimensional Mordell conjecture.* In *Arithmetic geometry* (Storrs, Connecticut 1984), 341–353, edited by G. Cornell and J. H. Silverman, Springer-Verlag, New York, 1986.
36. A. Wiles, *Modular elliptic curves and Fermat's last theorem.* Ann. of Math. (2) **141** (1995), no. 3, 443–551.
37. T. Womack, *Best known conductors for elliptic curves of given rank*, webpage copied to `www.math.harvard.edu/~elkies/womack2001.html`
38. D. Zagier, *Large integral points on elliptic curves.* Math. Comp. **48** (1987), no. 177, 425–436.

Binary GCD Like Algorithms for Some Complex Quadratic Rings

Saurabh Agarwal and Gudmund Skovbjerg Frandsen

BRICS*, Department of Computer Science, University of Aarhus
IT-Parken, Åbogade 34, DK-8200, Aarhus N, Denmark.
{saurabh, gudmund}@daimi.au.dk

Abstract. On the lines of the binary gcd algorithm for rational integers, algorithms for computing the gcd are presented for the ring of integers in $\mathbb{Q}(\sqrt{d})$ where $d \in \{-2, -7, -11, -19\}$. Thus a binary gcd like algorithm is presented for a unique factorization domain which is not Euclidean (case $d = -19$). Together with the earlier known binary gcd like algorithms for the ring of integers in $\mathbb{Q}(\sqrt{-1})$ and $\mathbb{Q}(\sqrt{-3})$, one now has binary gcd like algorithms for all complex quadratic Euclidean domains. The running time of our algorithms is $O(n^2)$ in each ring. While there exists an $O(n^2)$ algorithm for computing the gcd in quadratic number rings by Erich Kaltofen and Heinrich Rolletschek, it has large constants hidden under the big-oh notation and it is not practical for medium sized inputs. On the other hand our algorithms are quite fast and very simple to implement.

1 Introduction

Greatest common divisor is one of the most fundamental concepts of number theory. Elementary number theory texts introduce gcd very early and also present an algorithm to compute it, the Euclid's algorithm. However, it is not possible to extend Euclid's algorithm to all number rings. The rings in which one can extend Euclid's algorithm are called Euclidean rings. A large amount of effort has been put in identifying Euclidean number rings. Franz Lemmermeyer's paper on Euclidean number rings [12] contains an almost complete list of all known Euclidean number rings.

In 1965 a different algorithm to compute gcd was presented by J. Stein [19]. Apart from being very simple to understand, this algorithm has the virtue of being efficiently implementable on a computer as the only operations used by the algorithm are addition, subtraction and division by 2. Since divisions by 2 can be performed by right shifts (on a computer), this algorithm essentially has no divisions at all. This algorithm is popularly known as the binary gcd algorithm. In this paper we present extensions of this algorithm to four complex quadratic rings.

* Basic Research in Computer Science (www.brics.dk), funded by the Danish National Research Foundation.

D.A. Buell (Ed.): ANTS 2004, LNCS 3076, pp. 57–71, 2004.

1.1 Overview of Results

We successfully generalize the binary gcd algorithm to compute the gcd in the ring of integers in $\mathbb{Q}(\sqrt{d})$ where $d \in \{-2, -7, -11, -19\}$. In each case the time complexity of the algorithm is $O(n^2)$ with small constants hidden under the big-oh notation. The only operations used in our algorithms are addition, subtraction and division by a small fixed number (2 or 3 or 5). One of the main result is an extension of the binary gcd algorithm to a unique factorization domain (UFD) which is not Euclidean (case $d = -19$). Our extension clearly indicates that the binary gcd like algorithms are not restricted to Euclidean rings.

1.2 Road Map

Section 2 contains some preliminaries. In Sect. 3, we review some other algorithms for computing the gcd. The main idea of our algorithm is presented in Sect. 4. In Sect. 5, the algorithms for computing the gcd in the ring of integers in $\mathbb{Q}(\sqrt{d})$ are presented where $d \in \{-2, -7, -11, -19\}$.

2 Preliminaries

The definitions/facts in this section are found in most books on algebra and/or algebraic number theory (for example see [10,6,7]). Complex quadratic number fields are of the form $\mathcal{Q} = \mathbb{Q}(\sqrt{d})$ where d is a negative square-free rational integer. Any $\alpha \in \mathcal{Q}$ is of the form $a + b\sqrt{d}$ where $a, b \in \mathbb{Q}$. For any $\alpha = a + b\sqrt{d}$, the norm of α is defined as $N(\alpha) = \alpha\bar{\alpha}$ where $\bar{\alpha} = a - b\sqrt{d}$ is the conjugate of α. If $\mathcal{Z}$ denotes the ring of algebraic integers in $\mathcal{Q}$, then

$$\mathcal{Z} = \begin{cases} \mathbb{Z} + \mathbb{Z}\left[\sqrt{d}\right] & \text{if } d \equiv 2, 3 \pmod 4 \\ \mathbb{Z} + \mathbb{Z}\left[\frac{1}{2}(-1 + \sqrt{d})\right] & \text{if } d \equiv 1 \pmod 4 \end{cases}$$

There are nine complex quadratic rings which are also UFD. These are the ring of integers in $\mathbb{Q}(\sqrt{d})$ where $d \in \{-1, -2, -3, -7, -11, -19, -43, -67, -163\}$ [18]. From now on we will assume that $\mathcal{Z}$ is one of these nine rings.

For all $\alpha \in \mathcal{Z}$, $N(\alpha)$ is a non-negative rational integer and $N(\alpha) = 0$ iff $\alpha = 0$. An element $u \in \mathcal{Z}$ is a unit iff $N(u) = 1$. Any two elements $\alpha, \beta \in \mathcal{Z}$ are called associates if $\alpha|\beta$ and $\beta|\alpha$. A non-zero non-unit element $\rho \in \mathcal{Z}$ is a prime if $(\rho|\alpha\beta) \Rightarrow (\rho|\alpha$ or $\rho|\beta)$. If $\rho \in \mathcal{Z}$ is a prime, then there exists a rational prime p such that $N(\rho) = p$ or p^2. In the former case ρ is not associate to any rational prime and in the latter case ρ is an associate to p. The quotient ring $\mathcal{Z}/\rho\mathcal{Z}$ is a finite field with $N(\rho)$ elements. If $N(\rho) = p$ for some odd rational prime p, then $\{-\frac{p-1}{2}, \ldots, 0, \ldots, \frac{p-1}{2}\}$ forms a complete set of coset representatives for $\mathcal{Z}/\rho\mathcal{Z}$ and if $N(\rho) = 2$, then $\{0, 1\}$ forms a complete set of coset representatives for $\mathcal{Z}/\rho\mathcal{Z}$.

Let $\alpha, \beta \in \mathcal{Z}$ and $\alpha\beta \neq 0$. Then a non-zero element $\mathfrak{g} \in \mathcal{Z}$ is said to be greatest common divisor (gcd) of α and β if

a. $\mathfrak{g}|\alpha$ and $\mathfrak{g}|\beta$, and
b. for any $\gamma \in \mathcal{Z}\backslash\{0\}$, if $\gamma|\alpha$ and $\gamma|\beta$, then $\gamma|\mathfrak{g}$.

For any $\alpha \neq 0$, gcd of α and 0 is defined to be α. In the literature, gcd of α and β is denoted by (α, β) and we also use this notation. In general (α, β) is not unique. However if $\mathfrak{g}_1 = (\alpha, \beta)$ and $\mathfrak{g}_2 = (\alpha, \beta)$, then $\mathfrak{g}_1$ and $\mathfrak{g}_2$ are associates. It is customary to overload the '=' operator for gcd. Thus a statement like $(\alpha, \beta) = (\gamma, \eta)$ means that gcd of α, β and gcd of γ, η are associates. The following facts about gcd are easily shown.

Lemma 1. *Let $\alpha, \beta \in \mathcal{Z}$ be arbitrary integers and $\rho \in \mathcal{Z}$ be any prime.*

a. *If $\rho|\alpha$ and $\rho|\beta$, then $(\alpha, \beta) = \rho\,(\alpha/\rho, \beta/\rho)$.*
b. *If $\rho|\alpha$ and $\rho \nmid \beta$, then $(\alpha, \beta) = (\alpha/\rho, \beta)$.*
c. *$(\alpha, \beta) = (\alpha + \lambda\beta, \beta)$ for all $\lambda \in \mathcal{Z}$.*

3 Related Work

The main aim of this section is to present some of the algorithms which can possibly be used or extended to compute the gcd in different number rings. There are many different known ways of computing the gcd and the list of algorithms discussed in this section is not exhaustive. In the rest of this section the term 'ring' will always mean number ring.

3.1 Euclidean Algorithm

The problem of computing the gcd is as old as number theory itself. Euclid gave an algorithm to compute the gcd of rational integers in 300 B.C. [9]. This algorithm is called the Euclidean Algorithm (EA). However EA cannot be extended to all rings. The rings in which one can extend EA are the Euclidean rings. A fairly complete list of all known Euclidean rings can be found in [12].

Let R be any Euclidean ring. Then by definition of Euclidean ring, there exists $\varphi\colon R \mapsto \mathbb{N} \cup \{0\}$ such that for any $a, b \in R$ there exits $q, r \in R$ such that $a = bq + r$ and $\varphi(r) < f(b)$. If $r \neq 0$, then $(a, b) = (r, b)$ and if $r = 0$, then $(a, b) = b$. A basic step in EA takes inputs a and b and finds such q and r. This step is called Euclidean division. EA repeats this step until $r = 0$ and then outputs the result. As φ is a non-negative integer-valued function, EA will always terminate. The analysis of EA in $\mathbb{Z}$ is well studied [2,9]. The running time of EA in $\mathbb{Z}$ is $O(n^2)$. One can establish that the time complexity of EA in imaginary Euclidean quadratic rings is $O(\mu(n)n)$ (see for example [14,12]) where $\mu(n)$ is the complexity of multiplying two n-bit integers and $\mu(n) = O(n \log n \log \log n)$ by [16]. A similar bound follows for some cyclotomic rings from the works of Hendrik Lenstra [13], Renate Scheidler and Hugh Williams [15].

There are two techniques of speeding up EA for large inputs. First technique is by D.H. Lehmer [11]. The basic idea behind this scheme is to perform single-precision arithmetic with leading digits of the input most of the time and fewer multiple-precision operations. André Weilert [21] and George Collins [4] have established that a similar scheme on Gaussian integers has a time complexity of $O(n^2)$. Second technique is controlled Euclidean descent by Arnold Schönhage [16]. For the five imaginary quadratic Euclidean rings, André Weilert has established that this technique has a time complexity of $O(\mu(n) \log n)$ [22,20].

3.2 Non-euclidean Algorithms

Erich Kaltofen and Heinrich Rolletschek [8] gave an $O(n^3)$ algorithm for computing the gcd in all complex quadratic UFDs. They transform the problem of computing the gcd to that of computing a short vector in a suitable four dimensional integer lattice. They have also given an $O(n^2)$ algorithm to compute the gcd in any quadratic ring. This algorithm is based on a result which states that, given α and β in some quadratic ring, one can always divide $l\alpha$ by β and have a remainder with norm smaller than $N(\beta)$ where l is a small rational integer. However the algorithm has large constants under the big-oh notation [8,21].

Henri Cohen has given a general algorithm for computing the extended gcd by reducing the problem to that of computing Hermite normal form of a suitable rational integer matrix [3]. However, he has not given the exact complexity of this algorithm in different rings.

The simplicity of the operations used in the binary gcd algorithm makes it the method of choice on real computers [2]. This algorithm has been generalized by André Weilert [21] to the ring of integers in $\mathbb{Q}(\sqrt{-1})$ and by Ivan Damgård and Gudmund Frandsen [5] to the ring of integers in $\mathbb{Q}(\sqrt{-3})$. In both the cases the running time of the algorithm is $O(n^2)$ with small constants hidden under the big-oh notation. There have been several variations and enhancements of the original binary gcd algorithm (see the notes at the end of chapter 4 in [2]). Our approach is quite similar to that of Jonathan Sorenson's [17] k-ary algorithm. The main reason we look at the possibility of extending the binary gcd algorithm is its simplicity and its speed. While the approaches like controlled Euclidean descent are expected to be asymptotically faster, they are impractical for smaller inputs.

4 Binary GCD Like Algorithms in Number Rings

The binary gcd algorithm for $\mathbb{Z}$ is shown in Alg. 1. The algorithm is slightly modified to include negative integers. The algorithm is based on the following three facts:

a. If $2|a$ and $2|b$, then $(a,b) = 2\left(\frac{a}{2}, \frac{b}{2}\right)$.
b. If $2|a$ and $2 \nmid b$, then $(a,b) = \left(\frac{a}{2}, b\right)$.
c. If $2 \nmid a$ and $2 \nmid b$, then $(a,b) = \left(\frac{a-b}{2}, b\right) = \left(\frac{a+b}{2}, b\right)$.

Thus given two non-zero odd $a, b \in \mathbb{Z}$ with $|a| \geq |b|$, one can find $c \in \mathbb{Z}$ in $O(\log |a|)$ time such that $|c| \leq \frac{|a|}{2}$ and $(a,b) = (c,b)$. This fact forms the basis of

Algorithm 1 Binary Gcd algorithm for $\mathbb{Z}$ (a, b are inputs)

Find $i, j \geq 0$ such that $2^i|a$, $2^{i+1} \nmid a$, $2^j|b$ and $2^{j+1} \nmid b$.
$a = a/2^i$, $b = b/2^j$
Assert that $|a| \geq |b|$. Swap a and b if needed.
while true
    if $|a - b| \leq |a + b|$ then $c = a - b$
    else $c = a + b$
    if $c = 0$ then break
    Find $h \geq 1$ such that $2^h|c$ and $2^{h+1} \nmid c$.
    $c = c/2^h$
    if $|c| \geq |b|$ then $a = c$
    else $a = b$, $b = c$
return $2^{\min\{i,j\}}a$

the while loop in Alg. 1 and guarantees that the algorithm will terminate in at most $(\log|a||b| + 1)$ iterations of the while loop. This algorithm may not work for algebraic integers. This is because if α and β are algebraic integers such that $2 \nmid \alpha$ and $2 \nmid \beta$ then 2 need not divide $\alpha + \beta$ or $\alpha - \beta$. Thus the argument for termination of the above algorithm fails for algebraic integers.

Our aim is to create an algorithm similar to Alg. 1 for complex quadratic rings. As a first step we will generalize the binary gcd algorithm to use rational primes other than 2. This generalization can be seen as a special case of Sorenson's k-ary algorithm [17]. Suppose we have $a, b \in \mathbb{Z}$ such that a and b are co-prime to an odd prime p, then $p|(a + lb)$ for some $l \in \mathbb{Z}_p$ where $\mathbb{Z}_p = \mathbb{Z}/p\mathbb{Z}$ is the finite field of residues modulo p. The set $\{-\frac{p-1}{2}, \ldots, 0, \ldots, \frac{p-1}{2}\}$ forms a complete set of coset representatives for $\mathbb{Z}_p$ and hence we can always choose l such that $|l| \leq \frac{p-1}{2}$. Therefore by using p in place of 2 in Alg. 1, we have a p-ary algorithm for calculation of gcd in $\mathbb{Z}$ as shown in Alg. 2. In Alg. 2 assuming $|a| \geq |b|$, it takes $O(\log|a|)$ time to compute a suitable c in steps 5 and 6. Thus we can replace a with c such that $(a, b) = (c, b)$ and $|c| \leq \left(\frac{1}{2} + \frac{1}{2p}\right)|a|$ and the complexity of Alg. 2 is the same as the complexity of the binary gcd algorithm.

Let $\mathcal{Z}$ be an imaginary quadratic number ring. If ρ is a prime in $\mathcal{Z}$, then $\mathcal{F} = \mathcal{Z}/\rho\mathcal{Z}$ is a field. Thus if α and β are any two integers co-prime to ρ, then there exists a $\lambda \in \mathcal{F}$ such that $\alpha + \lambda\beta$ is divisible by ρ. By Lemma 1 we know that $(\alpha, \beta) = (\alpha + \lambda\beta, \beta)$. Thus by replacing $\{-\frac{p-1}{2}, \ldots, 0, \ldots, \frac{p-1}{2}\}$ with $\mathcal{F}$ in step 5 of Alg. 2 one can construct an algorithm similar to Alg. 2 for computing the gcd $\mathcal{Z}$. If $N(\alpha + \lambda\beta) \leq fN(\rho\alpha)$ for some $f < 1$, then one can show that such an algorithm will terminate. However for a fixed choice of $\mathcal{F}$, it is not guaranteed that there will exist a $\lambda \in \mathcal{F}$ satisfying the above termination condition. For example consider the ring of integers in $\mathbb{Q}(\sqrt{-2})$. In this ring $\zeta = \sqrt{-2}$ is a prime of norm 2. If α and β are any two integers in this ring and co-prime to ζ, then $\alpha \pm \beta$ is divisible by ζ. If γ is the norm-wise smaller of $\{\alpha + \beta, \alpha - \beta\}$, then $N(\gamma) \leq 2N(\alpha)$ (it follows using Lemma 2 which is mentioned later). Thus $N(\gamma/\zeta) \leq N(\alpha)$ and the above argument for termination breaks down. However

Algorithm 2 p-ary Gcd Algorithm for $\mathbb{Z}$ ($p \in \mathbb{Z}$ is a fixed odd prime and inputs are $a, b \in \mathbb{Z}$)

Find $i, j \geq 0$ such that $p^i | a$, $p^{i+1} \nmid a$, $p^j | b$ and $p^{j+1} \nmid b$.
$a = a/p^i$, $b = b/p^j$
Assert that $|a| \geq |b|$. Swap a and b if needed.
while true
  Find $l \in \left\{\frac{p-1}{2}, \dots, 0, \dots, \frac{p-1}{2}\right\}$ such that $(a + lb) \equiv 0 \pmod p$
  $c = a + lb$
  if $c = 0$ then break
  Find $h \geq 1$ such that $p^h | c$ and $p^{h+1} \nmid c$.
  $c = c/p^h$
  if $|c| \geq |b|$ then $a = c$
  else $a = b$, $b = c$
return $p^{\min\{i,j\}} a$

one can get around this problem in some imaginary quadratic rings as follows (one can get around the termination problem in this particular ring in another way which is mentioned in conclusion).

Suppose p is a rational prime. Then either p splits or ramifies or remains inert in a quadratic ring $\mathcal{Z}$. Suppose p splits into ρ and $\bar{\rho}$. In this situation ρ and $\bar{\rho}$ are not associates and are co-prime to each other. Since $N(\rho) = N(\bar{\rho}) = p$, $\{-\frac{p-1}{2}, \dots, 0, \dots, \frac{p-1}{2}\}$ or $\{0, 1\}$ forms a complete set of coset representatives for *both* $\mathcal{Z}/\rho\mathcal{Z}$ and $\mathcal{Z}/\bar{\rho}\mathcal{Z}$ depending on if p is odd or even. Now if α and β are any two integers co-prime to both ρ and $\bar{\rho}$, then we have two choices of λ (λ_1 and λ_2 such that $\rho|(\alpha + \lambda_1\beta)$ and $\bar{\rho}|(\alpha + \lambda_2\beta)$). The idea is now to use both primes and then choose λ which makes $\alpha + \lambda\beta$ small. However this trick is useful only when there is at most one choice of $\lambda \in \mathcal{F}$ which can result in $\alpha + \lambda\beta$ having a large norm. One can verify that this favorable situation occurs only when $N(\rho) \leq 5$.

5 GCD Algorithms for Complex Quadratic Rings

In this section we materialize the ideas presented in the last section and construct algorithms for computing the gcd in the ring of integers in $\mathcal{Q} = \mathbb{Q}(\sqrt{d})$ where $d \in \{-2, -7, -11, -19\}$. These algorithms can be seen as an instance of the abstract algorithm shown in Alg. 3. The main difference from Alg. 2 is that we use a pair of conjugate primes instead of one fixed prime. Let ρ and $\bar{\rho}$ be the primes used in Alg. 3. The equivalent of c from Alg. 2 is $C(\alpha, \beta)$ and is denoted by γ. For different rings, ρ and C are different.

In the rest of this section we will show how to choose ρ and C in different rings and prove the termination of the algorithm in each case. Note that if Alg. 3 terminates, it will terminate with correct answer if $(\alpha, \beta) = (C(\alpha, \beta), \beta)$. In all the rings, our choice of $C(\alpha, \beta)$ is $\alpha + l\beta$ for some rational integer l. Thus the

Algorithm 3 Gcd algorithm for $\mathcal{Z}$ (inputs are α, β and primes $\rho, \bar{\rho}$ are fixed)

Let $p_1 = \rho$, $p_2 = \bar{\rho}$
Find i_1, i_2, j_1 and j_2 such that $p_t^{i_t} | \alpha$, $p_t^{i_t+1} \nmid \alpha$, $p_t^{j_t} | \beta$ and $p_t^{j_t+1} \nmid \beta$ for $t = 1, 2$.
$\alpha = \alpha / p_1^{i_1} p_2^{i_2}$, $\beta = \beta / p_1^{j_1} p_2^{j_2}$
Assert that $N(\alpha) \geq N(\beta)$. Swap α and β if needed.
While true
    $\gamma = C(\alpha, \beta)$
    if $\gamma = 0$ then break
    Find (h_1, h_2) such that $p_t^{h_t} | \gamma$ and $p_t^{h_t+1} \nmid \gamma$ for $t = 1, 2$.
    $\eta = \gamma / p_1^{h_1} p_2^{h_2}$
    if $N(\eta) \geq N(\beta)$ then $\alpha = \eta$
    else $\alpha = \beta$, $\beta = \eta$
return $p_1^{\min\{i_1, j_1\}} p_2^{\min\{i_2, j_2\}} \alpha$

correctness follows from Lemma 1. The following lemma will be the major tool in proving the termination results.

Lemma 2. *Let $\mathcal{Z}$ be an imaginary quadratic ring. Let $\alpha, \beta \in \mathcal{Z}$ and $l, m \in \mathbb{Z}$. If $N(\alpha) \geq N(\beta)$, then*

a. $N(l\alpha + m\beta) \leq (|l| + |m|)^2 N(\alpha)$,
b. $\min\{N(l\alpha + m\beta), N(l\alpha - m\beta)\} \leq (l^2 + m^2) N(\alpha)$.

Proof. We note that for any integer α, $N(\alpha)$ is the same as the complex norm of α when α is viewed as a complex number. Proof of above statements now follows by elementary properties of complex norm. □

For $d \in \{-2, -7, -11\}$, we show that $N(\eta) \leq f N(\alpha)$ for some $f < 1$ and the termination of the algorithm is trivial. For $d = -19$, we show that in at most two iterations of the while loop, the product $N(\alpha)N(\beta)$ will decrease by a factor $f > 1$ and hence the algorithm will terminate. The time complexity of Alg. 3 is the same in all rings and we discuss this in Sect. 5.5. In the rest of this section we will compare algebraic integers with respect to the norm. Thus whenever we say that $\alpha < \beta$, it means that $N(\alpha) < N(\beta)$. Similarly, other concepts based on comparison like min and max are defined.

5.1 GCD Algorithm for $\mathcal{Z} = \mathbb{Z}\left[\frac{1}{2}(-1 + \sqrt{-19})\right]$

The algebraic integers in this ring are of the form $a + b\omega$ where $\omega = \frac{-1+\sqrt{-19}}{2}$ and $a, b \in \mathbb{Z}$. Here ω and $\bar{\omega}$ are non-associate primes of norm 5. If $\alpha, \beta \in \mathcal{Z}$ are co-prime to both ω and $\bar{\omega}$, then $\omega | (\alpha + l\beta)$ and $\bar{\omega} | (\alpha + m\beta)$ for some $l, m \in \{-2, -1, 1, 2\}$. Based on this observation, we construct a gcd algorithm in this ring by taking $\{\rho, \bar{\rho}\} = \{\omega, \bar{\omega}\}$ and defining $C(\alpha, \beta)$ as follows. If $\alpha + \beta$ is divisible by ω or $\bar{\omega}$ then return $\alpha + \beta$. If $\alpha - \beta$ is divisible by ω or $\bar{\omega}$ then return $\alpha - \beta$. Otherwise, let $\lambda = \min\{\alpha + 2\beta, \alpha - 2\beta\}$ and $\Lambda = \max\{\alpha + 2\beta, \alpha - 2\beta\}$. If λ is divisible by ω or $\bar{\omega}$, then return λ. Else return Λ with the assurance that it is divisible by $\omega\bar{\omega}$. Showing that such an algorithm terminates is slightly non-trivial. In the next two lemma we show that either after one iteration or after two iterations of the while loop, the product of the norms of the two integers decreases by a factor of at least 1.16. Thus the algorithm will terminate in at most $2(\log_{1.16} N(\alpha)N(\beta)) + 1)$ iterations of the while loop.

Lemma 3. *Let $\alpha, \beta \in \mathcal{Z}$ be the algebraic integers at the start of the while loop in Alg. 3 with $\{\rho, \bar{\rho}\} = \{\omega, \bar{\omega}\}$ and $C(\alpha, \beta)$ as described above. If $C(\alpha, \beta) \neq \min\{\alpha + 2\beta, \alpha - 2\beta\}$, then after one iteration of the while loop, the quantity $N(\alpha)N(\beta)$ decreases by a factor of at least $\frac{5}{4}$.*

Proof. Without loss of generality assume that $N(\alpha) \geq N(\beta)$ and $\min\{\alpha+2\beta, \alpha-2\beta\} = \alpha - 2\beta$. If $\gamma = C(\alpha, \beta) \neq \min\{\alpha + 2\beta, \alpha - 2\beta\}$, then either $\gamma = \alpha \pm \beta$ or $\gamma = \alpha + 2\beta$.

Suppose $\gamma = \alpha \pm \beta$. Then from Lemma 2, $N(\gamma) \leq (1+1)^2 N(\alpha) \leq 4N(\alpha)$. By the description of $\gamma = C(\alpha, \beta)$, note that γ is divisible by ω or $\bar{\omega}$. Thus $\eta = \frac{\gamma}{\omega^i \bar{\omega}^j}$ where i and j are non-negative rational integers with $i + j \geq 1$. As $i + j \geq 1$, $N(\omega^i \bar{\omega}^j) \geq 5$ and hence,

$$N(\eta) \leq \frac{N(\gamma)}{N(\omega^i \bar{\omega}^j)} \leq \frac{N(\gamma)}{5} \leq \frac{4}{5} N(\alpha) \ .$$

Since η will replace α, $N(\alpha)N(\beta)$ will decrease by a factor of at least $\frac{5}{4}$ in this case.

Now suppose $\gamma = \alpha + 2\beta$. Again using Lemma 2, $N(\gamma) \leq (1 + 2)^2 N(\alpha) \leq 9N(\alpha)$. From the description of $C(\alpha, \beta)$, note that $\omega\bar{\omega} | \gamma$. Thus $\eta = \frac{\gamma}{\omega^i \bar{\omega}^j}$ for some rational integers $i, j \geq 1$. Thus,

$$N(\eta) \leq \frac{N(\gamma)}{N(\omega^i \bar{\omega}^j)} \leq \frac{N(\gamma)}{25} \leq \frac{9}{25} N(\alpha) \ .$$

Hence the product of norms will decrease by a factor of at-least $\frac{25}{9} > \frac{5}{4}$ in this case. □

Lemma 4. *Let $\alpha, \beta \in \mathcal{Z}$ be the algebraic integers at the start of the while loop in Alg. 3 with $\{\rho, \bar{\rho}\} = \{\omega, \bar{\omega}\}$ and $C(\alpha, \beta)$ as described above. If $C(\alpha, \beta) = \min\{\alpha+2\beta, \alpha - 2\beta\}$, then $N(\alpha)N(\beta)$ never increases and either after one iteration or after two iterations of the while loop, the quantity $N(\alpha)N(\beta)$ decreases by a factor of at least* 1.16.

Proof. From the description of $C(\alpha, \beta)$ we know that $\gamma = C(\alpha, \beta)$ is divisible by ω or $\bar{\omega}$. If γ is divisible by both ω and $\bar{\omega}$, or γ is divisible by ζ^i where $\zeta = \omega$ or $\zeta = \bar{\omega}$ and $i \geq 2$, then on the lines of proof of Lemma 3, it follows that the product of norms will decrease by a factor of at least 5. Thus we need to consider only the case where γ is divisible by either ω or $\bar{\omega}$ and exactly once. Let γ be divisible by ζ where $\zeta = \omega$ or $\bar{\omega}$.

A complex quadratic integer can also be viewed as a point on the complex plane. On the complex plane, $N(\alpha)$ is same as the square of distance of α from origin *i.e*, $N(\alpha) = |\alpha|^2$. It is an elementary result in geometry that

$$\begin{aligned}|\alpha + 2\beta|^2 &= |\alpha|^2 + |2\beta|^2 + 2|\alpha||2\beta|\cos\theta \text{ and}\\ |\alpha - 2\beta|^2 &= |\alpha|^2 + |2\beta|^2 - 2|\alpha||2\beta|\cos\theta\end{aligned}$$

where θ is the angle between α and β. Without loss of generality we assume that $0 \leq \theta \leq \frac{\pi}{2}$. Then $N(\alpha - 2\beta) \leq N(\alpha + 2\beta)$ and the algorithm will choose $\gamma = \alpha - 2\beta$. Also assume that $|\alpha| \geq |\beta|$. Thus $|\alpha| = k|\beta|$ where $k \in \mathbb{R}$ and $k \geq 1$. Thus $N(\alpha) = k^2 N(\beta) \geq N(\beta)$ and hence

$$N(\gamma) = |\alpha - 2\beta|^2 = (k^2 + 4 - 4k\cos\theta)|\beta|^2 \ .$$

As $\eta = \gamma/\zeta$,

$$N(\eta) = \frac{N(\gamma)}{N(\zeta)} = \frac{k^2 + 4 - 4k\cos\theta}{5}|\beta|^2 \ .$$

Since the algorithm will replace α by η, the product of norms will change by a factor f, where

$$f = \frac{N(\eta)}{N(\alpha)} = \frac{N(\eta)}{k^2 N(\beta)} = \frac{1}{5} + \frac{4}{5k^2} - \frac{4\cos\theta}{5k} \ . \tag{5.1}$$

As $k \geq 1$ and $0 \leq \theta \leq \frac{\pi}{2}$, $f \leq 1$ and hence the algorithm will not increase the product of norms by replacing α by η. However, as f can get arbitrarily close to 1, we are not assured that there will be substantial decrease in the product of norms. However intuitively we can see that if θ is away from $\frac{\pi}{2}$, then f will be small. The idea of the proof is to show that when $\theta \approx \frac{\pi}{2}$, then the angle between η and β is away from $\frac{\pi}{2}$. Thus while there may not be substantial decrease in the product of norms in this iteration, there will be enough decrease in the next iteration. Concretely, we now show that the product of norms will decrease by a factor of 1.16 as long as $k \geq 1.1$ or $\theta \leq 0.444\pi$. Then we show that if $1 \leq k < 1.1$ and $0.444\pi < \theta \leq \frac{\pi}{2}$, then the acute angle between η and β (if the angle between η and β is obtuse then angle between η and $-\beta$ is acute) is less than 0.444π. Thus in the next iteration either by Lemma 3 or by the arguments in the previous lines, the product of norms will decrease by a factor of at least 1.16.
Case 1 : Suppose that $k \geq 1.1$. Then from (5.1)

$$f \leq \frac{1}{5} + \frac{4}{5 * 1.21} < 0.862 \ .$$

On the other hand if $\theta \leq 0.444\pi$. Then $\cos\theta > 0.175$ and hence from (5.1),

$$f < \frac{1}{5} + \frac{4}{5k^2} - \frac{4}{5k}(0.175) \ .$$

The expression on right takes maximum value at $k = 1$ (in the range $1 \leq k < \infty$ as $k < 1$ is not possible). Evaluating, $f < 0.86$. Thus whenever $k \geq 1.1$ or

$0 \leq \theta \leq 0.444\pi$, the product of norms will decrease by a factor of at least $\frac{1}{0.862} > 1.16$.

Case 2 : Now let $0.444\pi < \theta \leq \frac{\pi}{2}$ and $1 \leq k < 1.1$. Since $k < 1.1$, $|\alpha| = k|\beta| < |2\beta|$. The situation is something like as shown in Fig. 1. In the figure, the dashed cone indicates the region in which all integers making an angle (acute) greater than 0.444π with $\pm\beta$ can lie.

On the complex plane, $\omega = \frac{-1+\sqrt{-19}}{2} = \sqrt{5}e^{\iota\omega_\theta}$ and $\bar{\omega} = \frac{-1-\sqrt{-19}}{2} = \sqrt{5}e^{-\iota\omega_\theta}$, where $\omega_\theta = \pi - \tan^{-1}\sqrt{19}$. Thus after dividing γ by ω or $\bar{\omega}$ we get η, which is γ reduced in length by $\sqrt{5}$ and rotated by angle $\pm\omega_\theta$. Now we wish to show that η is outside the dashed cone.

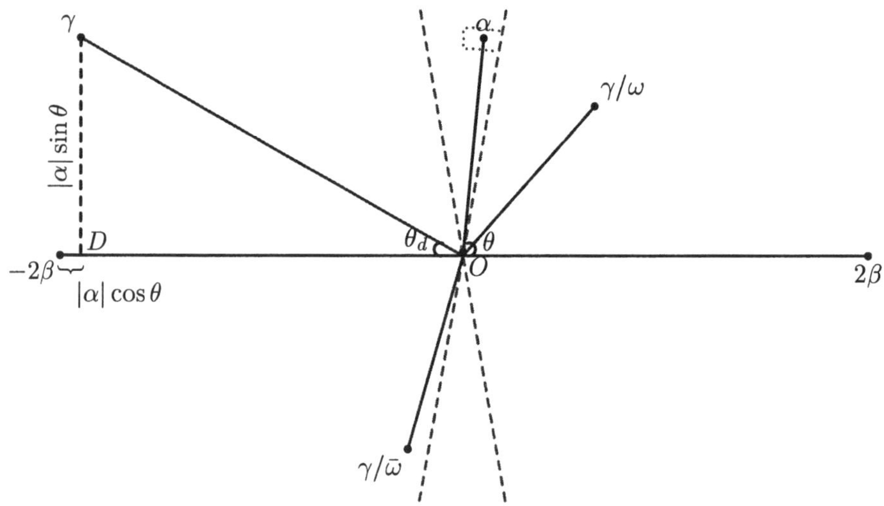

Fig. 1. A typical situation when $1 \leq k < 1.1$ and $0.444\pi < \theta \leq \frac{\pi}{2}$. The dashed conical region contains points which can make an acute angle greater than 0.444π with $\pm\beta$. α lies in this cone and inside the dotted region which is got by placing requirement $|\beta| \leq |\alpha| < 1.1|\beta|$. Once $\gamma = C(\alpha, \beta) = \alpha - 2\beta$ is divided by ω or $\bar{\omega}$, the resulting point gets reduced in length by $\sqrt{5}$ and rotated by $\pm\omega_\theta$ where $\omega_\theta = \pi - \tan^{-1}\sqrt{19}$. The lengths of line segment $\overline{\gamma D}$ and $\overline{(-2\beta)D}$ are $|\alpha|\sin\theta$ and $|\alpha|\cos\theta$ respectively, which are got by completing the parallelogram $O\alpha\gamma(-2\beta)$.

To show that η lies outside the dashed the cone, we have to find out the extremal values of θ_d subject to $1 \leq k < 1.1$ and $0.444\pi < \theta \leq \frac{\pi}{2}$. From the figure,

$$\tan\theta_d = \frac{|\alpha|\sin\theta}{2|\beta| - |\alpha|\cos\theta} = \frac{k\sin\theta}{2 - k\cos\theta} \ .$$

Since $k \cos\theta < 2$, θ_d is always acute. Analytically or graphically one can verify that in the region under consideration θ_d attains its maximum value at $k = 1.1$ and $\theta = 0.444\pi$ and attains its minimum value at $k = 1$ and $\theta = \frac{\pi}{2}$. Evaluating we get, $\theta_{d_{\min}} = \tan^{-1}\left(\frac{1}{2}\right) > 0.147\pi$ and $\theta_{d_{\max}} = \tan^{-1}\left(\frac{1.1\sin(0.444\pi)}{2-1.1\cos(0.444\pi)}\right) < 0.172\pi$. Thus we have,

$$0.147\pi < \theta_d < 0.172\pi \ .$$

Let ϕ be the angle between η and β. If it lies inside the dashed cone, then we should have $0.444\pi < \phi < 0.556\pi$ or $-0.556\pi < \phi < -0.444\pi$. As mentioned already, after division by ω or $\bar{\omega}$, γ will be rotated by $\pm\omega_\theta = \pm(\pi - \tan^{-1}\sqrt{19})$. By direct computation, $0.571\pi < \omega_\theta < 0.572\pi$. Now suppose that γ is rotated by ω_θ. Then using above extremal values of θ_d and ω_θ one sees that $0.718\pi < \phi < 0.744\pi$. Thus in this case η cannot lie in the cone. Similarly if γ is rotated by $-\omega_\theta$, then $-0.425\pi < \phi < -0.399\pi$. Thus again η cannot lie in the cone.

Thus in two iterations the product of norms will decrease by a factor of 1.16. We wish to note here that by a similar reasoning one can verify that if $\vartheta = (\alpha + 2\beta)/\zeta$ where $\zeta = \omega$ or $\zeta = \bar{\omega}$, then ϑ also lies outside the forbidden cone. While this is not crucial to the proof, it becomes crucial later when we use approximate norms. □

5.2 GCD Algorithm for $\mathcal{Z} = \mathbb{Z}\left[\frac{1}{2}(-1 + \sqrt{-11})\right]$

The integers in this ring are of form $\alpha = a + b\omega$ where $\omega = \frac{-1+\sqrt{-11}}{2}$ and $a, b \in \mathbb{Z}$. Here ω and $\bar{\omega}$ are non-associate primes of norm 3. Since there are no integers of norm 2 in this ring, ω and $\bar{\omega}$ are the smallest non-rational prime. As $\{-1, 0, 1\}$ forms a complete set of coset representatives for $\mathcal{Z}/\omega\mathcal{Z}$ and $\mathcal{Z}/\bar{\omega}\mathcal{Z}$, if $\alpha, \beta \in \mathcal{Z}$ are co-prime to both ω and $\bar{\omega}$, then ω divides one of $\{\alpha+\beta, \alpha-\beta\}$ and $\bar{\omega}$ divides one of $\{\alpha + \beta, \alpha - \beta\}$. Based on this observation, we construct a gcd algorithm in this ring by taking $\{\rho, \bar{\rho}\} = \{\omega, \bar{\omega}\}$ and defining $C(\alpha, \beta)$ as follows. Let $\lambda = \min\{\alpha+\beta, \alpha-\beta\}$ and $\Lambda = \max\{\alpha+\beta, \alpha-\beta\}$. If λ is divisible by either ω or $\bar{\omega}$, we pick λ. Else we pick Λ and we now know that Λ is divisible by *both* ω and $\bar{\omega}$. The following lemma shows that the product of norms decreases by a factor of at least $\frac{3}{2}$ in each iteration of the while loop and hence the algorithm will terminate in at most $(\log_{\frac{3}{2}} N(\alpha\beta) + 1)$ iterations of the while loop.

Lemma 5. *Let $\alpha, \beta \in \mathcal{Z}$ be the algebraic integers at the start of the while loop in Alg. 3 with $\{\rho, \bar{\rho}\} = \{\omega, \bar{\omega}\}$ and taking $C(\alpha, \beta)$ as described above. Then in each iteration of the while loop, the quantity $N(\alpha)N(\beta)$ decreases by a factor of at least $\frac{3}{2}$.*

Proof is similar to the proof of Lemma 3.

5.3 GCD Algorithm for $\mathcal{Z} = \mathbb{Z}\left[\frac{1}{2}(-1 + \sqrt{-7})\right]$

The integers in this ring are of form $\alpha = a + b\omega$ where $\omega = \frac{-1+\sqrt{-7}}{2}$ and $a, b \in \mathbb{Z}$. Here ω and $\bar{\omega}$ are non-associate primes of norm 2. If $\alpha, \beta \in \mathcal{Z}$ are co-prime to both ω and $\bar{\omega}$, then $\omega\bar{\omega}|(\alpha+\beta)$ and $\omega\bar{\omega}|(\alpha-\beta)$. Based on this observation, we choose $\{\rho, \bar{\rho}\} = \{\omega, \bar{\omega}\}$ and $C(\alpha,\beta) = \min\{\alpha+\beta, \alpha-\beta\}$ for Alg. 3 for this ring. Next lemma shows that the product of norms decreases by a factor of 2 in each iteration of the while loop and hence the algorithm will terminate in at most $(\log N(\alpha\beta)+1)$ iterations of the while loop.

Lemma 6. *Let $\alpha, \beta \in \mathcal{Z}$ be the algebraic integers at the start of the while loop in Alg. 3 with $\{\rho, \bar{\rho}\} = \{\omega, \bar{\omega}\}$ and $C(\alpha,\beta) = \min\{\alpha+\beta, \alpha-\beta\}$. Then in each iteration of the while loop, the quantity $N(\alpha)N(\beta)$ decreases by a factor of at least 2.*

Proof is similar to the proof of Lemma 3.

5.4 GCD Algorithm for $\mathcal{Z} = \mathbb{Z}\left[\sqrt{-2}\right]$

The integers in this ring are of form $\alpha = a + b\sqrt{-2}$ where $a, b \in \mathbb{Z}$. In this ring $\rho = 1 + \sqrt{-2}$ and $\bar{\rho} = 1 - \sqrt{-2}$ are non-associate primes of norm 3. The gcd algorithm in this ring is Alg. 3 with choices of ρ and $\bar{\rho}$ as above, and $C(\alpha,\beta)$ is the same as in the case of $d = -11$. The fact that the algorithm terminates also follows from a lemma similar to Lemma 5.

5.5 Runtime

Let $s(x)$ denote number of bits required to represent x. Let $\alpha = a + b\omega$ (where $\omega = \frac{-1+\sqrt{d}}{2}$ for $d \in \{-7, -11, -19\}$ and $\omega = \sqrt{-2}$ for $d = -2$). For any given ring ω is constant. Hence we represent input as a pair (a, b) where $a, b \in \mathbb{Z}$. Let $\alpha = a+b\omega$ and $\beta = c+d\omega$ be inputs to Alg. 3. If $2^{n-1} \leq \max\{|a|, |b|, |c|, |d|\} < 2^n$, then $s(\alpha) \leq 2n + O(1)$ and $s(\beta) \leq 2n + O(1)$. The main focus of this section is to show that the time complexity of Alg. 3 is $O(n^2)$ in all the rings, with small constants hidden under the big-oh notation. The following lemma is easily shown.

Lemma 7. *If $\alpha \in \mathcal{Z}$, then $s(\alpha) \leq s(N(\alpha)) \leq 2s(\alpha) + O(1)$.*

First consider the steps 2-3 of Alg. 3. Divisions are by a fixed prime and easily performed. For $d \in \{-7, -11, -19\}$, $\frac{\alpha}{\omega} = b - \frac{a}{p} - \frac{a}{p}\omega$ and $\frac{\alpha}{\bar{\omega}} = -b + \frac{a-b}{p}\omega$ where $p = N(\omega) = 2$ or 3 or 5 in the respective ring. For $d = -2$, $\frac{\alpha}{1+\sqrt{-2}} = \frac{a+2b}{3} + \frac{b-a}{3}\sqrt{-2}$ and $\frac{\alpha}{1-\sqrt{-2}} = \frac{a-2b}{3} + \frac{b+a}{3}\sqrt{-2}$. Thus each single division step takes at most $O(n)$ time. Since every single division decreases the product of norms of α and β by a factor of p (where $p = 2$ or 3 or 5), if a total of τ divisions are performed, then using Lemma 7 it follows that $\tau = O(n)$. Thus steps 2-3 take at most $O(n^2)$ time. As the while loop contains norm computations, the complexity of step 4 can be ignored. The following lemma will be useful before we consider the while loop in Alg. 3.

Lemma 8. *Let $\alpha, \beta \in \mathcal{Z}$ be the algebraic integers at the start of the while loop in Alg. 3. Then the algorithm halts in at most $k \cdot (s(\alpha) + s(\beta)) + O(1)$ iterations of the while loop for some positive real number k.*

Proof. Let $\mathcal{Z} = \mathbb{Z}\left[\frac{1}{2}(-1+\sqrt{-7})\right]$. Then from Lemma 3 we know that in each iteration of the while loop the quantity $N(\alpha\beta)$ decreases by a factor of 2. Thus the algorithm will halt in at most $(\log N(\alpha\beta) + 1)$ iterations of the while loop. But,

$$\log N(\alpha\beta) + 1 \leq s(N(\alpha)) + s(N(\beta)) + 1 \leq 2(s(\alpha) + s(\beta)) + O(1) \ .$$

Similarly we can show this in every ring. The exact values of constant k are shown in the fourth column of Table 1. □

From Lemma 8 it follows that the number of iterations of the while loop in the Alg. 3 are $O(n)$ for every ring. In each iteration of the while loop the operations performed are addition, subtraction, division by 2 or 3 or 5 and computation of norm. Addition and subtraction can be done in $O(n)$ time. A single divisibility test and a single division step can also be done in $O(n)$ time. There are now two issues. First, the number of divisions performed in each iteration and second, the computation of norm. Let us assume that there are t_i divisions in the ith iteration and the norm can be computed in $T(n)$ time. Since there are $O(n)$ iterations of the while loop, the combined time complexity of all the iterations of the while loop is $O(n(n + T(n) + \Sigma_i t_i))$.

We will now estimate $\Sigma_i t_i$. We know that product of norms will decrease by a factor of at least $p^{t_i - 2}$ in each iteration where $p = 2$ or 3 or 5. Hence $\prod_i p^{t_i-2} \leq N(\alpha)N(\beta)$. Since there are only $O(n)$ number of iterations, $\prod_i p^{t_i} \leq p^{O(n)} N(\alpha\beta)$ and hence $\Sigma_i t_i = O(n)$. Thus the combined time complexity of all the iterations of the while loop is $O(n^2 + nT(n))$. Since startup to while loop takes $O(n^2)$ time, the time complexity of Alg. 3 is $O(n^2 + nT(n))$.

Let us now concentrate on $T(n)$. Norm computation involves a fixed number of multiplications and additions. Hence $T(n)$ is of the order of multiplication of two n bit numbers. However it is not necessary to compute the norm *exactly*. The algorithm uses the norm to compare integers. When norms are very close, it does not make much difference which one the algorithm picks. If the algorithm uses approximate norms to compare integers, then it may have to do more iterations to get the gcd, but the number of iterations still remain $O(n)$. For details we refer to the earlier known binary gcd like algorithms for ring of integers in $\mathbb{Q}(\sqrt{-1})$ [21] and $\mathbb{Q}(\sqrt{-3})$ [5]. The approximate norm can be computed in $O(n)$ time [5] and hence the complexity of Alg. 3 becomes $O(n^2)$.

The fact that the approximate norm will terminate the algorithm is quite easy to show for the cases $d \in \{-2, -7, -11\}$. For $d = -19$ one needs the fact mentioned at the end of the proof of Lemma 4. The effect of approximation on the number of iterations of the while loop in Alg. 3 is shown in Table 1. In Table 1, the second column is the factor by which the product of norms would decrease if we computed the norm accurately. The third column lists the factor by which product of norms would decrease if we compute an approximate norm along the lines of lemma 1 in [5] by using only 16 most significant bits in multiplications. The number of iterations performed is quantified by the constant k of Lemma 8. The fourth column lists the value this constant when the norm is computed accurately. The revised values of this constant with approximate norm are listed in the last column. As a conclusion to this section we have the following theorem

Theorem 1. *The running time of Alg. 3 is $O(n^2)$ where n is the number of bits in the representation of input and where norm is computed approximately.*

Table 1.

Ring	f	f'	k	k'
$\mathbb{Z}\left[\sqrt{-2}\right]$	1.5	1.49	3.42	3.48
$\mathbb{Z}\left[\frac{1}{2}(-1+\sqrt{-7})\right]$	2	1.99	2	2.02
$\mathbb{Z}\left[\frac{1}{2}(-1+\sqrt{-11})\right]$	1.5	1.49	3.42	3.48
$\mathbb{Z}\left[\frac{1}{2}(-1+\sqrt{-19})\right]$	1.16	1.15	18.7	19.84

6 Conclusion

We have presented binary gcd like algorithms for computing the gcd for four complex quadratic rings. The main contribution of the paper is that this approach can be made to work for unique factorization domains which are not Euclidean. Our algorithms are quite simple to implement and are practical in use. Its an open question if it is possible to extend these algorithms to the remaining three imaginary quadratic unique factorization domains.

The two step analysis used for $\mathbb{Z}\left[\frac{1}{2}(-1+\sqrt{-19})\right]$ is not restricted to conjugate primes. Using similar analysis one can show that algorithms similar to Alg. 3 can be constructed for $\mathbb{Z}\left[\sqrt{-2}\right]$ using the prime $\sqrt{-2}$ and for $\mathbb{Z}\left[\frac{1}{2}(-1+\sqrt{-3})\right]$ using the prime 2. More details are available in [1].

References

1. Saurabh Agarwal. Binary gcd like algorithms in some number rings. Department of Computer Science, University of Aarhus, 2004.
2. Eric Bach and Jeffrey Shallit. *Algorithmic number theory. Vol. 1*. Foundations of Computing Series. MIT Press, Cambridge, MA, 1996. Efficient algorithms.
3. Henri Cohen. Hermite and Smith normal form algorithms over Dedekind domains. *Math. Comp.*, 65(216):1681–1699, 1996.
4. George E. Collins. A fast Euclidean algorithm for Gaussian integers. *J. Symbolic Comput.*, 33(4):385–392, 2002.

5. Ivan Bjerre Damgård and Gudmund Skovbjerg Frandsen. Efficient algorithms for gcd and cubic residuosity in the ring of Eisenstein integers. In *Fundamentals of Computation Theory—FCT 2003 (Malmö, Sweden, August 2003)*, volume 2751 of *Lecture Notes in Comput. Sci.*, pages 109–117. Springer, Berlin, 2003.
6. Thomas W. Hungerford. *Algebra*, volume 73 of *Graduate Texts in Mathematics*. Springer-Verlag, New York, 1980. Reprint of the 1974 original.
7. Kenneth Ireland and Michael Rosen. *A classical introduction to modern number theory*, volume 84 of *Graduate Texts in Mathematics*. Springer-Verlag, New York, second edition, 1990.
8. Erich Kaltofen and Heinrich Rolletschek. Computing greatest common divisors and factorizations in quadratic number fields. *Math. Comp.*, 53(188):697–720, 1989.
9. Donald E. Knuth. *The art of computer programming. Vol. 2*. Addison-Wesley Publishing Co., Reading, Mass., second edition, 1981. Seminumerical algorithms.
10. Serge Lang. *Algebra*. Addison-Wesley Publishing Company, third edition, 1993.
11. D. H. Lehmer. Euclid's algorithm for large numbers. 45:227–233, 1938.
12. Franz Lemmermeyer. The Euclidean algorithm in algebraic number fields. *Exposition. Math.*, 13(5):385–416, 1995. An updated version is available at the webpage, `http://www.rzuser.uni-heidelberg.de/~hb3/prep.html`.
13. H. W. Lenstra, Jr. Euclid's algorithm in cyclotomic fields. *J. London Math. Soc. (2)*, 10(4):457–465, 1975.
14. Heinrich Rolletschek. On the number of divisions of the Euclidean algorithm applied to Gaussian integers. *J. Symbolic Comput.*, 2(3):261–291, 1986.
15. Renate Scheidler and Hugh C. Williams. A public-key cryptosystem utilizing cyclotomic fields. *Des. Codes Cryptogr.*, 6(2):117–131, 1995.
16. A. Schönhage. Schnelle Berechnung von Kettenbruchentwilungen. *Acta Informatica*, 1:139–144, 1971.
17. Jonathan Sorenson. Two fast GCD algorithms. *J. Algorithms*, 16(1):110–144, 1994.
18. H. M. Stark. A complete determination of the complex quadratic fields of class-number one. *Michigan Math. J.*, 14:1–27, 1967.
19. J. Stein. Computational problems associated with Racah algebra. *J. Comput. Phys.*, (1):397–405, 1967.
20. André Weilert. Effiziente Algorithmen zur Berechnung von Idealsummen in Quadratischen Ornungen. Dissertation, Universitaet Bonn, 2000.
21. André Weilert. $(1+i)$-ary GCD computation in $\mathbb{Z}[i]$ as an analogue to the binary GCD algorithm. *J. Symbolic Comput.*, 30(5):605–617, 2000.
22. André Weilert. Asymptotically fast GCD computation in $\mathbb{Z}[i]$. In *Algorithmic number theory (Leiden, 2000)*, volume 1838 of *Lecture Notes in Comput. Sci.*, pages 595–613. Springer, Berlin, 2000.

On the Complexity of Computing Units in a Number Field

V. Arvind and Piyush P. Kurur

Institute of Mathematical Sciences
C.I.T Campus,Chennai, India 600 113
{arvind,ppk}@imsc.res.in

Abstract. Given an algebraic number field K, such that $[K : \mathbb{Q}]$ is constant, we show that the problem of computing the units group $\mathcal{O}_K^*$ is in the complexity class SPP. As a consequence, we show that *principal ideal testing* for an ideal in $\mathcal{O}_K$ is in SPP. Furthermore, assuming the GRH, the class number of K, and a presentation for the class group of K can also be computed in SPP. A corollary of our result is that solving PELL'S EQUATION, recently shown by Hallgren [12] to have a quantum polynomial-time algorithm, is also in SPP.

1 Introduction

The computation of units in a number field is a fundamental problem in computational number theory and is considered an important algorithmic task in the area. It has been the subject of considerable research in the last two decades and several algorithmic results as well some complexity-theoretic results have been pioneered. Much of this research (e.g. [17,7,6]) is based on ideas developed by Buchmann in [3,4].

In the present paper we are interested in the following problems in computational number theory, from a structural complexity perspective. Let K be a number field given by its minimal polynomial.

1. Computing a fundamental system of units that generates the units group $\mathcal{O}_K^*$ in $\mathcal{O}_K$.
2. Computing a presentation (i.e. a set of generators and relators) for the class group $Cl(K)$ of K and the class number $h(K)$.
3. Testing if a given ideal A of $\mathcal{O}_K$ is a principal ideal.

From a purely complexity theory perspective, earlier research on these problems was by McCurley [16], and Buchmann and Williams [7]. This was followed by the Thiel's work [17] where it is shown that the problem of principal ideal testing is in NP. Furthermore, *assuming the Generalized Riemann Hypothesis,* it is shown in [17] that principal ideal testing and verifying the class number are in NP $\cap$ coNP.

Our interest to further investigate the computational complexity of these problems is motivated by the recent exciting work of Hallgren [12] where it

D.A. Buell (Ed.): ANTS 2004, LNCS 3076, pp. 72–86, 2004.

is shown that computing a solution to PELL'S EQUATION is in BQP (the class of problems that have polynomial-time quantum algorithms). Hallgren's main result is that given a *quadratic* number field K, its regulator R_K can be computed by a polynomial-time quantum algorithm. The regulator is the solution to PELL'S EQUATION. For quadratic fields, Hallgren also indicates how principal ideal testing and computing the class group are problems in BQP. Hallgren's results, however, do not appear to generalize to number fields of larger degree. Thus, it remains an open problem if these problems are in BQP for number fields of degree more than two.

How does the class BQP relate to standard complexity classes defined using classical Turing machines? Fortnow and Rogers [11] show that BQP is contained in the counting complexity class AWPP (definitions in Section 1.1). Thus, in a sense, we can think of BQP as a *counting class.* Counting classes is an area of research in structural complexity theory motivated by Valiant's class #P (see e.g. [10]). Intuitively, counting complexity classes are defined by suitable restrictions on the number of accepting and rejecting paths in nondeterministic Turing machines. In the rest of this section we give formal definitions followed by a summary of our results.

1.1 SPP and Other Counting Complexity Classes

Let $\Sigma = \{0, 1\}$ be the finite alphabet. Let lg denote logarithm to base 2. Let FP denote the class of polynomial-time computable functions and NP denotes all languages accepted by polynomial-time nondeterministic Turing machines. Let $\mathbb{Z}$ denote integers. A function $f : \Sigma^* \to \mathbb{Z}$ is said to be *gap-definable* if there is an NP machine M (i.e. a nondeterministic polynomial time Turing machine M) such that, for each $x \in \Sigma^*$, $f(x)$ is the difference between the number of accepting paths and the number of rejecting paths of M on input x. Let GapP denote the class of gap-definable functions [10]. For each NP machine M let gap_M denote the GapP function defined by it.

A language L is in UP if there is an NP machine M accepting L such that M has at most one accepting path on any input. The class UP was defined by Valiant and it captures the complexity of 1-way functions. The complexity class SPP is defined as follows. A language L is in SPP if there is an NP machine M such that $x \in L$ implies $\mathrm{gap}_M(x) = 1$ and $x \notin L$ implies $\mathrm{gap}_M(x) = 0$. In this case we say that L is *accepted* by the machine M. Note that the class SPP is essentially a GapP analogue of the class UP and UP $\subseteq$ SPP.

We say that f is in GapP^A, for oracle $A \in \Sigma^*$, if there is an NP^A machine M^A such that, for each $x \in \Sigma^*$, $f(x)$ is the difference between the number of accepting paths and the number of rejecting paths of M^A on input x. For an oracle A, we can now define the class SPP^A.

The class PP is defined as follows: a language L is in PP if there is an $f \in$ GapP such that $x \in L$ if and only if $f(x) > 0$. PP is a hard counting class: by Toda's theorem we know that PH $\subseteq \mathrm{P}^{\mathrm{PP}}$. We say that a language $A \in \Sigma^*$ is *low* for PP if $\mathrm{PP}^A = $ PP. Characterizing the class of languages low for PP is an intriguing open question in structural complexity.

In [10] it is shown that every language in SPP is low for PP. Additionally, SPP has nice closure properties [10]: $P^{SPP} = SPP^{SPP} = SPP$. Another class that is low for PP [14] is BPP (the class of languages with polynomial-time randomized algorithms with error probability bounded by, say, 1/3.) Subsequently, the complexity class AWPP was introduced[1] in [9]. The class AWPP generalizes both BPP and SPP, and it is shown that every language in AWPP is low for PP. To complete the picture relating these classes, Fortnow and Rogers in [11] show that BQP is contained in AWPP. It is interesting to note that NP $\cap$ coNP is *not* known to be low for PP. Here is a diagram that shows the containments between the complexity classes discussed here.

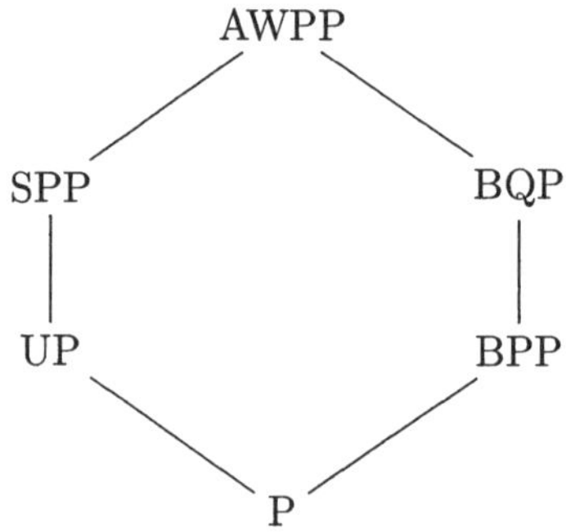

Although no containment is known between BQP and SPP, it is interesting to compare these classes in terms of natural problems they contain. Important problems known to be in SPP are Graph Isomorphism and the hidden subgroup problem for permutation groups [1]. These problems have resisted efficient deterministic or randomized algorithms, but are considered potential candidates for quantum algorithms. On the other hand, FP^{SPP} contains Integer Factoring and Discrete Log that have polynomial-time quantum algorithms.[2]

1.2 The New Results and the Methods

We now state the main results of the paper.

(a) Given a number field K (by its minimal polynomial as input), the problem of computing a fundamental system of units is in FP^{SPP}, assuming that K is a constant degree extension of $\mathbb{Q}$. As a consequence finding the regulator of K upto polynomially many bits of approximation is also in FP^{SPP}. As a corollary the PELL'S EQUATION problem is in FP^{SPP}.
(b) Given a constant-degree number field K and an ideal A of the ring $\mathcal{O}_K$, testing if A is a principal ideal is in SPP.

[1] For the definition see [9].

[2] In fact, these problems are even in FP^{UP}. Also, as $P^{SPP} = SPP$, notice that the class FP^{SPP} is essentially SPP: for $f \in FP^{SPP}$ and input x, the bits of $f(x)$ can be computed in SPP. A similar closure property holds for BQP.

(c) Given a constant-degree number field K (by its minimal polynomial as input), the problem of computing the class group of K (by finding a generator-relator presentation for it) and finding the class number of K is in $\mathrm{FP}^{\mathrm{SPP}}$, assuming GRH.

In particular, PELL'S EQUATION is also in $\mathrm{FP}^{\mathrm{SPP}}$. Thus, we add to the list of natural problems that are in both SPP and BQP. A brief outline of the methods used to show the above results is given below.

Let M be an oracle Turing machine. For a language A in NP, we say that M^A makes *UP-like queries* to A if there is an NP machine N accepting A such that on all inputs x, $M^A(x)$ makes *only* such queries y for which $N(y)$ has *at most* one accepting path. Effectively, it is like M having access to a UP oracle. We state a useful variant of a result from [15].

Theorem 1 ([15]). *Let M be a nondeterministic polynomial-time oracle machine with oracle $A \in \mathrm{NP}$ such that M^A makes UP-like queries to A then the function $h(x) = \mathrm{gap}_{M^A}(x)$ is in* GapP.

Next, we recall an important property of the class SPP shown in [10].

Theorem 2 ([10]). *If L is in SPP^A for some oracle $A \in \mathrm{SPP}$ then $L \in \mathrm{SPP}$. I.e. $\mathrm{SPP}^{\mathrm{SPP}} = \mathrm{SPP}$.*

The following lemma, which is a straightforward consequence of Theorem 1 and of Theorem 2, is in a form useful for this paper.

Lemma 1.

- *Suppose L is in SPP^A accepted by the nondeterministic polynomial-time oracle machine M^A with oracle $A \in \mathrm{NP}$ (i.e. $x \in L$ implies that $\mathrm{gap}_{M^A}(x) = 1$, and $x \notin L$ implies that $\mathrm{gap}_{M^A}(x) = 0$), such that the machine M^A makes UP-like queries to A, then L is in* SPP.
- *Suppose a function $f : \Sigma^* \to \Sigma^*$ is in FP^A (i.e. f is computed by a polynomial-time oracle transducer M^A) where $A \subset \mathrm{NP}$, such that the machine M^A makes UP-like queries to A, then f is in $\mathrm{FP}^{\mathrm{SPP}}$.*

Lemma 1 is a crucial tool in obtaining the $\mathrm{FP}^{\mathrm{SPP}}$ upper bounds. For computing a fundamental system of units in $\mathrm{FP}^{\mathrm{SPP}}$ we first show that a bound $B \in \mathbb{Q}$ can be computed in $\mathrm{FP}^{\mathrm{SPP}}$ such that the regulator R_K of K lies between B and $2B$. Once such a bound is computed, we again apply an algorithm based on Lemma 1 to compute a *canonical* fundamental system of units in $\mathrm{FP}^{\mathrm{SPP}}$. This notion of canonical fundamental system of units is developed and explained in Sections 2 and 3, where we show how to transform an arbitrary fundamental system of units to the canonical set.

Once we have the $\mathrm{FP}^{\mathrm{SPP}}$ upper bound for computing fundamental units, we can design an SPP algorithm for principal ideal testing. If we assume the generalized Riemann hypothesis then, by a result of Bach [2], we can apply our SPP algorithm for principal ideal testing and give an $\mathrm{FP}^{\mathrm{SPP}}$ algorithm for computing the class group $Cl(K)$.

1.3 Comparison with Previous Results

As mentioned, Thiel [17] has shown that principal ideal testing is in NP. Thiel also shows, assuming the GRH, that principal ideal testing and verifying class number are in NP ∩ coNP. On the other hand, our results on principal ideal testing and computing a fundamental system of units are *unconditional*, but applicable to only number fields of constant degree. The $\mathrm{FP}^{\mathrm{SPP}}$ upper bound for the class group problem depends on the GRH.

No containment relation is known between SPP and NP ∩ coNP or BQP and NP ∩ coNP. Furthermore, we remark here that NP ∩ coNP is not known to be low for PP. Thus, the results of Thiel [17] are incomparable to our results.

An important computational aspect in all our results is the notion of *compact representation* as explained by Thiel [17], based on Buchmann's earlier papers [3,4]. We need compact representations to succinctly express units as well as the generating element of a principal ideal in $\mathcal{O}_K$.

2 Compact Representation

Let K be a number field of degree n and let $\mathcal{O}$ be the ring of integer of K. Let D be the discriminant of K. For an element $\alpha \in K$ by $\mathrm{N}(\alpha)$ we mean the norm $\mathrm{N}_{\mathbb{Q}}^{K}(\alpha)$. Without loss of generality we assume that the input to the algorithm is $\mathcal{O}$ presented as a $\mathbb{Z}$-module with basis $\omega_1, \ldots \omega_n$ and constants c_{ijk} such that $\omega_i \omega_j = \sum_k c_{ijk}\omega_k$. For, computing the maximal order from a given order is reducible to the problem of finding the square free part of an integer which can be done in $\mathrm{FP}^{\mathrm{SPP}}$, as factoring integers is in $\mathrm{FP}^{\mathrm{SPP}}$. By size of $\mathcal{O}$ we mean $\sum size(c_{ijk})$. The constants c_{ijk} will be the called the explicit data for K.

Fractional ideals $\mathfrak{a}$ of $\mathcal{O}$ will be presented by giving a $\mathbb{Z}$-basis for $\mathfrak{a}$. Let $\alpha_i = \sum_j a_{ij}\omega_j$, $1 \leq i \leq n$, be a basis of $\mathfrak{a}$ then by $\mathrm{HNF}(\mathfrak{a})$ we mean the Hermite normal form of the matrix (a_{ij}). Once ω_i's are fixed, for every ideal $\mathfrak{a}$, $\mathrm{HNF}(\mathfrak{a})$ is unique. Since the Hermite normal form of a matrix can be computed in polynomial time this gives a polynomial time algorithm for testing whether two ideals are equal.

Let $\sigma_1, \ldots, \sigma_r$ be all the r real embeddings and $\sigma_{r+1}, \overline{\sigma}_{r+1}, \ldots, \sigma_{r+s}, \overline{\sigma}_{r+s}$ be all the $2s$ complex embeddings of K. Define the $r+s$ absolute values on K as follows.

$$|\alpha|_i = \begin{cases} |\sigma_i(\alpha)| & \text{if } 1 \leq i \leq r \\ |\sigma_i(\alpha)|^2 & \text{if } r+1 \leq i \leq r+s \end{cases}$$

For $\alpha \in K$, by height of α, denoted by $\mathrm{H}(\alpha)$, we mean $\max\{|\alpha|_i : 1 \leq i \leq r+s\}$.

Lemma 2. *Given $\mathcal{O}$, we have $\mathrm{H}(\omega_i) \leq n2^{size(\mathcal{O})}$, and $\lg D \leq n(2\lg n + size(\mathcal{O}))$.*

Proof. Let l be such that for $1 \leq i \leq n$ $\mathrm{H}(\omega_l) \geq \mathrm{H}(\omega_i)$. Then we have

$$\mathrm{H}(\omega_l\omega_l) \leq \sum_k |c_{llk}|\,\mathrm{H}(\omega_k) \leq n2^{size(\mathcal{O})}\mathrm{H}(\omega_l).$$

Hence $\mathrm{H}(\omega_l) \leq n2^{size(\mathcal{O})}$. Also since $D = \sum_{g \in S_n} \prod_{i=1}^{n} \sigma_{g(i)}(\omega_i)$ we have $D \leq n!.\mathrm{H}(\omega_l)^n$.

Fix a basis for $\mathcal{O}$. For any $\alpha \in \mathcal{O}$ there is a unique set of integers $a_1, \ldots, a_n$ such that $\alpha = \sum a_i \omega_i$. By giving the vector of integers a_i's the algebraic integer α is completely specified. Following [17] this is the *standard representation* of α which is unique for a fixed $\mathbb{Z}$-basis for $\mathcal{O}$. The size of α in standard representation is $size_s(\alpha) = \sum size(a_i)$.

The following result from [17] describes a *compact representation* of algebraic integers.

Theorem 3. *For $\alpha \in \mathcal{O}$ there exists $k \leq \lg(\lg(D) + (n-1)\lg \mathrm{H}(\alpha)) + 2$ and $\gamma, \alpha_i \in \mathcal{O}$ and $d_i \in \mathbb{Z}$, $1 \leq i \leq k$, with $\mathrm{H}(\gamma) \leq \mathrm{N}(\alpha)^{2/n}$, $\mathrm{H}(\alpha_i) \leq D^{\frac{3}{4}(m+2)}$ and $0 < d_i \leq \sqrt{D}$ such that*

$$\alpha = \gamma \prod_{i=1}^{k} \left(\frac{\alpha_i}{d_i}\right)^{2^{k-i}}.$$

Moreover, for $1 \leq j \leq k$ the ideal $\prod_{i=1}^{j} \left(\frac{\alpha_i}{d_i}\right)^{2^{k-i}} \mathcal{O}$ is a reduced ideal.

A product of this form can be presented as a tuple $\langle k, \gamma, \langle \alpha_i, d_i \rangle_{i=1}^{k} \rangle$. For a given compact representation of α the size of the representation is the sum of the sizes of the integers d_i and the sizes of algebraic integers γ and α_i's in their standard representation. Compact representations are not unique for a given α even for a given $\mathbb{Z}$-basis. Let $size_c(\alpha)$ denotes the maximum of sizes of all compact representation of α. Using Lemma 2 and [17, Corollary 15] we have the following theorem on compact representations.

Theorem 4. [17] *For nonzero $\alpha \in \mathcal{O}$*

$$size_s(\alpha) \leq (n \lg \mathrm{H}(\alpha) \cdot size(\mathcal{O}))^{O(1)}.$$
$$size_c(\alpha) \leq \left(n^2 \lg^2(n) \cdot size(\mathcal{O}). \lg(size(\mathcal{O})) \cdot \mathrm{N}(\alpha) \cdot \lg\lg \mathrm{H}(\alpha)\right)^{O(1)}$$

Furthermore, given the compact representation $\langle k, \gamma, \langle \alpha_i, d_i \rangle_{i=1}^{k} \rangle$ of α, there is a polynomial time algorithm that computes the Hermite Normal Form for the ideal $\alpha\mathcal{O}$.

Conversely the following proposition from [17] gives a bound on the height of the algebraic number based on the size of their representation.

Proposition 1. *Let α be any algebraic integer of K, a number field of discriminant D. Then we have:*

1. *For all j, $\mathrm{H}(\alpha) \leq n^2 2^{size_s(\alpha) + size(\mathcal{O})}$.*
2. *For all j, $\ln \mathrm{H}(\alpha) \leq \ln \mathrm{N}(\alpha) + n^3 \lg n \cdot size_c(\alpha) \cdot size(\mathcal{O}) \cdot 2^{size_c \alpha}$.*

3 Minimal Bases for Lattices

For a set of linearly independent vectors $\mathbf{a}_1, \mathbf{a}_2, \dots \mathbf{a}_n$ let $\mathbf{a}_1^*, \mathbf{a}_2^*, \dots, \mathbf{a}_n^*$ denote the corresponding GSO (Gram-Schmidt) basis. Given a lattice $\Lambda = \sum_{i=1}^{n} \mathbb{Z}\mathbf{b}_i$ in $\mathbb{R}^n$ with basis $\mathbf{b}_i = \sum_{j=1}^{n} b_{ij}\mathbf{e}_i$. Let $M = (\mu_{ij})$ be the matrix that transforms the GSO basis given by $\mathbf{b}_i^*$'s to the basis given by the $\mathbf{b}_i$'s. We say that the basis is *proper* if for every $i < j \leq n$ we have $-\frac{1}{2} \leq \mu_{ij} < \frac{1}{2}$. The following holds for any lattice.

Lemma 3. *Given a lattice Λ with basis $\mathbf{a}_1, \mathbf{a}_2, \dots, \mathbf{a}_n$, a new basis $\mathbf{b}_1, \mathbf{b}_2, \dots \mathbf{b}_n$ can be computed in polynomial time such that $\mathbf{b}_i$'s form a proper basis and $\mathbf{b}_i^* = \mathbf{a}_i^*$ (i.e. the GSO basis of both vectors are the same).*

Proof. Here is the algorithm.

$\mathbf{b}_1 := \mathbf{a}_1$;
1 **for** $i = 2$ *to* n **do**
  Let $\mathbf{a}_i = \mathbf{a}_i^* + \sum_{j=1}^{i-1} \mu_{ij}\mathbf{a}_j^*$;
  $\mathbf{b}_i := \mathbf{a}_i$;
2 **for** $j = i - 1$ ***downto*** 1 **do**
    if $\mu_{ij} \geq \frac{1}{2} \vee \mu_{ij} < -\frac{1}{2}$ **then**
      Let n be the nearest integer to μ_{ij};
3      $\mathbf{b}_i := \mathbf{b}_i - n\mathbf{b}_j$;
    end
  end
end

The invariant for the loop in step 1 is $-\frac{1}{2} \leq \mu_{kj} < \frac{1}{2}$ for all $1 \leq j < k \leq i-1$. If the invariant is violated at i for some j then the loop in step 2 fixes it. For a given k note that step 3 does not affect any of the μ_{ij} for $j > k$. It is also clear that step 3 does not affect the GSO of the basis.

Given the vector space $W = U \oplus V$ such that U and V are orthogonal, for $\mathbf{w} \in W$, if $\mathbf{w} = \mathbf{u} + \mathbf{v}$, $\mathbf{u} \in U$ and $\mathbf{v} \in V$, then $\mathbf{w}/U$ denotes the vector $\mathbf{v}$ (i.e. the component of $\mathbf{w}$ orthogonal to the space U). For a lattice Λ, Λ/V is the lattice $\{\mathbf{v}/V : \mathbf{v} \in \Lambda\}$. If $\mathbf{b}_1, \mathbf{b}_2 \dots \mathbf{b}_n$ forms a basis for Λ then any vector of Λ/V can be expressed as an integer linear combination of $\mathbf{b}_i/V$'s.

Given a lattice Λ, a basis $\mathbf{b}_1, \mathbf{b}_2, \dots, \mathbf{b}_n$ is called a *minimal basis* if it is proper and it satisfies the following conditions:

1. $\mathbf{b}_1$ is a ℓ_1-shortest vector in Λ.
2. For all i, if V_{i-1} is the span of the vectors $\mathbf{b}_1, \mathbf{b}_2, \dots, \mathbf{b}_{i-1}$ then $\mathbf{b}_i/V_{i-1}$ is the vector of least ℓ_1 norm in the lattice Λ/V_{i-1}.

To find a canonical basis for the lattice Λ one can define a total order on the set of all minimal basis of Λ and choose the least basis under that order. For two vectors $\mathbf{u} = \sum u_i\mathbf{e}_i$ and $\mathbf{v} = \sum v_i\mathbf{e}_i$, $\mathbf{u} \prec \mathbf{v}$ if $\|\mathbf{u}\|_1 < \|\mathbf{v}\|_1$ or if $\|\mathbf{u}\|_1 = \|\mathbf{v}\|_1$ then there is an $1 \leq i \leq n$ such that $u_j = v_j$ for all $1 \leq j < i$ and $u_i < v_i$.

Consider two minimal basis $A = \{\mathbf{a}_i\}_{i=1}^n$ and $B = \{\mathbf{b}_i\}_{i=1}^n$ for the lattice Λ. Let $A^* = \{\mathbf{a}_i^*\}_{i=1}^n$ and $B^* = \{\mathbf{b}_i^*\}_{i=1}^n$ be their GSO basis respectively. For two such minimal basis $A \prec B$ if there is an i such that for all $j < i$, $\mathbf{a}_j = \mathbf{b}_j$ and $\mathbf{a}_i^* \prec \mathbf{b}_i^*$.

Theorem 5. *On the set of minimal bases, the relation $\prec$ forms a total order.*

Proof. Suppose $\prec$ does not form a total order then we have two minimal basis A and B and an index i such that $\mathbf{a}_j = \mathbf{b}_j$ for all $j < i$ and $\mathbf{a}_i^* = \mathbf{b}_i^*$ yet $\mathbf{a}_i \neq \mathbf{b}_i$. Expressing $\mathbf{a}_i$ and $\mathbf{b}_i$ in the respective GSO basis we have

$$\mathbf{a}_i = \sum_{j=1}^{i} \alpha_j \mathbf{a}_j^* \qquad \mathbf{b}_i = \sum_{j=1}^{i} \beta_j \mathbf{b}_j^*.$$

Let k be the index such that for all $j > k$, $\alpha_j = \beta_j$ and $\alpha_k \neq \beta_k$. Clearly $k < i$. Since the basis A and B are proper we have α_j and β_j lie in the interval $\left[-\frac{1}{2}, \frac{1}{2}\right)$. Consider the vector $\mathbf{u} = \mathbf{a}_i - \mathbf{b}_i$. Since $\mathbf{a}_j^* = \mathbf{b}_j^*$ for all $1 \leq j \leq i$ we have $\mathbf{u} = (\alpha_k - \beta_k)\mathbf{a}_k^* + \mathbf{u}'$ where $\mathbf{u}'$ lies in the vector space V_{k-1} that is spanned by $\{\mathbf{a}_j\}_{j=1}^{k-1}$. But then $\| \mathbf{u}/V_{k-1} \|_1 = |\alpha_k - \beta_k| \, \| \mathbf{a_k}^* \|_1 < \| \mathbf{a_k}/V_{k-1} \|_1$ which contradicts the fact that A and B are minimal.

Given an algorithm for finding the $\prec$-least vector of a lattice, Theorem 5 suggests the following algorithm for finding $\prec$-minimum basis for a lattice Λ.

Input: A set of linearly independent vectors $\mathbf{b}_1, \mathbf{b}_2, \ldots, \mathbf{b}_n$.
Output: The $\prec$-minimum basis for Λ.
Let Λ be the lattice $\mathbb{Z}[\mathbf{b}_1, \mathbf{b}_2, \ldots, \mathbf{b}_n]$;
Find the $\prec$-least element $\mathbf{u}$ of the lattice Λ;
Let $\{\mathbf{u}_k\}_{k=1}^n$ be a basis for Λ such that $\mathbf{u}_1 = \mathbf{u}$;
Let $\mathbf{u}_k^* = \mathbf{u}_k - \frac{\langle \mathbf{u}, \mathbf{u}_k \rangle}{\langle \mathbf{u}, \mathbf{u} \rangle} \mathbf{u}$ $1 < k \leq n$;
Find the $\prec$-minimum basis for the lattice Λ^* generated by $\{\mathbf{u}_k^*\}_{k=2}^n$ recursively;
Let this basis be $\{\mathbf{b}_i' = \sum_{j=2}^n x_{ij} \mathbf{u}_j^*\}_{i=2}^n$;
Consider the basis $\{\mathbf{a}_i\}_{i=1}^n$ defined by $\mathbf{a}_1 = \mathbf{u}$ and $\mathbf{a}_i = \sum_{j=2}^n x_{ij} \mathbf{u}_j$, $2 \leq i \leq n$,
Converting this basis to a proper basis using Lemma 3 completes that algorithm;

Algorithm 1: Computing $\prec$-minimal basis

To complete this section we give an algorithm for finding the $\prec$-minimum element of a lattice Λ. We apply Lenstra's algorithm for integer linear programming [13] that runs in polynomial time if the number of variables is constant.

Lemma 4. *If $\Lambda \subset \mathbb{R}^n$ is a lattice of rank r then assuming n is a constant there is a polynomial time algorithm for finding the $\prec$-minimum element of Λ.*

Proof. Let $\Lambda = \mathbb{Z}[\mathbf{b}_1, \mathbf{b}_2, \ldots, \mathbf{b}_r]$ where the basis $\mathbf{b}_i$'s are given by $\mathbf{b}_i = \sum b_{ij} \mathbf{e}_j$. The algorithm goes in two steps. The first step is to find a vector with least ℓ_1 norm. In the subsequent steps this solution is refined further till we get the $\prec$-minimum vector.

To find a vector with least ℓ_1 norm we have the following integer programming problem in r variables $x_1, x_2, \ldots, x_r$: Minimize the expression $f(\mathbf{x}) = \sum_{i=1}^{n} |\sum_{j=1}^{r} b_{ji}x_j|$ given $f(\mathbf{x}) > 0$. Although this is not an integer linear programming problem one can use the algorithm of Lenstra here as follows: For the 2^n different vector $\mathbf{c} \in \{-1,1\}^n$, solve the following integer linear programming problem and pick the best among those 2^n different solutions.

$$\text{Minimize } \sum_{i=1}^{r}\left(\sum_{j=1}^{n} b_{ij}c_j\right)x_i,$$

under the constraints

$$\sum_{i=1}^{r}\left(\sum_{j=1}^{n} b_{ij}c_j\right)x_i > 0 \tag{1}$$

$$0 \leq c_j \sum_{i=1}^{r} b_{ij}x_i \leq B, \qquad 1 \leq j \leq n. \tag{2}$$

The first constraints expresses the fact that the solution should be nonzero. The second set of constraints express the fact that we are choosing the right c_i's.

The B in the equation is an upper bound on the ℓ_∞ of the shortest vector. The ℓ_1 norm of any particular vector in the basis will be a suitable value for B.

Having obtained a solution for the ℓ_1-shortest vector say $\mathbf{u}$ one has to refine the solution to get a $\prec$-minimum solution. Let $\mathbf{u}_j$ denote a ℓ_1-shortest vector which agrees with the $\prec$-minimum vector on all coordinates less than or equal to j, then $\mathbf{u}_n$ is the desired solution. Let $\mathbf{u}_0 = \mathbf{u}$. Having got the vector $\mathbf{u}_j$ to compute $\mathbf{u}_{j+1}$ we minimize $\sum_{i=1}^{r} x_i b_{ij+1}$ under the constraints $\| \sum_{i=1}^{r} x_i \mathbf{b}_i \|_1 = \| \mathbf{u}_j \|_1$ and $\sum_{i=1}^{r} x_i b_{ik} = u_{jk}$ for $1 \leq k \leq j$. Here u_{jk} denotes the component of $\mathbf{u}_j$ in the direction $\mathbf{e}_k$. One can use the same trick as before to convert this to an integer linear programming problem and use Lenstra's algorithm. Since the dimension n is bounded, the running time is polynomial.

Combining Lemma 4 and algorithm 1 we have the following result.

Theorem 6. *Given a lattice a basis $\{\mathbf{b}_i\}_{i=1}^{r}$ of a rank r lattice $\Lambda \subseteq \mathbb{R}^n$ there is a polynomial time algorithm to compute the $\prec$-minimal basis of Λ assuming n to be a constant*

4 Units of a Number Field

Let K be a number field of degree n and let $\mathcal{O}$ be the set of algebraic integers of K. If K has r real embeddings and $2s$ complex embeddings then by Dirichlet's theorem (see, e.g. [8]) there exists a set of $m = r + s - 1$ units $\{\varepsilon_i\}_{i=1}^{m}$, called a fundamental system of units, such that every unit of $\mathcal{O}$ can be expressed as $\zeta\varepsilon_1^{x_1}\ldots\varepsilon_m^{x_m}$, $x_i \in \mathbb{Z}$, where ζ is a root of unity in K. Consider the map $\mathrm{Log} : K \mapsto \mathbb{R}^m$ defines as follows:

$$\mathrm{Log}\,(\alpha) = \langle \ln|\alpha|_1, \ln|\alpha|_2, \ldots, \ln|\alpha|_m \rangle$$

From Dirichlet's theorem it follows that the set $\mathrm{Log}(\mathcal{O}^*)$ is a lattice in $\mathbb{R}^m$ with basis $\mathrm{Log}(\varepsilon_i)_{i=1}^m$. Often it is necessary to work with vectors in this lattice whose coordinates are in general irrationals. We will use rational approximations of these vectors instead.

An important algorithmic task that will be useful in the $\mathrm{FP}^{\mathrm{SPP}}$ algorithm is to compute a canonical fundamental system of units from a given fundamental system of units $U = \{\varepsilon_i\}_{i=1}^m$. In this section we give a polynomial time algorithm for the above task assuming that the degree $[K : \mathbb{Q}]$ is constant. The next theorem is a re-statement of [17, Lemma 16] using the bound in Lemma 2.

Theorem 7. [17] *There exists a fundamental system of units $\{\varepsilon_i\}_{i=1}^m$ for $\mathcal{O}$ such that $size_c(\varepsilon_i) = (n.size(\mathcal{O}))^{O(1)}$ for all $1 \leq i \leq m$.*

Consider the fundamental system of units $\{\eta_i\}_{i=1}^m$ that corresponds to $\prec$-minimal basis of the lattice $\mathrm{Log}(\mathcal{O}^*)$. We have the following observation.

Lemma 5. *For all $1 \leq i \leq m$, $size_c(\eta_i) = (n.size(\mathcal{O}))^{O(1)}$.*

Proof. The basis $\mathbf{b}_i = \mathrm{Log}(\eta_i)$ is the $\prec$-minimal basis of $\Lambda = \mathrm{Log}(\mathcal{O}^*)$. Let $\mathbf{b}_i^*$ denote the corresponding GSO basis and let $V_i = \mathrm{Span}\{\mathbf{b}_i\}_{i=1}^m$. Let $\{\varepsilon_i\}_{i=1}^m$ be any fundamental system of units satisfying the condition in Theorem 7 then $\mathbf{a}_i = \mathrm{Log}(\varepsilon_i)$ spans the lattice Λ. Without loss of generality we may assume that $\mathbf{a}_i \notin V_i$. Since $\mathbf{b}_i$'s form the $\prec$-minimal basis of Λ we have $\| \mathbf{b}_i^* \|_\infty \leq \| \mathbf{b}_i^* \|_1 \leq \| \mathbf{a}_i/V_{i-1} \|_1 \leq m \| \mathbf{a}_i \|_\infty$. Hence we have $\| \mathbf{b}_i \|_\infty \leq \| \mathbf{b}_i^* \|_\infty + \frac{1}{2}\sum_{j=1}^{i-1} \| \mathbf{b}_j^* \| \leq m.A.(i-1)/2$, where A is an upper bound on $\| \mathbf{a}_i \|_\infty$. From Theorem 7 and Proposition 1 we have $A \leq (n \cdot size(\mathcal{O}) + 2^{size_c(\varepsilon_i)})^{O(1)}$ where i is such that $\mathrm{Log}(\varepsilon_i)$ has the largest ℓ_∞ norm. Hence for every i we have $\ln \mathrm{H}(\eta_i) \leq (n \cdot size(\mathcal{O}) + 2^{size_c(\varepsilon_i)})^{O(1)}$. Together with Theorem 4 we have the result.

We now describe how a canonical fundamental system of units can be computed given an arbitrary fundamental system of units. The following theorem, based on a remark in [17], will be useful.

Theorem 8. [17] *Assuming that $[K : \mathbb{Q}]$ is constant, there is a polynomial time algorithm that takes as input a principal ideal $\mathfrak{a} = \alpha\mathcal{O}$ by its Hermite Normal Form and a good rational approximation for $\mathrm{Log}(\alpha)$, and outputs a compact representation for $\zeta\alpha$ where ζ is a root of unity in K.*

Remark 1. The point to note in the above theorem is that only $\mathrm{Log}(\alpha)$ is given (as a rational approximation) and not the compact representation of α. Also notice that $\mathrm{Log}(\alpha)$ is unique only upto multiplication by roots of unity in K. The theorem promises that one such element $\zeta\alpha$, which depends only on $\mathrm{HNF}(\mathfrak{a})$ and $\mathrm{Log}(\alpha)$, is computable in polynomial time.

Theorem 9. *Assuming $[K : \mathbb{Q}]$ is a constant, there is a polynomial time deterministic algorithm that takes as input a fundamental system of units (as compact representations) and outputs another fundamental system of units $\{\eta_i\}_{i=1}^m$ (as compact representations) corresponding to the $\prec$-minimal basis for $\mathrm{Log}(\mathcal{O}^*)$. Furthermore, $\{\eta_i\}_{i=1}^m$ is canonical in the sense that it does not depend on the input fundamental system of units.*

Proof. Given a fundamental system of units $\{\varepsilon_i\}_{i=1}^m$ compute $\{\text{Log}(\varepsilon_i)\}_{i=1}^m$ to the desired approximation. Compute the $\prec$-minimal basis for the lattice generated by $\{\text{Log}(\varepsilon_i)\}_{i=1}^m$ using algorithm in Theorem 6. Let it be $\{\mathbf{b}_i\}_{i=1}^m$. Use algorithm in Theorem 8 to compute compact representations of units $\{\eta_i\}_{i=1}^m$ corresponding to the vectors $\{\mathbf{b}_i\}_{i=1}^m$. Since the basis $\{\mathbf{b}_i\}_{i=1}^m$ is unique (upto approximation) and since all the algorithms involved are polynomial time deterministic algorithm the output generated is independent of the fundamental system of units that was given as input.

Given any set of units $\{\varepsilon_i\}$, we now analyze the approximation of $\text{Log}(\varepsilon_i)$ required in order to accurately compute the canonical fundamental system of units.

Let $\{\eta_i\}$ the canonical fundamental system of units and let $\mathbf{a}_i = \text{Log}(\eta_i)$. Let $\mathbf{b}_i = \text{Log}(\varepsilon_i)$. Consider the matrices $A = (a_{ij})$ and $B = (b_{ij})$ (recall that for a vector $\mathbf{v}$, v_i denotes its ith component). Since $\mathbf{a}_i$'s and $\mathbf{b}_i$'s span the same lattice $\Lambda = \text{Log}(\mathcal{O}^*)$, there is a unimodular transformation $U \in \text{SL}_m(\mathbb{Z})$ such that $A = UB$. Note that the determinant of B is the regulator which is at least 0.2 and hence it can be shown that each entries of U is of size bounded by a polynomial in the sized of entries in A and B. Let B_q denote the q bit approximations of B and let $A_q = UB_q$. We have

$$\| A_q - A \|_\infty = \| U(B_q - B) \|_\infty \leq m \| U \|_\infty 2^{-q}.$$

If we take q large enough so that $\| A_q - A \|_\infty$ is small enough for us to recover back the compact representation of η_i's we are through. It is easy to see that a q that is bounded by a polynomial in the sizes of entries of A and B is sufficient for this purpose.

Lemma 6. *In the algorithm of Theorem 9, it suffices to approximate* $\text{Log}(\varepsilon_i)$ *to an error of* 2^{-q} *where* $q \leq (\lg(\| A \|_\infty) \lg(\| B \|_\infty))^{O(1)}$.

5 Computing Units is in FP^{SPP}

In this section we give an FP^{SPP} algorithm for computing a fundamental system of units for a number field K. The algorithm is in two stages. In the first stage it computes a number B such that the regulator R_K lies in the range $[B, 2B)$. Notice that, having computed such a bound B, we can test in deterministic polynomial time if an arbitrary set of m algebraic numbers is a fundamental system of units. Given this value of B, in the second stage the FP^{SPP} algorithm computes a fundamental system of units of K. The first stage is described in the following lemma.

Lemma 7. *Given a number field K, there is an* FP^{SPP} *algorithm to compute a constant B such that the regulator R_K of K, lies in the interval $[B, 2B)$.*

Proof. We give a polynomial time algorithm that makes UP-like queries to an NP oracle. Consider the following NP language:

$$A = \{\langle x, \mathcal{O}_K \rangle \mid \text{there is a subgroup of index } y \text{ in } \mathcal{O}_K^* \; : \; x \leq yR_K < 2x\}.$$

We consider the following nondeterministic procedure that accepts A.

Input: A rational x and basis for the ring of integers, $\mathcal{O}$ of a number field K
Output: "Yes" if there is a subgroup of $\mathcal{O}^*$ of index y such that $x \leq yR_K < 2x$; "No" otherwise.

1 Guess the polynomial sized compact representations of m units of $\mathcal{O}$ say $\{\varepsilon\}_{i=1}^m$;
Compute rational approximations of $\{\mathrm{Log}\,(\varepsilon_i)\}_{i=1}^m$ and check if they form a linearly independent set. If not reject;
Compute the volume of the parallelepiped formed by the $\{\mathrm{Log}\,(\varepsilon_i)\}_{i=1}^m$ and check if it lies in the interval $[x, 2x)$. If not reject;

2 Use the algorithm in Theorem 9 to compute a canonical fundamental system of units say η_i's;
Check the whether the compact representations obtained in step 2 is same as the guessed compact representations. If yes accept else reject;

We now explain Step 1. First guess m (polynomial sized) compact representations of m algebraic integers $\{\alpha_i\}_{i=1}^m$. Applying Theorem 4 it is possible to compute $\alpha_i\mathcal{O}$ and check whether $\alpha_i\mathcal{O} = \mathcal{O}$ in polynomial time.

In the above NP machine if x is such that $x \leq R_K < 2x$ then there will be only one accepting path. This is because any set of m units that was guessed in step 1 will indeed be a fundamental system of units. For each of these accepting paths, step 2 will give a unique compact representation of a fundamental system of units— those units that in the Log map gives the $\prec$-minimum basis of $\mathrm{Log}\,(\mathcal{O}^*)$. Hence the only path that will accept is that which guessed that unique compact representation of units corresponding to the $\prec$-minimal basis of $\mathrm{Log}\,(\mathcal{O}^*)$.

It is known that the regulator of any number field is at least 0.2 [8]. We now describe the procedure that computes the required bound B:

Input: A $\mathbb{Z}$-basis for the ring of integers, $\mathcal{O}$, of a number field K
Output: A rational B such that $B \leq R_K < 2B$

```
B := 0.2;
while true do
    if ⟨B, O_K⟩ ∈ A then return B;
    B := 2B;
end
```

Since this procedure makes UP-like queries to the NP language A we can convert it into a $\mathrm{FP}^{\mathrm{SPP}}$ algorithm by Lemma 1.

Lemma 8. *Given a constant B such that $B \leq R_K < 2B$, a fundamental system of units can be computed in* $\mathrm{FP}^{\mathrm{SPP}}$.

Proof. First, consider the following nondeterministic polynomial time machine M. The machine M first guesses a set of m algebraic integers in their compact representation and then verifies in polynomial time that the guessed algebraic integers indeed form a fundamental system of units by first checking whether they are indeed units (check if $\alpha\mathcal{O} = \mathcal{O}$) and then calculating the volume of the parallelepiped (in the Log map) formed by the vectors corresponding to the guessed units, by a determinant computation. If this volume does not lie between B and $2B$, the machine M rejects on this computation path. Otherwise, applying

Theorem 9 along with the guessed fundamental system of units as input, the machine M now computes a canonical fundamental system of units and checks if it coincides with the guessed fundamental system of units. If they do coincide the machine M accepts along this computation path. It is clear from the above description that the nondeterministic machine M has a unique accepting path. Applying Lemma 1, we can now design from M an $\mathrm{FP}^{\mathrm{SPP}}$ algorithm that will compute a fundamental system of units, if it is additionally given B such that $B \leq R_K < 2B$.

Now, combining Lemmas 7 and 8 we immediately obtain the following.

Theorem 10. *There is a* $\mathrm{FP}^{\mathrm{SPP}}$ *algorithm to compute a fundamental system of units of the ring of integers of a number field* K *assuming that the degree* $[K : \mathbb{Q}]$ *is a constant.*

6 Principal Ideal Testing is in SPP

Given a number field K and a $\mathbb{Z}$-basis for its ring of integer $\mathcal{O}$, the *principal ideal testing* problem (denoted by PrI) problem is to check if an $\mathfrak{a}$ of $\mathcal{O}$ is a principal ideal. We show that this problem is in SPP.

Theorem 11. *Given a number field* K *with ring of integers* $\mathcal{O}$ *and the* $\mathbb{Z}$ *basis of a ideal* $\mathfrak{a}$*, checking whether* $\mathfrak{a}$ *is principal is in* SPP*, assuming* $[K : \mathbb{Q}]$ *is a constant.*

Proof. Without loss of generality we can assume that $\mathfrak{a}$ is an integral ideal. First compute a fundamental system of units $\{\varepsilon_i\}_{i=1}^{m}$ of $\mathcal{O}$ using the algorithm in Theorem 10. Guess the compact representation of an algebraic integer α. Check if $\alpha O = \mathfrak{a}$ if not reject. Check if $\mathrm{Log}\,(\alpha)$ lies in the fundamental parallelepiped $\{\mathbf{x} \in \mathbb{R}^m : \mathbf{x} = \sum \alpha_i \mathrm{Log}\,(\varepsilon_i)\,, \alpha_i \in [0,1]\}$. If not reject. Next, apply Theorem 8 to obtain a compact representation of α' from $\mathrm{Log}\,(\alpha)$ and $\mathfrak{a}$, such that $\mathrm{Log}\,(\alpha) = \mathrm{Log}\,(\alpha')$. If the compact representations of α and α' coincide we accept on that computation path and reject otherwise. The correctness of the easily follows from Theorem 8 and the fact that for every $\alpha \in \mathcal{O}$ there is a unique associate in the fundamental parallelepiped.

6.1 Computing the Class Number

Finally, we show that if we assume the generalized Riemann hypothesis (GRH) then finding the class number and a presentation for the class group are in $\mathrm{FP}^{\mathrm{SPP}}$. It is shown by Bach [2] that if the GRH is true then the class group of any number field K is generated by the ideal classes of all non-inert prime ideals of norm less $L = 12\ln^2|D|$. Let $\mathfrak{p}_1, \ldots, \mathfrak{p}_N$ be the (polynomially many) ideal classes of all non-inert prime ideals of norm less $L = 12\ln^2|D|$. We can compute these ideals $\mathfrak{p}_i$ in polynomial time as explained, for example in [8, Section 6.2.5] or [5].

Our goal is to compute the class number and a generator-relator presentation of the class group of K. Let G_i denote the subgroup of $Cl(K)$ generated by $\{\mathfrak{p}_1, \ldots, \mathfrak{p}_i\}$ and let $G_0 = \{id\}$. For each i let t_i be the least positive integer such that $\mathfrak{p}_i^{t_i} \in G_{i-1}$. Let s_{ij}, $1 \leq i \leq N$ and $1 \leq j < i$ be integers such that $0 \leq s_{ij} < t_j$ and such that $\mathfrak{p}_i^{t_i} \sim \prod_j^{i-1} \mathfrak{p}_j^{s_{ij}}$. The set of relators defined by

$$R = \left\{ \mathfrak{p}_i^{t_i} = \prod_{j=1}^{i-1} \mathfrak{p}_j^{s_{ij}} : 1 \leq i \leq N \right\}$$

together with the generator set $\{\mathfrak{p}_1, \ldots, \mathfrak{p}_N\}$ gives a generator-relator presentation of $Cl(K)$. Furthermore, notice that the s_{ij}'s are unique in the range $0 \leq s_{ij} < t_j$. Also, t_1 is the order of the ideal class of $\mathfrak{p}_1$ in $Cl(K)$. We will describe an $\mathrm{FP}^{\mathrm{SPP}}$ procedure for computing the set R inductively as follows.

Assume that the set of relators

$$R_i = \left\{ \mathfrak{p}_j^{t_j} = \prod_{k=1}^{j-1} \mathfrak{p}_k^{s_{jk}} : 1 \leq j \leq i \right\}$$

is already computed (where $R_0 = \emptyset$). It suffices to give an $\mathrm{FP}^{\mathrm{SPP}}$ procedure for computing R_{i+1}. To this end, we define an $\mathrm{NP}^{\mathrm{SPP}}$ language A as follows: A consists of the set of tuples $\langle x, y, \{\mathfrak{a}_j\}_{j=1}^{i}, \{(m_j, n_j)\}_{j=1}^{i-1} \rangle$ such that there is a $x \leq t < y$ and $m_j \leq s_j < n_j$ such that $\mathfrak{a}_i^t \sim \prod_{j=1}^{i-1} \mathfrak{a}_j^{s_j}$, where $\mathfrak{a}_j$'s are ideals in $\mathcal{O}_K$ given by their HNFs. It is easy to see that the language A is in $\mathrm{NP}^{\mathrm{SPP}}$: guess t and the s_j's and verify the class group identity by applying the SPP algorithm for principal ideal testing in Theorem 11.

The following code is a polynomial-time oracle computation (with oracle A) that computes R_{i+1} from R_i.

Let $T := 1$;
while *true* **do**
  if $\langle T, 2T, \{\mathfrak{p}_j\}_{j=1}^{i+1}, \{(1, t_j)\}_{j=1}^{i} \rangle \in A$ **then** break **else** $T := 2T$
end
Do a binary search for t_{i+1} in the range $[T, 2T)$ using A as oracle.;
Next, do a binary search to compute the s_j's. (* These s_j's are actually the s_{i+1j} in the definition of R. *)

It is easy to see that in the above algorithm only UP-like queries are made to A. More precisely, the queries will be of the form $\langle x, y, \{\mathfrak{p}_j\}_{j=1}^{i}, \{(m_j, n_j)\}_{j=1}^{i-1} \rangle$, with parameters such that the $\mathrm{NP}^{\mathrm{SPP}}$ machine for A will have at most one accepting path. The queries are UP-like as R_i is a set of relators for G_i. Now, using closure properties of SPP and Lemma 1 we can transform the above algorithm to an $\mathrm{FP}^{\mathrm{SPP}}$ procedure that inductively computes the generator-relator presentation of the class group $Cl(K)$. Observe that the class number is given by $\prod_{i=1}^{N} t_i$. Hence we have the following theorem.

Theorem 12. *Assuming the GRH, the class number and a generator-relator presentation for the class group of a constant-degree number field can be computed in* $\mathrm{FP}^{\mathrm{SPP}}$.

Acknowledgment. We thank the referees for their useful comments.

References

1. V. Arvind and P. P. Kurur. Graph Isomorphism is in SPP. In 43rd *Annual Symposium of Foundations of Computer Science*, pages 743–750. IEEE, November 2002.
2. E. Bach. Explicit bounds for primality testing and related problems. *Mathematics of Computation*, 55:355–380, 1990.
3. J. Buchmann. On computation of units and class number by a generalization of Lagrange's algorithm. *Journal of Number Theory*, 26:8–30, 1987.
4. J. Buchmann. On the period length of the generalized Lagrange algorithm. *Journal of Number Theory*, 26:31–37, 1987.
5. J. Buchmann and H. W. Lenstra Jr. Computing maximal orders and decomposing primes in number fields. preprint.
6. J. Buchmann and H. C. Williams. On the infrastructure of the principal ideal class of an algebraic number field of unit rank one. *Mathematics of Computation*, 50:569–579, 1988.
7. J. Buchmann and H. C. Williams. On the existence of a short proof for the value of the class number and regulator of a real quadratic field. *Number Theory and Applications*, 265:327–345, 1989.
8. H. Cohen. *A Course in Computational Algebraic Number Theory.* Springer-Verlag, Berlin, 1993.
9. S. Fenner, L. Fortnow, S. A. Kurtz, and L. Li. An oracle builder's toolkit. In *SCT: Annual Conference on Structure in Complexity Theory*, pages 120–131, 1993.
10. S. A. Fenner, L. Fortnow, and S. A. Kurtz. Gap-definable counting classes. In *Structure in Complexity Theory Conference*, pages 30–42, 1991.
11. L. Fortnow and J. D. Rogers. Complexity limitations on quantum computation. In *IEEE Conference on Computational Complexity*, pages 202–209, 1998.
12. S. Hallgren. Polynomial-time quantum algorithms for pell's equation and the principal ideal problem. In *Proceedings of the* 34th *ACM Symposium on Theory of Computing*, pages 653–658. IEEE, 2002.
13. H. W. Lenstra Jr. Integer programming with a fixed number of variables. *Mathematics of Operations Research*, 8:538–548, 1983.
14. J. Kobler, U. Schoning, S. Toda, and J. Toran. Turing machines with few accepting computations and low sets for PP. *Journal of Computer and System Sciences*, 44(2):272–286, 1992.
15. J. Köbler, U. Schöning, and J. Torán. *The Graph Isomorphism Problem: Its Structural Complexity.* Birkhauser, 1993.
16. K. S. McCurley. Cryptographic key distribution and computation in class groups. In *NATO Advanced Science Institute Series C*, volume 256, pages 459–479. Kluwer, Dordrecht, 1989.
17. C. Thiel. Under the assumption of the Generalized Riemann Hyothesis verifying the class number belongs to NP ∩ co-NP. In *Algorithmic Number Theory, First International Symposium, ANTS-I*, volume 877 of *Lecture Notes in Computer Science*, pages 234–247. Springer, 1994.

Implementing the Arithmetic of $C_{3,4}$ Curves

Abdolali Basiri[1,3], Andreas Enge[2], Jean-Charles Faugère[1], and Nicolas Gürel[2]

[1] Laboratoire d'Informatique de Paris 6 (CNRS/UMR 7606)
4 place Jussieu, 75252 Paris Cedex 05, France
{basiri,jcf}@calfor.lip6.fr

[2] INRIA Futurs & Laboratoire d'Informatique (CNRS/FRE 2653)
École polytechnique, 91128 Palaiseau Cedex, France*
{enge,gurel}@lix.polytechnique.fr

[3] Department of Mathematics and Computer Sciences
Damghan University of Sciences, Damghan, Iran
basiriab2@yahoo.com

Abstract. We provide explicit formulae for realising the group law in Jacobians of superelliptic curves of genus 3 and $C_{3,4}$ curves. It is shown that two distinct elements in the Jacobian of a $C_{3,4}$ curve can be added with 150 multiplications and 2 inversions in the field of definition of the curve, while an element can be doubled with 174 multiplications and 2 inversions. In superelliptic curves, 10 multiplications are saved.

1 Introduction

The interest in the arithmetic of low genus algebraic curves has been spurred by the fact that their Jacobians provide attractive groups to implement discrete logarithm based cryptosystems. While the attention first focused on the simplest types of curves, namely elliptic and hyperelliptic ones, it is shifting towards more complicated ones in the form of superelliptic and more generally $C_{a,b}$ curves. Attacks on high [5,6] and medium genus hyperelliptic cryptosystems [9,20] make it seem advisable to concentrate on curves of genus at most 3, which leaves superelliptic cubics and $C_{3,4}$ curves. There are a number of generic algorithms for computing $\mathcal{L}$-spaces of arbitrary curves, thus implementing the arithmetic of their Jacobian groups, like [13,15], to cite only the most recent ones. For superelliptic and $C_{a,b}$ curves, faster special purpose algorithms have been developed, relying either on Gröbner basis computation [1] or LLL on polynomials [8,12,3]. None of these articles provide a precise count of the number of operations for the arithmetic of non-hyperelliptic curves of low genus.

In [2], the present authors identify special Jacobian elements, called "typical", that, while admitting a special simplified representation by two polynomials (cf. Section 2), yet cover the major part of the Jacobian. (In fact, in the cryptographic context of a large finite base field and randomly chosen elements, one

* Projet TANC, Pôle Commun de Recherche en Informatique du Plateau de Saclay, CNRS, École polytechnique, INRIA, Université Paris-Sud

D.A. Buell (Ed.): ANTS 2004, LNCS 3076, pp. 87–101, 2004.

does not expect to ever encounter non-typical elements.) Then two algorithms for adding typical elements are developed. The first algorithm is inspired by Cantor's algorithm for hyperelliptic curves. The second one is obtained by applying the FGLM algorithm for changing the ordering of a Gröbner basis, and yields explicit formulae for the group law in terms of operations with polynomials. These two algorithms for superelliptic cubics generalise readily to $C_{3,4}$ curves. A first quick implementation in the superelliptic case required about 250 multiplications in the underlying field for the algorithm à la Cantor, and about 200 multiplications for the formulae on the polynomial level, thus improving by a factor of 3 on our implementation of the algorithm of [8].

Flon and Oyono use a slightly different approach in [7] to obtain explicit formulae for the group law on typical elements.[1] The formulae they give for superelliptic cubics require 156 resp. 174 field multiplications (depending on whether two distinct elements are added or an element is doubled) and 2 inversions in the field. For $C_{3,4}$ curves, they announce 177 resp. 198 multiplications and again 2 inversions, without providing further details.

In the present article we show how a carefully optimised implementation of our formulae allows to save up to 16 multiplications for superelliptic curves and 27 multiplications for $C_{3,4}$ curves compared to [7]. We explain in detail how we obtained a straight line program, suited for a straightforward implementation in constrained environments such as smartcards, from the formulae manipulating polynomials, and for the first time we provide details for $C_{3,4}$ curves.

In the following section, we give a concise overview of superelliptic cubics and $C_{3,4}$ curves and relate without proof the algorithm developed in [2]. After a few remarks on the underlying field arithmetic in Section 3, we present in Section 4 algorithms for computing with low degree polynomials. These algorithms serve as a toolbox for transforming the formulae on the polynomial level into formulae on the coefficient level, which we describe together with an exact operation count in Section 5.

2 Jacobians of $C_{3,4}$ Curves

Let K be a perfect field. A *superelliptic curve of genus* 3 or *Picard curve* over K is a smooth affine curve C of the form $C : Y^3 = f(X)$ with $f \in K[X]$ monic of degree 4; a $C_{3,4}$ *curve* is more general and may additionally have a term of the form $h(X)Y$ with $h \in K[X]$ of degree at most 2. The place at infinity of the function field extension $K(C)/K(X)$ corresponding to such a curve is totally ramified and rational over K, whence the K-rational part of the Jacobian of C is isomorphic as a group to the ideal class group of the coordinate ring $K[C] = K[X,Y]/(Y^3 + hY - f)$. By the Riemann–Roch theorem, each ideal class contains a unique ideal $\mathfrak{a}$ of minimal degree $\#(K[C]/\mathfrak{a})$ not exceeding 3, and this ideal is called the *reduced* representative of the class. In the following,

[1] Technically speaking, they compute a minimum with respect to the $C_{3,4}$ order in the ideal itself and end up in the inverse class, while we compute a minimum in the inverse class and end up with a reduced representative of the ideal itself.

we shall only consider ideals not containing two distinct prime ideals above a rational prime of $K[X]$. Such an ideal can be written as

$$\mathfrak{a} = (u, Y - v) \text{ with } u,\, v \in K[X],\ u \text{ monic, } \deg v < \deg u \text{ and } u | v^3 + hv - f;$$

its degree is exactly $\deg u$.

Most ideal classes (a proportion of $1 - \frac{1}{q}O(1)$ in the case of a finite ground field $\mathbb{F}_q$) are represented by such an ideal satisfying furthermore $\deg u = 3$ and $\deg v = 2$; we call these ideals "typical". Conversely, it is shown in [2], that for superelliptic curves of genus 3 a typical ideal is automatically reduced, and the proof carries over to $C_{3,4}$ curves. In the remainder of this article, we shall only consider typical ideals and the cases where ideals and polynomials behave in a typical way (for instance, the remainder of a polynomial of degree at least $n+1$ upon division by a polynomial of degree n is supposed to be $n-1$). Whenever these assumptions do not hold (which one does not expect to happen in practice), one may have recourse to a slower generic algorithm. Alternatively, one may develop specific formulae, that actually turn out to be simpler than in the common case.

The product of two ideal classes, represented by ideals $\mathfrak{a}_i = (u_i, Y - v_i)$, $\deg u_i = 3$, $\deg v_i = 2$, $i \in \{1, 2\}$, is obtained in two steps. The *composition* step corresponds to ideal multiplication and yields $\mathfrak{a} = (u, Y - v) = \mathfrak{a}_1\mathfrak{a}_2$. Here, $u = u_1u_2$ is monic of degree 6, and v of degree 5 is computed by interpolation. In the *addition* case, where $u_1 \neq u_2$ (and typically u_1 and u_2 are coprime), v is obtained by Chinese remaindering as follows:

$$s_1 = u_1^{-1} \bmod u_2;\ t = s_1(v_2 - v_1) \bmod u_2;\ v = v_1 + tu_1. \tag{1}$$

In the *doubling* case $\mathfrak{a}_1 = \mathfrak{a}_2$, a Hensel lift yields

$$\begin{aligned} &s_3 = (3v_1^2 + h)^{-1} \bmod u_1;\ w_1 = \frac{v_1^3 + hv_1 - f}{u_1};\ t = -s_3w_1 \bmod u_1; \\ &v = v_1 + tu_1. \end{aligned} \tag{2}$$

The *reduction* step takes as input an ideal $\mathfrak{a} = (u, Y - v)$ with u of degree 6 and v of degree 5, and outputs an equivalent ideal $\mathfrak{a}' = (u', Y - v')$ of degree 3, which by [2] is the reduced representative of its class. Let e be the minimum with respect to the $C_{3,4}$ order of the ideal $(u, Y^2 + vY + v^2 + h)$ in the class of $\mathfrak{a}^{-1}$, that is, e is the element whose pole at infinity has minimal multiplicity. It is shown in [2] for superelliptic curves and easily generalised to $C_{3,4}$ curves that

$$e = tY^2 + \varphi Y + \psi,$$

where the polynomial t, of degree 2, is obtained by executing two steps of the extended Euclidian algorithm on u and v, and $\varphi = tv \bmod u$; otherwise said, there is a linear polynomial s such that $su + tv = \varphi$ of degree 3. Moreover, $\psi = t(v^2 + h) \bmod u$. Then the reduced ideal $\mathfrak{a}' = \frac{e}{u}\mathfrak{a} = (u', Y - v')$ is computed as follows:

$$\begin{aligned}
\lambda &:= \frac{t^2 f - \varphi\psi}{u} \\
\mu &:= \frac{\varphi^2 - t\psi + t^2 h}{u} \\
u' &= \frac{f\left(t(t^2 f - 3\varphi\psi) + \varphi^3\right) + \psi^3 \; + \; h\left(t\psi(th - 2\psi) + \varphi(t^2 f + \varphi\psi)\right)}{u^2} \\
v' &= \mu^{-1}\lambda \bmod u'
\end{aligned} \tag{3}$$

Here, all divisions by u are exact, that is with remainder zero.

3 Field Arithmetic

In the previous section, we have shown how to realise the arithmetic of $C_{3,4}$ curves using a representation by polynomials. The main goal of this article is to adopt a lower level point of view and to provide formulae for the coefficients of the output polynomials in terms of those of the input polynomials. Thus, the operations we consider as elementary are operations in the field defining the curve. Our main motivation being potential cryptographic applications, we have (not too small) finite fields in mind, but the final formulae will hold for any field. However, for certain optimisations to work, we exclude fields of characteristic 2, 3 and 5. (As a side note, a superelliptic cubic is singular in characteristic 3, while in characteristic 2, it has a special structure, which might make it less attractive for cryptography, cf. [10].) There is only one division by 5 in our formulae; if need be, it could be removed at the expense of a few extra multiplications.

There are a thousand and one ways of organising the computations, so an optimality criterion is needed. Naturally, this should be the running time of a group operation in the Jacobian, which will ultimately depend on the concrete implementation and the concrete environment. A reasonable and theoretically tractable approximation is the number of elementary field operations. In many situations (over not too small finite fields, for instance, but not over the rational numbers) additions and subtractions take a negligible time compared to multiplications and inversions (divisions being realised as an inversion followed by a multiplication). Notice that multiplications by small natural numbers can be realised by a few additions, so they come for free.

Moreover, we do not count divisions by small constants (precisely, 2, 3 and 5) either. For instance, in a prime finite field $\mathbb{F}_p$, represented by $\{0, \dots, p-1\}$, division of an element a by 2 is trivial: either a is even, then it may be divided by 2 as an integer; or it is odd, then $a + p$ is even and $\frac{a+p}{2}$ is a representative of the result in $\{0, \dots, p-1\}$. Slightly less straightforward, a division of a by 3 may be realised by first computing the remainder of a upon division by 3. Since $4 \equiv 1 \pmod 3$, this is a matter of splitting a in blocks of two bits using bit masks and shifts and adding these base 4 digits, much as the test for divisibility by 9 of a number in decimal notation. Then, one adds the appropriate multiple of p and divides by 3. For the remainder modulo 5, an alternating sum of base 4 digits may be used. In a general finite field, represented as a vector space over $\mathbb{F}_p$, divisions by small constants can be carried out coordinate wise.

Finally, this leaves us with two variables to minimise, the number of multiplications and the number of inversions, and there is a trade off between these two. Depending on the library for finite field arithmetic, an inversion usually takes between three and ten times as long as a multiplication. We therefore tried to eliminate as many inversions as possible, as long as this introduced only a few extra multiplications. For instance, two independent inversions of field elements a and b may be replaced by one inversion and three multiplications as follows:

$$u = (a \cdot b)^{-1},\ a^{-1} = u \cdot b,\ b^{-1} = u \cdot a.$$

4 Polynomial Arithmetic

Once the algorithm of Section 2 exhibited, the remaining task is no more connected to geometry, but rather to the topic of symbolic computation. The rational formulae given there are expressed in terms of operations with polynomials; it remains to phrase them in terms of their coefficients. A straightforward implementation of the polynomial arithmetic involved is of course trivial; the problem of minimising the number of field operations, however, appears to be hard. The only feasible approach we have found consists of performing local optimisations on pieces of the formulae.

In this section, we review different approaches and algorithms of polynomial arithmetic useful for this task. All of them are well-known in the symbolic computation community; however, textbooks often focus on the asymptotic behaviour of the algorithms and do not treat the very small instances we are interested in. While commenting on our choices for the concrete case of $C_{3,4}$ curves, we hope that the following overview will be helpful in further situations.

4.1 Multiplication

Let $M(m,n)$ denote the number of field multiplications carried out for multiplying two polynomials with m resp. n coefficients, that is, of degree $m-1$ and $n-1$; and let $M(n) := M(n,n)$. If useful, we indicate the employed algorithm by a subscript. For instance, the "naïve" method has $M(m,n) = mn$. A trivial improvement arises when one or both polynomials are monic; in the latter case, the equation
$(X^{m-1}+f(X))(X^{n-1}+g(X)) = X^{m+n-2}+f(X)X^{n-1}+g(X)X^{m-1}+(fg)(X)$
shows that at the expense of a few additions, only $M(m-1,n-1)$ multiplications are needed.

A substantial improvement is obtained by Karatsuba's multiplication [16]. Using the relation $(aX+b)(cX+d) = a \cdot cX^2 + \big((a+b)\cdot(c+d) - ac - bd\big)X + b \cdot d$, it achieves $M_{\mathrm{K}}(2) = 3$, and, by recursively splitting the polynomials in half,

$$M_{\mathrm{K}}(2^m, 2^n) = 2^{m-n}3^n \text{ for } m \geq n.$$

Generalisations by Toom [21] and Cook [4] to the product of two polynomials with three coefficients yield $M_{\mathrm{TC}}(3) = 5$, and the analogous recursive strategy may be applied.

4.2 Exact Division

The simplest way of computing the quotient and remainder of $v = v_n X^n + v_{n-1} X^{n-1} + \cdots$ divided by $u = u_m X^m + u_{m-1} X^{m-1} + \cdots$ is the schoolbook method: invert u_m; the first term of the quotient, $u_m^{-1} v_n X^{n-m}$, is then computed with one multiplication; multiply it back by u, subtract from v; and continue in the same way.

If the remainder is of no interest or known to be zero, then it is not necessary to multiply back by all of u. In fact, it suffices to consider only the leading terms of u and v, while the lower ones determine the remainder. The k leading coefficients of the quotient are obtained with one inversion and

$$1 + \cdots + k \tag{4}$$

multiplications. If moreover u is monic, then the inversion and the multiplications by u_m^{-1} are saved, resulting in

$$1 + \cdots + (k-1) \tag{5}$$

multiplications. Letting $k = n - m + 1$ yields the full quotient. As observed by Jebelean [14] in the case of integer division, if the division is exact, that is, the remainder is zero, then it is also possible to work with only the trailing coefficients of u and v from right to left; in fact, his algorithm amounts to the division of the reciprocal polynomials of u and v. The number of operations is unaffected, but the algorithm deploys its benefits when used to work from both sides simultaneously as suggested by Schönhage in [19], using the k lowest and $n - m + 1 - k$ highest coefficients for some k. For given u and v, the value $k = \lfloor \frac{n-m+1}{2} \rfloor$ is optimal. If the effort for computing u and v is to be taken into account, other choices may be preferable; for instance, we shall use $k = 1$ most of the time.

4.3 Short Product

As seen in Section 4.2 on exact divisions, one is sometimes interested in only the trailing (or leading) coefficients of a polynomial. If this polynomial is the result of a multiplication, then these coefficients are obtained by what is known as a "short product". In the case of trailing coefficients, it can be seen as the product of truncated power series, returning upon input of two polynomials u and v of degree $n - 1$ their product modulo X^n. Instead of computing the full product of u and v and then truncating, Mulders suggested the following algorithm [17]: choose a cutoff point $k \geq \frac{n}{2}$, and write

$$u = u_0 + u_1 X^k \text{ and } v = v_0 + v_1 X^k.$$

Then $uv \bmod X^n$ is computed recursively as

$$u_0v_0 + \big((u_0v_1 \bmod X^{n-k}) + (u_1v_0 \bmod X^{n-k})\big)X^k$$

by a full and two short products. If $S(n)$ denotes the number of field multiplications required for computing a short product modulo X^n, we obtain the recursive formula

$$S(n) = M(k) + 2S(n-k).$$

Hanrot and Zimmermann [11] showed that if the full product is computed by Karatsuba's algorithm, then the optimal cutoff point k is the largest power of 2 not exceeding n. For instance, a short product modulo X^3 is computed by a full Karatsuba product of order 2 and two further field multiplications, resulting in altogether $S(3) = 5$ multiplications. This is the same number as for the full product by the Toom–Cook approach, but fewer additions and no division by 3 are required. It is to be expected that with Toom–Cook as the basic multiplication method, the optimal cutoff point will be once or twice a power of 3; then $S(4)$ becomes $S(4) = M_{\mathrm{TC}}(3) + 2S(1) = 7$. To generalise to products of polynomials of different degrees, let for $d \geq m \geq n$ the value $S(m, n; d)$ denote the number of multiplications required to compute the product modulo X^d of a polynomial of degree $m-1$ with one of degree $n-1$. Then for a cutoff point $k \leq n$ we obtain

$$S(m, n; d) = M(k) + S(m-k, k; d-k) + S(n-k, k; d-k).$$

For instance, $S(4, 3; 4)$ with Toom–Cook and $k = 3$ becomes

$$S(4, 3; 4) = M_{\mathrm{TC}}(3) + S(1, 3; 1) + S(0, 3; 1) = 5 + 1 + 0 = 6.$$

4.4 Interpolation

Karatsuba and Toom–Cook multiplication essentially work by evaluating the factors at small arguments (say, 0, 1, −1, 2, $\cdots$), multiplying the values and interpolating the result. Additionally, they treat the values at "∞", that is the leading coefficients, separately. Of course, this approach can be extended to higher degree polynomials as long as the base field has a sufficiently large characteristic so that the interpolation points are different. Evaluating a (low degree) polynomial in small integers requires only additions and multiplication by (small) constants, interpolation uses also divisions by (small) integers. So in our model, these steps cost nothing. One thus obtains a complexity of

$$M(m, n) = m + n - 1.$$

But the interpolation approach is not limited to simple multiplications; it can be extended to arbitrary polynomial and even rational formulae as long as the result is a polynomial, that is, all divisions are exact. For instance, the polynomial u' as computed in (3) is monic of degree 3, so it can be reconstructed by computing the values of $u' - X^3$ in 0, 1 and -1. Each value requires as many *field multiplications* as there are *polynomial multiplications* in the numerator of the formula, after adding parentheses and suitably reusing common subexpressions, and additionally a squaring and an inversion of the corresponding values of u in the denominator and a multiplication by the inverses. As mentioned in Section 3, the different inversions may be pooled into only one if one is willing to spend more time with multiplications.

4.5 Extended Euclidian Algorithm

The classical extended Euclidian algorithm, upon input of two polynomials r_{-1} and r_0 with $\deg r_{-1} \geq \deg r_0$, computes the greatest common divisor d of r_{-1} and r_0 together with multipliers a and b such that

$$d = ar_{-1} + br_0.$$

It proceeds by iterated divisions with remainder, until the remainder vanishes, and thus requires a certain number of inversions. The greatest common divisor is only defined up to multiplication by constants, and it is possible to modify the algorithm to use pseudodivisions without field inversions (this is essentially the subresultant algorithm). Keeping track of the multipliers u and v then requires extra multiplications. Assume by induction that remainders r_{i-1} and r_i and multipliers a_{i-1}, a_i, b_{i-1} and b_i are given such that

$$r_{i-1} = a_{i-1}r_{-1} + b_{i-1}r_0 \text{ and } r_i = a_i r_{i-1} + b_i r_0;$$

the initial values being $a_{-1} = b_0 = 1$, $a_0 = b_{-1} = 0$.

Let ℓ be the function that to a polynomial associates its leading coefficient. Then the pseudodivision of r_{i-1} by r_i yields a quotient q_{i+1} and a remainder r_{i+1} such that

$$\ell(r_i)^{\deg r_{i-1} - \deg r_i + 1} r_{i-1} = q_{i+1} r_i + r_{i+1},$$

$a_{i+1} = \ell(r_i)^{\deg r_{i-1} - \deg r_i + 1} a_{i-1} - q_{i+1} a_i$ and $b_{i+1} = \ell(r_i)^{\deg r_{i-1} - \deg r_i + 1} b_{i-1} - q_{i+1} b_i$ satisfy $r_{i+1} = a_{i+1} r_{-1} + b_{i+1} r_0$.

We analyse the most common case in more detail, where $\deg r_0 = \deg r_{-1} - 1$ and the remainder degrees drop by 1 in each step, that is, all the q_i have degree 1. Letting $r_{i-1} = \alpha X^k + \beta X^{k-1} + \cdots$ and $r_i = \gamma X^{k-1} + \delta X^{k-2} + \cdots$, the next quotient is computed as $q_{i+1} = \gamma \cdot \alpha X + (\gamma \cdot \beta - \delta \cdot \alpha)$. This requires 3 multiplications in the general case, 1 multiplication if r_{i-1} or r_i are monic and comes for free if both of them are monic.

The next remainder is obtained as $r_{i+1} = \ell(r_i)^2 \cdot r_{i-1} - q_{i+1} \cdot r_i$, using interpolation with $1 + 2 \deg r_i$ multiplications (or just $\deg r_i$ if r_i is monic).

Now, $\ell(r_i)^2$ is known, and computing a_{i+1} is free for $i = 0$, and requires 1 multiplication for $i = 1$ and $M(1, \deg a_{i-1}+1) + M(2, \deg a_i + 1) = i - 1 + M(2, i)$ multiplications for $i \geq 2$. If furthermore r_0 is monic, then $a_2 = -q_1$ comes also for free, and $a_3 = \ell(r_2)^2 + q_1 \cdot q_2$ requires only $M(2) = 3$ multiplications.

Obtaining b_{i+1} is free for $i = 0$ and requires $M(2) = 3$ multiplications for $i = 1$ and $M(1, \deg b_{i-1} + 1) + M(2, \deg b_i + 1) = i + M(2, i+1)$ multiplications for $i \geq 2$.

The following table summarises the number of multiplications carried out to compute the greatest common divisor, a and b, depending on the degree n of r_0, that is, the number of division steps, and on the monicity of r_{-1} and r_0. It is assumed that polynomials are multiplied using interpolation as explained in Section 4.4.

n	generic			r_{-1} monic			r_0 monic		
	gcd	a	b	gcd	a	b	gcd	a	b
1	6	0	0	4	0	0	2	0	0
2	14	1	3	12	1	3	7	0	3
3	24	5	9	22	5	9	16	3	9
4	36	11	17	34	11	17	27	9	17

4.6 Modular Division by Linear Algebra

One ingredient in our formulae for superelliptic arithmetic is the computation of $v' = \mu^{-1}\lambda \pmod{u}$, where μ is of degree 1, λ of degree 2, and u monic of degree 3. This computation can be carried out by first determining $\mu^{-1} \bmod u$ by the extended Euclidian algorithm as described in Section 4.5 and division by the greatest common divisor, a constant in the base field; then multiplying by λ and finally reducing modulo u. In our implementation, these steps require 22 multiplications and one inversion. In this section, we describe a different approach, saving 2 multiplications. The problem to be solved is much less generic than those of the previous sections; the proposed solution, relying on linear algebra, is quite general, however, and may also be applied to different constellations of degrees.

Write $\mu = \mu_1 X + \mu_0$, $\lambda = \lambda_2 X^2 + \lambda_1 X + \lambda_0$ and $v' = x_2 X^2 + x_1 X + x_0$ with unknown x_2, x_1, x_0. Then, by degree considerations, there is a further unknown value γ such that $\mu v + \gamma u = \lambda$. Comparing coefficients, we obtain a system of four linear equations in four variables, in which the equation $\gamma = -\mu_1 x_2$ can be substituted immediately. Performing Gaussian elimination yields the solution

$$\begin{aligned} x_2 &= \alpha^{-1}\beta \\ x_0 &= \mu_0^{-1}(\lambda_0 + \mu_1 u_0 x_2) \\ x_1 &= \mu_0^{-1}(\lambda_1 + \mu_1(u_1 x_2 - x_0)) \end{aligned}$$

with

$$\begin{aligned} \alpha &= (\mu_1^2 u_1 + \mu_0^2 - (\mu_0\mu_1)u_2)\mu_0 - (\mu_1 u_0)\mu_1^2 \\ \beta &= -\lambda_1(\mu_0\mu_1) + \lambda_0\mu_1^2 + \lambda_2\mu_0^2. \end{aligned}$$

α and β are computed with 11 multiplications. Then α and μ_0 are inverted simultaneously with 3 multiplications and one inversion as described in Section 3. The computation of x_2, x_0 and x_1 then requires $1 + 2 + 3 = 6$ multiplications (reusing the expression $\mu_1 u_0$ needed for α), for a total of 20 multiplications and one inversion.

5 Explicit Formulae

In this section, we develop explicit formulae for the composition and reduction steps, counting precisely the number of field multiplications and inversions. To keep the presentation readable, we use the building blocks of polynomial arithmetic of Section 4; expanding these to formulae involving only field operations is a trivial task.

We use two little tricks to speed up the computations. First, it is possible to complete the fourth power in f; after the linear change of variables $X \mapsto X - \frac{f_3}{4}$ we may assume that $f = X^4 + f_2X^2 + f_1X + f_0$. The impact of this observation is not very high, since f is hardly used in the formulae (it is mainly implicitly present in the relations $v^3 + vh \equiv f \pmod{u}$).

Second, the composition step involves the extended Euclidian algorithm, and the resulting greatest common divisor is normalised to be 1. This normalisation step requires an inversion, which can be saved by modifying the output of the composition to be (u, v, d) with d in the base field such that the real ideal product is given by $(u, Y - d^{-1}v)$. The composition step being of little interest *per se*, there is no harm done as long as this modification is taken into account in the reduction process.

5.1 Composition – Addition

Theorem 1. *On input of two ideals* $\mathfrak{a}_1 = \langle u_1, Y - v_1\rangle$ *and* $\mathfrak{a}_2 = \langle u_2, Y - v_2\rangle$ *such that* $u_i | v_i^3 + v_ih - f$ *and* $\gcd(u_1, u_2) = 1$, *and assuming the typical behaviour of the remainder degrees during the Euclidian algorithm, polynomials* u *and* v *and a field element* d *such that* $\mathfrak{a}_1\mathfrak{a}_2 = \langle u, Y - d^{-1}v\rangle$ *can be computed with* 37 *multiplications.*

Proof. We first compute $u = u_1u_2$ by Toom–Cook multiplication with 5 field multiplications. Then, implementing (1), we determine s_1 of degree 2 and d such that $s_1u_1 \equiv d \pmod{u_2}$ by applying the extended Euclidian algorithm to u_2 and $u_1 - u_2$. According to the table in Section 4.5, with $n = 2$ and r_{-1} monic this needs $12 + 3 = 15$ multiplications. We then compute $t_1 = s_1(v_2 - v_1)$ with 5 multiplications and the quotient q of the result by u_2 with 1 multiplication according to (5). The polynomial $t = t_1 - q \cdot u_2$ of degree 2 is then obtained by interpolation on three points with 3 multiplications for the values of $q \cdot u_2$. Finally, $v = d \cdot v_1 + u_1 \cdot t$ is computed with $M(1, 3) + M(3) = 8$ multiplications.

5.2 Composition – Doubling

With the preparations of Section 4, the composition part of doubling is as straightforward as addition, but it requires noticeably more operations.

Theorem 2. *On input of an ideal* $\mathfrak{a}_1 = \langle u_1, Y - v_1\rangle$ *such that* $u_1 | v_1^3 + v_1h - f$, *and assuming the typical behaviour of the remainder degrees during the Euclidian algorithm, polynomials* u *and* v *and a field element* d *such that* $\mathfrak{a}_1^2 = \langle u, Y - d^{-1}v\rangle$ *can be computed with* 61 *multiplications.*

Proof. We use the notation introduced in (2). First, we compute $u = u_1^2$ and v_1^2 with 5 multiplications each. To obtain w_1, we start with the four highest coefficients of $v_1^3 + v_1 h - f = v_1^2 \cdot (v_1 + h) - f$ using $S(4,3;4) = 6$ multiplications (see Section 4.3). An exact division by u_1 yields w_1 with 6 multiplications as shown in (5). The next step is to determine s_3 of degree 2 and d in the base field such that $s_3 \cdot (3v_1^2 + h) \equiv d \pmod{u_1}$. By the table in Section 4.5, this requires 19 multiplication since $n = 3$ and r_{-1} is monic.

We then reduce w_1, which is of degree 3, modulo the monic u_1 by subtracting the appropriate multiple of u_1, obtained with 3 multiplications; multiply by s_3 with $M(3) = 5$ multiplications and reduce again modulo u_1, which takes 1 multiplication for the quotient and 3 multiplications for the remainder, yielding t in a total of 12 multiplications.

Finally, v is obtained multiplying v_1 by d and t by u with $M(3,1)+M(3) = 8$ multiplications.

Notice that for adding as well as for doubling, the composition step is not more costly on $C_{3,4}$ than on superelliptic curves.

5.3 Reduction

Theorem 3. *On input of an ideal* $\mathfrak{a} = \langle u, Y - d^{-1}\tilde{v} \rangle$ *such that* $u | (d^{-1}\tilde{v})^3 + d^{-1}\tilde{v}h - f$, *$u$ monic of degree* 6, *$\tilde{v}$ of degree* 5, *the reduced representative* $\mathfrak{a}' = \langle u', Y - v' \rangle$ *in the ideal class of* $\mathfrak{a}$ *can be computed with* 113 *multiplications and* 2 *inversions. In the superelliptic case,* 10 *multiplications may be saved.*

Proof. To facilitate keeping track of the total number of multiplications, from time to time we provide their balance, having the number for the superelliptic case precede that for the general $C_{3,4}$ case.

0/0

We use the notation of Section 2. Let furthermore $v = d^{-1}\tilde{v}$, and denote the coefficient of a polynomial in front of X^i by a subscript i. The first step of the algorithm consists of finding the minimum $e = tY^2 + \varphi Y + \psi$ of $\mathfrak{a}$ with respect to the $C_{3,4}$ order, where the polynomial t, of degree 2, is obtained by executing two steps of the extended Euclidian algorithm on u and v, $\varphi = tv \bmod u$ is of degree 3 and $\psi = tv^2 + th \bmod u$ is of degree 5. Notice that e is defined only up to multiplication by constants; we shall compute the representative with leading coefficient 1 for the $C_{3,4}$ order, that is, with ψ monic. Inspection of (3) shows that then also u' will be monic, and no further normalisation will be needed.

To avoid inverting d, we shall use $\tilde{v}$ in the place of v, and correct the polynomials later on. Thus, we determine $\tilde{t}$ of degree 2, $\tilde{\varphi}$ of degree 3 and a linear polynomial ζ such that $\tilde{\varphi} = \tilde{t}\tilde{v} \bmod u = t \cdot \tilde{v} - \zeta \cdot u$. Using the algorithm of Section 4.5, the computation of $\tilde{t}$ and ζ requires 17 multiplications. By carrying out the Euclidian algorithm symbolically and simplifying the resulting formulae by hand, one may save 2 squarings as follows. The δ_i designate temporary variables.

$$\begin{aligned}
\delta_1 &= u_4 \cdot \tilde{v}_5 - \tilde{v}_3 \\
\delta_2 &= u_5 \cdot \tilde{v}_5 - \tilde{v}_4 \\
\delta_3 &= \delta_1 \cdot \tilde{v}_5 - \delta_2 \cdot \tilde{v}_4 \\
\tilde{t}_2 &= \delta_3 \cdot \tilde{v}_5 \\
\zeta_1 &= t_2 \cdot \tilde{v}_5 \\
\delta_4 &= \delta_3 \cdot u_5 \\
\delta_5 &= \delta_2 \cdot \tilde{v}_3 + \delta_4 - \tilde{v}_5 \cdot (\tilde{v}_5 \cdot u_3 - \tilde{v}_2) \\
\tilde{t}_1 &= \delta_5 \cdot \tilde{v}_5 \\
\zeta_0 &= \tilde{v}_5 \cdot (t_1 - \delta_2 \cdot \delta_3) \\
\tilde{t}_0 &= (\delta_5 - \delta_4) \cdot \delta_2 + \delta_1 \cdot \delta_3
\end{aligned}$$

15/15

The polynomial $\tilde{\varphi}$ is obtained via interpolation from $\tilde{t}$, $\tilde{v}$, u and ζ with 8 multiplications. Then, we compute polynomials $\tilde{\psi}$ and ξ such that

$$\tilde{\psi} = \tilde{\varphi}\tilde{v} \bmod u = \tilde{\varphi} \cdot \tilde{v} - \xi \cdot u,$$

the correction by the additional term th in the definition of ψ being postponed. Computing the 3 leading coefficients of $\tilde{\varphi}\tilde{v}$ takes $S(3) = 5$ multiplications, and the three coefficients of the quotient ξ by the monic u are obtained with 3 multiplications. Then $\tilde{\psi}$ may be computed by interpolation on six points with 12 multiplications.

43/43

Since we worked with $\tilde{v}$ instead of v, the polynomials $\tilde{t}$, $\tilde{\varphi}$ and $\tilde{\psi}$ have to be adjusted by powers of d, at the same time as making $\tilde{\psi}$ monic. We profit from the inversion of $\tilde{\psi}_5$ by computing simultaneously the inverse of u_0, which will be needed later on, with 3 multiplications and one field inversion as described in Section 3. Then, we obtain the minimum e via

$$t = (\psi_5^{-1} d \cdot d) \cdot \tilde{t},\ \varphi = (\psi_5^{-1} \cdot d) \cdot \tilde{\varphi},\ \psi = \psi_5^{-1} \cdot \tilde{\psi} + th.$$

As will become clear in the following, we do in fact not need the coefficients φ_1, ψ_1 and ψ_2, whence this step can be carried out with 11 multiplications in the superelliptic case. When $h \neq 0$, the computation of $t_0 h_0$ requires an additional multiplication, and ψ_3 and ψ_4 need the two leading terms of th, obtained with $S(2) = 3$ multiplications.

57/61

Define the polynomials λ and μ of degree 2 and 1, respectively, as in the equations before (3). In the following, we shall perform polynomial arithmetic "from both sides" as described in Section 4.2. All constant coefficients are computed separately. For instance, λ_0 and μ_0 are obtained with 7 multiplications if $h = 0$, including those by the already computed u_0^{-1}. For $C_{3,4}$ curves, μ_0 requires additionally the computation of $t_0^2 h_0$, which is obtained with one extra multiplication since $t_0 h_0$ has already been used for ψ_0.

For $\mu_1 = -t_2$, there is nothing to do. Taking into account that $f_4 = 1$, $f_3 = 0$ and $\psi_5 = 1$, the numerator of λ starts with

$$(t_2^2 - \varphi_3)X^8 + (2t_1 \cdot t_2 - \varphi_2 - \varphi_3 \cdot \psi_4)X^7 + \cdots,$$

and these coefficients are computed with 3 multiplications. The two leading terms of the quotient by u require another multiplication, so that the total number of multiplications for λ and μ becomes 11 in the superelliptic and 12 in the $C_{3,4}$ case.

68/73

The polynomial u' of the result is computed via (3). For $h = 0$, the constant coefficient u_0' is easily seen to be computable with 7 multiplications, reusing values like φ_0^2 already needed for λ_0 or μ_0. If $h \neq 0$, then the term

$$h_0 \cdot \big(t_0\psi_0 \cdot (t_0 h_0 - 2\psi_0) + \varphi_0 \cdot (t_0^2 f_0 + \varphi_0\psi_0)\big)$$

requires only the 3 additional multiplications marked with a dot.

75/83

For the high degree part of u', we compute the leading terms of the numerator $X^{15} + \alpha X^{14} + \beta X^{13} + \cdots$ of (3) as

$$\begin{aligned}
\alpha &= t_2 \cdot (\lambda_2 - 2(\varphi_3 + h_2)) + 3\psi_4, \\
\beta &= 3(t_1 \cdot \lambda_2 + \psi_3 + \psi_4^2) - t_2 \cdot (3\varphi_2 + 2h_1) + (\varphi_3^2 - 3t_2 \cdot \psi_4) \cdot \varphi_3 \\
&\quad + h_2 \cdot \big(t_2^2 \cdot (h_2 + \varphi_3) + (\varphi_3^2 - 3t_2\psi_4) - t_2\psi_4 - 2t_1\big),
\end{aligned}$$

where we have used the relation $\lambda_2 = t_2^2 - \varphi_3$. These quantities are obtained with 7 multiplications in the superelliptic case and 9 multiplications in the case of $C_{3,4}$ curves. Then the leading coefficients of u' are given by

$$u_3' = 1,\ u_2' = \alpha - 2u_5,\ u_1' = \beta - u_5 \cdot (2u_2' + u_5) - 2u_4$$

with 1 multiplication.

83/93

Finally, v' is computed as $v' = \mu^{-1}\lambda \bmod u$ with 20 multiplications and one inversion as described in Section 4.6.

103/113

In the following table, we summarise the number of multiplications carried out by our algorithm for adding or doubling divisors, totalling the efforts for the composition and the reduction step. We distinguish between ordinary multiplications and squarings in the base field. While we did not pursue this distinction in the present article due to space restrictions, separating these two numbers is a simple exercise. The number of inversions is always 2.

	superelliptic			$C_{3,4}$		
	mult.	sqr.	m.+s.	mult.	sqr.	m.+s.
addition	129	11	140	139	11	150
doubling	143	21	164	153	21	174

6 Concluding Remarks

Formulae for the arithmetic of hyperelliptic curves of genus 3 are reported in [18]. They require 76 field multiplications and one inversion for adding two distinct elements, and 71 multiplications and one inversion for doubling an element. While our formulae for superelliptic and $C_{3,4}$ curves need more operations, the factor of only about 2 shows that $C_{3,4}$ curves constitute a reasonable alternative to hyperelliptic curves for cryptographic use.

Availability of the Formulae

The MAGMA code of our formulae can be downloaded from the web at the address

`http://www.lix.polytechnique.fr/Labo/Andreas.Enge/C34.html`

Acknowledgement. Thanks to Pierrick Gaudry for his comments on our work.

References

1. Seigo Arita. Algorithms for computations in Jacobian group of C_{ab} curve and their application to discrete-log based public key cryptosystems. *IEICE Transactions*, J82-A(8):1291–1299, 1999. In Japanese. English translation in the proceedings of the Conference on The Mathematics of Public Key Cryptography, Toronto 1999.
2. Abdolali Basiri, Andreas Enge, Jean-Charles Faugère, and Nicolas Gürel. The arithmetic of Jacobian groups of superelliptic cubics. Rapport de Recherche 4618, INRIA, November 2002. Available at http://www.inria.fr/rrrt/rr-4618.html, to appear in *Mathematics of Computation*.
3. Mark L. Bauer. The arithmetic of certain cubic function fields. *Mathematics of Computation*, 73(245):387–413, 2004.
4. Stephen A. Cook. *On the Minimum Computation Time of Functions*. PhD thesis, Harvard University, 1966.
5. Andreas Enge. Computing discrete logarithms in high-genus hyperelliptic Jacobians in provably subexponential time. *Mathematics of Computation*, 71(238):729–742, 2002.

6. Andreas Enge and Pierrick Gaudry. A general framework for subexponential discrete logarithm algorithms. *Acta Arithmetica*, 102(1):83–103, 2002.
7. Stéphane Flon and Roger Oyono. Fast arithmetic on Jacobians of Picard curves. Preprint, Cryptology ePrint Archive 2003/079, available at http://eprint.iacr.org/2003/079, 2003.
8. S. D. Galbraith, S. M. Paulus, and N. P. Smart. Arithmetic on superelliptic curves. *Mathematics of Computation*, 71(237):393–405, 2002.
9. Pierrick Gaudry. An algorithm for solving the discrete log problem on hyperelliptic curves. In Bart Preneel, editor, *Advances in Cryptology — EUROCRYPT 2000*, volume 1807 of *Lecture Notes in Computer Science*, pages 19–34, Berlin, 2000. Springer-Verlag.
10. Pierrick Gaudry and Nicolas Gürel. An extension of Kedlaya's point counting algorithm to superelliptic curves. In Colin Boyd, editor, *Advances in Cryptology — ASIACRYPT 2001*, volume 2248 of *Lecture Notes in Computer Science*, pages 480–494, Berlin, 2001. Springer-Verlag.
11. G. Hanrot and P. Zimmermann. A long note on Mulders' short product. *Journal of Symbolic Computation*, 37(3):391–401, 2004.
12. Ryuichi Harasawa and Joe Suzuki. Fast Jacobian group arithmetic on C_{ab} curves. In Wieb Bosma, editor, *Algorithmic Number Theory — ANTS-IV*, volume 1838 of *Lecture Notes in Computer Science*, pages 359–376, Berlin, 2000. Springer-Verlag.
13. Floran Heß. Computing Riemann–Roch spaces in algebraic function fields and related topics. *Journal of Symbolic Computation*, 33(4):425–445, 2002.
14. Tudor Jebelean. An algorithm for exact division. *Journal of Symbolic Computation*, 15:169–180, 1993.
15. Kamal Khuri-Makdisi. Linear algebra algorithms for divisors on an algebraic curve. *Mathematics of Computation*, 73(245):333–357, 2004.
16. А. Карацуба and Ю. Офман. *Умножение многозначных чисел на автоматах. Доклады Академии Наук СССР*, 145(2):293–294, 1962. English translation: Multiplication of Multidigit Numbers on Automata, *Soviet Physics — Doklady*, 7(7):595–596, 1963.
17. Thom Mulders. On short multiplications and divisions. *Applicable Algebra in Engineering, Communication and Computing*, 11:69–88, 2000.
18. Jan Pelzl, Thomas Wollinger, Jorge Guajardo, and Christof Paar. Hyperelliptic curve crypstosystems: Closing the performance gap to elliptic curves. In Colin D. Walter, Çetin K. Koç, and Christof Paar, editors, *Cryptographic Hardware and Embedded Systems — CHES 2003*, volume 2779 of *Lecture Notes in Computer Science*, pages 351–365, Berlin, 2003. Springer-Verlag.
19. Arnold Schönhage and Ekkehart Vetter. A new approach to resultant computations and other algorithms with exact division. In Jan van Leeuwen, editor, *Algorithms — ESA'94*, volume 855 of *Lecture Notes in Computer Science*, pages 448–459, Berlin, 1994. Springer-Verlag.
20. Nicolas Thériault. Index calculus attack for hyperelliptic curves of small genus. In Chi Sung Laih, editor, *Advances in Cryptology — ASIACRYPT 2003*, volume 2894 of *Lecture Notes in Computer Science*, pages 75–92, Berlin, 2003. Springer-Verlag.
21. А. Л. Тоом. О сложности схемы из функционалных элементов, реализирующей умножение целых чисел. *Доклады Академии Наук СССР*, 150(3):496–498, 1963. English translation: The complexity of a scheme of functional elements realizing the multiplication of integers, *Soviet Mathematics*, 4:714–716, 1963.

Pseudocubes and Primality Testing

P. Berrizbeitia[1], S. Müller[2*], and H.C. Williams[2]

[1] Departamento de Matematicas P. y A., Universidad Simón Bolívar, Apartado 89000, Caracas 1080-A, Venezuela
pedrob@usb.ve

[2] Centre for Applied Cryptographic Research, University of Calgary, 2500 University Drive NW, Calgary, Alberta, Canada, T2N 1N4
(smueller,williams)@math.ucalgary.ca

Abstract. The recent ideas of Agrawal, Kayal, and Saxena have produced a milestone in the area of deterministic primality testing. Unfortunately, their method, as well as their successors are mainly of theoretical interest, as they are much too slow for practical applications.
Via a totally different approach, Lukes et al. developed a test which is conjectured to prove the primality of N in time only $O((\lg N)^{3+o(1)})$. Their (plausible) conjecture concerns the distribution of pseudosquares. These are numbers which locally behave like perfect squares but are nevertheless not perfect squares.
While squares are easy to deal with, this naturally gives rise to the question of whether the pseudosquares can be replaced by more general types of numbers. We have succeeded in extending the theory to the cubic case. To capture pseudocubes we rely on interesting properties of elements in the ring of Eisenstein integers and suitable applications of cubic residuacity. Surprisingly, the test itself is very simple as it can be formulated in the integers only. Moreover, the new theory appears to lead to an even more powerful primality testing algorithm than the one based on the pseudosquares.

1 Introduction

1.1 Motivation

The aim of this paper is both theoretical and practical. It has been known since the 1930s [17] that the theory of so-called pseudosquares yields a very powerful machinery for primality testing of large integers N. In fact, assuming some reasonable heuristics [17] this gives a deterministic primality test in time $O((\lg N)^{3+o(1)})$ which should be compared to the times $O((\log N)^{10.5+o(1)})$, resp. $O((\log N)^{6+o(1)})$ required by AKS and successors, [1,15], or for their random time counterparts [2,5,6], which require time $O((\log N)^{4+o(1)})$.

The above heuristics have been confirmed for numbers to 2^{80}. Based on the investigations made so far, this makes the pseudosquare approach the most efficient primality testing algorithm for primes of about 100 binary digits, without

* Research supported under APART [Austrian Programme for Advanced Research and Technology] by the Austrian Academy of Sciences.

D.A. Buell (Ed.): ANTS 2004, LNCS 3076, pp. 102–116, 2004.

requiring any unproven assumptions. Specifically, for primes of this size, only 73 modular exponentiations plus some trial divisions are required to prove primality, see [4]. It should be noted here that this is similar to the amount of time that typically would be spent on any probabilistic method, like Miller-Rabin, which however unlikely, might return the wrong answer.

Pseudosquares are integers which behave locally like squares modulo all primes in a certain range, while they are themselves not squares. A natural problem is to extend the theory of these pseudosquares to more general types of numbers. Numbers that 'behave' like an xth power without being a perfect xth power are of interest here. However, even extending the theory to the case $x = 3$ has proved to be challenging. It has not even been clear how such numbers are to be defined, nor how to investigate their theoretical behaviour, let alone their practical applications. Indeed, in the 1980s D. Lehmer posed a question tantamount to this to one of the authors (Williams), and it has remained a challenging open problem since.

1.2 The Main Results

In this note we succeed in giving a positive answer to the above in two respects. We establish the theoretical foundations and properties for pseudocubes and develop practical applications. The theory involves a number of new results by making use of cubic residuacity and cubic reciprocity.

While the quadratic theory is very simple and well-understood, the extension of it's important features to the cubic case is not easily apparent. Surprisingly, while the theory is more involved we have managed to describe the cubic counterpart in a clear and simple manner. Although the theoretical setting is the ring of the Eisenstein integers, the applications can easily be described and put into practice in the ring of integers only. Growth estimates of the pseudocubes suggest that the primality testing algorithms based on the new theory eventually will even become more efficient than those based on the pseudosquares.

While the primality test that we develop is quite efficient for any given N in a certain range, it must be borne in mind that determining that range requires a lengthy precomputation – the process of determining (or bounding from below) M_x, which is likely not possible in polynomial time. Once this has been performed, however, we can apply our test of complexity $O((\log N)^3)$ to any N in the range.

2 Background and Sketch of the Problem

2.1 Pseudosquares

Definition 1. *Let p be an odd prime. The pseudosquare L_p is defined as follows.*

1. $L_p \equiv 1 \bmod 8$
2. *the Legendre symbol* $\left(\frac{L_p}{q}\right) = 1$ *for all odd primes* $q \leq p$,

3. L_p is the least positive nonsquare integer satisfying (1) and (2).

The main result of the paper will be a result analogous to Theorem 1 (see [17]) in terms of pseudocubes.

Theorem 1. *If*

- $N < L_p$,
- $q^{(N-1)/2} \equiv \pm 1 \bmod N$ *for all primes* q, *s.t.* $2 \leq q \leq p$,
- $q^{(N-1)/2} \equiv -1 \bmod N$ *for at least one prime* q, *s.t.* $2 \leq q \leq p$,

then N *is a prime or a power of a prime.*

The above is a simplified and somewhat weaker version of the theorem in [17], which as an additional feature includes trial division. To extend this theory, we investigate the underlying properties. The difficult part is to establish that no composite integers N survive the test. Suppose that P_i, $i \geq 2$ are different primes dividing N. Then the conditions of the theorem imply the following, where $\nu_2(m)$ is the highest power of 2 dividing m,

$$\nu_2(P_i - 1) = \nu_2(N-1) \text{ for all } P_i | N, \tag{2.1}$$

$$\text{if } \left(\frac{c}{P_1 P_2}\right) \neq 1 \text{ then } c^{(N-1)/2} \not\equiv \pm 1 \bmod P_1 P_2. \tag{2.2}$$

Both of these lines immediately imply the above assertion, $c^{(N-1)/2} \not\equiv \pm 1 \bmod N$, which proves N does not fulfill Solovay-Strassen, and hence is not prime. The correctness of these two properties is established via Definition 1 through the use of quadratic reciprocity. We give a short summary to collect the main features necessary for cubic reciprocity.

2.2 The Ring of Eisenstein Integers

We first fix notation and recall some useful facts for which a general reference is [13] (see also [10,14]). Let $\omega = e^{2\pi i/3} = (-1+\sqrt{-3})/2$, a primitive cube root of unity and let D denote the ring $\mathbb{Z}[\omega]$. For any $\alpha = a + b\omega \in D$ we shall denote the norm and the trace by $N(\alpha) = \alpha\overline{\alpha} = a^2 - ab + b^2$ and $Tr(\alpha) = \alpha + \overline{\alpha} = 2a - b$, respectively, where $\overline{\alpha}$ denotes complex conjugation.

D is a unique factorization domain, in which there are three types of primes: the inert rational primes $q \equiv 2 \bmod 3$ of norm q^2, the primes π of prime norm $p = \pi\overline{\pi} \equiv 1 \bmod 3$, and the prime $1-\omega$ which lies over 3 as $3 = -\omega^2(1-\omega)^2$. The units in D are the six roots of unity $\pm\omega^i$ for $i = 0, 1, 2$. An element $\alpha \in D$ is called primary, if $\alpha \equiv 2 \bmod 3$. With the exception of $1-\omega$ (having no primary associate), every prime in D has exactly one primary associate.

An easy way to find the decomposition of $p \equiv 1 \bmod 3$ as $p = \pi\overline{\pi}$ can be obtained from the quadratic partition of p (see [21]), resp. $4p$. Indeed, any prime $p \equiv 1 \bmod 3$ can be written as

$$4p = L^2 + 27M^2, \quad L \equiv 1 \bmod 3,$$

with L and M^2 unique integers. From this, it follows that

$$\pi = \frac{L+3M}{2} + 3M\omega.$$

Conversely, to any $\pi = a + b\omega$, primary, we associate the pair of integers (L, M) defined as $L = 2a - b$ and $M = b/3$. We note that the pair of integers associated in this way to $\overline{\pi}$ is $(L, -M)$.

2.3 The Cubic Symbol and the Cubic Reciprocity Law

For any $\alpha \in D$ and any prime $\pi \in D$ with $N(\pi) \neq 3$ the cubic residue character $\left(\frac{\alpha}{\pi}\right)_3$ is defined as follows,

- if $\pi | \alpha$ then $\left(\frac{\alpha}{\pi}\right)_3 = 0$,
- if $\pi \not| \alpha$ then $\left(\frac{\alpha}{\pi}\right)_3$ is the unique element $\omega^i \in \{1, \omega, \omega^2\}$ satisfying $\alpha^{(N(\pi)-1)/3} \equiv \omega^i \bmod \pi$,
- if $\alpha, \beta \in D$ with $3 \not| N(\beta)$ we define $\left(\frac{\alpha}{\beta}\right)_3 = \prod_{i=1}^k \left(\frac{\alpha}{\pi_i}\right)_3$, where $\beta = \prod_{i=1}^k \pi_i$ and all $\pi_i \in D$ are prime. Clearly, if π and γ are primary, then $-\pi\gamma$ is primary. If π is primary, then $\pi = \pm\pi_1 \ldots \pi_t$, where the π_i are (not necessarily distinct) primary primes.

We note the following properties.

1. The symbol is bimultiplicative, that is $\left(\frac{\alpha\beta}{\pi}\right)_3 = \left(\frac{\alpha}{\pi}\right)_3 \left(\frac{\beta}{\pi}\right)_3$ and $\left(\frac{\alpha}{\pi\eta}\right)_3 = \left(\frac{\alpha}{\pi}\right)_3 \left(\frac{\alpha}{\eta}\right)_3$
2. Conjugation acts on the symbol as follows: $\overline{\left(\frac{\alpha}{\pi}\right)_3} = \left(\frac{\alpha}{\pi}\right)_3^2 = \left(\frac{\alpha^2}{\pi}\right)_3$ and $\overline{\left(\frac{\alpha}{\pi}\right)_3} = \left(\frac{\overline{\alpha}}{\overline{\pi}}\right)_3$

The Cubic Reciprocity Law (CRL) is often formulated in terms of primary primes in D. By the properties just stated, this then automatically yields the more general form, as follows. Let α and β be primary elements of D of coprime norm. Then

$$\left(\frac{\alpha}{\beta}\right)_3 = \left(\frac{\beta}{\alpha}\right)_3.$$

The Supplement to the CRL treats the case for ω and $1 - \omega$, which, not being primary, are not covered by the general CRL. Suppose that $N(\alpha) \neq 3$. If $\alpha = a + 3b\omega$ is primary (including the case $b = 0$) and $a = 3m - 1$ then

$$\left(\frac{1-\omega}{\alpha}\right)_3 = \omega^{2m} \text{ and } \left(\frac{\omega}{\alpha}\right)_3 = \omega^{m+b}.$$

2.4 Extending the Quadratic to the Cubic Case: Problems

If P_1 and P_2 ($P_1 \neq P_2$) are prime divisors of N, then, since $P_1P_2 \leq N$, by the definition of L_p, there is some prime $q \leq p$ such that $\left(\frac{P_1P_2}{q}\right) \neq 1$. By letting $c = -1, 2$, or q, accordingly, quadratic reciprocity gives $\left(\frac{c}{P_1P_2}\right) \neq 1$, which is (2.2), and we may assume this value is -1. Analogous results will be obtained when working in D, where the basis c will be replaced by a suitable $\alpha \in D$, and the L_p by its relevant cubic counterpart. For this section suppose we know $N = \nu\overline{\nu}$ and π_i is any primary prime in D lying over P_i dividing N. In trying to extend the quadratic theory we encounter a number of difficulties, for example:

Using an argument similar to the above (see [17]), it is easy to verify (2.1). This then immediately implies that if N is composite, it has to have at least three distinct prime divisors. The same is not true for the cubic analogue.

Any odd integer N is only divisible by primes $P_i \equiv 1 \bmod 2$, but a priori it is not known if, when $N \equiv \sigma \bmod 3$, for $\sigma \in \{1, -1\}$, that also each $P_i \equiv \sigma \bmod 3$.

The theory of the pseudosquares gives us elements for which the value of the quadratic character (mod N or P_i respectively), is different from 1. However, we note that if in the CRL both α and β are rational then the assertion is trivially true and follows from the fact that $\left(\frac{n}{q}\right)_3 = 1$ for any integer n and any $q \equiv -1 \bmod 3$. Unfortunately this implies that in extending the theory we require complex elements in the ring of Eisenstein integers.

Straightforward generalization of the definition of pseudosquares and of Theorem 1 into the setting of the Eisenstein integers would involve some challenges. Firstly, there are several ways that pseudocubes may be defined as certain elements lying over rational primes. Unfortunately none of these approaches seems to be practical. Secondly, N would need to be decomposed as $N = \nu\overline{\nu}$ if $N \equiv 1 \bmod 3$. Similarly, each prime $q \leq p$ which is 1 mod 3, would need to be decomposed as $\lambda\overline{\lambda} = q$. Finally, the testing condition formulated utilizing the theory of the cubic symbol would lead to the testing conditions $\lambda^{(N-1)/3} \equiv \left(\frac{\lambda}{\nu}\right)_3 \bmod \nu$, and $\overline{\lambda}^{(N-1)/3} \equiv \left(\frac{\overline{\lambda}}{\nu}\right)_3 \bmod \nu$. Below, we will avoid this quite involved setting and develop a practical approach which can be verified in the integers only.

3 Pseudocubes

3.1 Definition and Fundamental Properties

Definition 2. *For any non negative real number x, the pseudocube M_x is defined as the smallest positive integer satisfying the following properties.*

1. $M_x \equiv \pm 1 \bmod 9$,
2. $M_x^{(q-1)/3} \equiv 1 \bmod q$ *for all primes* $q \leq x$, $q \equiv 1 \bmod 3$.
3. M_x *is relatively prime to all primes* $q \leq x$.
4. M_x *is not a cube of an integer.*

Although condition (3) is not necessary in the definition, it is natural to have it included and the reasons for doing so will become apparent below.

We also introduce the following notation:

Definition 3. *For any prime q we choose α_q, a primary prime in D lying over q. So $\alpha_q = q$ if $q \equiv -1 \bmod 3$ and $\alpha_q \overline{\alpha}_q = q$ if $q \equiv 1 \bmod 3$. So there are two possibilities for α_q in the latter case. Also, we denote by λ_q the quotient $\alpha_q/\overline{\alpha}_q$.*

The quadratic reciprocity law allows us to give an alternative definition of the pseudosquares. Similarly, the CRL allows us to give the following alternative definition for the pseudocubes.

Proposition 1. *M_x is the smallest integer $\equiv \pm 1 \bmod 9$ which satisfies*

1. $\left(\frac{\alpha_q}{M_x}\right)_3 = 1$ *for all primes $q \leq x$, $q \neq 3$.*
2. *M_x is not a cube of an integer.*
3. $\left(\frac{\omega}{M_x}\right)_3 = 1$.

Proof. (1) Let q be a prime less or equal than x and $N = M_x$. If $q \equiv -1 \bmod 3$, then, by definition and CRL, $(\frac{\alpha_q}{N})_3 = (\frac{q}{N})_3 = \left(\frac{N}{q}\right)_3 = 1$. If $q \equiv 1 \bmod 3$ then, by definition of a pseudocube we must have $N^{(q-1)/3} \equiv 1 \bmod q$, hence also modulo α_q, so $\left(\frac{N}{\alpha_q}\right)_3 = 1$. The CRL gives then that $(\frac{\alpha_q}{N})_3 = 1$, as desired.

(2) is part of the definition of a pseudocube.

The final assertion follows from the supplement to the CRL since $M_x \equiv \pm 1 \bmod 9$. □

3.2 Useful Lemmas

If, α and β are complex integers in D then knowledge of $(\alpha/\beta)_3$ reveals no information concerning $(\alpha/\overline{\beta})_3$. In this context the following lemma which gives a result for the product $(\alpha/\beta)_3(\alpha/\overline{\beta})_3$ is useful, see also [12,8].

Lemma 1. *Suppose α, β are distinct primary complex elements in D. Then* $\left(\frac{\alpha/\overline{\alpha}}{\beta}\right)_3 = \left(\frac{\alpha}{N(\beta)}\right)_3$.

Proof. $\left(\frac{\alpha/\overline{\alpha}}{\beta}\right)_3 = \left(\frac{\alpha}{\beta}\right)_3 \left(\frac{\overline{\alpha}}{\beta}\right)_3^{-1} = \left(\frac{\alpha}{\beta}\right)_3 \left(\frac{\alpha}{\overline{\beta}}\right)_3$. □

As for primes $p \equiv -1 \bmod 3$, $N(p) = p^2$, the cubic symbol $\left(\frac{\alpha}{p}\right)_3$ involves exponents $(p^2 - 1)/3$. A simplification involving only the exponent $(p+1)/3$ is given in [8]. In fact, $\left(\frac{\alpha}{p}\right)_3 \equiv (\frac{\overline{\alpha}}{\alpha})^{(p+1)/3} \bmod p$. Also, from Lemma 1, the analogous result for $p = \pi\overline{\pi} \equiv 1 \bmod 3$ reads $\left(\frac{\alpha}{p}\right)_3 \equiv (\frac{\alpha}{\overline{\alpha}})^{(p-1)/3} \bmod \pi$.

The last two lines motivate the following definition, which is analogous to the one given in the biquadratic setting [13,7]. The convenience of this notation will become more apparent later on.

Definition 4. *For any integer N, $3 \nmid N$, we let $N^* = N$ if $N \equiv 1 \bmod 3$, and $N^* = -N$, when $N \equiv -1 \bmod 3$.*

We note that $3 | N^* - 1$ and $(NM)^* = N^* M^*$.

We are now ready for the notion of an Eisenstein pseudoprime w.r.t. the base q. Recall the notation given in Definition 3.

Definition 5. *Let q be a prime not dividing N. We say that N is an Eisenstein pseudoprime w.r.t the base q if N is composite and*

$$\left(\frac{\alpha_q}{N}\right)_3 \equiv \lambda_q^{(N^*-1)/3} \bmod N.$$

We next show that the definition does not depend on the choice of the primary element α_q. This only has to be verified when $q \equiv 1 \bmod 3$. The other choice of a primary element would be $\overline{\alpha}_q$, so the properties of the symbol would lead to the inverse on the left hand side. But the corresponding λ_q would also lead to the inverse on the right hand side, and this establishes our claim.

We summarize the above in the following which verifies that Definition 5 is really a notion of pseudoprime.

Proposition 2. *If N is a prime then it fulfills the condition in Definition 5 w.r.t the base q for any prime $q \neq N$.*

Proof. Suppose $N \equiv 1 \bmod 3$. Let ν be a primary prime lying over N. So $N = \nu\overline{\nu}$. By Lemma 1, $\left(\frac{\alpha_q}{N}\right)_3 = \left(\frac{\alpha_q/\overline{\alpha}_q}{\nu}\right)_3 = \left(\frac{\lambda_q}{\nu}\right)_3 \equiv {\lambda_q}^{(N-1)/3} = {\lambda_q}^{(N^*-1)/3} \bmod \nu$. Since ν was arbitrary the same equality holds modulo $\overline{\nu}$, hence modulo N.

Suppose $N \equiv -1 \bmod 3$. Then it remains prime in D. So in this case $\left(\frac{\alpha_q}{N}\right)_3 \equiv {\alpha_q}^{(N^2-1)/3} \equiv ({\alpha_q}^{1-N})^{(-N-1)/3} \equiv {\lambda_q}^{(N^*-1)/3} \bmod N$. □

Remark 1. Note that if q is a prime $-1 \bmod 3$ then any composite N relatively prime to q is a Eisenstein pseudoprime w.r.t. the base q. This is because on one hand $\left(\frac{\alpha_q}{N}\right)_3 = \left(\frac{q}{N}\right)_3 = 1$, and on the other hand, $\lambda_q = 1$, so the equality trivially holds.

3.3 The Main Theorem

Theorem 2. *Let N be odd, $N < (M_x)^{2/3}$, N not a prime or a prime power. Then, there is a prime $q \leq x$, $q \equiv 1 \bmod 3$ such that N is not an Eisenstein pseudoprime w.r.t. the base q.*

The proof of the theorem will be based on the lemma below. We first introduce some notation that will facilitate the statement and proof of the lemma.

- We will say that N satisfies Hypothesis (*) if N is odd, not a prime power, and

$$\left(\frac{\alpha_q}{N}\right)_3 \equiv \lambda_q^{(N^*-1)/3} \bmod N \text{ holds for all prime } q \leq x, q \neq 3.$$

- Let $\nu_3(N^* - 1) = s$ (that is, the exact power of 3 dividing $N^* - 1$). Let $t_{N^*} = (N^* - 1)/3^s$. Note that t_{N^*} is an integer and $3 \nmid t_{N^*}$. Let P be any prime divisor of N and denote $t_{P^*} = (P^* - 1)/3^s$.

Lemma 2. *Suppose N satisfies hypothesis (*), $N < M_x$, and N is not a cube of an integer. Then,*

1. $\nu_3(P^* - 1) = \nu_3(N^* - 1)$
2. *for all primes q, $q \neq 3$, $q \leq x$ we have:*

$$\left(\frac{\alpha_q}{N}\right)_3^{t_{P^*}} = \left(\frac{\alpha_q}{P}\right)_3^{t_{N^*}}.$$

Proof. We first claim that $\nu_3(P^* - 1) \geq \nu_3(N^* - 1)$. If $s = \nu_3(N^* - 1) = 1$ then the inequality holds trivially so we assume that s is at least 2, so $N^* \equiv 1 \bmod 9$. Since $N < M_x$, then by Proposition 1 there is a prime $q \leq x$, such that $\left(\frac{\alpha_q}{N}\right)_3 \neq 1$. Since N satisfies hypothesis (*) then $\lambda_q^{(N^*-1)/3} \equiv \omega, \omega^2 \bmod N$, hence mod π, for any prime π in D dividing N. It follows that the element $\lambda_q^{(N^*-1)/3^s}$ has order 3^s mod π, so 3^s divides $N(\pi) - 1$. In both cases, $P \equiv \pm 1 \bmod 3$, this leads to 3^s divides $P^* - 1$, which proves our first claim. It follows that t_{P^*} is an integer.

We will prove part 2) of the lemma and then use that to prove part 1).

Consider the following two equalities, which hold for all primes q less or equal to x and different from 3:

$$\left(\frac{\alpha_q}{N}\right)_3 \equiv \lambda_q^{(N^*-1)/3} \equiv \lambda_q^{3^{s-1}t_{N^*}} \bmod N \tag{3.1}$$

(which is hypothesis (*)), and

$$\left(\frac{\alpha_q}{P}\right)_3 \equiv \lambda_q^{(P^*-1)/3} \equiv \lambda_q^{3^{s-1}t_{P^*}} \bmod P \tag{3.2}$$

(which holds by Proposition 2 because P is prime).

Since P divides N then both equalities hold mod P. Raising the first equality to the t_{P^*} and the second to the t_{N^*} we obtain

$$\left(\frac{\alpha_q}{N}\right)_3^{t_{P^*}} \equiv \left(\frac{\alpha_q}{P}\right)_3^{t_{N^*}} \bmod P. \tag{3.3}$$

Since $P \neq 3$ then this leads to the equality, which is part 2) of the lemma.

To prove 1) it suffices to show that t_{P^*} is not a multiple of 3. If $P^* \not\equiv 1 \bmod 9$ then $\nu_3(P^* - 1) = 1$, and since we proved that $\nu_3(P^* - 1) \geq \nu_3(N^* - 1)$ it follows that $s = \nu_3(N^* - 1) = 1$, so $t_{P^*} = (P^* - 1)/3 \not\equiv 0 \bmod 3$, since $P^* \not\equiv 1 \bmod 9$.

Alternatively, if $P^* \equiv 1 \bmod 9$, then since P is a divisor of N then P is also $< M_x$. By Proposition 1, there must be a prime $q \leq x$ ($q \neq 3$) such that $\left(\frac{\alpha_q}{P}\right)_3$ is not 1. Since t_{N^*} is not 1 mod 3, then, for this particular α_q the right hand side of the equality in 2) is not 1, hence the left side cannot be 1, which shows t_{P^*} cannot be a multiple of 3. □

Proof of Theorem 2. Under the hypothesis in the theorem, N must have more than one prime divisor. Let P and Q be two different prime divisors of N. We assume that N satisfies hypothesis (*). Applying the lemma to both P and Q we obtain:

$$\left(\frac{\alpha_q}{N}\right)_3^{t_{P^*}} = \left(\frac{\alpha_q}{P}\right)_3^{t_{N^*}}$$

$$\left(\frac{\alpha_q}{N}\right)_3^{t_{Q^*}} = \left(\frac{\alpha_q}{Q}\right)_3^{t_{N^*}}$$

Using the fact that $(t_P^*)^2 = (t_N^*)^2 = (t_Q^*)^2 \equiv 1 \bmod 3$, one can deduce from the two equalities that

$$\left(\frac{\alpha_q}{P}\right)_3^{t_{P^*}t_{Q^*}} = \left(\frac{\alpha_q}{Q}\right)_3 \tag{3.4}$$

holds for all primes q less or equal to x.

It follows that for $i = 1$ or 2 we get

$$\left(\frac{\alpha_q}{P^iQ}\right)_3 = \left(\frac{\alpha_q}{P}\right)_3^i \left(\frac{\alpha_q}{Q}\right)_3 = \left(\frac{\alpha_q}{P}\right)_3^{i+t_{P^*}t_{Q^*}}.$$

This last exponent is $\equiv 0 \bmod 3$ for $i \equiv -t_{P^*}t_{Q^*} \bmod 3$, which in particular implies $(P^iQ)^* \equiv 1 \bmod 9$. Also,

$$\left(\frac{\alpha_q}{P^iQ}\right)_3 = 1 \text{ for all } q \text{ up to } x \text{ for } i = 1 \text{ or } 2. \tag{3.5}$$

But $P^iQ \leq P^2Q \leq PN < N^{3/2} < M_x$, which gives a contradiction to the minimality of the pseudocube M_x. It follows that hypothesis (*) fails to be true for some q up to x and, finally, by Remark 1 this has to be the case for some $q \not\equiv -1 \bmod 3$. □

4 Practical Considerations and Estimates on Pseudocubes

4.1 A More Practical Version

Recall that while Theorem 1 tests for the Solovay-Strassen condition on the left hand side, it does not require computing the quadratic symbol on the right hand side. Instead, it is required that the -1 appears on the right hand side at least once. Unfortunately this requirement on the existence of a basis which is a quadratic nonresidue makes the algorithm a random time deterministic test. In fact, this requirement on the -1 makes the test essentially a Solovay-Strassen test. It is known [11, Exercise 3.24], that if N is odd and if there exists an integer

b with $b^{(N-1)/2} \equiv -1 \bmod N$, then any integer a with $a^{(N-1)/2} \equiv \pm 1 \bmod N$ also satisfies $a^{(N-1)/2} \equiv \left(\frac{a}{N}\right) \bmod N$.

On the other hand, since any witness to the Solovay-Strassen test (or to the Euler pseudoprime test) is also a witness to the Miller-Rabin test (or the strong pseudoprime test), then the symbol $\left(\frac{q}{N}\right)$ need not be computed. So to do something equally effective in our case, namely, to avoid the computation of the cubic symbol $\left(\frac{\alpha_q}{N}\right)_3$ we will first give the notion of a strong cubic pseudoprime.

Definition 6. *Let q, α_q, λ_q, N^*, and t_{N^*} be as above. We say that N is a strong cubic pseudoprime w.r.t. base q if N is composite and either*

$$\lambda_q^{t_{N^*}} \equiv 1 \bmod N,$$

or there is some i, $0 \le i \le s-1$, such that

$$\lambda_q^{t_{N^*}3^i} \equiv \omega^j \bmod N, \text{ for } j = 1 \text{ or } -1.$$

The following result is an extension of a well known theorem of Selfridge [20] to the cubic case.

Theorem 3. *If N is a strong cubic pseudoprime w.r.t. the base q, then N is an Eisenstein pseudoprime w.r.t. q.*

Proof. Let $\lambda_q^{t_{N^*}} \equiv 1 \bmod N$. Because $3 \nmid t_{N^*}$, this implies $\lambda_q^{(P^*-1)/3} \equiv 1 \bmod P$ for any $P|N$. So $\left(\frac{\lambda_q}{\pi}\right)_3 = 1$ for π lying over P. Now $\left(\frac{\alpha_q}{P}\right)_3 = \left(\frac{\lambda_q}{\pi}\right)_3 = 1$, and hence $\left(\frac{\alpha_q}{N}\right)_3 = 1$, as desired.

Now assume $\lambda_q^{t_{N^*}3^i} \equiv \omega^j \pmod N$ for some i and j, $0 \le i \le s-1$, $j \not\equiv 0 \bmod 3$. If μ_P is the least positive integer such that $\lambda_q^{\mu_P} \equiv 1 \bmod P$ for any prime $P|N$, then $\mu_P | 3^{i+1} t_{N^*}$, but $\mu_P \nmid 3^i t_{N^*}$. But μ_P divides $P^* - 1$ by Proposition 2, so that $P^* = 1 + 3^{i+1} k_P$ for some integer k_P.

Note that if N is a product of (not necessarily different) primes P, then

$$k_{N^*} = 3^{s-i-1} t_{N^*} \equiv \sum_{P|N} k_P \bmod 3. \tag{4.1}$$

This follows since $N^* = \prod_{P|N} P^*$, which gives $1 + 3^s t_{N^*} = 1 + 3^{i+1} \sum_{P|N} k_P \bmod 3^{2i+2}$, and hence, $3^{s-i-1} t_{N^*} \equiv \sum_{P|N} k_P \bmod 3^{i+1}$.

Again, by Proposition 2, $\left(\frac{\alpha_q}{P}\right)_3^{t_{N^*}} \equiv \lambda_q^{3^i k_P t_{N^*}} \bmod P$, and by hypothesis, $\lambda_q^{3^i k_P t_{N^*}} \equiv \omega^{j k_P} \bmod N$. From these two equations it follows that

$$\left(\frac{\alpha_q}{P}\right)_3^{t_{N^*}} = \omega^{j k_P}.$$

Multiplying over all primes P dividing N we obtain $\left(\frac{\alpha_q}{N}\right)_3^{t_{N^*}} = \omega^{j \sum_{P|N} k_P} = \omega^{j 3^{s-i-1} t_{N^*}}$, where the last step follows from (4.1). Finally, since $t_{N^*}^2 \equiv 1 \bmod 3$, we get $\left(\frac{\alpha_q}{N}\right)_3 = \omega^{j 3^{s-i-1}} \equiv \lambda_q^{t_{N^*} 3^{s-1}} = \lambda_q^{(N^*-1)/3} \bmod N$. □

As a corollary of Theorem 2 and Theorem 3 we obtain the following:

Theorem 4. *Let N be as in Theorem 2. Then there is a prime q, $q \equiv 1 \bmod 3$, $q \leq x$, such that N is not a strong cubic pseudoprime w.r.t. the base q.*

For practical purposes in determining the primality of a certain $N < (M_x)^{2/3}$, the use of Theorem 4 is faster, since it requires no calculation of the cubic residuacity symbol. We will show that the cost of a round of the strong cubic pseudoprime test is practically about the same as the cost of one round of the classic strong pseudoprime test.

4.2 Evaluation

For any prime $q \equiv 1 \bmod 3$, $q \leq p$, the primary primes α_q can easily be found via Cornaccia-Smith (see e.g. [9,11]) or by trial, since q is small. This gives $\alpha_q = a + b\omega$, from where we get $\lambda_q \equiv \alpha_q/\overline{\alpha}_q \bmod N$. The evaluation of powers of $\lambda_q = c + d\omega$ modulo N according to Definition 6 can done very efficiently in the integers only, as shown in the following lemma. The result is an element in D, reduced modulo N, and one only needs to verify whether this is $1, \omega$, or ω^2, as desired.

Lemma 3. *Given $\lambda_q = c + d\omega$, the evaluation of $\lambda_q^m \bmod N$ can be achieved in $\log(m)$ squarings and $\log(m)$ multiplications, modulo N.*

Proof. We define the integers $T_j = T_j(q)$ as $Tr(\lambda_q^j) = \lambda_q^j + \overline{\lambda}_q^j \equiv \lambda_q^j + 1/\lambda_q^j \bmod N$. They satisfy the recurrence relations

$$T_{2j} = T_j^2 - 2, \quad T_{2j+1} = T_j T_{j+1} - T_1 \tag{4.2}$$

with the initial values $T_0 = 2, T_1 \equiv \lambda_q + 1/\lambda_q \bmod N$. Thus, if we have $T_j \bmod N$ and $T_{j+1} \bmod N$, then we may compute either the pair $T_{2j} \bmod N$, $T_{2j+1} \bmod N$, or the pair $T_{2j+1} \bmod N$, $T_{2j+2} \bmod N$, with each choice taking one multiplication modulo N, and one squaring modulo N. This way, starting from T_0, T_1 we can recursively arrive at any pair T_k, T_{k+1}. Which type of the two moves has to be made can be read off from the binary representation of m in the target pair $m, m+1$, see e.g. [11].

Finally, the desired value of $\lambda_q^m \bmod N$ can be obtained via the following:

- $(T_m, T_{m+1}) \equiv (2, T_1) \bmod N$, iff $\lambda_q^m \equiv \overline{\lambda}_q^m \equiv 1 \bmod N$,
- $(T_m, T_{m+1}) \equiv (-1, T) \bmod N$, where $2T \equiv -T_1 \pm 3d \bmod N$, iff $\lambda_q^m \equiv \omega, \omega^2 \bmod N$.

This can be seen as follows. We have $(\lambda - \overline{\lambda})(\lambda^m - \overline{\lambda}^m) = 2T_{m+1} - T_1 T_m$ where, for brevity we write λ for λ_q, and $\overline{\lambda}$ for $\overline{\lambda}_q$.

If $(T_m, T_{m+1}) \equiv (2, T_1) \bmod N$, then $(T_1^2 - 4)(\lambda^m - \overline{\lambda}^m)/(\lambda - \overline{\lambda}) \equiv 0 \bmod N$, which implies $\lambda^m \equiv \overline{\lambda}^m \bmod N$, since we may assume $\gcd(N, T_1^2 - 4) = 1$. Since $T_m \equiv 2 \bmod N$, this gives $\lambda^m \equiv \overline{\lambda}^m \equiv 1 \bmod N$, as required. The converse is immediate.

If $(T_m, T_{m+1}) \equiv (-1, T) \bmod N$ then $\lambda^m + \overline{\lambda}^m \equiv -1 \bmod N$ and $(\lambda - \overline{\lambda})(\lambda^m - \overline{\lambda}^m) \equiv 2T_{m+1} - T_1 T_m \equiv 2T + T_1 \equiv \pm 3d \bmod N$. Now, $3d \equiv (\omega - \omega^2)^2 d \equiv (\omega - \omega^2)(\lambda - \overline{\lambda}) \bmod N$, and therefore $\lambda^m - \overline{\lambda}^m \equiv \pm(\omega - \omega^2) \bmod N$. Hence, $2\lambda^m \equiv -1 \pm (\omega - \omega^2) \bmod N$, as desired. The converse follows again immediately. □

Remark 2. In many cases the evaluation can be somewhat simplified via the following. Suppose $N \equiv 1 \bmod 3$ and we attempt to compute the square root of $-3 \bmod N$, under the assumption that N is prime. If any such method fails to return the correct answer, N is disclosed as composite. Otherwise, we obtain the value of $\omega \bmod N$ as $(-1 - \sqrt{-3})/2 \bmod N$. Now, exponentiation of any $\lambda_q = c + d\omega \bmod N$ is simply the exponentiation of a rational integer modulo N. There are several ways for computing the square root of -3 modulo a prime N. The simplest requiring one, resp. two exponentiations, when $N \equiv 3 \bmod 4$, resp. $N \equiv 5 \bmod 8$. In the other cases, a quadratic nonresidue modulo N is required, and in practice this is always easily found, see e.g. [3].

4.3 Growth-Rate Estimates

The natural question that arises is whether there is any advantage to using pseudocubes instead of pseudosquares for testing primality. It is a difficult question to answer, mainly because the actual rate of growth of pseudosquares or pseudocubes is not known. In this section we will try to make such a comparison by using heuristic estimates of the size of both the pseudosquares and the pseudocubes.

Let p_i denote the ith prime, and q_i denote the ith prime $\equiv 1 \bmod 3$. Recall that the test for the pseudosquares involves all primes $\leq p$, where p is the smallest prime p_n such that $N < L_{p_n}$. The cubic test only requires the primes $q \equiv 1 \bmod 3$, $q \leq p$ which are about half of all the primes. On the other hand, the p in this case is the smallest prime q_n such that $N < M_{q_n}^{2/3}$. Since the running time for each round requires about the same time for both of the tests, we conclude that the test based on the pseudocubes will be 'better' than the test based on the pseudosquares, only if

$$M_{q_n}^{2/3} > L_{p_n}. \tag{4.3}$$

In [17] it is conjectured that the pseudosquares L_{p_n} should have a growth rate of the form

$$L_{p_n} \approx c_1 2^n \log p_n \tag{4.4}$$

which implies a growth rate for L_p of the form $2^{(p/\log p)(1+o(1))}$. This estimate was derived from the conjecture that the solutions of

$$x \equiv 1 \bmod 8, \quad \left(\frac{x}{p_i}\right) = 1 \quad (i = 1, 2, ..., n) \tag{4.5}$$

are equidistributed in the region $0 < x < 8p_2p_3...p_n$.

Making the analogous heuristic assumption that the solutions of

$$x \equiv \pm 1 \bmod 18, \quad x^{(q_i-1)/3} \equiv 1 \bmod q_i \quad (i = 1, 2, ..., n) \tag{4.6}$$

are equidistributed in the region $0 < x < 18q_1...q_n$ we would expect that

$$M_{q_n} \approx 9 \prod_{i=1}^{n} q_i / S,$$

where S is the number of solutions of (4.6). Since

$$S = \prod_{i=1}^{n} (q_i - 1)/3,$$

we get

$$M_{q_n} \approx 3^{n+2} \prod_{i=1}^{n} q_i / (q_i - 1).$$

Now, Mertens' Theorem for arithmetic progressions states that

$$\prod_{p_i \leq x, p_i \equiv l \bmod k} \left(1 - \frac{1}{p_i}\right) \approx c(\log x)^{-1/\phi(k)},$$

where the constant c and the error term can be made explicit, see [22]. From this, we get

$$M_{q_n} \approx c_2 3^n (\log q_n)^2 \tag{4.7}$$

for a constant c_2. To compare their relative growth rates, we get from (4.4) and (4.7),

$$\frac{(M_{q_n})^{2/3}}{L_{p_n}} \approx \frac{c_2^{2/3} 3^{2n/3} (\log q_n)^{4/3}}{c_1 2^n (\log p_n)} \geq c \left(\frac{3^{2/3}}{2}\right)^n > 1$$

for n sufficiently large. This suggests the following.

Conjecture. Under the same heuristic estimates (4.5) and (4.6) for the pseudosquares and the pseudocubes we have $M_{q_n}^{2/3} > L_{p_n}$ for sufficiently large n.

Remark 3. In [18] character sum techniques have been used to make rigorous some of the required statements about the distribution of numbers, as in (4.5). Schinzel showed that, under the ERH, for every $\epsilon > 0$ and for p_n sufficiently large, $(1 - \epsilon)\sqrt{p_n} < \log L_{p_n} < (2 \log 2 + \epsilon) p_n / \log p_n$. He also obtained a much weaker result unconditionally. Unfortunately, in spite of using the powerful ERH, his results are far from being as precise as the conjectured (4.4).

4.4 Some Data

A complete list of pseudocubes M_{q_k} up to $q_k = 313$ can be found in [19]. Further numerical computations were obtained in [16], which we give below, as they have never been published.

q_k	M_{q_k}
331	75 017 625 272 879 381
337	75 017 625 272 879 381
349	75 017 625 272 879 381
367	> 100 000 000 000 000 000

Thus, on comparing these data with the table in [17], we see that the above heuristic that the pseudocubes will eventually outgrow the pseudosquares is supported by the existing numerical evidence. However, we have yet to reach the point where we get $M_{q_n}^{2/3} > L_{p_n}$. Further computations of M_{q_n} may provide more information concerning this.

References

1. M. Agrawal, N. Kayal, and N. Saxena. PRIMES is in P. *Annals of Mathematics*, To appear.
2. R. Avanzi and P. Mihailescu. Efficient 'quasi - deterministic' primality test improving AKS. Available at http://www-math.uni-paderborn.de/ preda/, 2003.
3. E. Bach and J. Shallit. *Algorithmic number theory. Vol. 1.* Foundations of Computing Series. MIT Press, Cambridge, MA, 1996. Efficient algorithms.
4. D. Bernstein. Doubly focused enumeration of locally square polynomial values. Available at http://cr.yp.to/focus.html, 2003.
5. D. Bernstein. Proving primality in essentially quartic random time. Available at http://cr.yp.to/papers.html/quartic, 2003.
6. P. Berrizbeitia. Sharpening "Primes is in P" for a large family of numbers. Available at http://arxiv.org/abs/math.NT/0211334, 2002.
7. P. Berrizbeitia and T. Berry. Biquadratic reciprocity and a Lucasian primality test. *Math. Comp.*, Posted July 1, 2003.
8. P. Berrizbeitia and T. G. Berry. Cubic reciprocity and generalised Lucas-Lehmer tests for primality of $A \cdot 3^n \pm 1$. *Proc. Amer. Math. Soc.*, 127(7):1923–1925, 1999.
9. H. Cohen. *A course in computational algebraic number theory*, volume 138 of *Graduate Texts in Mathematics*. Springer-Verlag, Berlin, 1993.
10. D. A. Cox. *Primes of the form $x^2 + ny^2$*. A Wiley-Interscience Publication. John Wiley & Sons Inc., New York, 1989. Fermat, class field theory and complex multiplication.
11. R. Crandall and C. Pomerance. *Prime numbers*. Springer-Verlag, New York, 2001. A computational perspective.
12. A. Guthmann. Effective primality tests for integers of the forms $N = k \cdot 3^n + 1$ and $N = k \cdot 2^m 3^n + 1$. *BIT*, 32(3):529–534, 1992.
13. K. Ireland and M. Rosen. *A classical introduction to modern number theory*, volume 84 of *Graduate Texts in Mathematics*. Springer-Verlag, New York, second edition, 1990.

14. F. Lemmermeyer. *Reciprocity laws.* Springer Monographs in Mathematics. Springer-Verlag, Berlin, 2000. From Euler to Eisenstein.
15. H. W. J. Lenstra and C. Pomerance. Primality Testing with Gaussian Periods. Manuscript, 2003.
16. R. Lukes. *A Very Fast Electronic Number Sieve.* PhD thesis, University of Manitoba, Winnipeg, 1995.
17. R. F. Lukes, C. D. Patterson, and H. C. Williams. Some results on pseudosquares. *Math. Comp.*, 65(213):361–372, S25–S27, 1996.
18. A. Schinzel. On pseudosquares. In *New trends in probability and statistics, Vol. 4 (Palanga, 1996)*, pages 213–220. VSP, Utrecht, 1997.
19. A. J. Stephens and H. C. Williams. An open architecture number sieve. In *Number theory and cryptography (Sydney, 1989)*, volume 154 of *London Math. Soc. Lecture Note Ser.*, pages 38–75. Cambridge Univ. Press, Cambridge, 1990.
20. H. C. Williams. Primality testing on a computer. *Ars Combin.*, 5:127–185, 1978.
21. H. C. Williams. An M^3 public-key encryption scheme. In *Advances in cryptology—CRYPTO '85 (Santa Barbara, Calif., 1985)*, volume 218 of *Lecture Notes in Comput. Sci.*, pages 358–368. Springer, Berlin, 1986.
22. K. S. Williams. Mertens' theorem for arithmetic progressions. *J. Number Theory*, 6:353–359, 1974.

Elliptic Curves with a Given Number of Points

Reinier Bröker and Peter Stevenhagen

Mathematisch Instituut, Universiteit Leiden,
Postbus 9512, 2300 RA Leiden, The Netherlands
{reinier,psh}@math.leidenuniv.nl

Abstract. We present a non-archimedean method to construct, given an integer $N \geq 1$, a finite field $\mathbf{F}_q$ and an elliptic curve $E/\mathbf{F}_q$ such that $E(\mathbf{F}_q)$ has order N.

1 Introduction

A classical theorem of Hasse from 1934 states that for an elliptic curve E defined over the finite field $\mathbf{F}_q$ of q elements, the order of the group $E(\mathbf{F}_q)$ of $\mathbf{F}_q$-rational points is an integer in the *Hasse interval*

$$\mathcal{H}_q = [q + 1 - 2\sqrt{q}, q + 1 + 2\sqrt{q}]$$

around q. If E is given in some standard way, say by a Weierstrass equation over $\mathbf{F}_q$, there are several algorithms that compute the order of $E(\mathbf{F}_q)$. The 1985 algorithm by Schoof [8,9] runs in time polynomial in $\log q$, and in small characteristic p there are even faster p-adic algorithms due to Satoh [7] and Kedlaya [5].

The situation is rather different in the case of the following problem, which can be seen as an 'inverse problem' to the point counting problem.

Problem. *Given an integer $N \geq 1$, find a finite field $\mathbf{F}_q$ and an elliptic curve $E/\mathbf{F}_q$ for which the number of $\mathbf{F}_q$-rational points equals N.*

As with other inverse problems, such as in Galois theory, this is mathematically a natural question to ask. In this particular case, an efficient solution to the problem would also be desirable in view of the need in current applications to construct elliptic curves having point groups satisfying various smoothness requirements with respect to their order. It is one of the reasons why we focus on the order N, and do not specify the finite field $\mathbf{F}_q$ as being part of the input. In addition, we will use the freedom with respect to the choice of a base field $\mathbf{F}_q$ to our advantage.

A necessary condition for our problem to be solvable for given N is clearly that N is contained in *some* Hasse interval $\mathcal{H}_q$, so we would like the union $\bigcup_q \mathcal{H}_q$ over all prime powers q to contain *all* positive integers. It is easy to see that the contribution to the union coming from the 'true' prime powers q that are not primes is negligible: it is contained in a zero density subset of $\mathbf{Z}_{\geq 1}$. For this reason, we may and will restrict in the sequel to the case where the base field $\mathbf{F}_q$

D.A. Buell (Ed.): ANTS 2004, LNCS 3076, pp. 117–131, 2004.

is the prime field coming from a prime number q. In this particular case, all integers in $\mathcal{H}_q$ actually do occur as the group order of $E(\mathbf{F}_q)$ for some elliptic curve E, so $N \in \mathcal{H}_q$ is sufficient to guarantee the existence of a solution. (For arbitrary prime powers q there are often not enough supersingular curves to realize all orders congruent to 1 modulo the characteristic.)

For the equality $\mathbf{Z}_{\geq 1} = \bigcup_{q \text{ prime}} \mathcal{H}_q$ we need to show that the primes are not too far apart, i.e., that the gap between consecutive primes q and q' is roughly bounded by $4\sqrt{q}$ for large q. This is more than what is currently known to be true: even under assumption of the Riemann hypothesis the gap between consecutive primes can only be shown to be of order $O(\sqrt{q}(\log q)^2)$. However, from a practical, algorithmic point of view there are always *lots* of primes q for which a large integer N is contained in $\mathcal{H}_q$. Indeed, by the prime number theorem, we expect 1 out of every $\log N$ integers around N to be prime, so for large N the set of primes q having $N \in \mathcal{H}_q$ is on average of size $4\sqrt{N}/\log N$, and finding such q is never a problem in practice.

Once we have found a prime $q > 3$ for which we have $N \in \mathcal{H}_q$ (we now require $N > 1$), there is the following *naive algorithm* to find an elliptic curve having exactly N rational points over $\mathbf{F}_q$. Suppose that we are not in the easy cases where we have $N = q + 1$ (then any supersingular curve over $\mathbf{F}_q$ will do) or where one of the few curves with j-invariant 0 or 1728 has the right number of points. Then we try

$$E_a : y^2 = x^3 + ax - a \qquad \text{with} \quad j(E_a) = 1728\frac{4a}{4a+27}$$

for random $a \in \mathbf{F}_q^* \setminus \{-\frac{27}{4}\}$ as the Weierstrass equation of the desired curve until we find a curve having N points. More precisely, we write $N = q + 1 - t$ and check whether for our a the point $(1,1) \in E_a(\mathbf{F}_q)$ is annihilated by $N = q+1-t$ or $q+1+t$. If it is, we check whether the number of $\mathbf{F}_q$-rational points is indeed $q+1\pm t$. For order $N = q+1-t$ we are done, for order $q+1+t$ not E_a itself but its quadratic twist has N points. Even though the distribution of the group orders of elliptic curves over $\mathbf{F}_q$ is not quite uniform, we expect to examine $O(\sqrt{q}) = O(\sqrt{N})$ curves E_a before we hit a curve having exactly N points. As the amount of time spent per a is usually very small, and certainly polynomial in $\log N$, this yields a probabilistic algorithm with expected running time $O(N^{\frac{1}{2}+o(1)})$. It is quite practical for small values of N, but becomes unwieldy for $N \gg 10^{15}$.

In the next section we briefly describe a classical deterministic algorithm based on complex multiplication methods which, although not asymptotically faster than the naive algorithm, can be improved in various ways. Our first improvement is a p-adic approach to complex multiplication based on the recent work of Couveignes and Henocq [3]. It is described in section 3, and illustrated by the explicit computation in section 4 of an 'ANTS 6 curve' having 2004061320040618 rational points. Our second improvement, in section 5, consists of using 'small' modular functions in this p-adic context to push the limits of what is feasible by p-adic methods. Although the resulting algorithm is still far from polynomial, its power is illustrated in section 6 by the computation of an elliptic curve having 10^{30} rational points.

2 Complex Multiplication

A deterministic algorithm to produce an elliptic curve E over the prime field $\mathbf{F}_q$ having $N \in \mathcal{H}_q$ points is provided by the theory of complex multiplication. One writes $N = q+1-t$ and observes that the Frobenius endomorphism $F_q : E \to E$ on the desired curve $E/\mathbf{F}_q$ satisfies the quadratic relation $F_q^2 - tF_q + q = 0$ of discriminant $\Delta = t^2 - 4q < 0$ in $\mathrm{End}(E) = \mathrm{End}_{\overline{\mathbf{F}}_q}(E)$. Assume we are not in the supersingular case $t = 0$. Then our observation gives rise to an embedding

$$\begin{aligned} \mathrm{End}(E) &\longrightarrow K = \mathbf{Q}(\sqrt{\Delta}) \\ F_q &\longmapsto \pi_q = (t + \sqrt{\Delta})/2 \end{aligned}$$

that maps F_q to a prime element π_q of trace t and norm q in the quadratic order $\mathcal{O}_\Delta = \mathbf{Z}[(\Delta + \sqrt{\Delta})/2] \subset K$ of discriminant Δ. By the *Deuring lifting theorem* [6, Chapter 13, Section 5], there exist a number field $H \supset K$ and an elliptic curve $\tilde{E}/H$ such that

1. there exists $\phi_q \in \mathrm{End}(\tilde{E})$ satisfying $\phi_q^2 - t\phi_q + q = 0$;
2. the prime q splits completely in $H/\mathbf{Q}$, and for every prime $\mathfrak{q}|q$ in H the reduced curve $\tilde{E}$ mod $\mathfrak{q}$ is an elliptic curve having $q + 1 - t$ points.

In fact, the reduction of the endomorphism $\phi_q \in \mathrm{End}(\tilde{E})$ above modulo a prime $\mathfrak{q}|q$ in H yields the Frobenius endomorphism of the curve $\tilde{E}$ mod $\mathfrak{q}$ over $\mathbf{F}_q$.

The smallest field $H \supset K$ over which a curve $\tilde{E}$ satisfying 1 and 2 can be defined is the Hilbert class field of K. More explicitly, from a list of reduced binary quadratic forms $[a, b, c]$ of discriminant $D = \mathrm{disc}(K)$, which can be viewed as a list of elements of the *class group* $\mathrm{Cl}(D)$ of discriminant D, we can form the *class polynomial*

$$F_D = \prod_{[a,b,c] \in \mathrm{Cl}(D)} \left(X - j\Big(\frac{-b + \sqrt{D}}{2a}\Big) \right) \in \mathbf{Z}[X]$$

of discriminant D. Here $j : \mathbf{H} \to \mathbf{C}$ denotes the well-known elliptic modular function on the complex upper half plane. Using sufficiently accurate complex approximations of the zeroes $j(\frac{-b+\sqrt{D}}{2a})$ of F_D, one may *exactly* determine F_D as it has integral coefficients. Once we have F_D, we are essentially done. Indeed, any zero of the irreducible polynomial $F_D \in \mathbf{Z}[X]$ generates the Hilbert class field H of K over K, and modulo q the polynomial $\bar{F}_D \in \mathbf{F}_q[X]$ splits into linear factors. (In fact, this property of F_D is an excellent check for the correctness of any algorithm to compute F_D.) The zeroes of $\bar{F}_D$ in $\mathbf{F}_q$ are the j-invariants of the elliptic curves over $\mathbf{F}_q$ having endomorphism ring isomorphic to the ring of integers $\mathcal{O}_D$ of K. If $\bar{j} \in \mathbf{F}_q$ is one of these zeroes, we write down a curve with this j-invariant, and check whether it has $q + 1 - t$ points. If it hasn't, we have found a curve with $q+1+t$ points, and (for $j \neq 0, 1728$) its quadratic twist has $N = q + 1 - t$ points.

This deterministic algorithm, although relatively simple, is not much faster than the naive algorithm, as for large D the class polynomial F_D is of degree

$h(D) = O(D^{\frac{1}{2}+o(1)})$ and has coefficients of size $O(D^{\frac{1}{2}+o(1)})$, making the total running time of order $O(D^{1+o(1)})$. However, if we have the freedom to pick q on input N, it is often possible to find primes q for which the associated discriminant $\Delta = \Delta(q) = (q-1-N)^2 - 4N$ is not only of size $N^{\frac{1}{2}+o(1)}$ (this is simply done by picking $q \in [N+1-2\sqrt{N}, N+1+2\sqrt{N}]$ close to the end points of the interval), but moreover has a large square factor. This leads to a field discriminant D of K which is quite a bit smaller than Δ, and makes the method feasible in cases where the naive method would take too long. As we currently cannot even *prove* the existence of a single prime in $q \in [N+1-2\sqrt{N}, N+1+2\sqrt{N}]$, we certainly cannot prove the existence of q for which $\Delta(q)$ has large square factors and D is of order substantially smaller than $O(N^{\frac{1}{2}+o(1)})$.

3 A Non-Archimedean Approach

The key feature of the complex multiplication method in the previous section is the computation of the class polynomial $F_D \in \mathbf{Z}[X]$ of the order $\mathcal{O}_D$ for suitable D. As the zeroes modulo q of F_D are j-invariants of curves $E/\mathbf{F}_q$ for which either E or its quadratic twist has exactly N points, this immediately solves our problem. In this section we achieve the computation of F_D in an other way, using p-adic instead of complex approximations of the zeroes of F_D. Working in a non-archimedean setting has the advantage that we no longer have to cope with the problem of rounding errors that arises in the complex case. It does require a p-adic substitute for the complex analytic method to evaluate j in CM-points of $\mathbf{H}$ using Fourier expansions, and this is provided by the recent work of Couveignes and Henocq [3] explained in this section.

Let $N = q+1-t$ and $\Delta = t^2 - 4q$ be as in the previous section, and $D < -4$ the discriminant of $\mathbf{Q}(\sqrt{\Delta})$. We first construct an elliptic curve E over a finite field $\mathbf{F}_p$ which has CM with $\mathcal{O}_D$. As we want p to be as small as possible, we let s be the smallest positive integer of the same parity as D for which $p = (s^2 - D)/4$ is prime. For $D \equiv 1 \bmod 8$ such p does not exist for parity reasons unless $D = -7$, and we pick the smallest positive s for which $p = (s^2 - 4D)/4$ is prime instead. In practice we expect s to be small, at most a power of $\log |D|$, so that p is of the same order of magnitude as D. Unfortunately, even under GRH proven upper bounds [3] for s are much weaker.

As there is a prime element $\pi_p = (s + \sqrt{D})/2$ (or $\pi_p = (s + 2\sqrt{D})/2$) of norm p and trace $s > 0$ in the order $\mathcal{O}_D = \mathbf{Z}[\pi_p]$ (or $\mathcal{O}_{4D} = \mathbf{Z}[\pi_p]$), there exists an ordinary elliptic curve over $\mathbf{F}_p$ having CM by $\mathbf{Z}[\pi_p]$ and $p + 1 \pm s$ points over $\mathbf{F}_p$. We can find such a curve $E/\mathbf{F}_p$ by applying the naive algorithm, as we saw in section 2 that $|D|$ is much smaller than N. We have $\mathrm{End}(E) = \mathcal{O}_D$ for $D \not\equiv 1 \bmod 8$. For $D \equiv 1 \bmod 8$ the ring $\mathrm{End}(E)$ is either equal to $\mathbf{Z}[\pi_p] = \mathcal{O}_{4D}$ or to $\mathcal{O}_D \supsetneq \mathbf{Z}[\pi_p]$. We are in the second case if all 2-torsion of E is $\mathbf{F}_p$-rational, and in the first if it isn't. As for $D \equiv 1 \bmod 8$ the class polynomials F_D and F_{4D} have the same degree and both generate the Hilbert class field H of K, they are both fine for our purposes. Alternatively, in case we have $\mathrm{End}(E) = \mathcal{O}_{4D}$ there is a unique point P of order 2 in $E(\mathbf{F}_p)$, and we can replace E by the 2-isogenous curve $E/\langle P\rangle$ to achieve $\mathrm{End}(E) = \mathcal{O}_D$. We assume that we have $\mathrm{End}(E) = \mathcal{O}_D$ in the sequel. Note that D is by definition fundamental.

By the Deuring lifting theorem, there exists a curve $\tilde{E}/H$ and a prime $\mathfrak{p}|p$ such that $\tilde{E}$ reduces modulo $\mathfrak{p}$ to E and such that we have $\mathrm{End}(\tilde{E}) \cong \mathrm{End}(E)$. As p splits completely in $H/\mathbf{Q}$ the curve $\tilde{E}$ is actually defined over $\mathbf{Q}_p$. In fact, $\tilde{E}$ is the *unique* elliptic curve over $\mathbf{Q}_p$ with reduction E and endomorphism ring $\mathrm{End}(\tilde{E}) \cong \mathrm{End}(E) \cong \mathcal{O}_D$. It is this *canonical lift* $\tilde{E}$ of E that we want to compute, as its j-invariant is a zero of our class polynomial.

Let $\mathrm{Ell}_D(\mathbf{F}_p)$ be the set of $\overline{\mathbf{F}}_p$-isomorphism classes of elliptic curves over $\mathbf{F}_p$ with endomorphism ring $\mathcal{O}_D$. The j-invariants of the elements in $\mathrm{Ell}_D(\mathbf{F}_p)$ are the zeroes of $(F_D \bmod p)$, so $\mathrm{Ell}_D(\mathbf{F}_p)$ is finite of order $h(D) = \#\mathrm{Cl}(D)$. It can be identified with the similarly defined set $\mathrm{Ell}_D(\mathbf{Q}_p)$ of $\overline{\mathbf{Q}}_p$-isomorphism classes of elliptic curves over $\mathbf{Q}_p$ with endomorphism ring $\mathcal{O}_D$. The j-invariants of the elements in $\mathrm{Ell}_D(\mathbf{Q}_p)$ are the zeroes of F_D. Let $\mathbf{C}_p$ be the completion of an algebraic closure of $\mathbf{Q}_p$, and write $X_D(\mathbf{C}_p)$ for the set of isomorphism classes of elliptic curves over $\mathbf{C}_p$ with the property that their reduction is an element of $\mathrm{Ell}_D(\mathbf{F}_p)$. Then $X_D(\mathbf{C}_p)$ is a p-adic analytic space, as the j-invariant identifies it with a subset of $\mathbf{C}_p$. It consists of $h(D)$ discs of p-adic radius 1. Each disc contains exactly one element of $\mathrm{Ell}_D(\mathbf{Q}_p)$, and this is the subset of $\mathbf{C}_p$ we want to compute. It consists of the (j-invariants of the) isomorphism classes of elliptic curves in $X_D(\mathbf{C}_p)$ having CM with $\mathcal{O}_D$.

Let $I \subset \mathcal{O}_D$ be an $\mathcal{O}_D$-ideal prime to p and $E/\mathbf{F}_p$ an elliptic curve in $\mathrm{Ell}_D(\mathbf{F}_p)$. Then there is a separable isogeny $E \to E_I$ which has the subgroup $E[I]$ of I-torsion points of E as its kernel. In this way, we obtain a bijection $\bar{\rho}_I : \mathrm{Ell}_D(\mathbf{F}_p) \to \mathrm{Ell}_D(\mathbf{F}_p)$ that sends the isomorphism class of E to that of E_I. We obtain an action of the group $I(p)$ of $\mathcal{O}_K$-ideals prime to p on $\mathrm{Ell}_D(\mathbf{F}_p)$, and since principal $\mathcal{O}_K$-ideals act trivially, this action factors via the quotient map $I(p) \twoheadrightarrow \mathrm{Cl}(D)$. This makes $\mathrm{Ell}_D(\mathbf{F}_p)$ into a principal homogeneous $\mathrm{Cl}(D)$-space.

The fundamental idea in [3] is that the action of $I(p)$ on $\mathrm{Ell}_D(\mathbf{F}_p)$ admits a natural lift to an action of $I(p)$ on $X_D(\mathbf{C}_p)$. More precisely, for $I \in I(p)$ the map $\rho_I : X_D(\mathbf{C}_p) \to X_D(\mathbf{C}_p)$ is a p-adic analytic map that lifts $\bar{\rho}_I$ in the sense that on $\mathrm{Ell}_D(\mathbf{Q}_p) \subset X_D(\mathbf{C}_p)$, the restriction of ρ_I is the standard Galois action that factors via $\mathrm{Cl}(D)$. If $I = (\alpha)$ is principal with $\alpha \in \mathcal{O}_D \setminus \mathbf{Z}$, then $\rho_\alpha = \rho_I$ stabilizes each disk around a CM-point in $\mathrm{Ell}_D(\mathbf{Q}_p)$, and has the CM-point in the disk as its unique fixed point.

It is shown in [3] that the derivative of the map ρ_α in a point of $\mathrm{Ell}_D(\mathbf{Q}_p)$ equals $\alpha/\bar{\alpha}$, and this can be used to compute the j-invariant of a CM-point in $\mathrm{Ell}_D(\mathbf{Q}_p)$ starting from an arbitrary point in the disk using a Newton iteration process. If E_1 is any lift of E to $\mathbf{C}_p$, we put

$$(1) \qquad j(E_{k+1}) = j(E_k) - \frac{j(\rho_\alpha(E_k)) - j(E_k)}{(\alpha/\bar{\alpha}) - 1} \qquad \text{for } k \in \mathbf{Z}_{\geq 1}.$$

If α is small with respect to p, then $(\alpha/\bar{\alpha}) - 1$ is a unit in $\mathbf{Z}_p$ and the sequence (1) converges to the j-invariant of the canonical lift $\tilde{E}$. In each step, the p-adic precision of the approximation is doubled.

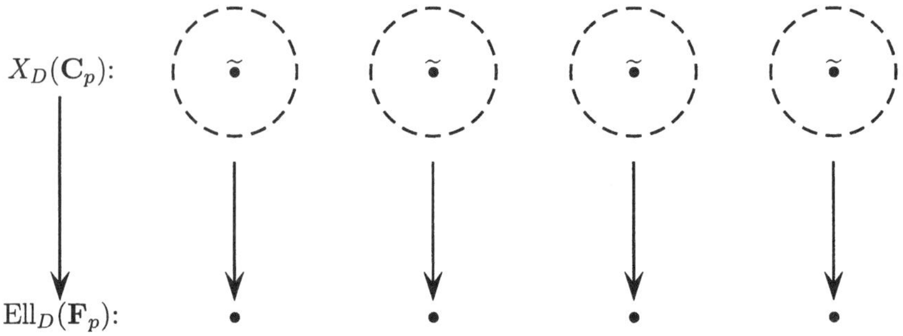

For the definition of ρ_I on the isomorphism class of E' in $X_D(\mathbf{C}_p)$, we note that the subgroup $E[I]$ of I-torsion points of the reduction $E/\mathbf{F}_p$ of E' lifts canonically to a subgroup $E'[I]$ of E', and we put $\rho_I(E') = E'_I$, with $E \to E'_I$ the isogeny with kernel $E'[I]$. This provides a lift of $E[I]$ to a *group scheme* over the p-adic disk in $X_D(\mathbf{C}_p)$ lying over $[E] \in \mathrm{Ell}_D(\mathbf{F}_p)$. More algorithmically, if E is given by a Weierstrass model, the subgroup $E[I]$ can be described by a separable polynomial $\bar{f}_I \in \mathbf{F}_p[X]$ having the x-coordinates of the affine points in $E[I]$ as its zeroes. If I has norm n, then $p \nmid n$ and $\bar{f}_I$ divides the n-th division polynomial of E in $\mathbf{F}_p[X]$. Choosing a Weierstrass model for E' reducing to that for E, we can lift $\bar{f}_I$ uniquely by Hensel's lemma to a factor f_I of the n-th division polynomial of E', and this factor describes the subgroup $E'[I]$. Note that ρ_I is the identity on $X_D(\mathbf{C}_p)$ if I is generated by an integer, and that, consequently, $\rho_{\bar{I}}$ is the inverse of ρ_I.

For the explicit computation of $\rho_\alpha(E')$, we take an element $\alpha = a + b\pi_p$ in $\mathcal{O}_D \setminus \mathbf{Z}$ that is sufficiently smooth, i.e., a product of $\mathcal{O}_D$-ideals $L = (\ell, a + b\pi_p)$ of small prime norm $\ell \neq p$. Such an element is found by sieving in the set

$$\{a + b\pi_p : a, b \in \mathbf{Z}, b \neq 0, (a, b) = 1\}.$$

Again, the smoothness properties are in practice much better than what can be rigorously proved [3]. For the computation of $\rho_L(E')$, we first compute the action of $\bar{\rho}_L$ on the reduction $E = \bar{E}' \in \mathrm{Ell}_D(\mathbf{F}_p)$, i.e., the j-invariant of $E_L = \bar{E}'_L$. The kernel $E[L]$ of the isogeny $E \to E_L$ is a cyclic subgroup of order ℓ of $E[\ell]$ that is an eigenspace of the Frobenius morphism with eigenvalue $-b/a \in \mathbf{F}_\ell$. The corresponding polynomial $\bar{f}_L \in \mathbf{F}_p[X]$ can be computed by the techniques used by Atkin and Elkies to improve Schoof's original point counting algorithm, see [9]. These techniques also yield a Weierstrass model for E_L; we only need the j-invariant $j(E_L) \in \mathbf{F}_p$ of $E_L = \bar{\rho}_L(E)$. From the decomposition of ρ_α into 'prime degree' maps ρ_L, we obtain a *cycle* of isogenies

$$E \xrightarrow{\bar{\rho}_{L_1}} E_{L_1} \xrightarrow{\bar{\rho}_{L_2}} E_{L_1L_2} \longrightarrow \ldots \xrightarrow{\bar{\rho}_{L_t}} E_{(\alpha)} = E. \tag{2}$$

To compute the action of L on the lift E' of E, we do *not* lift $\bar{f}_L$ to some precision to a divisor of the ℓ-th division polynomial of E', and then find a Weierstrass model and the j-invariant for E_L using Vélu's formulas [11]. As the division polynomial has degree $(\ell^2-1)/2$ for $\ell > 2$, this would be rather time-consuming. Instead, we exploit the ℓ-th modular polynomial $\Phi_\ell(X,Y) \in \mathbf{Z}[X,Y]$, which is of degree $\ell+1$ in each of the variables, and which we can precompute for a number of small primes ℓ. As there is an isogeny $E' \to E'_L$ of degree ℓ, we know that $j(E'_L)$ is a zero of $\Phi_\ell(j(E'),Y) \in \mathbf{Z}_p[Y]$. Since we know the j-invariant of the reduction of E'_L, we also know which root to approximate in $\mathbf{Z}_p$, and this reduces the lifting process to a simple Hensel lift of a zero of a polynomial of degree $\ell+1$.

For our Newton process in (1), we start from $E \in \mathrm{Ell}_D(\mathbf{F}_p)$, and compute the cycle of $\mathbf{F}_p$-isogenies in (2). We then lift E arbitrarily to a curve E_1 over $\mathbf{Q}_p$, the '1 digit precision' approximation of the canonical lift. Now we compute lifts over $\mathbf{Q}_p$ of our $\mathbf{F}_p$-isogenies in 2 digit precision, using the modular polynomials, and use the value of $\rho_\alpha(E_1)$ obtained to update E_1 as in (1) to a 2 digit precision approximation E_2 of the canonical lift. We continue this process of making (not really closed) cycles over $\mathbf{Q}_p$, doubling the precision of the computation at each step, until we have the canonical lift with high enough accuracy (see [2, Chapter 7, Section 6] for an estimate of the required accuracy).

If we know the canonical lift $\tilde{E}$, we compute its Galois conjugates again via the modular polynomials. For this, we need small primes L that generate the class group. Under GRH, this can be done using primes L not exceeding the Bach bound $6\log^2(|D|)$, see [2, Chapter 5, Section 5]. In practice, this is never a problem. If we have all the conjugates to the required precision, we find by simple expansion of the product below the class polynomial

$$F_D = \prod_{[I]\in \mathrm{Cl}(D)} (X - j(\tilde{E}_I)) \in \mathbf{Z}[X].$$

4 An Elliptic Curve for ANTS 6

We illustrate the working of our p-adic method by computing a tailor made elliptic curve for ANTS 6 having exactly $N = 2004061320040618$ points.

First we look for a small discriminant, so we write $N = q + 1 - t$ for various primes q and search for a large square dividing $\Delta = t^2 - 4q$. In this example, the choice

$$N = 2004061230508291 + 1 + 89532326$$

yields $\Delta = -2^3 \cdot 3 \cdot 619^2 \cdot 22567$, so we take $D = -2^3 \cdot 3 \cdot 22567 = -541608$ in this section. The corresponding class group $\mathrm{Cl}(D)$ has order 132.

Our goal is to compute the class polynomial F_D. We will do this p-adically, so we first find an elliptic curve over some $\mathbf{F}_p$ which has CM with $\mathcal{O}_D$. The smallest integer $s > 0$ for which $(s^2 - D)/4$ is prime is $s = 2$, so we have $D = 2^2 - 4p$ with $p = 135403$. We fix this value of p for the rest of this section.

We now apply the naive method to find a curve $E_a : y^2 = x^3 + ax - a$ over $\mathbf{F}_p$ with trace of Frobenius 2. We find that E_{1737} has trace 2 and take this as our base curve $E/\mathbf{F}_p$.

Next we determine which element $\alpha \in \mathcal{O}_D \setminus \mathbf{Z}$ we will use for the map ρ_α. The ideal $(\alpha) = (22539 + 4\pi_p)$ of norm $510353281 = 19^2 \cdot 29^2 \cdot 41^2$ factors as

$$(\alpha) = L_{19}^2 \cdot L_{29}^2 \cdot L_{41}^2 = (19, \pi_p + 6)^2 \cdot (29, \pi_p - 13)^2 \cdot (41, \pi_p - 13)^2.$$

We compute the action for the prime ideal L_{19}. The eigenvalue of the action of Frobenius on the 19-torsion is $-6 \in \mathbf{F}_{19}$. If we evaluate the modular polynomial $\Phi_{19}(X,Y)$ in $X = j(E) = 41556 \in \mathbf{F}_p$, we get a polynomial which has two roots over $\mathbf{F}_p$, namely 19533 and 54827. From this we deduce that L_{19} sends $j(E)$ to one of these two roots; we don't know which one yet.

We just *guess* that the correct j-invariant is $54827 \in \mathbf{F}_p$. Following Elkies [9], we now compute the eigenspace S of the 19-torsion corresponding to this isogeny. We get the x-coordinates of the points on E in S as zeroes of

$$\bar{f}_{L_{19}} = X^9 + 29873X^8 + 49874X^7 + 131130X^6 + 49222X^5 + 46538X^4 + 111513X^3 + 68602X^2 + 126444X + 20947 \in \mathbf{F}_p[X]$$

Since we know that the eigenvalue for L_{19} is -6, we can now just check whether

$$(X^p, Y^p) = -6 \cdot (X, Y)$$

holds for points in S, i.e., we compute both (X^p, Y^p) and $-6 \cdot (X,Y)$ in the ring

$$\mathbf{F}_p[X,Y]/(\bar{f}_{L_{19}}(X), Y^2 - X^3 + 1737X - 1737).$$

Note that the $\cdot$ means adding *on the curve*! In this example, it turns out that (X^p, Y^p) and $-6 \cdot (X,Y)$ are not the same. It follows that the correct j-invariant of the L_{19}-isogenous curve is the other value $19533 \in \mathbf{F}_p$.

The action of L_{19} on the curve with j-invariant 19533 is now easier to compute: the modular polynomial has again two roots, but one of the roots has to be $j(E)$. This root corresponds to the action of $\bar{L}_{19}$, so we pick the other root. If we compute the entire cycle corresponding to L, we get:

$$41556 \xrightarrow{L_{19}} 19533 \xrightarrow{L_{19}} 100121 \xrightarrow{L_{29}} 86491 \xrightarrow{L_{29}} 40349 \xrightarrow{L_{41}} 32517 \xrightarrow{L_{41}} 41556.$$

We now lift $E/\mathbf{F}_p$ to $E_1/\mathbf{Q}_p$ by lifting the coefficients of the Weierstrass equation arbitrarily. The polynomial $\Phi_{19}(j(E_1), Y) \in \mathbf{Z}_p[Y]$ has exactly two roots, one of which reduces to 19533 modulo p. We compute this root, which is the value of the L_{19}-action on $j(E_1)$, to two p-adic digits of precision. We continue to lift the whole cycle to two p-adic digits of precision, and update $j(E_1)$ according to formula (1) to obtain $j(E_2)$. Starting from $j(E_2)$, we now lift the cycle to four p-adic digits of precision, compute $j(E_3)$ from this, and so on. We obtain

$$\begin{aligned} j(\tilde{E}) &= 41556 + O(p) \\ &= 41556 - 17953p + O(p^2) \\ &= 41556 - 17953p - 51143p^2 - 17793p^3 + O(p^4) \\ &= 41556 - 17953p - 51143p^2 - 17793p^3 + 45123p^4 + 52596p^5 + 18237p^6 \\ &\quad + 42211p^7 + O(p^8) \\ &= 41556 - 17953p - 51143p^2 - 17793p^3 + 45123p^4 + 52596p^5 + 18237p^6 \\ &\quad + 42211p^7 + 45716p^8 + 58788p^9 + 18836p^{10} - 4101p^{11} - 60004p^{12} \\ &\quad - 24668p^{13} + 27527p^{14} - 58942p^{15} + O(p^{16}). \end{aligned}$$

From the estimate in [2, Chapter 7, Section 6], we know that we need approximately 3000 decimals of accuracy. As we have $10^{3000} \approx p^{585}$, we compute $j(\tilde{E})$ up to 585 p-adic digits. The class group $\mathrm{Cl}(D) \cong \mathbf{Z}/2\mathbf{Z} \times \mathbf{Z}/66\mathbf{Z}$ is generated by a prime of norm 2 and a prime of norm 29, so we find all 132 conjugates of $j(\tilde{E})$ under $\mathrm{Gal}(H/K)$ up to 585 p-adic digits using the modular polynomials Φ_2 and Φ_{29}. In the end, we expand the polynomial of degree 132 to find the class polynomial

$$F_D = \prod_{[I] \in \mathrm{Cl}(D)} (X - j(\tilde{E}_I)) \in \mathbf{Z}[X]$$

We now compute a root of $F_D \in \mathbf{F}_q$, and note that we can check our computations so far by testing whether F_D splits completely in $\mathbf{F}_q[X]$. One of the 132 roots is $j = 5215470850369 \in \mathbf{F}_q$, and an elliptic curve with this j-invariant is $E_a : y^2 = x^3 + ax - a$ with $a = 27j/(4(1728 - j)) = 1460967812073632 \in \mathbf{F}_q$. We know that E_a has CM with $\mathcal{O}_D$ and that its trace of Frobenius equals $\pm t$. To test whether it actually equals t, we look at the order of $P = (1,1) \in E_a(\mathbf{F}_q)$. We see that $(q+1+t)P \neq O = (q+1-t)P$, so E_a must have trace equal to t. We conclude that the curve defined by

$$y^2 = x^3 + 1460967812073632x + 543093418434659$$

over $\mathbf{F}_{2004061230508291}$ has exactly 2004061320040618 rational points.

5 Using Class Invariants

A serious drawback of the complex multiplication method is that it requires the computation of the class polynomial F_D, which grows rapidly in size with D, and is already sizable for moderately small discriminants. Around 1900, Weber [12] computed generating polynomials for Hilbert class fields using the values at CM-points of modular functions of higher level instead of the j-function. These techniques have become important again in an algorithmic context, and many complications from Weber's days are now well understood [4,10]. The classical theory of class invariants is firmly rooted in complex analytic arguments, but much of it can be made to work in our non-archimedean setting. This section gives an indication of what can be done, leaving a fuller treatment to [1].

The complex multiplication method to compute the class polynomial F_D is based on the fact that its zeroes are the j-invariants of the elliptic curves in characteristic zero having complex multiplication with $\mathcal{O}_D$. In the complex analytic setting, we simply list the complex lattices (up to scaling) giving rise to such curves and compute their j-invariants to sufficient accuracy using the q-expansion of j. In the p-adic setting, we first compute one such curve in the finite set $\mathrm{Ell}_D(\mathbf{F}_p)$. Then we lift the action on $\mathrm{Ell}_D(\mathbf{F}_p)$ of the group $I(p)$ of $\mathcal{O}_D$-ideals prime to p to an action on the set $X(\mathbf{C}_p)$ of their $\mathbf{C}_p$-lifts in such a way that the induced action on the subset $\mathrm{Ell}_D(\mathbf{Q}_p)$ of canonical lifts is the standard Galois action coming from $\mathrm{Cl}(D)$. This enables us to compute the finite subset $\mathrm{Ell}_D(\mathbf{Q}_p) \subset X(\mathbf{C}_p) \xrightarrow{j} \mathbf{C}_p$ in $\mathbf{C}_p$ consisting of the zeroes of F_D. We use a

Newton type lift of our curve in $\mathrm{Ell}_D(\mathbf{F}_p)$ to $\mathrm{Ell}_D(\mathbf{Q}_p)$ together with the explicit Galois action of $\mathrm{Cl}(D)$ on $\mathrm{Ell}_D(\mathbf{Q}_p)$, which we handle by exploiting modular polynomials.

We now want to replace j by modular functions of higher level. These are elements of the modular function field $\mathcal{F} = \bigcup_{n\geq 1} \mathcal{F}_n$ as defined in [6, Chapter 6, Section 3]. Here $\mathcal{F}_n$ denotes the field of modular functions of level n over $\mathbf{Q}$. It can be viewed as the function field of the modular curve $X(n)$ over the cyclotomic field $\mathbf{Q}(\zeta_n)$. Over $\mathcal{F}_1 = \mathbf{Q}(j)$, it is generated by the Fricke functions of level n, which are normalized x-coordinates of n-torsion points on an elliptic curve with j-invariant j.

The modular functions f we use are *integral* over $\mathbf{Z}[j]$, so they are given as the zero of some irreducible polynomial $\Psi_f(X,j) \in \mathbf{Z}[j,X]$. If we specialize j to be the j-invariant of a curve $\tilde{E} \in \mathrm{Ell}_D(\mathbf{Q}_p)$, then the roots of the polynomial $\Psi_f(X, j(\tilde{E})) \in H[X]$, which has integral coefficients, lie in the ray class field H_n of conductor n of $K = \mathbf{Q}(\sqrt{D})$, with n the level of f. It is known that for many choices of 'small' f, one or more of these roots are *class invariants* that actually lie in the Hilbert class field $H = H_1 \subset H_n$ of K. If we can determine which roots end up in H, and compute the explicit Galois action of $\mathrm{Cl}(D)$ on such a root, we can compute its irreducible polynomial over K or $\mathbf{Q}$ just like we did this for j. In the complex analytic setting the tool to perform these tasks is Shimura's reciprocity law [4,10]. It tells us in which points $\tau \in K \subset \mathbf{H}$ a function f should be evaluated to obtain a class invariant, and describes the conjugates of $f(\tau)$ over K as the values of conjugates of f over $\mathbf{Q}(j)$ in certain other points $\tau' \in K$. These values can be approximated in $\mathbf{C}$ using the q-expansions of f and its conjugates. Once we have computed the irreducible polynomial of $f(\tau)$ over K or $\mathbf{Q}$, one can use the relation $\Psi_f(f(\tau), j(\tau)) = 0$ to obtain information on $j(\tau)$ itself.

In a p-adic setting, we cannot deal with j and f as functions on the complex upper half plane, and the expansion of modular functions as Fourier series in $q = e^{2\pi i\tau}$ has no non-archimedean analogue when dealing as we do with CM-curves, which have integral j-invariants. What we do have is an action from class field theory of the $\mathcal{O}_D$-ideals coprime to n on the roots of $\Psi_f(X, j(\tilde{E})) \in H[X]$ in H_n associated to $j(\tilde{E})$. This action factors via the class group $\mathrm{Cl}(n^2D)$ of the order of discriminant n^2D, which is a non-maximal order for $n > 1$.

Example 1. The modular function $j : \mathbf{H} \to \mathbf{C}$ has a holomorphic cube root $\gamma_2 : \mathbf{H} \to \mathbf{C}$ that is modular of level 3 and has $\Psi_{\gamma_2}(X,j) = X^3 - j$. It is the unique root of $X^3 - j$ having a rational q-expansion. If D is not divisible by 3 and we write $\mathcal{O}_D = \mathbf{Z}[\tau]$ with $\tau + \bar{\tau} = 0 \bmod 3$, then $\gamma_2(\tau)$ is a class invariant, and its 'size' is only one third of that of $j(\tau)$.

If $\tilde{E} : y^2 = x^3 + ax + b$ is in $\mathrm{Ell}_D(\mathbf{Q}_p)$ and $c_1, \ldots, c_4$ are the 4 roots of its 3-division polynomial, then

$$\frac{-48a}{2a - 3(c_1c_2 + c_3c_4)} \tag{3}$$

is a cube root of $j(\tilde{E})$. As there are 3 ways to divide the roots $c_1, \ldots, c_4$ in two sets of two roots each, formula (3) yields 3 distinct cube roots of j. Note that there is no obvious way to single out an expression in (3) as being 'the function' corresponding to γ_2.

If I is an $\mathcal{O}_D$-ideal prime to $3p$, the isogeny $\tilde{E} \to \tilde{E}_I$ induces an isomorphism $\tilde{E}[3] \xrightarrow{\sim} \tilde{E}_I[3]$ on the 3-torsion subgroups, and maps each possible cube root of $j(\tilde{E})$ in (3) to some well-defined cube root of $j(\tilde{E}_I)$. This is the Galois action of the Artin symbol of I, which maps $j(\tilde{E})$ to $j(\tilde{E}_I)$, on these cube roots. It provides an extension of the map $\rho_I : \mathrm{Ell}_D(\mathbf{Q}_p) \to \mathrm{Ell}_D(\mathbf{Q}_p)$ to the set of the cube roots of (the j-invariants of) the elements in $\mathrm{Ell}_D(\mathbf{Q}_p)$. As all these cube roots are p-adically integral, reduction of this map provides an extension of $\bar{\rho}_I : \mathrm{Ell}_D(\mathbf{F}_p) \to \mathrm{Ell}_D(\mathbf{F}_p)$ to the set of cube roots in $\bar{\mathbf{F}}_p$, provided we have $p \neq 3$. In a similar way, we have an extension of the map $\rho_I : X_D(\mathbf{C}_p) \to X_D(\mathbf{C}_p)$ to the set of cube roots.

Example 1 nicely illustrates that the cube roots of j are functions on the modular curve $X(3)$, the points of which can be viewed as isomorphism classes of elliptic curves with complete 3-level structure: in order to have a well defined value not only the j-invariant of the curve but also some 'ordering' of 3-torsion points is required.

As the field $\mathcal{F}_n$ is generated over $\mathbf{Q}(j)$ by Fricke functions, a modular function f of level n is always a rational expression in j and the roots of the n-th division polynomial of a curve with j-invariant j. As in Example 1, the action ρ_I of an $\mathcal{O}_D$-ideal I coprime to pn on $X_D(\mathbf{C}_p)$ naturally maps the roots of $\Psi_f(X, j)$ for $j \in X_D(\mathbf{C}_p)$ to the roots of $\Psi_f(X, \rho_I(j))$. This observation suffices to treat modular functions providing class invariants by p-adic methods similar to that in section 3.

Suppose f is an integral modular function of level n that is known to provide class invariants for $\mathcal{O}_D$ at certain CM-points for $\mathcal{O}_D$ in $\mathbf{H}$. Then we know that for every curve $\tilde{E} \in \mathrm{Ell}_D(\mathbf{Q}_p)$, certain roots of $\Psi_f(X, j(\tilde{E})) \in \mathbf{Q}_p[X]$ are class invariants, so they lie in $\mathbf{Q}_p$. If $E \in \mathrm{Ell}_D(\mathbf{F}_p)$ is the reduction of $\tilde{E}$, then $\Psi_f(X, j(E)) \subset \mathbf{F}_p[X]$ will have roots in $\mathbf{F}_p$, and we want to know which roots in $\mathbf{F}_p$ arise as the reduction of class invariants. Let β be a root, and assume for simplicity that $\Psi_f(X, j(E))$ is separable. (This is usually the case in practice as p is not too small, of size $O(D^{1+\varepsilon})$.) The roots of $\Psi_f(X, j(\tilde{E})) \in \mathbf{Q}_p[X]$ all lie in H_n, and they are class invariants exactly when they are fixed by the Galois group

$$\mathrm{Gal}(H_n/H) \cong (\mathcal{O}_D/n\mathcal{O}_D)^*/\mathcal{O}_D^* = \ker[\mathrm{Cl}(n^2D) \to \mathrm{Cl}(D)].$$

It follows that β arises as the reduction of a class invariant if it is fixed by the maps $\bar{\rho}_x : \mathrm{Ell}_D(\mathbf{F}_p) \to \mathrm{Ell}_D(\mathbf{F}_p)$ for a set of generators x of $(\mathcal{O}_D/n\mathcal{O}_D)^*/\mathcal{O}_D^*$. We can compute $\bar{\rho}_x(j(E))$ as before when x is sufficiently smooth, and we need the extension of the action to the roots of $\Psi_f(X, j(E))$. In theory this can be done by working with explicit Weierstrass models and an explicit description of f in terms of n-torsion points as in Example 1. In practice we work with *modular polynomials* for prime degree ℓ isogenies relating the roots of $\Psi_f(X, j_1)$ to that of $\Psi_f(X, j_2)$ when j_1 and j_2 are j-invariants of ℓ-isogenous curves. Such modular polynomials are in many ways similar to the modular polynomials $\Phi_\ell(X,Y)$ arising for the j-function, and they may be found using complex analytic methods involving q-expansions.

Suppose we find that a certain root $\beta \in \mathbf{F}_p$ of $\Psi_f(X, j(E))$ is the reduction of a class invariant $\tilde{\beta}$. Then we lift the *pair* (E,β) to the pair $(\tilde{E},\tilde{\beta})$ consisting of the canonical lift $\tilde{E}/\mathbf{Q}_p$ and the class invariant $\tilde{\beta}$. In theory this can be done by lifting $j(E)$ as in section 3, using cycles of smooth isogenies, and then compute the Hensel lift of β to the root $\tilde{\beta}$. This method makes use of the modular polynomials Φ_ℓ for j, which are big. In many cases the corresponding polynomials Φ_ℓ^f for f are much more pleasant to work with, so it is better to lift isogeny cycles not in terms of j-invariants but in terms of a root of $\Psi_f(X,j)$, and find the resulting j-invariant from its corresponding root.

All that remains is computing the conjugates of the pairs $(\tilde{E},\tilde{\beta})$ under $\mathrm{Cl}(D)$. This is done in exactly the same manner as before, using small primes that generate $\mathrm{Cl}(D)$. We are only interested in the conjugates of $\tilde{\beta}$, so we use the modular polynomials Φ_ℓ^f to compute these conjugates. In the end we expand the polynomial

$$F_D^f = \prod_{[I]\in \mathrm{Cl}(D)} (X - \tilde{\beta}^I) \in \mathbf{Z}[X] \quad (\text{or } \mathcal{O}_D[X].)$$

This polynomial splits again completely in $\mathbf{F}_q[X]$, and from a root in $\mathbf{F}_q$ we compute the corresponding j-value in $\mathbf{F}_q$.

Example 2. We take for f the Weber function $\mathfrak{f}$, which is classically defined on $\mathbf{H}$ in terms of the Dedekind η-function as $\mathfrak{f}(\tau) = \zeta_{48}^{-1}\eta(\frac{\tau+1}{2})/\eta(\tau)$. It is a modular function of level 48 of degree 72 over $\mathbf{Q}(j)$ with

$$\Psi_\mathfrak{f}(X,j) = (X^{24}-16)^3 - jX^{24} \in \mathbf{Z}[j,X].$$

For discriminants $D \equiv 1 \bmod 8$ with $3 \nmid D$, it yields class invariants when evaluated at appropriate points $\tau \in \mathbf{H}$. In other cases small powers of $\mathfrak{f}$ often have the same property [4].

In principle we can express $\mathfrak{f}$ in terms of x-coordinates of 48-torsion points, but there is no need to do this. It suffices to find, for $E/\mathbf{F}_p$ a curve having CM with $\mathcal{O}_D$, first a root $\beta \in \mathbf{F}_p$ of $\Psi_f(X,j(E))$ that is the reduction of a class invariant $\tilde{\beta}$, then a good approximation of the root $\tilde{\beta}$ of $\Psi_\mathfrak{f}(X,j(\tilde{E}))$ in $\mathbf{Q}_p$, and finally good approximations in $\mathbf{Q}_p$ of the conjugates of $\tilde{\beta}$ under $\mathrm{Cl}(D)$. These questions ultimately reduce to computing the action of an ideal L of prime norm $\ell \nmid pn$ on pairs (E',β'), with E' in $X_D(\mathbf{C}_p)$ or $\mathrm{Ell}_D(\mathbf{F}_p)$ and β' a root of $\Psi_f(X,j(E'))$ in $\mathbf{Q}_p$ or $\mathbf{F}_p$. For E' we know how to compute $j(\tilde{E}'_L)$, for $(\beta')^L$ we use the fact that it is a zero of the modular polynomial $\Phi_\ell^\mathfrak{f}(\beta',X)$ *and* of $\Psi_\mathfrak{f}(X,j(E'_L))$. Usually there are only two roots $\Phi_\ell^\mathfrak{f}(\beta',X)$ that we need to consider, the correct one and the image under the action of $\bar{L}$. It is of great help that the modular polynomials $\Phi_\ell^\mathfrak{f}$ are quite a bit smaller than the classical modular polynomials for j. For small ℓ they are really small, like

$$\Phi_5^\mathfrak{f}(X,Y) = (X^5-Y)(X-Y^5)+5XY$$
$$\Phi_7^\mathfrak{f}(X,Y) = (X^7-Y)(X-Y^7)+7(XY-X^4Y^4).$$

For $\ell = 13$ it takes at least two of these pages to write down Φ_ℓ, but we have

$$\Phi^{\mathfrak{f}}_{13}(X,Y) = (X^{13}-Y)(X-Y^{13}) + 5\cdot 13XY \\ +13(X^2Y^{12} + X^{12}Y^2 + 4X^{10}Y^4 + 4X^{10}Y^4 + 6X^6Y^8 + 6X^8Y^6).$$

6 An Elliptic Curve Having 10^{30} Points

We illustrate the power of the method presented in the previous section by constructing an elliptic curve having exactly $N = 10^{30}$ rational points.

Just as in section 4, we first look for a suitable discriminant. We write $N = q+1-t$, and by looking at $|t|$ slightly less than $2\sqrt{N} = 2\cdot 10^{15}$, we find for trace $t = 1999999999167682$ that the number $q = 10^{30} + t - 1$ is prime and that

$$\Delta = t^2 - 4q = -2^{12}\cdot 3^2\cdot 5^2\cdot 17^2\cdot 367^2\cdot 3943\cdot 23537$$

has a large square factor leading to $D = -92806391$. The corresponding class group $\mathrm{Cl}(D)$ has order 15610. Computing the Hilbert class polynomial for our D would require an accuracy of 313618 decimals, which is clearly not practical.

Instead, we notice that we have $D \equiv 1 \bmod 8$ and $D \not\equiv 0 \bmod 3$, so we can use the classical Weber function $\mathfrak{f}$ to compute a class polynomial for $H(K)$. We first compute an elliptic curve in $\mathrm{Ell}_D(\mathbf{F}_p)$ for a 'small' prime p. We have $D \equiv 1 \bmod 8$, and the smallest $s \in \mathbf{Z}_{>0}$ with $p = (s^2 - 4D)/4$ prime is $s = 132$, leading to $p = 92810747$. The first curve $E_a/\mathbf{F}_p$ of trace 132 we encounter is $E_{1086} : y^2 = x^3 + 1086x - 1086$ of j-invariant 37202456. As E has all three of its two-torsion points defined over $\mathbf{F}_p$, its endomorphism ring is $\mathcal{O}_D$, not $\mathcal{O}_{4D}$.

We now have to determine which root β of the polynomial $\Psi_{\mathfrak{f}}(X, j(E)) = (X^{24}-16)^3 - j(E)X^{24} \in \mathbf{F}_p[X]$ is the reduction of a *class invariant* $\tilde{\beta} \in \mathbf{Z}_p$. We are lucky since ± 21677132 are the only two roots in $\mathbf{F}_p$. Since $-\tilde{\beta}$ is also a class invariant, it does not matter which root we pick. We take $\beta = 21677132$.

For the smooth ideal inducing ρ_α we pick $(\alpha) = (-420+\pi_p)$, which factors as

$$(11, 8+2\pi_p)\cdot(17, 4+2\pi_p)^2\cdot(23, 16+2\pi_p)^2\cdot(31, 13+2\pi_p)\cdot(41, 30+2\pi_p).$$

Just as in section 4, we compute the cycle in $\mathbf{F}_p$ for the *j-invariants*:

$$37202456 \xrightarrow{L_{11}} 4967239 \xrightarrow{L_{17}} \cdots \xrightarrow{L_{31}} 21402782 \xrightarrow{L_{41}} 37202456.$$

Using this cycle, we can also compute the cycle for β. For instance, the modular polynomial $\Phi^{\mathfrak{f}}_{11}(\beta, Y) \in \mathbf{F}_p[Y]$ has two roots: 32604444 and 60476019. In order to determine which root to take, we note that β is a root of $\Psi_{\mathfrak{f}}(X, j(E_{L_{11}})) = (X^{24}-16)^3 - j(E_{L_{11}})X^{24} \in \mathbf{F}_p[X]$. We find that 60476019 is the root we need. Continuing like this, we get the following cycle for β in $\mathbf{F}_p$:

$$21677132 \xrightarrow{L_{11}} 60476019 \xrightarrow{L_{17}} \cdots \xrightarrow{L_{31}} 53004472 \xrightarrow{L_{41}} 21677132.$$

Just as in section 4, we lift $E/\mathbf{F}_p$ to $E_1/\mathbf{Q}_p$ by lifting the coefficients of its Weierstrass equation. We could now compute the canonical lift $\tilde{E}$ in two p-adic digits just like we did in section 4, and use that information to compute $\tilde{\beta} \in \mathbf{Z}_p$ in two p-adic digits accuracy. Indeed, $\tilde{\beta}$ is a root of $\Psi_{\mathfrak{f}}(X, j(\tilde{E})) \in \mathbf{Z}_p[X]$, and since we have $(\tilde{\beta} \bmod p) = \beta$, this root is just a Hensel lift of β. This approach has the disadvantage that we have to use modular polynomials for the j-function.

Instead, we lift β to a root β_1 of $\Psi_{\mathfrak{f}}(X, j(E_1)) \in \mathbf{Z}_p[X]$. Now we lift the cycle that we had for $\beta \in \mathbf{F}_p$ to a cycle for $\beta_1 \in \mathbf{Z}_p$ by applying the small modular polynomials for $\mathfrak{f}$ once more. Since we know that $\beta_1^{(\alpha)}$ is a root of $\Psi_{\mathfrak{f}}(X, j(E_1^{(\alpha)}))$, we can compute

$$j(E_1^{(\alpha)}) = \frac{((\beta_1^{(\alpha)})^{24} - 16)^3}{(\beta_1^{(\alpha)})^{24}}$$

and use this value to update $j(E_1)$ as in formula (1) to a value of the j-invariant of the canonical lift $\tilde{E}$ that is accurate to two p-adic digits. Knowing $j(\tilde{E}) \bmod p^2$, we can lift $\beta \in \mathbf{F}_p$ to a root of $\Psi_{\mathfrak{f}}(X, j(\tilde{E}))$ in $\mathbf{Z}_p$ that is accurate to two p-adic digits. We continue this process of doubling the precision until we have $\tilde{\beta}$ with sufficient accuracy. The first four cycles yield:

$$\begin{aligned}
\tilde{\beta} &= 21677132 + O(p) \\
&= 21677132 + 28966941p + O(p^2) \\
&= 21677132 + 28966941p + 7010373p^2 + 31182954p^3 + O(p^4) \\
&= 21677132 + 28966941p + 7010373p^2 + 31182954p^3 - 33808617p^4 \\
&\quad +27519307p^5 - 31601027p^6 - 36195013p^7 + O(p^8) \\
&= 21677132 + 28966941p + 7010373p^2 + 31182954p^3 - 33808617p^4 \\
&\quad +27519307p^5 - 31601027p^6 - 36195013p^7 - 8331811p^8 \\
&\quad -33957007p^9 - 18191700p^{10} + 5895954p^{11} - 42670221p^{12} \\
&\quad +23637278p^{13} - 40784695p^{14} + 7754196p^{15} + O(p^{16}).
\end{aligned}$$

We expect to need $\lceil 313618/72 \rceil = 4356$ decimals of accuracy, so we compute $\tilde{\beta}$ upto 550 p-adic digits. The class group $\mathrm{Cl}(D)$, which is cyclic of order 15610, is generated by a prime of norm 11. We can thus compute all the conjugates of $\tilde{\beta}$ under $\mathrm{Gal}(H/K)$ to 550 p-adic digits using the modular polynomial $\Phi_{11}^{\mathfrak{f}}$. In the end, we expand the polynomial of degree 15610 to find the class polynomial

$$F_D^{\mathfrak{f}} = \prod_{[I] \in \mathrm{Cl}(D)} (X - \tilde{\beta}^I) \in \mathbf{Z}[X],$$

which we reduce modulo q to compute a root $\gamma \in \mathbf{F}_q$. From γ, we compute the corresponding j-value and write down a curve $\bar{E}$ with that j-invariant. We then know that $\bar{E}$ has CM with $\mathcal{O}_D$, but we still need to check whether the trace really equals t. This turns out not to be the case, so we conclude that the quadratic twist of $\bar{E}$, given by

$$y^2 = x^3 + 669397215131271955483581235905x + 363369366443977510319399421188$$

over $\mathbf{F}_q$, with $q = 10^{30} + 1999999999167681$, has exactly 10^{30} rational points. Checking that this curve indeed has the required number of points is an easy matter for the current point counting algorithms.

References

[1] Bröker, R.: *Thesis*, Universiteit Leiden, in preparation.

[2] Cohen, H.: *A course in computational algebraic number theory*, Graduate Texts in Mathematics **138**, (1993).

[3] Couveignes, J.-M. and Henocq, T.: *Action of modular correspondences around CM points* in Algorithmic Number Theory Symposium V, Lecture Notes in Computer Science **2369**, (2002), 234–243.

[4] Gee, A.: *Class invariants by Shimura's reciprocity law*, Journal de Théorie des Nombres de Bordeaux **11**, (1999), 45–72.

[5] Kedlaya, K.: *Counting Points on Hyperelliptic Curves using Monsky-Washnitzer Cohomology*, Journal Ramanujan Mathematical Society **16**, (2002), 323–338.

[6] Lang, S.: *Elliptic functions*, 2nd edition Graduate Texts in Mathematics **112**, (1987).

[7] Satoh, T.: *The canonical lift of an ordinary elliptic curve over a finite field and its point counting*, Journal Ramanujan Mathematical Society **15**, (2000), 247-270.

[8] Schoof, R.: *Elliptic Curves over Finite Fields and the Computation of Square Roots mod p*, Mathematics of Computation **44**, (1985), 483–494.

[9] Schoof, R.: *Counting points on elliptic curves over finite fields*, Journal de Théorie des Nombres de Bordeaux **7**, (1995), 219–254.

[10] Stevenhagen, P.: *Hilbert's 12th problem, complex multiplication and Shimura reciprocity* in Class field theory—its centenary and prospect (Tokyo, 1998) Adv. Stud. Pure Math. **30**, Math. Soc. Japan, (2001), 161–176.

[11] Vélu, J.: *Isogénies entre courbes elliptiques*, C.R. Math. Acad. Sc. Paris **273**, (1971), 238–241.

[12] Weber, H.: *Lehrbuch der Algebra*, Vol. III. Chelsea reprint, original edition, 1908.

Rational Divisors in Rational Divisor Classes

N. Bruin*[1] and E.V. Flynn**[2]

[1] Department of Mathematics, Simon Fraser University
Burnaby, BC, Canada V5A 1S6
nbruin@cecm.sfu.ca

[2] Department of Mathematical Sciences, University of Liverpool
Liverpool L69 3BX, United Kingdom
evflynn@liverpool.ac.uk

Abstract. We discuss the situation where a curve $\mathcal{C}$, defined over a number field K, has a known K-rational divisor class of degree 1, and consider whether this class contains an actual K-rational divisor. When $\mathcal{C}$ has points everywhere locally, the local to global principle of the Brauer group gives the existence of such a divisor. In this situation, we give an alternative, more down to earth, approach, which indicates how to compute this divisor in certain situations. We also discuss examples where $\mathcal{C}$ does not have points everywhere locally, and where no such K-rational divisor is contained in the K-rational divisor class.

1 Introduction

The following result is typically proved as a direct consequence of the local to global principle of the Brauer group (see, for example, [2] or p.30 of [3]).

Lemma 1. *Let $\mathcal{C}$ be a curve defined over a number field K with points everywhere locally. Then any K-rational degree 1 divisor class $\mathcal{D}$ contains a K-rational divisor D.*

Such divisor classes have relevance to the application of second descents [4], as well as the application of the Brauer-Manin obstruction to higher genus curves, where such a class $\mathcal{D}$ is used to obtain an embedding $P \mapsto [P] - \mathcal{D}$ from $\mathcal{C}(K)$ to $J(K)$, the Mordell-Weil group of the Jacobian. This embedding can sometimes be used to find information about $\mathcal{C}(K)$.

Our intention here is to describe, in a concise and explicit manner, how the problem of finding a rational divisor in a rational divisor class corresponds to finding a rational point on a certain algebraic variety. We give an example of how this description can be used in practice to find a rational divisor explicitly, given a rational divisor class.

* Partially supported by an NSERC grant.
** Supported by EPSRC grant GR/R82975/01.

D.A. Buell (Ed.): ANTS 2004, LNCS 3076, pp. 132–139, 2004.

2 The Brauer-Severi Variety of a Divisor Class

Let $\mathcal{C}$ be a curve defined over a field K and let $\overline{K}$ be a separable closure of K. We say a divisor class $\mathcal{D} \in \mathrm{Pic}_{\mathcal{C}}(\overline{K})$ is *rational* if it is fixed under the action of $\mathrm{Gal}(\overline{K}/K)$. This means that for any divisor $D \in \mathcal{D}$ and $\sigma \in \mathrm{Gal}(\overline{K}/K)$, we have that ${}^{\sigma}D$ and D are linearly equivalent. In the language of Galois modules and Galois cohomology, we have

$$\mathrm{Pic}_{\mathcal{C}}(\overline{K})^K = H^0(K, \mathrm{Pic}_{\mathcal{C}}(\overline{K})) = \{\mathcal{D} \in \mathrm{Pic}_{\mathcal{C}}(\overline{K}) : \mathcal{D} \text{ is a rational divisor class}\}.$$

This group is different from $\mathrm{Pic}_{\mathcal{C}}(K)$, which simply consists of the linear equivalence classes in $\mathrm{Div}_{\mathcal{C}}(K)$. There is an obvious embedding $\mathrm{Pic}_{\mathcal{C}}(K) \subset \mathrm{Pic}_{\mathcal{C}}(\overline{K})^K$ and we identify $\mathrm{Pic}_{\mathcal{C}}(K)$ with its image.

For $D \in \mathrm{Div}_{\mathcal{C}}(\overline{K})$, we adopt the standard notation

$$\mathcal{L}(D) = \{f \in \overline{K}(\mathcal{C}) : (f) \geq -D\}. \tag{1}$$

This is a finite dimensional vector space over $\overline{K}$. We write $l(D)$ for its dimension. The Riemann-Roch theorem asserts that, for any divisor $D \in \mathrm{Div}_{\mathcal{C}}(\overline{K})$ and any canonical divisor κ of $\mathcal{C}$, we have $l(D) - l(\kappa - D) = \deg D - g + 1$. Furthermore $l(D)$ only depends on the equivalence class of D and we write $l([D]) = l(D)$.

We write

$$\mathcal{V}_D(\overline{K}) = \mathbb{P}\mathcal{L}(D) \tag{2}$$

for the complete linear system of D. Via the map $f \mapsto (f) + D$, we see that this set is in bijection with the set of effective divisors linearly equivalent to D:

$$\mathcal{V}_D(\overline{K}) \simeq \mathcal{V}_{[D]}(\overline{K}) := \{D' \in \mathrm{Div}_{\mathcal{C}}(\overline{K}) : D' \geq 0 \text{ and } D' \sim D\}. \tag{3}$$

Let $\mathcal{D}$ be a rational divisor class with $l(\mathcal{D}) > 0$. Then $\mathcal{V}_{\mathcal{D}}(\overline{K})$ is fixed under $\mathrm{Gal}(\overline{K}/K)$ and has the structure of a Galois set. Generalizing the above notation, for an extension L of K, we write $\mathcal{V}_{\mathcal{D}}(L)$ for the effective divisors of $\mathcal{C}$ defined over L and in $\mathcal{D}$.

The functor $\mathcal{V}_{\mathcal{D}}$ is represented by a scheme over K, which is called the *Brauer-Severi variety* associated to $\mathcal{D}$.

It follows that the following are equivalent.

- $\mathcal{V}_{\mathcal{D}}(K) \neq \emptyset$
- $\mathcal{V}_{\mathcal{D}}(K) \simeq \mathbb{P}^{(l(\mathcal{D})-1)}(K)$.
- $\mathcal{D}$ contains a rational divisor.

In some cases it is easy to see that the conditions above hold. For instance, if $\mathcal{D}$ is a rational divisor class with $l(\mathcal{D}) = 1$, then over $\overline{K}$, there is a unique effective divisor D representing $\mathcal{D}$. Consequently, D is fixed under $\mathrm{Gal}(\overline{K}/K)$ and therefore is itself rational.

If a curve has a rational point P_0 then the property above is sufficient to deduce that any rational divisor class contains rational divisors. We use that for any divisor D and point P the following inequalities hold:

$$l(D) \leq l(D + P) \leq l(D) + 1. \tag{4}$$

It follows from (4) that for any divisor class $\mathcal{D}$ there exists an integer n such that $l(\mathcal{D}+[nP_0]) = 1$. The argument above shows that the rational divisor class $\mathcal{D}+[nP_0]$ contains a rational divisor D and therefore $D - nP_0 \in \mathcal{D}$.

In particular, if a curve $\mathcal{C}$ over a number field K has points everywhere locally and $l(\mathcal{D}) > 0$ then $\mathcal{V}_\mathcal{D}$ has rational points everywhere locally. Hence, Lemma 1 is equivalent to the assertion that the local-to-global principle applies to the Brauer-Severi varieties $\mathcal{V}_\mathcal{D}$. For $l(\mathcal{D}) = 2$, we have that $\mathcal{V}_\mathcal{D}$ is a curve of genus 0, and the Hasse-Minkowski theorem confirms that such varieties obey a local-to-global principle.

In fact, without assuming local solvability of $\mathcal{C}$, the geometry of $\mathcal{V}_\mathcal{D}$ still allows deductions to be made about representability of divisor classes by rational divisors. For instance, any $\mathcal{D}$ with $l(\mathcal{D}) = 2$ is representable by a divisor over a quadratic extension of K, since any curve of genus 0 has quadratic points.

We now sketch how one can proceed, given a rational divisor class $\mathcal{D}$, to derive an explicit model of $\mathcal{V}_\mathcal{D}$ in such a way that finding a rational point on it allows the construction of a representing rational divisor. Suppose that $\mathcal{D}$ is represented by a divisor D over some finite extension $L = K(\alpha)$ of K of degree, say, d.

1. Determine a basis $f_1, \dots, f_{l(D)} \in L(\mathcal{C})$ of $\mathcal{L}(D)$.
2. Over L, we have $\mathcal{V}_\mathcal{D} \simeq \mathbb{P}^{l(D)-1}$ via the inverse of the map
$$(t_1 : \cdots : t_{l(D)}) \mapsto D + (t_1 f_1 + \cdots + t_{l(D)} f_{l(D)}).$$
This establishes a model of $\mathcal{V}_\mathcal{D}$ over L, with $(t_1 : \ldots : t_{l(D)})$ as projective coordinates. Putting $t_1 = 1$ yields an affine chart $(t_2, \dots, t_{l(D)})$ of $\mathcal{V}_\mathcal{D}$ over L.
3. In order to descend our model of $\mathcal{V}_\mathcal{D}$ over L to a model over K, we compute a representation of $D_t = D+(f_1+t_2f_2+\cdots t_{l(D)}f_{l(D)}) \in \mathrm{Div}_\mathcal{C}(L(t_2,\dots,t_{l(D)}))$ corresponding to the generic point $(1 : t_2 : \ldots : t_{l(D)})$ on $\mathcal{V}_\mathcal{D}$ as a scheme over L.
Let $\mathcal{C}$ be given as a plane curve with coordinates X and Y in general position over K. The effective divisor D_t can be described by the equations $g(X) = 0, Y = h(X)$, with $g, h \in L(t_2, \dots, t_{l(D)})[X]$, with g monic, $\deg(g) = \deg(D)$ and $\deg(h) = \deg(D) - 1$, since we have taken X and Y such that degeneracies do not occur.
4. We substitute $t_i = \sum_{j=0}^{d-1} t_{i,j}\alpha^j$ and write
$$g(X) = \sum_{k=0}^{d-1} g_k(X)\alpha^j, \text{ where } g_k(X) \in K(\{t_{i,j}\}_{i,j})[X],$$
and similarly for $h(X)$. A point $(1 : t_1 : \cdots : t_{l(D)})$ corresponds to a divisor over K precisely when that divisor can be described by equations not involving α. Thus, we are led to consider the equations obtained by insisting that $g_k(X) = h_k(X) = 0$ for $k = 1, \dots, d-1$ as polynomials in X. Those equations define an affine chart of $\mathcal{V}_\mathcal{D}$ over K, with coordinates $t_{i,j}$ over K. To get a model of $\mathcal{V}_\mathcal{D}$, one can take the projective closure.

5. If one finds a point $(t_{i,j})$, one can reconstruct g, h from these values and obtain a description of a K-rational divisor in $\mathcal{D}$. Equivalently, one can reconstruct $(1 : t_2 : \cdots : t_{l(D)})$ and thus obtain a K-rational specialization of D_t.

Of course, the procedure described above only applies to divisor classes satisfying $l(\mathcal{D}) > 0$. In general, one should select some divisor D_0 over K, preferrably of minimal positive degree, and an integer n such that $l(\mathcal{D} + [nD_0])$ is minimally positive. One can then apply the procedure to that divisor class and derive a suitable representative of $\mathcal{D}$ from the result, or conclude that none exists.

3 Finding a Rational Divisor on a Curve of Genus 2

We will give an example of this for a genus 2 curve

$$\mathcal{C} : Y^2 = f_6X^6 + f_5X^5 + \cdots + f_0 = F(X), \text{ with } F(X) \in K[X] \tag{5}$$

and a rational divisor class $\mathcal{B}$ of degree 3, represented by an effective divisor defined over a quadratic extension of K.

We assume that $\mathcal{B}$ is a rational divisor class with $\mathcal{B} = [P_1 + P_2 + P_3]$ and $P_1, P_2, P_3 \in \mathcal{C}(K(\sqrt{d}))$, where the P_i are not all Weierstrass points. It follows that $l(\mathcal{B}) = 2$. We can arrive at an equation for $\mathcal{V}_\mathcal{B}$ in the by following the outline in Section 2. We will give an account that can be read independently, but point out the correspondences with the general algorithm.

Let $G_t(x) \in K(\sqrt{d})[t][x]$ be the cubic in x such that $y = G_t(x)$ passes through the points P_1, P_2, P_3 for all values of t. For any value $t \in K(\sqrt{d})$ we have

$$\frac{(x - x_1)(x - x_2)(x - x_3)}{y - G_t(x)} \in \mathcal{L}(\mathcal{B}).$$

This allows us to write down a basis of $\mathcal{L}(\mathcal{B})$ for Step 1 in Section 2. We find $f_1 = 1$, $f_2 = \frac{(x-x_1)(x-x_2)(x-x_3)}{y-G_0(x)}$. We get the coordinates $(1 : t)$ on $\mathcal{L}(\mathcal{B})$.

For any value of t, we have that the identity

$$\{y = G(x)\} \cap \mathcal{C} = P_1 + P_2 + P_3 + (x_1, y_1) + (x_2, y_2) + (x_3, y_3) \simeq 3\mathcal{O}, \tag{6}$$

where $\mathcal{O}$ is any canonical divisor[1] of $\mathcal{C}$. The divisor $(x_1, y_1) + (x_2, y_2) + (x_3, y_3)$ can be described by $\{C_t(x) = 0, y = H_t(x)\}$, where $C_t(x), H_t(x) \in K(\sqrt{d})[t][x]$ are a monic cubic and a quadratic in x respectively. This is the required description for Step 3 of Section 2.

Substituting $t = t_1 + \sqrt{d}\, t_2$, we get that the coefficients c_i of $C_{t_1+\sqrt{d}\,t_2}$ are of the form $c_i = c_{i,0}(t_1, t_2) + \sqrt{d}\, c_{i,1}(t_1, t_2)$, with $c_{i,j} \in K[t_1, t_2]$, and similarly for H_t. Setting the $\sqrt{d}$-components to 0 yields 6 equations in t_1, t_2 over K, which

[1] The divisor $\mathcal{O} = \infty^+ + \infty^-$, consisting of the intersection of $\mathcal{C}$ with $X = \infty$, is a popular choice among people computing with curves of genus 2.

describe the genus 0 curve $\mathcal{V}_\mathcal{B}$. This corresponds to Step 4 of Section 2, where we put $t_2 = t_{2,1} + \sqrt{d}\,t_{2,2}$.

If $\mathcal{C}$ has points everywhere locally, then Lemma 1 asserts that $\mathcal{V}_\mathcal{B}$ has a rational point. Let $C_{t_0}(x), H_{t_0}(x)$ correspond to such a point. Then the divisor class $\mathcal{B}$ is represented by the rational divisor $B = 3\mathcal{O} - \{C_{t_0}(x) = 0, y = H_{t_0}(x)\}$. This corresponds to Step 5 of Section 2.

Of course, since there exists $\mathcal{O}$, which is of degree 2 and defined over K, the above can be applied to the rational divisor class $\mathcal{D} = [P_1 + P_2 + P_3 - \mathcal{O}]$ of degree 1, which is represented by the rational divisor $D = B - \mathcal{O}$.

We illustrate the above ideas with a detailed worked example. We first observe how a K-rational divisor class can arise naturally, in such a way that the contained K-rational divisor is not immediately apparent.

Let $\mathcal{C}$ be the genus 2 curve, defined over $\mathbb{Q}$,

$$\mathcal{C} : Y^2 = F(X) = -X^6 - X^5 - 2X^4 - 2X^3 + X^2 - 2X + 2, \tag{7}$$

which is easily checked to have points in $\mathbb{R}$ and every $\mathbb{Q}_p$. One can perform a 2-descent on the Mordell-Weil group $\mathcal{J}(\mathbb{Q})$ of the Jacobian, as described in [6], using the map

$$\mu : \mathcal{J}(\mathbb{Q}) \to \mathbb{Q}(\theta)^*/\mathbb{Q}^*(\mathbb{Q}(\theta)^*)^2 : [\sum(x_i, y_i)] \mapsto \prod(x_i - \theta), \tag{8}$$

where θ is a root of $F(X)$. One of the steps of the 2-descent on $\mathcal{J}(\mathbb{Q})$ is the computation of the kernel of μ, generated by $2\mathcal{J}(\mathbb{Q})$ and $[P_1 + P_1' - \mathcal{O}]$, where $P_1 = (\frac{1}{2} + \frac{1}{5}\alpha, \frac{7}{40} + \frac{12}{25}\alpha)$ and $\alpha = \sqrt{-55}$, with P_1' denoting the conjugate of P_1 with respect to $\mathbb{Q}(\alpha)/\mathbb{Q}$. Let P_1^- be the hyperelliptic involute of P_1. Then general properties of μ (see [6] or p.55 of [3]) guarantee that $[P_1^- + P_1' - \mathcal{O}] \in 2\mathcal{J}(\mathbb{Q}(\alpha))$, and computing preimages under the multiplication by 2 map on $\mathcal{J}(\mathbb{Q}(\alpha))$, one finds $\mathcal{D}_1 = [(1+\sqrt{-5}, 2\alpha) + (1-\sqrt{-5}, 2\alpha) - \mathcal{O}]$ which satisfies $[P_1^- + P_1' - \mathcal{O}] = 2\mathcal{D}_1$. Clearly $\mathcal{D}_1' = -\mathcal{D}_1$, since conjugation merely negates the y-coordinates. Let

$$\begin{aligned} &P_1 = (\tfrac{1}{2} + \tfrac{1}{5}\alpha, \tfrac{7}{40} + \tfrac{12}{25}\alpha), P_2 = (1 + \sqrt{-5}, 2\alpha), P_3 = (1 - \sqrt{-5}, 2\alpha), \\ &\mathcal{D} = [P_1] + \mathcal{D}_1 = [P_1 + P_2 + P_3 - \mathcal{O}], \text{ where } \alpha = \sqrt{-55}. \end{aligned} \tag{9}$$

Then

$$\mathcal{D}' = [P_1'] + \mathcal{D}_1' = [P_1'] - \mathcal{D}_1 = [P_1'] - 2\mathcal{D}_1 + \mathcal{D}_1 = [P_1'] - [P_1^- + P_1' - \mathcal{O}] + \mathcal{D}_1 = \mathcal{D},$$

so that $\mathcal{D}$ is defined over $\mathbb{Q}$. We now have a naturally occurring divisor class $\mathcal{D}$ of degree 1, which is defined over $\mathbb{Q}$, but whose naturally occurring representative $P_1 + P_2 + P_3 - \mathcal{O}$ is not itself defined over $\mathbb{Q}$. This is a common outcome of an application of 2-descent on $\mathcal{J}(\mathbb{Q})$ for genus 2 curves. Note that such $\mathcal{D}$ are of some interest, as they allow an embedding of $\mathcal{C}(\mathbb{Q})$ into $\mathcal{J}(\mathbb{Q})$ via the map $P \mapsto [P] - \mathcal{D}$, even when no obvious member of $\mathcal{C}(\mathbb{Q})$ is available.

Since our curve has points everywhere locally, we know that $\mathcal{D}$ does contain an actual $\mathbb{Q}$-rational divisor. We now illustrate how this can be found in practice. One first finds the general $Y = G(X)$, through P_1, P_2, P_3, where $G(X)$ is cubic in X, and where there is a free parameter t, since we have one less than the number of points required to define the cubic uniquely. This parametrized family of cubics is

$$\begin{aligned} G(X) = &tX^3 + (\tfrac{3}{2} - \tfrac{5t}{2} - \tfrac{2\alpha}{5} - \tfrac{t\alpha}{5})X^2 \\ &+(-3 + 7t + \tfrac{4\alpha}{5} + \tfrac{2t\alpha}{5})X + 9 - 3t - \tfrac{2\alpha}{5} - \tfrac{6t\alpha}{5}. \end{aligned} \tag{10}$$

Computing $G(X)^2 - F(X)$, where $F(X)$ is as in (7), and removing the cubic factor $(X - X(P_1))(X - X(P_2))(X - X(P_3))$ leaves the residual cubic

$$\begin{aligned} C(X) = &10(1 + t^2)X^3 + (35 + 2\alpha + 30t - 8t\alpha - 25t^2 - 2t^2\alpha)X^2 \\ &+(-50 - 4\alpha - 60t + 16t\alpha + 70t^2 + 4t^2\alpha)X \\ &+30 + 12\alpha + 180t - 8t\alpha - 30t^2 - 12t^2\alpha. \end{aligned} \tag{11}$$

Let $(x_1, y_1) + (x_2, y_2) + (x_3, y_3) - \mathcal{O} \in \mathcal{D}$. Then x_1, x_2, x_3 are the roots of $C(X)$, for some choice of $t \in \mathbb{Q}(\alpha)$. Furthermore, the $y_i = -G(x_i)$, where $G(X)$ is as in (10). We now compute the quadratic $Y = H(X)$ which passes through $(x_1, -G(x_1)), (x_2, -G(x_2)), (x_3, -G(x_3))$, namely

$$\begin{aligned} H(X) = &((15 - 60t - 15t^2 - 4\alpha - 4t\alpha + 4t^2\alpha)X^2 \\ &+(-30 + 120t + 30t^2 + 8\alpha + 8t\alpha - 8t^2\alpha)X \\ &+90 - 60t - 90t^2 - 4\alpha - 24t\alpha + 4t^2\alpha)/(10 + 10t^2). \end{aligned} \tag{12}$$

We have now parametrized divisors of the form $(x_1, y_1) + (x_2, y_2) + (x_3, y_3) - \mathcal{O}$ which are in $\mathcal{D}$. Our requirement for $(x_1, y_1) + (x_2, y_2) + (x_3, y_3) - \mathcal{O}$ to be $\mathbb{Q}$-rational is the same as requiring that the ratios of the coefficients of $C(x)$ are $\mathbb{Q}$-rational (giving six polynomials in t, which can be reduced to three polynomials on the assumption that a given cofficient of $C(x)$ is nonzero; however, we shall prefer to write out all six polynomials in full), and that the actual coefficients of $H(x)$ are also $\mathbb{Q}$-rational (giving three polynomials in t). This gives in total nine polynomials in t total that must be $\mathbb{Q}$-rational. Let $t = t_1 + t_2\alpha$. Then on computing the coefficients on α in our nine expressions, we find nine quartics in t_1, t_2, all with a common factor of $t_1^2 + 55t_2^2 + 15t_2 + 1$. Then

$$t_1^2 + 55t_2^2 + 15t_2 + 1 = 0 \tag{13}$$

is our desired curve of genus 0, which has points everywhere locally, and hence globally. The solution of smallest height is $t_1 = 1/7, t_2 = -1/7$, corresponding to $t = 1/7 - \alpha/7$. Substituting this into $C(X)$ and $H(X)$, and removing from $C(X)$ a factor of $-5(5 + 2\alpha)/49$ (permissible, since the roots of $C(X)$ are unaffected), gives

$$C(X) = 2X^3 + X^2 + 2X + 2, \quad H(X) = -\frac{1}{2}X^2 + X - 1. \tag{14}$$

These $C(X), H(X)$ define $(x_1, y_1) + (x_2, y_2) + (x_3, y_3) - \mathcal{O}$, which is our desired $\mathbb{Q}$-rational divisor in our given $\mathbb{Q}$-rational divisor class $\mathcal{D}$. We summarize the above as follows.

Example 1. *Let $\mathcal{C}$ be the curve (7) and $\mathcal{D}$ be the $\mathbb{Q}$-rational divisor class in (9). Then $\mathcal{D}$ contains the $\mathbb{Q}$-rational divisor $(x_1, y_1) + (x_2, y_2) + (x_3, y_3) - \mathcal{O}$, where x_1, x_2, x_3 are the roots of $C(X)$ and $y_i = H(x_i)$ for $i = 1, 2, 3$, with $C(X), H(X)$ as in (14).*

We have written a Maple file at [1] which performs the above computation for any curve of genus 2.

4 Examples Where the Rational Divisor Class Contains No Rational Divisor

Suppose now that our genus 2 curve is defined over a number field K and is of the form

$$\begin{aligned} &\mathcal{H} : Y^2 = kF_1(X)F_2(X), \text{ with } F_1(X) = a_3X^3 + a_2X^2 + a_1X + a_0, \\ &k \in K^*, \text{ each } a_i = g_i + h_i\sqrt{d} \in K(\sqrt{d}), \text{ and } F_1(X), F_2(X) \text{ conjugate.} \end{aligned} \tag{15}$$

We shall also assume that $F_1(X)$ is not defined over K. Let e_1, e_2, e_3 denote the roots of $F_1(X)$. This is another situation where we have a naturally occurring degree 1 divisor class $\mathcal{D} = [(e_1, 0)+(e_2, 0)+(e_3, 0)-\mathcal{O}]$ which is defined over K, even though the given representative is defined over $K(\sqrt{d})$ and not generally over K. If $\mathcal{H}$ has points everywhere locally, we know that $\mathcal{D}$ must contain a divisor defined over K, but we do not make that assumption in this section. The parametrized familiy of cubics through $(e_1, 0), (e_2, 0), (e_3, 0)$ is simply $Y = G(X) = tF_1(X)$, where $t = t_1+t_2\sqrt{d}$ and t_1, t_2 are K-rational parameters. Replacing Y by $tF_1(X)$ in $Y^2 - kF_1(X)F_2(X)$ and removing the known factor $F_1(X)$ gives the residual cubic $C(X) = t^2F_1(X) - kF_2(X)$. Let $(x_1, y_1) + (x_2, y_2) + (x_3, y_3) - \mathcal{O} \in \mathcal{D}$. Then x_1, x_2, x_3 are the roots of $C(X)$, for some choice of $t \in K(\sqrt{d})$. Furthermore, each $y_i = -G(x_i)$; we compute the quadratic $Y = H(X)$ passing through the $(x_i, -G(x_i))$, namely

$$\begin{aligned} H(X) = 2k\ell\big((g_2h_3 - g_3h_2)X^2 + (g_1h_3 - g_3h_1)X + g_0h_3 - g_3h_0\big), \text{ where} \\ \ell = (-k + t_2^2d - t_1^2)(g_3t_2 + h_3t_1)d - (k + t_2^2d - t_1^2)(g_3t_1 + h_3t_2d)\sqrt{d}. \end{aligned} \tag{16}$$

Now, suppose that $(x_1, y_1) + (x_2, y_2) + (x_3, y_3) - \mathcal{O}$ is defined over K. Then the ratios of coefficients of $C(X)$ must be in K, giving the six equations

$$(k+t_2^2d-t_1^2)(-k+t_2^2d-t_1^2)(g_ih_j-g_jh_i) = 0, \text{ for } \{i, j\} \subset \{0, 1, 2, 3\},\ i < j. \tag{17}$$

Furthermore, the coefficients of $H(X)$ must be in K, giving the three equations

$$(k + t_2^2d - t_1^2)(g_3t_1 + h_3t_2d)(g_ih_3 - g_3h_i) = 0, \text{ for } i = 0, 1, 2. \tag{18}$$

Inspecting (17), we note that we cannot have all $g_ih_j - g_jh_i = 0$, since then our original curve $\mathcal{H}$ would have zero discriminant. So, one of the first two factors in (17) must be 0, giving that $\text{Norm}(t) = \pm k$. If $\text{Norm}(t) = -k$ then a similar argument (we have placed the details in the file [1]) applied to the equations (18) shows that either $\mathcal{H}$ has zero discriminant (which is not permitted) or that F_1, F_2 are each defined over K (which is also not permitted). In summary, if $\mathcal{D}$ contains a K-rational divisor then we must have $\text{Norm}(t) = k$ for some t.

Conversely, if $\text{Norm}(t) = k$ for some $t \in K(\sqrt{d})$, then the K-rational divisor $(x_1, y_1) + (x_2, y_2) + (x_3, y_3) - \mathcal{O}$ in $\mathcal{D}$ is defined by

$$C_0(X) = tF_1 - t'F_2(X), \quad H_0(X) = -4k^2 d(g_3 t_2 + h_3 t_1) \sum_{i=0}^{2} (g_i h_3 - g_3 h_i) X^i. \quad (19)$$

We summarize this as follows.

Lemma 2. *Let $\mathcal{H} : Y^2 = kF_1(X)F_2(X)$ be a curve of genus 2 of the type (15), defined over a number field K, where $F_1(X)$ is defined over $K(\sqrt{d})$ and not over K. Let $\mathcal{D}$ be the K-rational divisor class $[(e_1, 0) + (e_2, 0) + (e_3, 0) - \mathcal{O}]$, where e_1, e_2, e_3 are the roots of $F_1(X)$. Then $\mathcal{D}$ contains a K-rational divisor if and only if $\text{Norm}(t) = k$ for some $t \in K(\sqrt{d})$, in which case the required divisor is $(x_1, y_1) + (x_2, y_2) + (x_3, y_3) - \mathcal{O}$, where x_1, x_2, x_3 are the roots of $C_0(X)$ and each $y_i = H_0(x_i)$, with C_0, H_0 as in (19).*

In this situation, our genus 0 curve is simply the curve $t_1^2 - dt_2^2 = k$. Of course, any choice of K, d, k (such as $\mathbb{Q}, 2, 5$), where k is not a norm in $K(\sqrt{d})$, will give an example where $\mathcal{D}$ does not contain a K-rational divisor.

References

1. N. Bruin and E.V. Flynn. Maple programs for computing rational divisors in rational divisor classes. Available at ftp://ftp.liv.ac.uk/pub/genus2/maple/ratdiv
2. D. Coray and C. Manoil. On large Picard groups and the Hasse principle for curves and K3 surfaces. *Acta. Arith.*, **LXXVI.2** (1996), 165–189.
3. J.W.S. Cassels and E.V. Flynn. *Prolegomena to a Middlebrow Arithmetic of Curves of Genus 2*. LMS–LNS 230. Cambridge University Press, Cambridge, 1996.
4. J.W.S. Cassels. Second descents for elliptic curves. *J. reine angew. Math.* **494** (1998), 101–127.
5. V. Scharaschkin. Local Global Problems and the Brauer-Manin Obstruction. PhD Thesis, University of Michigan, 1999.
6. M. Stoll. Implementing 2-descent for Jacobians of hyperelliptic curves. *Acta Arith.* **98** (2001), 245–277.

Conjectures about Discriminants of Hecke Algebras of Prime Level

Frank Calegari[***1] and William A. Stein[†2]

[1] Harvard University
fcale@math.harvard.edu
http://www.math.harvard.edu/~{}fcale
[2] Harvard University,
was@math.harvard.edu
http://modular.fas.harvard.edu/

Abstract. In this paper, we study p-divisibility of discriminants of Hecke algebras associated to spaces of cusp forms of prime level. By considering cusp forms of weight bigger than 2, we are are led to make a precise conjecture about indexes of Hecke algebras in their normalisation which implies (if true) the surprising conjecture that there are no mod p congruences between non-conjugate newforms in $S_2(\Gamma_0(p))$, but there are almost always many such congruences when the weight is bigger than 2.

1 Basic Definitions

We first recall some commutative algebra related to discriminants, then introduce Hecke algebras of spaces of cusp forms.

1.1 Commutative Algebra

In this section we recall the definition of discriminant of a finite algebra and note that the discriminant is nonzero if and only if no base extension of the algebra contains nilpotents.

Let R be a ring and let A be an R-algebra that is free of finite rank as an R-module. The *trace* of $x \in A$ is the trace, in the sense of linear algebra, of left multiplication by x.

Definition 1 (Discriminant). *Let $\omega_1, \ldots, \omega_n$ be an R-basis for A. Then the* discriminant $\mathrm{disc}(A)$ *of A is the determinant of the $n \times n$ matrix $(\mathrm{tr}(\omega_i \omega_j))$.*

The discriminant is only well-defined modulo squares of units in R. When $R = \mathbf{Z}$ the discriminant is well defined, since the only units are ± 1.

We say that A is *separable over* R if for every extension R' of R, the ring $A \otimes R'$ contains no nilpotents.

*** Supported in part by the American Institute of Mathematics
† Supported in part by a National Science Foundation Postdoctoral Fellowship

D.A. Buell (Ed.): ANTS 2004, LNCS 3076, pp. 140–152, 2004.

Proposition 1. *Suppose R is a field. Then A has nonzero discriminant if and only if A is separable over R.*

Proof. For the convenience of the reader, we summarize the proof in [Mat86, §26]. If A contains a nilpotent then that nilpotent is in the kernel of the trace pairing, so the discriminant is 0. Conversely, if A is separable then we may assume that R is algebraically closed. Then A is an Artinian reduced ring, hence isomorphic as a ring to a finite product of copies of R, since R is algebraically closed. Thus the trace form on A is nondegenerate.

1.2 The Discriminant Valuation

We next introduce Hecke algebras attached to certain spaces of cusp forms of prime level p, define the discriminant valuation as the exponent of the largest power of p that divides the discriminant, and observe that there are eigenform congruences modulo p exactly when the discriminant valuation is positive. We then present an example to illustrate the definitions.

Let Γ be a congruence subgroup of $\mathrm{SL}_2(\mathbf{Z})$. In this paper, we will only consider $\Gamma = \Gamma_0(p)$ for p prime. For any positive integer k, let $S_k(\Gamma)$ denote the space of holomorphic weight k cusp forms for Γ. Let

$$\mathbf{T} = \mathbf{Z}[\ldots, T_n, \ldots] \subset \mathrm{End}(S_k(\Gamma))$$

be the associated Hecke algebra, which is generated by Hecke operators T_n for all integers n, including $n = p$ (we will sometimes write U_p for T_p). Then $\mathbf{T}$ is a commutative ring that is free as a module over $\mathbf{Z}$ of rank equal to $\dim S_k(\Gamma)$. We will also sometimes consider the image $\mathbf{T}^{\mathrm{new}}$ of $\mathbf{T}$ in $\mathrm{End}(S_k(\Gamma)^{\mathrm{new}})$.

Definition 2 (Discriminant Valuation). *Let p be a prime, k a positive integer, and suppose that $\Gamma = \Gamma_0(p)$. Let $\mathbf{T}$ be the corresponding Hecke algebra. Then the* discriminant valuation *of Γ in weight k is*

$$d_k(\Gamma) = \mathrm{ord}_p(\mathrm{disc}(\mathbf{T})).$$

We expect that $d_k(\Gamma)$ is finite for the following reason. The Hecke operators T_n, with n not divisible by p, are diagonalizable since they are self adjoint with respect to the Petersson inner product. When $k = 2$ one knows that U_p is diagonalizable since the level is square free, and when $k > 2$ one expects this (see [CE98]). If $\mathbf{T}$ contains no nilpotents, Proposition 1 implies that the discriminant of $\mathbf{T}$ is nonzero. Thus $d_k(\Gamma)$ is finite when $k = 2$ and conjectured to be finite when $k > 2$.

Let p be a prime and suppose that $\Gamma = \Gamma_0(p)$. A *normalised eigenform* is an element $f = \sum a_n q^n \in S_k(\Gamma)$ that is an eigenvector for *all* Hecke operators T_ℓ, including those that divide p, normalised so that $a_1 = 1$. The quantity $d_k(\Gamma)$ is of interest because it measures mod p congruences between normalised eigenforms in $S_k(\Gamma)$.

Proposition 2. *Assume that $d_k(\Gamma)$ is finite. The discriminant valuation $d_k(\Gamma)$ is positive (i.e., the discriminant is divisible by p) if and only if there is a congruence in characteristic p between two normalized eigenforms in $S_k(\Gamma)$. (The two congruent eigenforms might be Galois conjugate.)*

Proof. It follows from Proposition 1 that $d_k(\Gamma) > 0$ if and only if $\mathbf{T} \otimes \overline{\mathbf{F}}_p$ is not separable. The Artinian ring $\mathbf{T} \otimes \overline{\mathbf{F}}_p$ is not separable if and only if the number of ring homomorphisms $\mathbf{T} \otimes \overline{\mathbf{F}}_p \to \overline{\mathbf{F}}_p$ is less than

$$\dim_{\overline{\mathbf{F}}_p} \mathbf{T} \otimes \overline{\mathbf{F}}_p = \dim_{\mathbf{C}} S_k(\Gamma).$$

Since $d_k(\Gamma)$ is finite, the number of ring homomorphisms $\mathbf{T} \otimes \overline{\mathbf{Q}}_p \to \overline{\mathbf{Q}}_p$ equals $\dim_{\mathbf{C}} S_k(\Gamma)$. The proposition follows from the fact that for any ring R, there is a bijection between ring homomorphisms $\mathbf{T} \to R$ and normalised eigenforms with q-expansion in R.

The same proof also shows that a prime ℓ divides the discriminant of $\mathbf{T}$ if and only if there is a congruence mod ℓ between two normalized eigenforms in $S_k(\Gamma)$

Example 1. If $\Gamma = \Gamma_0(389)$ and $k = 2$, then $\dim_{\mathbf{C}} S_2(\Gamma) = 32$. Let f be the characteristic polynomial of T_2. One can check that f is square free and 389 exactly divides the discriminant of f. This implies that $d_2(\Gamma) = 1$ and that T_2 generates $\mathbf{T} \otimes \mathbf{Z}_{389}$ as an algebra over $\mathbf{Z}_{389}$. (If T_2 only generated a subring of $\mathbf{T} \otimes \mathbf{Z}_{389}$ of finite index > 1, then the discriminant of f would be divisible by 389^2.)

Modulo 389 the characteristic polynomial f is congruent to

$$\begin{aligned}&(x+2)(x+56)(x+135)(x+158)(x+175)^2(x+315)(x+342)(x^2+387)\\&(x^2+97x+164)(x^2+231x+64)(x^2+286x+63)(x^5+88x^4+196x^3+\\&113x^2+168x+349)(x^{11}+276x^{10}+182x^9+13x^8+298x^7+316x^6+\\&213x^5+248x^4+108x^3+283x^2+x+101)\end{aligned}$$

The factor $(x+175)^2$ indicates that $\mathbf{T} \otimes \mathbf{F}_{389}$ is not separable over $\mathbf{F}_{389}$ since the image of $(\overline{f}/(x+175))(T_2)$ in $\mathbf{T} \otimes \mathbf{F}_{389}$ is nilpotent (it is nonzero but its square is 0). There are 32 eigenforms over $\mathbf{Q}_2$ but only 31 mod 389 eigenforms, so there must be a congruence. There is a newform F in $S_2(\Gamma_0(389), \overline{\mathbf{Z}}_{389})$ whose a_2 term is a root of

$$x^2 + (-39 + 190 \cdot 389 + 96 \cdot 389^2 + \cdots)x + (-106 + 43 \cdot 389 + 19 \cdot 389^2 + \cdots).$$

There is a congruence between F and its $\mathrm{Gal}(\overline{\mathbf{Q}}_{389}/\mathbf{Q}_{389})$-conjugate.

2 Computing Discriminants

In this section we sketch the algorithm that we use for computing the discriminants mentioned in this paper.

This algorithm was inspired by a discussion of the second author with Hendrik Lenstra. We leave the details of converting the description below into standard matrix operations to the reader. Also, the modular symbols algorithms needed to compute Hecke operators are quite involved.

Let $\Gamma = \Gamma_0(p)$, and let $k \geq 2$ be an integer. The following sketches an algorithm for computing the discriminant of the Hecke algebra $\mathbf{T}$ acting on $S_k(\Gamma)$.

1. For any given n, we can explicitly compute a matrix that represents the action of Hecke operators T_n on $S_k(\Gamma)$ using modular symbols. We use the second author's MAGMA [BCP97] packages for computing with modular symbols, which builds on work of many people (including [Cre97] and [Mer94]).
2. Using the Sturm bound, as described in the appendix to [LS02], find an integer b such that $T_1, \ldots, T_b$ generate $\mathbf{T}$ as a $\mathbf{Z}$-module. (The integer b is $\lceil (k/12) \cdot [\mathrm{SL}_2(\mathbf{Z}) : \Gamma] \rceil$.)
3. Find a subset B of the T_i that form a $\mathbf{Q}$-basis for $\mathbf{T} \otimes_{\mathbf{Z}} \mathbf{Q}$. (This uses Gauss elimination.)
4. View $\mathbf{T}$ as a ring of matrices acting on $\mathbf{Q}^d$, where $d = \dim(S_k(\Gamma))$ and try random sparse vectors $v \in \mathbf{Q}^d$ until we find one such that the set of vectors $C = \{T(v) : T \in B\}$ are linearly independent.
5. Write each of $T_1(v), \ldots, T_b(v)$ as $\mathbf{Q}$-linear combinations of the elements of C.
6. Find a $\mathbf{Z}$-basis D for the $\mathbf{Z}$-span of these $\mathbf{Q}$-linear combinations of elements of C. (This basis D corresponds to a $\mathbf{Z}$-basis for $\mathbf{T}$, but is much easier to find that directly looking for a $\mathbf{Z}$-basis in the space of $d \times d$ matrices that $\mathbf{T}$ is naturally computed in.)
7. Unwinding what we have done in the previous steps, find the trace pairing on the elements of D, and deduce the discriminant of $\mathbf{T}$ by computing the determinant of the trace pairing matrix.

A very time-consuming step, at least in our implementation, is computing D from $T_1(v), \ldots, T_b(v)$ expressed in terms of C, and this explains why we embed $\mathbf{T}$ in $\mathbf{Q}^d$ instead of viewing the elements of $\mathbf{T}$ as vectors in $\mathbf{Q}^{d \times d}$.

An implementation by the second author of the above algorithm is included with the MAGMA computer algebra system. The relevant source code is in the file `Geometry/ModSym/linalg.m` in the `package` directory (or ask the second author of the apper to send you a copy `linalg.m`). We illustrate the use of MAGMA to compute discriminants below, which were run under MAGMA V2.10-21 for Linux on a computer with an Athlon 2800MP processor (2.1Ghz).

```
> M := ModularSymbols(389,2, +1);
> S := CuspidalSubspace(M);
> time D  := DiscriminantOfHeckeAlgebra(S);
Time: 0.750
> D;
62967005472006188288017473632139259549820493155023510831l\
04000000
```

```
> Factorisation(D);
[ <2, 53>, <3, 4>, <5, 6>, <31, 2>, <37, 1>, <389, 1>, ...]
> M := ModularSymbols(997,2, +1); S := CuspidalSubspace(M);
> time D  := DiscriminantOfHeckeAlgebra(S);
Time: 55.600
```

The reason for the $+1$ in the construction of modular symbols is so that we compute on a space that is isomorphic as a $\mathbf{T}$-module to one copy of $S_2(\Gamma_0(p))$, instead of two copies.

3 Data about Discriminant Valuations

In this section we report on our extensive computations of $d_k(\Gamma_0(p))$. We first note that there is only one $p < 50000$ such that $d_2(\Gamma_0(p)) > 0$. Next we give a table of values of $d_4(\Gamma_0(p))$, which seems to exhibit a nice pattern.

3.1 Weight Two

Theorem 1. *The only prime $p < 60000$ such that $d_2(\Gamma_0(p)) > 0$ is $p = 389$, with the possible exception of* 50923 *and* 51437.

Computations in this direction by the second author have been cited in [Rib99], [MS01], [OW02], and [MO02]. For example, Theorem 1 is used for $p < 1000$ in [MS01] as a crucial step in proving that if E is an elliptic curve over $\mathbf{Q}(\mu_p)$, with $17 \leq p < 1000$, then not all elements of $E(\overline{\mathbf{Q}})[p]$ are rational over $\mathbf{Q}(\mu_p)$.

Proof. This is the result of a large computer computation. The rest of this proof describes how we did the computation, so the reader has some idea how to replicate or extend the computation. The computation described below took about one week using a cluster equipped with 10 Athlon 2000MP processors. The computations are nontrivial; we compute spaces of modular symbols, supersingular points, and Hecke operators on spaces of dimensions up to 5000.

The aim is to determine whether or not p divides the discriminant of the Hecke algebra of level p for each $p < 60000$. If T is an operator with integral characteristic polynomial, we write $\mathrm{disc}(T)$ for $\mathrm{disc}(\mathrm{charpoly}(T))$, which also equals $\mathrm{disc}(\mathbf{Z}[T])$. We will often use that

$$\mathrm{disc}(T) \bmod p = \mathrm{disc}(\mathrm{charpoly}(T) \bmod p).$$

We ruled out the possibility that $d_k(\Gamma_0(p)) > 0$ for most levels $p < 60000$ by computing characteristic polynomials of Hecke operators using an algorithm that the second author and D. Kohel implemented in MAGMA ([BCP97]), which is based on the Mestre-Oesterle method of graphs [Mes86] (or contact the second author for an English translation). Our implementation is available as the "Module of Supersingular Points" package that comes with MAGMA. We computed $\mathrm{disc}(T_q)$ modulo p for several small primes q, and in most cases found a prime q such that this discriminant is nonzero. The following table summarises how often we used each prime q (note that there are 6057 primes up to 60000):

q	number of $p < 60000$ where q smallest s.t. $\mathrm{disc}(T_q) \neq 0 \bmod p$
2	5809 times
3	161 (largest: 59471)
5	43 (largest: 57793)
7	15 (largest: 58699)
11	15 (the smallest is 307; the largest 50971)
13	2 (they are 577 and 5417)
17	3 (they are 17209, 24533, and 47387)
19	1 (it is 15661)

The numbers in the right column sum to 6049, so 8 levels are missing. These are

$$389, 487, 2341, 7057, 15641, 28279, 50923, \text{ and } 51437.$$

(The last two are still being processed. 51437 has the property that $\mathrm{disc}(T_q) = 0$ for $q = 2, 3, \ldots, 17$.) We determined the situation with the remaining 6 levels using Hecke operators T_n with n composite.

p	How we rule level p out, if possible
389	p does divide discriminant
487	using charpoly(T_{12})
2341	using charpoly(T_6)
7057	using charpoly(T_{18})
15641	using charpoly(T_6)
28279	using charpoly(T_{34})

Computing T_n with n composite is very time consuming when p is large, so it is important to choose the right T_n quickly. For $p = 28279$, here is a trick we used to quickly find an n such that $\mathrm{disc}(T_n)$ is not divisible by p. This trick might be used to speed up the computation for some other levels. The key idea is to efficiently discover which T_n to compute. Computing T_n on the full space of modular symbols is difficult, but using projections we can compute T_n on subspaces of modular symbols with small dimension more quickly (see, e.g., [Ste00, §3.5.2]). Let M be the space of mod p modular symbols of level $p = 28279$, and let $f = \gcd(\mathrm{charpoly}(T_2), \mathrm{deriv}(\mathrm{charpoly}(T_2)))$. Let V be the kernel of $f(T_2)$ (this takes 7 minutes to compute). If $V = 0$, we would be done, since then $\mathrm{disc}(T_2) \neq 0 \in \mathbf{F}_p$. In fact, V has dimension 7. We find the first few integers n so that the charpoly of T_n on V has distinct roots, and they are $n = 34$, 47, 53, and 89. We then computed charpoly(T_{34}) directly on the whole space and found that it has distinct roots modulo p.

3.2 Some Data about Weight 4

The following are the valuations $d = d_4(\Gamma_0(p))$ at p of the discriminant of the Hecke algebras associated to $S_4(\Gamma_0(p))$ for $p < 500$. This data suggests a pattern, which motivates Conjecture 1 below.

p	2	3	5	7	11	13	17	19	23	29	31	37	41	43	47	53	59
d	0	0	0	0	0	2	2	2	2	4	4	6	6	6	6	8	8
p	61	67	71	73	79	83	89	97	101	103	107	109	113	127	131	137	139
d	10	10	10	12	12	12	14	16	16	16	16	18	18	20	20	22	24
p	149	151	157	163	167	173	179	181	191	193	197	199	211	223	227	229	233
d	24	24	26	26	26	28	28	30	30	32	32	32	34	36	36	38	38
p	239	241	251	257	263	269	271	277	281	283	293	307	311	313	317	331	337
d	38	40	40	42	42	44	44	46	46	46	48	50	50	52	52	54	56
p	347	349	353	359	367	373	379	383	389	397	401	409	419	421	431	433	439
d	56	58	58	58	60	62	62	62	65	66	66	68	68	70	70	72	72
p	443	449	457	461	463	467	479	487	491	499							
d	72	74	76	76	76	76	78	80	80	82							

4 Speculations

Motivated by the promise of a pattern suggested by the table in Section 3.2, we computed $d_k(\Gamma_0(p))$ for many values of k and p. Our observations led us to the following results and conjectures.

Theorem 2. *Suppose p is a prime and $k \geq 4$ is an even integer. Then $d_k(\Gamma_0(p)) > 0$ unless*

$$\begin{aligned}(p,k) \in \{&(2,4),(2,6),(2,8),(2,10),\\ &(3,4),(3,6),(3,8),\\ &(5,4),(5,6),(7,4),(11,4)\},\end{aligned}$$

in which case $d_k(\Gamma_0(p)) = 0$.

Proof. From [Rib91], mod p eigenforms on $\Gamma_0(p)$ of weight k arise exactly from mod p eigenforms on $\Gamma_0(1)$ of weight $(k/2)(p+1)$. Moreover, there is an equality of dimensions of vector spaces:

$$\dim S_{(k/2)(p+1)}(\Gamma_0(1)) + \dim S_{(k/2)(p+1)-(p-1)}(\Gamma_0(1)) = \dim S_k(\Gamma_0(p)).$$

Thus the dimension of $S_k(\Gamma_0(p))$ is bigger than the number of mod p eigenforms whenever $\dim S_{(k/2)(p+1)-(p-1)}(\Gamma_0(1))$ is non-zero. The cases of dimension zero correspond exactly to the finite list of exceptions above, for which one can explicitly calculate that $d_k(\Gamma_0(p)) = 0$.

Note that for $k = 2$, however, there is a canonical identification of spaces

$$S_{(p+1)}(\Gamma_0(1), \overline{\mathbf{F}}_p) \simeq S_2(\Gamma_0(p), \overline{\mathbf{F}}_p),$$

described geometrically in [Gro90]. For $k = 4$, the data suggests that the discriminants $d_4(\Gamma_0(p))$ are significantly larger than zero for large p, and the table above suggests a formula of the form $2 \cdot \lfloor p/12 \rfloor$ (Not entirely co-incidentally, this is the difference in dimension of the spaces $S_4(\Gamma_0(p))$ and $S_{2(p+1)}(\Gamma_0(1)))$. This exact formula is not correct, however, as evidenced by the case when $p = 139$. If we consider the Hecke algebra $\mathbf{T}_4$ for $p = 139$ in more detail, however, we observe that $\mathbf{T}_4 \otimes \mathbf{Q}_{139}$ is *ramified* at 139, and in particular contains two copies of the field $\mathbf{Q}_{139}(\sqrt{139})$. Just as in the case when $k = 2$ and $p = 389$, there is a "self congruence" between the associated ramified eigenforms and their Galois conjugates. For all other p in the range of the table, there is no ramification, and all congruences take place between distinct eigenforms. Such congruences are measured by the *index* of the Hecke algebra, which is defined to be the index of $\mathbf{T}$ in its normalisation $\widetilde{\mathbf{T}}$. If we are only interested in mod p congruences (rather than mod ℓ congruences for $\ell \neq p$), one can restrict to the index of $\mathbf{T} \otimes \mathbf{Z}_p$ inside its normalisation. There is a direct relation between the discriminant and the index. Suppose that $\mathbf{T} \otimes \mathbf{Q}_p = \prod K_i$ for certain fields $K_i/\mathbf{Q}_p$ (We may assume here that $\mathbf{T}$ is not nilpotent, for otherwise both the discriminant and index are infinite). Then if $i_p(\Gamma) = \mathrm{ord}_p([\mathbf{T}, \widetilde{\mathbf{T}}])$, then

$$d_p(\Gamma) = 2 i_p(\Gamma) + \sum \mathrm{ord}_p(\Delta(K_i/\mathbf{Q}_p)).$$

If we now return to the example $k = 4$ and $p = 139$, we see that the discrepancy from the discriminant $d_p(\Gamma_0(139)) = 24$ to the estimate $2\lfloor 139/12 \rfloor = 22$ is exactly accounted for by the two eigenforms with coefficients in $\mathbf{Q}_{139}(\sqrt{139})$, which contribute 2 to the above formula. This leads us to predict that the index is exactly given by the formula $\lfloor p/12 \rfloor$. Note that for primes p this is exactly the dimension of $S_{p+3}(\Gamma_0(1))$. Similar computations lead to the following more general conjecture.

Let $k = 2m$ be an even integer and p a prime. Let $\mathbf{T}$ be the Hecke algebra associated to $S_k(\Gamma_0(p))$ and let $\widetilde{\mathbf{T}}$ be the integral closure of $\mathbf{T}$ in $\mathbf{T} \otimes \mathbf{Q}$ (which is a product of number fields).

Conjecture 1 Suppose $p > k - 1$. Then

$$\mathrm{ord}_p([\widetilde{\mathbf{T}} : \mathbf{T}]) = \left\lfloor \frac{p}{12} \right\rfloor \cdot \binom{m}{2} + a(p, m),$$

where

$$a(p,m) = \begin{cases} 0 & \text{if } p \equiv 1 \pmod{12}, \\ 3 \cdot \binom{\lceil \frac{m}{3} \rceil}{2} & \text{if } p \equiv 5 \pmod{12}, \\ 2 \cdot \binom{\lceil \frac{m}{2} \rceil}{2} & \text{if } p \equiv 7 \pmod{12}, \\ a(5,m) + a(7,m) & \text{if } p \equiv 11 \pmod{12}. \end{cases}$$

Here $\binom{x}{y}$ is the binomial coefficient "x choose y", and floor and ceiling are as usual. The conjecture is very false if $k \gg p$.

When $k = 2$, the conjecture specializes to the assertion that $[\widetilde{\mathbf{T}} : \mathbf{T}]$ is not divisible by p. A possibly more familiar concrete consequence of the conjecture is the following conjecture about elliptic curves. The modular degree of an elliptic curve E is the smallest degree of a surjective morphism $X_0(N) \to E$, where N is the conductor of E.

Conjecture 2. Suppose E is an elliptic curve of prime conductor p. Then p does not divide the modular degree m_E of E.

Using the algorithm in [Wat02], M. Watkins has computed modular degrees of a huge number of elliptic curves of prime conductor $p < 10^7$, and not found a counterexample. Looking at smaller data, there is only one elliptic curve E of prime conductor $p < 20000$ such that the modular degree of E is even as big as the conductor of E, and that is a curve of conductor 13723. This curve has equation $[1, 1, 1, -10481, 408636]$, modular degree $m_E = 16176 = 2^4 \cdot 3 \cdot 337$. The modular degree can be divisible by large primes. For example, there is a Neumann-Setzer elliptic curve of prime conductor 90687593 whose modular degree is 1280092043, which is over 14 times as big as 90687593. In general, for an elliptic curve of conductor N, one has the estimate $m_E \gg N^{7/6-\epsilon}$ (see [Wat04]).

5 Conjectures Inspired by Conjecture 1

First, some notation. Let p be an odd prime. Let $\Gamma = \Gamma_0(p)$, and let

$$S_k(R) := S_k(\Gamma)^{\text{new}} \otimes R.$$

The spaces S_k carry an action of the Hecke algebra $\mathbf{T}_k^{\text{new}}$, and a Fricke involution w_p. If $\frac{1}{2} \in R$, the space S_k can be decomposed into $+$ and $-$ eigenspaces for w_p. We call the resulting spaces S_k^+ and S_k^- respectively. Similarly, let M_k^+ and M_k^- be the $+1$ and -1 eigenspaces for w_p on the full spaces of new modular forms of weight k for $\Gamma_0(p)$.

It follows from [AL70, Lem. 7] (which is an explicit formula for the trace to lower level) and the fact that U_p and w_p both preserve the new subspace, that the action of the Hecke operator U_p on S_k is given by the formula

$$U_p = -p^{(k-2)/2} w_p.$$

This gives rise to two quotients of the Hecke algebra:

$$\mathbf{T}^+ = \mathbf{T}^{\text{new}}/(U_p + p^{(k-2)/2}) \quad \text{and} \quad \mathbf{T}^- = \mathbf{T}^{\text{new}}/(U_p - p^{(k-2)/2}).$$

where $\mathbf{T}^+$ and $\mathbf{T}^-$ act on S^+ and S^-, respectively. Recall that $\widetilde{\mathbf{T}}$ is the normalization (integral closure) of $\mathbf{T}$ in $\mathbf{T} \otimes \mathbf{Q}$. Let $\widetilde{\mathbf{T}}^{\text{new}}$ denote the integral closure of $\mathbf{T}^{\text{new}}$ in $\mathbf{T}^{\text{new}} \otimes \mathbf{Q}$.

Lemma 1. *There are injections*

$$\mathbf{T}^{\text{new}} \hookrightarrow \mathbf{T}^+ \oplus \mathbf{T}^- \hookrightarrow \widetilde{\mathbf{T}}^{\text{new}}.$$

We now begin stating some conjectures regarding the rings $\mathbf{T}^{\pm}$.

Conjecture 3. Let $k < p-1$. Then $\mathbf{T}^{+}$ and $\mathbf{T}^{-}$ are integrally closed. Equivalently, all congruences between distinct eigenforms in $S_k(\overline{\mathbf{Z}}_p)$ take place between + and − eigenforms.

Note that for $k = 2$, there cannot be any congruences between + and − forms because this would force $1 \equiv -1 \mod p$, which is false, because p is odd. Thus we recover the conjecture that $p \nmid [\widetilde{\mathbf{T}} : \mathbf{T}]$ when $k = 2$. Our further conjectures go on to describe explicitly the congruences between forms in S_k^{+} and S_k^{-}.

Let E_2 be the non-holomorphic Eisenstein series of weight 2. The q-expansion of E_2 is given explicitly by

$$E_2 = 1 - 24 \sum_{n=1}^{\infty} q^n \left(\sum_{d|n} d \right).$$

Moreover, the function $E_2^* = E_2(\tau) - pE_2(p\tau)$ is holomorphic of weight 2 and level $\Gamma_0(p)$, and moreover on q-expansions, $E_2^* \equiv E_2 \mod p$.

Lemma 2. *Let $p > 3$. Let $f \in M_k(\Gamma_0(p), \overline{\mathbf{F}}_p)$ be a Hecke eigenform. Then θf is an eigenform inside $S_{k+2}(\Gamma_0(p), \overline{\mathbf{F}}_p)$.*

Proof. One knows that $\partial f = \theta f - kE_2 f/12$ is of weight $k+2$. On q-expansions, $E_2 \equiv E_2^* \mod p$, and thus for $p > 3$,

$$\theta f \equiv \partial f + kE_2^* f/12 \pmod{p}$$

is the reduction of a weight $k+2$ form of level p. It is easy to see that θf is a cuspidal Hecke eigenform.

Let us now assume Conjecture 3 and consider the implications for $k = 4$ in more detail. The space of modular forms $M_2(\Gamma_0(p), \overline{\mathbf{F}}_p)$ consists precisely of S_2 and the Eisenstein series E_2^*. The map θ defined above induces maps:

$$\theta : S_2^{+}(\overline{\mathbf{F}}_p) \to S_4(\overline{\mathbf{F}}_p), \qquad \theta : M_2^{-}(\overline{\mathbf{F}}_p) \to S_4(\overline{\mathbf{F}}_p).$$

The images are distinct, since $\theta f = \theta g$ implies (with some care about a_p) that $f = g$.

Conjecture 4. Let $f \in S_2(\overline{\mathbf{Z}}_p)$ and $g \in S_4(\overline{\mathbf{Z}}_p)$ be two eigenforms such that $\theta f \equiv g \mod p$. Then the eigenvalue of w_p on f and g have opposite signs.

Assuming this, we get inclusions:

$$\theta S_2^{+}(\overline{\mathbf{F}}_p) \hookrightarrow S_4^{-}(\overline{\mathbf{F}}_p), \qquad \theta M_2^{-}(\overline{\mathbf{F}}_p) \hookrightarrow S_4^{+}(\overline{\mathbf{F}}_p).$$

Now we are ready to state our main conjecture:

Conjecture 5. There is an Hecke equivariant exact sequence:

$$0 \longrightarrow \theta S_2^+(\overline{\mathbf{F}}_p) \longrightarrow S_4^-(\overline{\mathbf{F}}_p) \longrightarrow S_4^+(\overline{\mathbf{F}}_p) \longrightarrow \theta M_2^-(\overline{\mathbf{F}}_p) \longrightarrow 0.$$

Moreover, the map $S_4^-(\overline{\mathbf{F}}_p) \to S_4^+(\overline{\mathbf{F}}_p)$ here is the largest such equivariant map between these spaces. Equivalently, a residual eigenform of weight 4 and level p occurs in both the $+$ and $-$ spaces if and only if it is not in the image of θ.

Let us give some consequences of our conjectures for the index of $\mathbf{T}^{\text{new}}$ inside its normalisation. Fix a residual representation $\overline{\rho} : \text{Gal}(\overline{\mathbf{Q}}/\mathbf{Q}) \to \text{GL}_2(\mathbf{F}_q)$ and consider the associated maximal ideal $\mathfrak{m}$ inside $\mathbf{T}_4$. If $\overline{\rho}$ lies in the image of θ then our conjecture implies that it is not congruent to any other eigenform. If $\overline{\rho}$ is not in the image of θ, then it should arise exactly from a pair of eigenforms, one inside $S_4^+(\overline{\mathbf{Q}}_p)$ and one inside $S_4^-(\overline{\mathbf{Q}}_p)$. Suppose that $q = p^r$. If there is no ramification in $\mathbf{T} \otimes \mathbf{Q}$ over p (this is often true), then the $+$ and $-$ eigenforms will both be defined over the ring $W(\mathbf{F}_q)$ of Witt vectors of $\mathbf{F}_q$. Since $U_p = p$ on S_4^- and $-p$ on S_4^+, these forms can be at most congruent modulo p. Thus the completed Hecke algebra $(\mathbf{T}_4)_{\mathfrak{m}}$ is exactly

$$\{(a,b) \in W(\mathbf{F}_q) \oplus W(\mathbf{F}_q), |a \equiv b \mod p\}.$$

One sees that this has index $q = p^r$ inside its normalisation. Thus the (log of the) total index is equal to $\sum r_i$ over all eigenforms that occur inside S_4^+ and S_4^-, which from our exact sequence we see is equal to

$$\dim S_4^- - \dim S_2^+.$$

Conjecture 1 when $k = 4$, would then follow from the equality of dimensions:

$$\dim S_4^-(\overline{\mathbf{F}}_p) - \dim S_2^+(\overline{\mathbf{F}}_p) = \left\lfloor \frac{p}{12} \right\rfloor.$$

We expect that something similar, but a little more complicated, should happen in general. In weight $2k$, there are mod p^{k-r} congruences exactly between forms in the image of θ^{r-1} but not of θ^r.

5.1 Examples

We write small s's and m's for dimensions below.

Let $p = 101$. Then $s_2^+ = 1$, $m_2^- = 7 + 1 = 8$, $s_4^- = 9$, $s_4^+ = 16$. We predict the index should be $9 - 1 = 8 = \lfloor 101/12 \rfloor$. In the table below, we show the characteristic polynomials of T_2 on S_4^- and S_4^+, and for weight 2, we take the characteristic polynomial of θT_2 (or the same, taking $F(x/2)$ where $F(x)$ is the characteristic polynomial of T_2). Note that we have to add the Eisenstein series, which has characteristic polynomial $x - 1 - 2$, which becomes $x - 6 \equiv x + 95 \mod 101$ under θ.

Factors of the Characteristic Polynomial of T_2 for $p = 101$.

$\theta S_2^+(\overline{\mathbf{F}}_{101})$	$S_4^-(\overline{\mathbf{F}}_{101})$	$S_4^+(\overline{\mathbf{F}}_{101})$	$\theta M_2^-(\overline{\mathbf{F}}_{101})$
(x)	(x)	$(x+46)$	$(x+95)$
	$(x+46)$	$(x+95)$	$(x^2+90x+78)$
	$(x^2+58x+100)$	$(x^2+58x+100)$	$(x^2+96x+36)$
	$(x^5+2x^4+27x^3$ $+49x^2+7x+65)$	$(x^2+90x+78)$	$(x^3+16x^2$ $+35x+72)$
		$(x^2+96x+36)$	
		$(x^3+16x^2+35x+72)$	
		$(x^5+2x^4+27x^3$ $+49x^2+7x+65)$	

Here are some further conjectures when $k > 4$.

Conjecture 6. Let p and k be such that $4 < k < p-1$. There is an Hecke equivariant exact sequence:

$$0 \longrightarrow \theta S_{k-2}^+(\overline{\mathbf{F}}_p) \longrightarrow S_k^-(\overline{\mathbf{F}}_p) \longrightarrow S_k^+(\overline{\mathbf{F}}_p) \longrightarrow \theta S_{k-2}^-(\overline{\mathbf{F}}_p) \longrightarrow 0.$$

Moreover, all forms not in the image of θ contribute maximally to the index (a factor of $p^{(k-2)/2}$). Thus the total index should be equal to

$$\frac{(k-2)}{2}(\dim S_k^+ - \dim S_{k-2}^-) \quad + \quad \text{the index at level } p \text{ and weight } k-2.$$

This is the sum

$$\sum_{n=2}^{k} \frac{(2n-2)}{2}(s_{2n}^+ - s_{2n-2}^-).$$

When $k = 4$, we need to add the Eisenstein series to S_2^- in our previous conjecture. Note that $s_k^+ - s_{k-2}^- = s_k^- - s_{k-2}^+$ for $k > 4$ (and with s_2^- replaced by m_2^- when $k = 2$). This follows from our conjectures, but can easily be proved directly. As an example, when $p = 101$, we have $s_2^+ = 1$, $s_4^- = 9$, $s_6^+ = 17$, $s_8^- = 26$, $s_{10}^+ = 34$, $s_{12}^- = 42$, $s_{14}^+ = 51$, and so we would predict the indexes I_k to be as given in the following table:

k	I_k?
2	0
4	$8 = 8 + 0$
6	$24 = 24 + 0$
8	$51 = 48 + 3$
10	$83 = 80 + 3$
12	$123 = 120 + 3$
14	$177 = 168 + 9$

This agrees with our conjectural formula, which says that the index should be equal in this case to

$$8\binom{k/2}{2} + 3\binom{\lceil k/6 \rceil}{2}.$$

it also agrees with computation.

References

[AL70] A. O. L. Atkin and J. Lehner, *Hecke operators on $\Gamma_0(m)$*, Math. Ann. **185** (1970), 134–160.

[BCP97] W. Bosma, J. Cannon, and C. Playoust, *The Magma algebra system. I. The user language*, J. Symbolic Comput. **24** (1997), no. 3-4, 235–265, Computational algebra and number theory (London, 1993). MR 1 484 478

[CE98] R. F. Coleman and B. Edixhoven, *On the semi-simplicity of the U_p-operator on modular forms*, Math. Ann. **310** (1998), no. 1, 119–127. MR 99b:11043

[Cre97] J. E. Cremona, *Algorithms for modular elliptic curves*, second ed., Cambridge University Press, Cambridge, 1997.

[Gro90] B. H. Gross, *A tameness criterion for Galois representations associated to modular forms (mod p)*, Duke Math. J. 61 (1990), no. 2, 445–517. MR 1 074 305

[LS02] J.-C. Lario and R. Schoof, *Some computations with Hecke rings and deformation rings*, Experiment. Math. **11** (2002), no. 2, 303–311, With an appendix by Amod Agashe and William Stein. MR 2004b:11072

[Mat86] H. Matsumura, *Commutative ring theory*, Cambridge University Press, Cambridge, 1986, Translated from the Japanese by M. Reid. MR 88h:13001 ed., Cambridge University Press, Cambridge, 1997.

[Mer94] L. Merel, *Universal Fourier expansions of modular forms*, On Artin's conjecture for odd 2-dimensional representations, Springer, 1994, pp. 59–94.

[Mes86] J.-F. Mestre, *La méthode des graphes. Exemples et applications*, Proceedings of the international conference on class numbers and fundamental units of algebraic number fields (Katata) (1986), 217–242.

[MO02] F. Momose and S. Ozawa, *Rational points of modular curves $X_{\mathrm{split}}(p)$*, Preprint (2002).

[MS01] L. Merel and W. A. Stein, *The field generated by the points of small prime order on an elliptic curve*, Internat. Math. Res. Notices (2001), no. 20, 1075–1082. MR 1 857 596

[OW02] K. Ono and W. McGraw, *Modular form congruences and selmer groups*, Preprint (2002).

[Rib91] K. A. Ribet, *Letter to Gerd Faltings*, Unpublished (1991).

[Rib99] K. A. Ribet, *Torsion points on $J_0(N)$ and Galois representations*, Arithmetic theory of elliptic curves (Cetraro, 1997), Springer, Berlin, 1999, pp. 145–166. MR 2001b:11054

[Ste00] W. A. Stein, *Explicit approaches to modular abelian varieties*, Ph.D. thesis, University of California, Berkeley (2000).

[Wat02] M. Watkins, *Computing the modular degree of an elliptic curve*, Experiment. Math. **11** (2002), no. 4, 487–502 (2003). MR 1 969 641

[Wat04] M. Watkins, *Explicit lower bounds on the modular degree of an elliptic curve*, Preprint (2004).

Montgomery Scalar Multiplication for Genus 2 Curves

Sylvain Duquesne

Université Montpellier II, Laboratoire I3S, UMR CNRS 5149
Place Eugène Bataillon CC 051, 34005 Montpellier Cedex, France
duquesne@math.univ-montp2.fr

Abstract. Using powerful tools on genus 2 curves like the Kummer variety, we generalize the Montgomery method for scalar multiplication to the jacobian of these curves. Previously this method was only known for elliptic curves. We obtain an algorithm that is competitive compared to the usual methods of scalar multiplication and that has additional properties such as resistance to timings attacks. This algorithm has very important applications in cryptography using hyperelliptic curves and more particularly for people interested in cryptography on smart cards.

1 Introduction

In this paper we are dealing with the scalar multiplication on the jacobian of curves defined over a field of large characteristic. One of the motivations is that this operation is the main part in cryptography based on the jacobian of curves which is becoming more and more popular. For elliptic curves, Montgomery [24] developed a method for certain curves (which are said to be in the Montgomery form) which allows faster scalar multiplication than the usual methods of exponentiation for groups. This method has the extra advantage that it is resistant to side-channel attacks which is very interesting for people who want to use elliptic curves in cryptography on smart cards. The aim of this paper is the generalization of this method to genus 2 curves. In the following, $\mathbf{K}$ will denote a field of characteristic $k \geq 7$. For cryptographic application, the base field we have in mind is $\mathbb{F}_p$ where p is prime.

2 The Montgomery Method for Scalar Multiplication on Elliptic Curves

Let E be an elliptic curve defined over $\mathbf{K}$ by the equation

$$y^2 = x^3 + a_4 x + a_6 \ .$$

Every elliptic curve defined over $\mathbf{K}$ is isomorphic to a curve given by such an equation which is called the short Weierstrass form. The set $E(\mathbf{K})$ of the points $P = (x, y)$ verifying this equation with x and y in $\mathbf{K}$, forms (together with the point at infinity) a group which will be denoted additively. The problem we are interested in is the following :

D.A. Buell (Ed.): ANTS 2004, LNCS 3076, pp. 153–168, 2004.

Scalar multiplication
Given a point $P \in E(\mathbf{K})$ and an integer n, compute nP as fast as possible.

Of course there are a lot of very old methods to do this, such as the classical double and add algorithm and its variants (like the sliding window method). To improve these algorithms one can choose other systems of coordinates (i.e. other means to represent points on the curve) [4]. For example, the best-known coordinates are projective ones. They are obtained by introducing a new coordinate, usually called Z, which is the lcm of the denominators of x and y. Of course, x and y are replaced by X and Y such that $x = X/Z$ and $y = Y/Z$. This choice of coordinates allows to avoid inversions, which are very costly operations. In the following, we will work with projective coordinates for reasons of efficiency. Nevertheless, the same work can be done with other systems of coordinates. The algorithm we will present here is slightly different from the usual exponentiation algorithms in the sense that the purpose is not to minimize the number of operations but rather the cost of each operation.

2.1 The Algorithm

The original idea of Montgomery [24] was to avoid the computation of the y-coordinate, so that one can hope that basic operations (doubling and adding) are easier to compute. Since for any x-coordinate, there are two corresponding points on the curve (x, y) and $(x, -y)$, this restriction is equivalent to identifying a point on the curve and its opposite. When trying to add two points $\pm P$ and $\pm Q$, one cannot decide if the result obtained is $\pm(P+Q)$ as required or $\pm(P-Q)$. Nevertheless, some operations remain possible like doubling since it is not difficult to decide if the result is $\pm 2P$ or the point at infinity. Unfortunately, doubling is not sufficient for a complete scalar multiplication: one really needs to perform some additions. In fact additions are possible if the difference $P-Q$ is known. The principle of the computation of nP is to use pairs of consecutive powers of P, so that the difference between the two components of the pair is always known and equals to P. The algorithm for scalar multiplication is as follows:

Algorithm 1. *Montgomery scalar multiplication algorithm on elliptic curves*
Input : $P \in E(\mathbf{K})$ *and* $n \in \mathbb{Z}$.
Output : x and z-coordinate of nP.

Step 1. Initialize $Q = (Q_1, Q_2) = (\mathcal{O}, P)$ *where* $\mathcal{O}$ *is the point at infinity.*

Step 2. If the bit of n is 0, $Q = (2Q_1, Q_1 + Q_2)$.

Step 3. If the bit of n is 1, $Q = (Q_1 + Q_2, 2Q_2)$.

Step 4. After doing that for each bit of n, return Q_1.

In fact, at each step, $Q = (kP, (k+1)P)$ for some k and we compute either $(2kP, (2k+1)P)$ or $((2k+1)P, (2k+2)P)$ in the following step, so that we always have $Q_2 - Q_1 = P$.

Let us note that contrary to double and add or sliding window methods, both an addition and a doubling are done for each bit of the exponent. It is the price to be paid to avoid the computation of the y-coordinate but we hope that the gain obtained thanks to this restriction will be sufficient to compensate for the large number of operations. That is the reason why the Montgomery form for elliptic curves has been introduced. The interested reader will find more details for this section in [24] or [26].

2.2 The Montgomery Form

An elliptic curve E is transformable into the Montgomery form if it is isomorphic to a curve given by an equation of the type

$$E_m : By^2 = x^3 + Ax^2 + x \ .$$

It is easy to prove ([26]) that E is transformable in the Montgomery form if and only if

- the polynomial $x^3 + a_4 x + a_6$ has at least one root α in $\mathbf{K}$,
- the number $3\alpha^2 + a_4$ is a square in $\mathbf{K}$.

Thus all elliptic curves are not transformable into the Montgomery form. Nevertheless, since the two coefficients can be chosen arbitrarily in $\mathbf{K}$, the number of curves in such a form is of the same order as for general elliptic curves (for example $O(p^2)$ if $\mathbf{K} = \mathbb{F}_p$).
Please note that the first condition means that there is at least one 2-torsion point on the curve E, so that the cardinality of the curve is even.

2.3 Formulas for Doubling and Adding

Let us now describe the arithmetic of curves in the Montgomery form.

Proposition 1. *Let* $\mathbf{K}$ *be a field of characteristic* $k \neq 2, 3$ *and let* E_m *be an elliptic curve defined over* $\mathbf{K}$ *in the Montgomery form. Let* $P = (X_p, Y_p, Z_p)$ *and* $Q = (X_q, Y_q, Z_q) \in E_m(\mathbf{K})$ *be given in projective coordinates. Assume that the difference* $P - Q = (x, y)$ *is known in affine coordinates. Then we obtain the* X *and* Z*-coordinates for* $P + Q$ *and* $2P$ *by the following formulas :*

$$\begin{aligned} X_{p+q} &= ((X_p - Z_p)(X_q + Z_q) + (X_p + Z_p)(X_q - Z_q))^2 \ , \\ Z_{p+q} &= x\,((X_p - Z_p)(X_q + Z_q) - (X_p + Z_p)(X_q - Z_q))^2 \ , \end{aligned}$$

$$\begin{aligned} 4X_pZ_p &= (X_p + Z_p)^2 - (X_p - Z_p)^2 \ , \\ X_{2p} &= (X_p + Z_p)^2 (X_p - Z_p)^2 \ , \\ Z_{2p} &= 4X_pZ_p \left((X_p - Z_p)^2 + \frac{A+2}{4} 4X_pZ_p \right) \ . \end{aligned}$$

In this way, both an addition and a doubling take 3 multiplications and 2 squares each so that the cost of this algorithm is about $10|n|_2$ multiplications where $|n|_2$ denotes the number of bits of n.
In the best case with usual scalar multiplication, one needs 4 multiplications and 4 squaring just for doubling and more for adding. Thus, for curves in the Montgomery form, this method is interesting. In practice, the gain obtained is about 10 percent (compared in [6] for 192 bits with a sliding window method of size 4 after a Koyama-Tsuruoka recoding [18] and using mixed jacobian modified coordinates [4]).

2.4 General Case

We are now interested in a Montgomery method for scalar multiplication for elliptic curves which cannot be transformed into the Montgomery form. In fact the method for scalar multiplication is the same, we just need to have formulas for doubling and adding. These formulas can be found in [1], [10] or [13]. Let us recall them.

Proposition 2. *Let* $\mathbf{K}$ *be a field of characteristic* $k \neq 2,3$ *and let* E *be an elliptic curve defined over* $\mathbf{K}$ *as described in Sect. 2. Let* $P = (X_p, Y_p, Z_p)$ *and* $Q = (X_q, Y_q, Z_q) \in E(\mathbf{K})$ *be given in projective coordinates. Assume that the difference* $P - Q = (x, y)$ *is known in affine coordinates. Then we obtain the* X *and* Z*-coordinates for* $P + Q$ *and* $2P$ *by the following formulas :*

$$\begin{aligned}
X_{p+q} &= -4a_6 Z_p Z_q (X_p Z_q + X_q Z_p) + (X_p X_q - a_4 Z_p Z_q)^2 \ , \\
Z_{p+q} &= x(X_p Z_q - X_q Z_p) \ , \\
X_{2p} &= \left(X_p^2 - a_4 Z_p^2\right)^2 - 8a_6 X_p Z_p^3 \ , \\
Z_{2p} &= 4Z_p \left(X_p^3 + a_4 X_p Z_p^2 + a_6 Z_p^3\right) \ .
\end{aligned}$$

Addition can be evaluated in 10 multiplications and doubling in 9. Thus, in this way, the scalar multiplication can be performed in about $19|n|_2$ multiplications in the base field. This method can even be optimized in most cases to $17|n|_2$ multiplications [7]. Of course, in this case, the algorithm is not interesting any more compared with the usual methods. However, it can be useful in some situations as we will see in the next section.

2.5 Use and Interest in Cryptography

In this section, we are dealing with elliptic curve cryptography. Elliptic curve cryptosystems were simultaneously introduced by Koblitz [14] and Miller [23]. They are becoming more and more popular because the key length can be chosen smaller than with RSA cryptosystems for the same level of security. This small key size is especially attractive for small cryptographic devices like smart cards. In all schemes (such as encryption/decryption or signature generation/verification) the dominant operation is the scalar multiplication of some point on the elliptic curve. Hence, the efficiency of this scalar multiplication is central in elliptic curve cryptography, and more generally in cryptography based on the discrete logarithm problem. In the case where the curve is in the Montgomery form, we saw in the previous sections that the Montgomery scalar multiplication method allows to compute the multiple of any given point on the curve faster than with the usual scalar multiplication algorithms. Unfortunately, we also saw that some elliptic curves cannot be transformed into the Montgomery form. This is for example the case for most of the standards. The reason is really simple: any curve which can be transformed in the Montgomery form has a 2-torsion point so that its cardinality is divisible by 2 and this is not ideal for cryptographic use since we prefer to use curves with prime order.

In the general case, the Montgomery method can also be applied but is much more time-consuming. Indeed, we need to perform both an addition and a doubling for each bit of the exponent. This is not the case for example in the classical double and add algorithm where we only have to perform an addition every two bits on average (and even fewer with the sliding window method). Nevertheless, this particularity allows to resist to side-channel attacks on smart cards which is not the case with other algorithms.

This kind of attacks uses observations like timings [16], power consumption [17] or electromagnetic radiation [28]. They are based on the fact that addition and doubling are two different operations. In this situation, it is easy to decide, for each bit of the exponent, if the algorithm (double and add for example) is performing either a doubling (if the bit is 0) or a doubling and an addition (if the bit is 1). Hence, it is easy to recover the whole exponent (which was the secret key). Of course, various countermeasures have been proposed to secure the elliptic curve scalar multiplication against side-channel attacks [5]. For example, if one wants to protect a double and add algorithm, one can perform extra, useless, additions when the bit of the exponent is 0. In this way, for each bit of the exponent we perform both an addition and a doubling so that bits of the exponent are indistinguishable, but this is of course time consuming.

With the Montgomery scalar multiplication method, we always have to perform both an addition and a doubling for each bit of the exponent, so that this method is resistant against side-channel attacks. Therefore it is always interesting even with 19 multiplications at each step for general curves.

Of course elliptic curves in the Montgomery form are very attractive for people interested in elliptic curve cryptosystems on smart cards since, on the one hand, the scalar multiplication method is the most efficient one known to date and, on the other hand, it is resistant to side-channel attacks. That is one of the reasons why we want to generalize this method to hyperelliptic curves of genus 2.

Finally, for some cryptosystems, the x-coordinate of nP is sufficient but others, like the elliptic curve signature scheme ECDSA, require the y-coordinates. To recover it, we use the following result from [27] in the case of a curve in the Montgomery form.

Proposition 3. *Suppose that $R = P + Q$ with $P = (x_1, y_1)$, $Q = (x', y')$ and $R = (x^+, y^+)$. Then, if $y_1 \neq 0$, one has*

$$y^+ = \frac{(x'x_1 + 1)(x' + x_1 + 2A) - 2A - (x' - x_1)^2 x^+}{2By_1} .$$

For general curves, it is also possible to recover the y-coordinate ([1]).

In order to generalize this method to genus 2 curves, let us first recall some lowbrow background on these curves.

3 Background on Genus 2 Curves

First, let us note that every genus 2 curve is hyperelliptic, so that, in the following we will not state that the curves we are interested in are hyperelliptic. Moreover, we will concentrate on imaginary hyperelliptic curves. Since the characteristic of the field $\mathbf{K}$ has been chosen different from 2 and 5, the hyperelliptic curves we are interested in are given by an equation of the form

$$\mathcal{C} : y^2 = f(x) = x^5 + f_3x^3 + f_2x^2 + f_1x + f_0 \text{ with } f_0, f_1, f_2, f_3 \in \mathbf{K} . \quad (1)$$

Contrary to elliptic curves, the set of points on genus 2 curves does not form a group. Thus, one can define the jacobian of $\mathcal{C}$, denoted $\mathcal{J}(\mathcal{C})$ which is the quotient of the group of divisors of degree 0 by the principal divisors. In the case of elliptic curves, this jacobian is isomorphic to the curve itself. More details on the definition of the jacobian can be found in [15]. Our purpose in this paper is to give an algorithm for scalar multiplication in this jacobian. There are mainly two means to represent elements in the jacobian. The first one is a consequence of the Riemann-Roch theorem and says that a divisor class can be represented by a couple of points ($P_1 = (x_1, y_1)$ and $P_2 = (x_2, y_2)$) on the curve which are conjugated over some quadratic extension of the base field $\mathbf{K}$. The second one makes explicit the correspondence of ideal classes and divisor classes and was introduced by Mumford [25]:

Theorem 1. *Let the function field be given via the irreducible polynomial $y^2 = f(x)$ where $f \in \mathbf{K}[x]$ and deg(f)=5. Each non trivial ideal class over $\mathbf{K}$ can be represented via a unique ideal generated by $u(x)$ and $y - v(x)$, $u, v \in \mathbf{K}[x]$, where u is monic, deg(v) <deg(u) ≤ 2 and $u|(v^2 - f)$.*

The correspondence between these representations is that $u(x) = (x - x_1)(x - x_2)$ and $v(x_i) = y_i$ with appropriate multiplicities. This Mumford representation was used by Cantor to develop his algorithm to compute the group law on jacobians of curves [2]. Several researchers such as Harley [12], or more recently Lange [19], [20], [21] made explicit the steps of Cantor's Algorithm and list the operations one really needs to perform. They obtained explicit formulas for the group law on the jacobian.

The basic objects are now defined and we can give an analog for genus 2 curves of the Montgomery form for elliptic curves.

4 A Montgomery-Like Form for Genus 2 Curves

4.1 Definition

In the following, we will say that a curve C is transformable into Montgomery-like form if it is isomorphic to a curve given by an equation of the type

$$By^2 = x^5 + f_4x^4 + f_3x^3 + f_2x^2 + x \ . \qquad (2)$$

It is easy to prove that a curve C as defined in Sect. 3 is transformable into Montgomery-like form if and only if

- the polynomial $f(x)$ has at least one root α in $\mathbf{K}$.
- the number $f'(\alpha)$ is a fourth power in $\mathbf{K}$.

Thus, as in the case of elliptic curves, not all the curves are transformable into the Montgomery-like form. Nevertheless, since the four coefficients can be chosen arbitrarily in $\mathbf{K}$, the number of such curves is about the same as for general genus 2 curves ($O(p^4)$ if $\mathbf{K} = \mathbb{F}_p$).
Please note that the first condition means that there is at least one 2-torsion element in the jacobian variety of the curve C, so that the cardinality of the jacobian is even.

4.2 The Kummer Surface

With elliptic curves, the main idea of the Montgomery method was to avoid the computation of the y-coordinate. At first sight an analog for genus 2 curves would be to avoid the computation of the polynomial v in the Mumford representation and keep only u. But this is not satisfying since it has no mathematical sense. In fact, with elliptic curves, working only with the x-coordinate means that we identify a point and its opposite. The analog for genus 2 curves is called the Kummer surface, where a divisor and its opposite are identified. The Kummer surface is a quartic surface in $\mathbb{P}^3$. We give here the definition of the Kummer surface and its properties without demonstrations for curves in Montgomery-like form. They were obtained using the same method as that in the book of Cassels and Flynn on genus 2 curves ([3] or [8]). The Kummer surface is the image of the map

$$\begin{aligned} \kappa : J(\mathbf{K}) &\longrightarrow \mathbb{P}^3(\mathbf{K}) \\ \{(x_1,y_1),(x_2,y_2)\} &\longmapsto \left(1, x_1+x_2, x_1x_2, \frac{F_0(x_1,x_2)-2By_1y_2}{(x_1-x_2)^2}\right) , \end{aligned}$$

with

$$F_0(x_1, x_2) = (x_1 + x_2) + 2f_2x_1x_2 + f_3(x_1 + x_2)x_1x_2 + 2f_4x_1^2x_2^2 + (x_1 + x_2)x_1^2x_2^2 \ .$$

In the following, for any divisor $\mathcal{A} \in \mathcal{J}(\mathcal{C}))$, we will denote

$$\kappa(\mathcal{A}) = (k_1(\mathcal{A}), k_2(\mathcal{A}), k_3(\mathcal{A}), k_4(\mathcal{A})) \ .$$

More precisely, the Kummer surface is the projective locus given by an equation K of degree four in the first three variables and of degree two in the last one. The exact equation can be found online [9]. In passing from the jacobian to the Kummer surface, we have lost the group structure (as was already the case with elliptic curves) but traces of it remain. For example, it is possible to double on the Kummer surface.
Nevertheless, for general divisors $\mathcal{A}$ and $\mathcal{B}$, we cannot determine the values of the $k_i(\mathcal{A}+\mathcal{B})$ from the values of the $k_i(\mathcal{A})$ and $k_i(\mathcal{B})$ since the latter do not distinguish between $\pm\mathcal{A}$ and $\pm\mathcal{B}$, and so not between $\mathcal{A}\pm\mathcal{B}$. However the values of the $k_i(\mathcal{A}+\mathcal{B})k_j(\mathcal{A}-\mathcal{B}) + k_i(\mathcal{A}-\mathcal{B})k_j(\mathcal{A}+\mathcal{B})$ are well determined. We have

Theorem 2. *There are explicit polynomials φ_{ij} biquadratic in the $k_i(\mathcal{A}), k_i(\mathcal{B})$ such that projectively*

$$k_i(\mathcal{A}+\mathcal{B})k_j(\mathcal{A}-\mathcal{B}) + k_i(\mathcal{A}-\mathcal{B})k_j(\mathcal{A}+\mathcal{B}) = \varphi_{ij}(\mathcal{A}, \mathcal{B}) \ . \tag{3}$$

Using these biquadratic forms, we can easily compute the $k_i(\mathcal{A}+\mathcal{B})$ if the $k_i(\mathcal{A}-\mathcal{B})$ are known. We can also compute the $k_i(2\mathcal{A})$ by puting $\mathcal{A} = \mathcal{B}$.

4.3 Formulas for Adding

Proposition 4. *Let* $\mathbf{K}$ *be a field of characteristic $k \neq 2, 3$ and let $\mathcal{C}$ be a curve of genus 2 defined over* $\mathbf{K}$ *in the Montgomery form as defined in Sect. 4. Let $\mathcal{K}$ denote the Kummer surface of $\mathcal{C}$. Let $\mathcal{A}$ and $\mathcal{B}$ be two divisors on the jacobian of $\mathcal{C}$, $\kappa(\mathcal{A}) = (k_1(\mathcal{A}), k_2(\mathcal{A}), k_3(\mathcal{A}), k_4(\mathcal{A}))$ and $\kappa(\mathcal{B}) = (k_1(\mathcal{B}), k_2(\mathcal{B}), k_3(\mathcal{B}), k_4(\mathcal{B}))$ their images in the Kummer surface. Assume that the difference $\mathcal{A}-\mathcal{B}$ is known and that the last coordinate of its image in the Kummer surface is one (remember we are in $\mathbb{P}^3(\mathbf{K})$). Then we obtain the Kummer coordinates for $\mathcal{A}+\mathcal{B}$ by the following formulas :*

$$\begin{aligned}
k_1(\mathcal{A}+\mathcal{B}) &= \varphi_{11}(\mathcal{A}, \mathcal{B}) \ , \\
k_2(\mathcal{A}+\mathcal{B}) &= 2\,(\varphi_{12}(\mathcal{A}, \mathcal{B}) - k_1(\mathcal{A}+\mathcal{B})k_2(\mathcal{A}-\mathcal{B})) \ , \\
k_3(\mathcal{A}+\mathcal{B}) &= k_1(\mathcal{A}-\mathcal{B})\varphi_{33}(\mathcal{A}, \mathcal{B}) \ , \\
k_4(\mathcal{A}+\mathcal{B}) &= 2(\varphi_{14}(\mathcal{A}, \mathcal{B}) - k_1(\mathcal{A}+\mathcal{B})k_4(\mathcal{A}-\mathcal{B})) \ ,
\end{aligned}$$

where the φ_{ij} are the biquadratic forms described in Sect. 4.2.

The expressions of the $\varphi_{ij}(\mathcal{A}+\mathcal{B})$ are available by anonymous ftp [9] but require a large number of operations in the base field to be computed. The main difficulty is to find expressions which require the least possible multiplications in $\mathbf{K}$. We now give more precisely these expressions for the φ_{ij} we are interested in. For clarity we will denote $\kappa(\mathcal{A}) = (k_1, k_2, k_3, k_4)$ and $\kappa(\mathcal{B}) = (l_1, l_2, l_3, l_4)$.

$$\begin{aligned}
\varphi_{11}(\mathcal{A},\mathcal{B}) &= ((k_4l_1 - k_1l_4) + (k_2l_3 - k_3l_2))^2 \ , \\
\varphi_{12}(\mathcal{A},\mathcal{B}) &= ((k_2l_3 + k_3l_2) + (k_1l_4 + k_4l_1))(f_3(k_1l_3 + k_3l_1) + (k_2l_4 + k_4l_2)) + \\
&\quad 2(k_1l_3 + k_3l_1)(f_2(k_1l_3 + k_3l_1) + (k_1l_2 + k_2l_1) - (k_3l_4 + k_4l_3)) + \\
&\quad 2f_4(k_1l_4 + k_4l_1)(k_2l_3 + k_3l_2) \ , \\
\varphi_{33}(\mathcal{A},\mathcal{B}) &= ((k_3l_4 - k_4l_3) + (k_1l_2 - k_2l_1))^2 \ , \\
\varphi_{14}(\mathcal{A},\mathcal{B}) &= (k_1l_1 - k_3l_3)(f_3((k_1l_4 + k_4l_1) - (k_2l_3 + k_3l_2)) + \\
&\qquad 2((k_1l_2 + k_2l_1) - (k_3l_4 + k_4l_3)) + \\
&\qquad f_2(k_4l_4 + k_2l_2) + 2f_4(k_1l_1 - k_3l_3)) + \\
&\quad (k_2l_2 - k_4l_4)((k_2l_3 + k_3l_2) - (k_1l_4 + k_4l_1) - f_2(k_1l_1 + k_3l_3)) \ .
\end{aligned}$$

4.4 Formulas for Doubling

Proposition 5. *Let $\mathbf{K}$ be a field of characteristic $k \neq 2, 3$ and let $\mathcal{C}$ be a curve of genus 2 defined over $\mathbf{K}$ in the Montgomery form as defined in Sect. 4. Let $\mathcal{K}$ denote the Kummer surface of $\mathcal{C}$. Let also $\mathcal{A}$ be a divisor on the jacobian of $\mathcal{C}$ and $\kappa(\mathcal{A}) = (k_1, k_2, k_3, k_4)$ its image in the Kummer surface. Then we obtain the Kummer coordinates for $2\mathcal{A}$ ($\kappa(2\mathcal{A}) = \delta_1, \delta_2, \delta_3, \delta_4$) by the following formulas :*

$$\begin{aligned}
\delta_1 &= 2\varphi_{14}(\mathcal{A},\mathcal{A}) \ , \\
\delta_2 &= 2\varphi_{24}(\mathcal{A},\mathcal{A}) + 2f_3K(\mathcal{A}) \ , \\
\delta_3 &= 2\varphi_{34}(\mathcal{A},\mathcal{A}) \ , \\
\delta_4 &= \varphi_{44}(\mathcal{A},\mathcal{A}) + 2K(\mathcal{A}) \ ,
\end{aligned}$$

where the φ_{ij} are the biquadratic forms described in Sect. 4.2 and K is the equation of the Kummer surface also described in Sect. 4.2 and such that $K(\mathcal{A}) = 0$.

This is just a consequence of Theorem 2. Let us note that in δ_2 and δ_4 we added a multiple of the equation of the Kummer surface in order to simplify the expressions as much as possible. We give now more precisely these expressions for the δ_i.

$$\begin{aligned}
\delta_1 &= 8(k_1^2 - k_3^2)(f_4(k_1^2 - k_3^2) + 2(k_1k_2 - k_3k_4)) + \\
&\quad 8(k_1k_4 - k_2k_3)(k_4^2 - k_2^2 + f_2(k_1k_4 - k_2k_3) + f_3(k_1^2 - k_3^2)) \ , \\
\delta_2 &= 8(k_1^2 + k_3^2 - f_3k_1k_3 - 3k_2k_4)(k_2^2 + k_4^2 - f_3(k_1^2 + k_3^2) + 4k_1k_3) + \\
&\quad 16(k_2k_4 + f_3k_1kl_3)(f_4(k_1k_2 + k_3k_4) + 2(k_2^2 + k_4^2) + f_2(k_1k_4 + k_2k_3)) +
\end{aligned}$$

$$32k_1k_3(4k_2k_4 + f_2(k_1k_2 + k_3k_4) + (f_2^2 + f_4^2)k_1k_3 + 8f_4(k_1k_4 + k_2k_3)) \ ,$$
$$\delta_3 = 8(k_1^2 - k_3^2)(f_2(k_1^2 - k_3^2) + 2(k_1k_4 - k_2k_3) + f_3(k_1k_2 - k_3k_4)) + 8(k_3k_4 - k_1k_2)(k_4^2 - k_2^2 + f_4(k_3k_4 - k_1k_2)) \ ,$$
$$\delta_4 = (k_2^2 + k_4^2)((k_2^2 + k_4^2) - 2f_3(k_1^2 + k_3^2) - 8k_1k_3) + (k_1^2 + k_3^2)(f_3k_1k_3 + f_4(k_1k_4 + k_2k_3) + 2k_2k_4 + f_2(k_1k_2 + k_3k_4) + (f_3^2 - 4f_2f_4)(k_1^2 + k_3^2)) - 8f_2f_4(k_1k_3)^2 \ .$$

5 The Montgomery Scalar Multiplication on Genus 2 Curves in Montgomery-Like Form

5.1 Algorithm

We give here an analog for genus 2 curves of the Montgomery method for scalar multiplication on elliptic curves described in Sect. 2.1. In the case of elliptic curves, Montgomery's method [24] avoids the computation of the y-coordinate. We saw that an equivalent method in genus 2 was to work on the Kummer surface. Of course we have the same restriction in the case of genus 2 curves, namely that it is not possible to add two divisors except if their difference is known. If $\mathcal{D}$ is some divisor, recalling that our goal is the computation of $n\mathcal{D}$ for some integer n, the principle is, as it was already the case for elliptic curves, to use pairs of consecutive powers of $\mathcal{D}$, so that the difference between the two components of the pair is always known and equal to $\mathcal{D}$. The algorithm for scalar multiplication is as follows:

Algorithm 2. *Montgomery scalar multiplication algorithm for genus 2 curves*
Input : $\mathcal{D} \in \mathcal{J}(\mathbf{K})$ *and* $n \in \mathbb{Z}$.
Output : $\kappa(n\mathcal{D})$, *the image in the Kummer surface of* $n\mathcal{D}$.

Step 1. Initialize $(\mathcal{A}, \mathcal{B}) = ((0,0,0,1), \kappa(\mathcal{D}))$ *where* $(0,0,0,1)$ *is the image in the Kummer surface of the neutral element on* $\mathcal{J}(\mathcal{C})$.

Step 2. If the bit of n *is 0,* $(\mathcal{A}, \mathcal{B}) = (2\mathcal{A}, \mathcal{A} + \mathcal{B})$.

Step 3. If the bit of n *is 1,* $(\mathcal{A}, \mathcal{B}) = (\mathcal{A} + \mathcal{B}, 2\mathcal{B})$.

Step 4. After doing that for each bit of n, *return* $\mathcal{A}$.

Note that, at each step, we always have $\mathcal{B} - \mathcal{A} = \kappa(\mathcal{D})$ so that the addition of $\mathcal{A}$ and $\mathcal{B}$ is possible.

5.2 Number of Operations

At each step of the algorithm, we perform both an addition and a doubling, hence we just have to count the number of operations required for each of them. In the following, M will denote a multiplication in $\mathbf{K}$ and S a squaring.

Table 1. Addition of $\mathcal{A}$ and $\mathcal{B}$ in $\mathcal{K}(\mathcal{C})$ if $\mathcal{A}-\mathcal{B}$ is known

expressions	operations
precomputations $\{k_i l_j\}_{i,j=1..4}$	$16M$
$\varphi_{11}(\mathcal{A},\mathcal{B})$	S
$\varphi_{12}(\mathcal{A},\mathcal{B})$	$6M$
$\varphi_{33}(\mathcal{A},\mathcal{B})$	S
$\varphi_{14}(\mathcal{A},\mathcal{B})$	$6M$
$\kappa(\mathcal{A}+\mathcal{B})$	$3M$
total	$31M+2S$

Remark 1. The 31 multiplications include 7 multiplications by coefficients of the curve.

Table 2. Doubling of $\mathcal{A}$ in $\mathcal{K}(\mathcal{C})$

expressions	operations
precomputations $\{k_i k_j\}_{i,j=1..4}$	$6M+4S$
$\delta_1(\mathcal{A},\mathcal{B})$	$5M$
$\delta_2(\mathcal{A},\mathcal{B})$	$11M$
$\delta_3(\mathcal{A},\mathcal{B})$	$5M$
$\delta_4(\mathcal{A},\mathcal{B})$	$5M$
total	$31M+5S$

Remark 2. The 31 multiplications include 16 multiplications by coefficients of the curve. Moreover the multiplications $f_3k_1k_3$, $f_3(k_1^2+k_3^2)$, $f_4(k_1k_4+k_2k_3)$ and $f_2(k_1k_2+k_3k_4)$ are not counted in δ_4 since they were already computed in δ_2. Finally, we of course assumed that f_2f_4, $f_3^2-4f_2f_4$ and $f_2^2+f_4^2$ were precomputed.

Hence, on a curve in the Montgomery form as in (2), the scalar multiplication using the Montgomery method requires $69|n|_2$ base field multiplications (assuming that a squaring is a multiplication), where $|n|_2$ is the number of bits of n.

5.3 Comparison with Usual Algorithms for Scalar Multiplication

To date, the best algorithms for scalar multiplication on genus 2 curves defined over a field of odd characteristic are obtained by using mixed weighted projective coordinates [21]. In this case, Lange needs 41 multiplications both for a mixed addition and for a doubling. Hence our formulas requires fewer base field operations. But, in the Montgomery algorithm, we must perform both an addition and a doubling for each bit of the exponent whereas one can use efficient algorithms (like the sliding window method) with Lange's formulas. Nevertheless, this algorithm is still interesting for many reasons.

- As was the case for elliptic curves and as explained in Sect. 2.5, the Montgomery algorithm is resistant to side-channel attacks, contrary to other algorithms for scalar multiplications. For this reason it will be of interest to people who need to implement hyperelliptic curves protocols on smart cards or systems sensitive to side-channel attacks. For example, if one wants to make safe algorithms using mixed weighted projective coordinates, one needs to perform an extra addition when the bit of the exponent is one. In this case, for each bit of the exponent, 82 base field operations are required and with only 69, our algorithm allows a gain of 16 percent, which is significant.
- This algorithm is very easy to implement, there are no precomputations (as in the sliding window method) and an element on the Kummer surface requires only 4 base field elements whereas weighted projective coordinates require 8 of them so that it is also interesting in terms of memory usage. This last remark will be an advantage for constrained environments like smart cards.
- It is very dependent of the coefficients of the curve. Indeed there are 23 multiplications by these coefficients but only 2 in Lange's formulas. Hence a good choice of the coefficients of the curve certainly allows better timings. This is the subject of the following section.

5.4 Some Special Cases

In order to decrease the number of base field operations for our algorithm, certain choices of coefficients of the curve are better to use. For example there are 6 multiplications by f_3 in the formulas given in Sects. 4.3 and 4.4 so that, if one chooses $f_3 = 0$ or 1, the total amount of multiplications necessary for each bit of the exponent is 63 instead of 69. In the following table, we summarize the gain obtained in each operation. Let us note that there is no gain for the calculation of φ_{11}, φ_{33} and precomputations.

Table 3. Gain obtained if ...

	$f_2 = 0$	f_2 small	$f_3 = 0$ or small	$f_4 = 0$	f_4 small
φ_{12}	1	1	1	2	1
φ_{14}	2	2	1	1	1
δ_1	1	1	1	1	1
δ_2	2	2	2	2	2
δ_3	1	1	1	1	1
δ_4	2	1	0	2	1
total	9	8	6	9	7

Remark 3. If two of these conditions on the coefficients are satisfied the gain obtained is just the sum of the gains.

Of course this kind of restriction implies that fewer curves are taken into account. For example, if $\mathbf{K} = \mathbb{F}_p$ and $f_3 = 0$, one can only choose three coefficients in $\mathbb{F}_p$

(namely f_2, f_4 and B) so that the number of such curves is $O(p^3)$. Thus we lose in generality. However, in cryptography, one only needs to find a curve such that the order of its jacobian is divisible by a huge prime number. For this, one needs enough choices of curves in order to be able to find a curve with this property and $O(p^3)$ choices are of course widely sufficient.
Let us now examine more precisely a particular case and compare our algorithm to usual ones. Let $\mathcal{C}$ be a genus 2 curve defined over $\mathbf{K}$ by an equation of the form

$$By^2 = x^5 + f_3x^3 + \varepsilon x^2 + x \text{ with } \varepsilon = 0 \text{ or } \pm 1 \text{ and } B \text{ and } f_3 \in \mathbf{K} . \quad (4)$$

There are $O(p^2)$ curves in this form (which is sufficient to find one of these with nice properties for use in cryptography). Here, our algorithm of scalar multiplication requires 52 multiplications for each bit of the exponent whereas with mixed weighted projective coordinates,

- a sliding window method with window size equal to 4 requires in average 48 multiplications,
- a classical double and add requires 61 multiplications on average,
- a side-channel attack resistant double and add requires 81 multiplications.

Thus, our algorithm is 15 percent faster than a double and add, not so far from the sliding window method (around 7 percent) and much more efficient if one wants the operation to be resistant to side-channel attacks. Indeed, in this case, we obtain a gain of 36 percent. Of course one can even be faster than the sliding window method by choosing a small coefficient f_3 but the number of such curves becomes small.

Remark 4. Another means to accelerate this algorithm would be to choose f_2, f_3 and f_4 one word long. For example, on a 32 bits processor, if we are working on some finite field of cryptographic size for genus 2 curves, a multiplication of a coefficient of the curve and an element of the base field is about three times faster than the usual multiplication in the base field. Hence, as there are 23 multiplications by coefficients of the curve, our algorithm will require the equivalent of 53 multiplications, which is not so bad.

5.5 Examples

In this section, the base field is the prime field $\mathbb{F}_{2^{80}+13}$ (so that cryptosystems based on genus 2 curves defined over this field have the same security level than those based on elliptic curves defined over some 160 bits prime field). Let $\mathcal{C}_1$, $\mathcal{C}_2$ and $\mathcal{C}_3$ be the genus 2 curves respectively defined by the equations

$$\begin{aligned} 442946377804223815965 77\, y^2 &= x^5 + 271937476575616687835 34\, x^4 \\ &\quad + 297559220985752192390 37\, x^3 \\ &\quad + 760478620401411267378 26\, x^2 + x \ , \\ 103779310474567225222 92\, y^2 &= x^5 + 773041988679881578656 77\, x^3 + x^2 + x \ , \\ 694188370684424938642 20\, y^2 &= x^5 + x^3 + x \ . \end{aligned}$$

We compared our algorithm on these curves with a sliding window of size 4, a classical double and add and a double and always add (used to resist against side-channel attacks). For these three algorithms, we of course always used the weighted projective coordinates as in [21] which are the more efficient ones. In the following table, we provide the timings obtained using GMP 4.1.2 on a Pentium IV 3.06 GHz. We carried out 1000 scalar multiplications in each case with various divisors and 160 bits exponents.

Table 4. Timings

	$\mathcal{C}_1$	$\mathcal{C}_2$	$\mathcal{C}_3$
Sliding window	13.4 ms	13.3 ms	12.9 ms
Double and add	16 ms	16 ms	15.5 ms
Double and always add	21.5 ms	21.5 ms	21 ms
Montgomery method	18.3 ms	13.6 ms	11.9 ms

6 Conclusion and Prospects

Thanks to the theory of the Kummer surface of a hyperelliptic curve of genus 2, we have generalized to genus 2 curves the method of Montgomery for scalar multiplication on elliptic curves. As Montgomery does for elliptic curves, we restrict to curves transformable into Montgomery-like form. However, there are no theoretical obstructions to generalize this method to all genus 2 curves. Indeed Propositions 4.3 and 4.4 remain valid but the total amount of multiplications to compute the biquadratic forms is really huge so that this method is not competitive with the classical ones. This is not so surprising since it was already the case for elliptic curves.

In fact, for people interested in cryptography, this restriction is not very important since the number of choices of curves remains largely the same. The only significant restriction is that the order of the jacobian of such curves is even and then cannot be prime. But working with a jacobian whose order is twice a prime is not less efficient than working with a prime order.

For elliptic curves, the standards are not transformable into the Montgomery form because of this restriction and it's really a shame because the Montgomery method for scalar multiplication is the most interesting one (the fastest, easy to implement, resistant to side-channel attacks). Up to now, there are no standards for genus 2 curves. If such standards exist one day, it would be useful to take the method that we developed into account.

Moreover, we have seen that, with some restrictions, we obtain very interesting timings for the scalar multiplication on the jacobian of genus 2 curves. It would be nice to verify (even if there is no reason for this) that these restrictions are not awkward for finding jacobians suitable for cryptography (i.e. with a large prime dividing the order). Unfortunately, algorithms for finding the order of the jacobian over $\mathbb{F}_p$ are still under development ([11], [22]).

Finally, it would be very interesting to study the case of the characteristic 2, since it is in that case that this method is the most efficient for elliptic curves.

For this, all the necessary mathematical objects, such as the Kummer surface, remain to be defined.

References

1. Brier, E., Joye, M.: Weierstrass Elliptic Curves and Side-Channel Attacks, Public Key Cryptography, Lecture Notes in Computer Science, **2274** (2002)
2. Cantor, D. G.: Computing on the Jacobian of a hyperelliptic curve. Math. Comp., **48** (1987) 95–101
3. Cassel, J. W. S., Flynn, E. V.: Prolegomena to a middlebrow arithmetic of curves of genus 2, London Mathematical Society Lecture Note Series, **230** (1996)
4. Cohen, H., Miyaji, A., Ono, T.: Efficient elliptic curve exponentiation using mixed coordinates, Asiacrypt'98, Lecture Notes in Computer Science, **1514** (1998) 51–65
5. Coron, J. S.: Resistance against differential power analysis for elliptic curve cryptosystems, CHES'99, Lecture Notes in Computer Science, **1717** (1999) 292–302
6. Doche, C., Duquesne, S.: Manual for the elliptic curve library, Arehcc report (2003)
7. Duquesne, S.: Improvement of the Montgomery method for general elliptic curves defined over $\mathbb{F}_p$, preprint (2003)
8. Flynn, E. V.: The group law on the Jacobian of a curve of genus 2, J. reine angew. Math., **439** (1993), 45–69
9. Flynn, E. V.: ftp site, ftp://ftp.liv.ac.uk/pub/genus2/kummer
10. Fischer, W., Giraud, C., Knudsen, E. W., Seifert, J. P.: Parallel scalar multiplication on general elliptic curves over $\mathbb{F}_p$ hedged against Non-Differential Side-Channel Attacks, preprint
11. Gaudry, P., Schost, E.: Construction of secure random curves of genus 2 over prime fields, Eurocrypt'04, Lecture Notes in Computer Science (2004)
12. Harley, R.: Fast arithmetic on genus 2 curves, available at http://cristal.inria.fr/~harley/hyper (2000)
13. Izu, T., Takagi, T.: A fast Elliptic Curve Multiplication Resistant against Side Channel Attacks, preprint
14. Koblitz, N.: Elliptic curve cryptosystems, Math. Comp., **48** (1987) 203–209
15. Koblitz, N.: Algebraic aspects of cryptography, Algorithms and Computation in Mathematics, **3** (1998)
16. Kocher, P. C.: Timing attacks on implementations of DH, RSA, DSS and other systems, CRYPTO'96, Lecture Notes in Computer Science, **1109** (1996), 104–113
17. Kocher, P. C., Jaffe, J., Jun, B.: Differential power analysis, CRYPTO'99, Lecture Notes in Computer Science, **1666** (1999) 388-397
18. Koyama, K., Tsuruoka, Y.: Speeding up elliptic cryptosystems by using a signed binary window method, Crypto'92, Lecture Notes in Computer Science, **740** (1993) 345–357
19. Lange, T.: Efficient Arithmetic on Genus 2 Hyperelliptic Curves over Finite Fields via Explicit Formulae, Cryptology ePrint Archive, **121** (2002)
20. Lange, T.: Inversion-Free Arithmetic on Genus 2 Hyperelliptic Curves, Cryptology ePrint Archive, **147** (2002)
21. Lange, T.: Weighted Coordinates on Genus 2 Hyperelliptic Curves, Cryptology ePrint Archive, **153** (2002)
22. Matsuo, K. Chao, J. Tsujii, S.: An improved baby step giant step algorithm for point counting of hyperelliptic curves over finite fields, ANTS-V, Lecture Notes in Computer Science, **2369** (2002) 461–474

23. Miller, V. S.: Use of elliptic curves in cryptography, Crypto'85, Lecture Notes in Computer Science, **218** (1986) 417–426
24. Montgomery, P. L.: Speeding the Pollard and elliptic curve methods of factorization, Math. Comp., **48** (1987) 243–164
25. Mumford, D.: Tata lectures on Theta II, Birkhäuser (1984)
26. Okeya, K., Kurumatani, H., Sakurai, K.: Elliptic curves with the Montgomery-form and their cryptographic applications, Public Key Cryptography, Lecture Notes in Computer Science, **1751** (2000) 238–257
27. Okeya, O., Sakurai, K.: Efficient Elliptic Curve Cryptosystems from a Scalar Multiplication Algorithm with Recovery of the y-Coordinate on a Montgomery-Form Elliptic Curve, Cryptographic Hardware and Embedded Systems, Lecture Notes in Computer Science, **2162** (2001) 126–141
28. Quisquater, J. J., Samyde, D.: ElectroMagnetic Analysis (EMA): Measures and Countermeasures for Smart Cards, e-smart 2001, Lecture Notes in Computer Science, **2140**, (2001), 200–210

Improved Weil and Tate Pairings for Elliptic and Hyperelliptic Curves

Kirsten Eisenträger[1*], Kristin Lauter[2], and Peter L. Montgomery[2]

[1] School of Mathematics, Institute for Advanced Study, Einstein Drive, Princeton, NJ 08540 eisentra@ias.edu

[2] Microsoft Research, One Microsoft Way, Redmond, WA 98052
klauter@microsoft.com, petmon@microsoft.com

Abstract. We present algorithms for computing the *squared* Weil and Tate pairings on elliptic curves and the *squared* Tate pairing on hyperelliptic curves. The squared pairings introduced in this paper have the advantage that our algorithms for evaluating them are deterministic and do not depend on a random choice of points. Our algorithm to evaluate the squared Weil pairing is about 20% more efficient than the standard Weil pairing. Our algorithm for the squared Tate pairing on elliptic curves matches the efficiency of the algorithm given by Barreto, Lynn, and Scott in the case of arbitrary base points where their denominator cancellation technique does not apply. Our algorithm for the squared Tate pairing for hyperelliptic curves is the first detailed implementation of the pairing for general hyperelliptic curves of genus 2, and saves an estimated 30% over the standard algorithm.

1 Introduction

The Weil and Tate pairings have been proposed for use in cryptography, including one round 3-way key establishment, identity-based encryption, and short signatures [9]. For a fixed positive integer m, the Weil pairing e_m is a bilinear map that sends two m-torsion points on an elliptic curve to an mth root of unity in the field. For elliptic curves, the Weil pairing is a quotient of two applications of the Tate pairing, except that the Tate pairing needs an exponentiation which the Weil pairing omits.

For cryptographic applications, the objective is a bilinear map with a specific recipe for efficient evaluation, and no clear way to invert. The Weil and Tate pairings provide such tools. Each pairing has a practical definition which involves finding functions with prescribed zeros and poles on the curve, and evaluating those functions at pairs of points.

For elliptic curves, Miller [10] gave an algorithm for the Weil pairing. (See also the Appendix B to [3], for a probabilistic implementation of Miller's algorithm which recursively generates and evaluates the required functions based on a

* The research for this paper was done while the first author was visiting Microsoft Research. We thank S. Galbraith for constructive suggestions.

D.A. Buell (Ed.): ANTS 2004, LNCS 3076, pp. 169–183, 2004.

random choice of points.) For Jacobians of hyperelliptic curves, Frey and Rück [7] gave a recursive algorithm to generate the required functions, assuming the knowledge of intermediate functions having prescribed zeros and poles.

For elliptic curves, we present an improved algorithm for computing the *squared* Weil pairing, $e_m(P,Q)^2$. Our deterministic algorithm does not depend on a random choice of points for evaluation of the pairing. Our algorithm saves about 20% over the standard implementation of the Weil pairing [3]. We use this idea to obtain an improved algorithm for computing the *squared* Tate pairing for elliptic and hyperelliptic curves. The Tate pairing is already more efficient to implement than the Weil pairing. Our new squared Tate pairing is more efficient than Miller's algorithm for the Tate pairing for elliptic curves, for another 20% saving. For pairings on special families of elliptic curves in characteristics 2 and 3, some implementation improvements were given in [8] and [1]. Another deterministic algorithm was given in [1]. In [2], an algorithm for the pairing on ordinary elliptic curves in arbitrary characteristic is given. Our squared pairing matches the efficiency of the algorithm in [2] in the case of arbitrary base points where their denominator cancellation technique does not apply.

For hyperelliptic curves, we use Cantor's algorithm to produce the intermediate functions assumed by Frey and Rück. We define a squared Tate pairing for hyperelliptic curves, and use the knowledge of these intermediate functions to implement the pairing and give an example. Our analysis shows that using the squared Tate pairing saves roughly 30% over the standard Tate pairing for genus 2 curves. Our algorithm for the pairing on hyperelliptic curves can be thought of as a partial generalization of the Barreto-Lynn-Scott algorithm for elliptic curves in the sense that we give a deterministic algorithm which is more efficient to evaluate than the standard one. It remains to be seen whether some denominator cancellation can also be achieved in the hyperelliptic case by choosing base points of a special form as was done for elliptic curves in [2]. For a special family of hyperelliptic curves, Duursma and Lee have given a closed formula for the pairing in [5], but ours is the first algorithm for the Tate pairing on general hyperelliptic curves, and we have implemented the genus 2 case. The squared Weil pairing or the squared Tate pairing can be substituted for the Weil or Tate pairing in many of the above cryptographic applications.

The paper is organized as follows. Section 2 provides background on the Weil pairing for elliptic curves and gives the algorithm for computing the squared Weil pairing. Section 3 does the same for the squared Tate pairing for elliptic curves. Section 4 presents the squared Tate pairing for hyperelliptic curves and shows how to implement it. Section 5 gives an example of the hyperelliptic pairing.

2 Weil Pairings for Elliptic Curves

2.1 Definition of the Weil Pairing

Let E be an elliptic curve over a finite field $\mathbb{F}_q$. In the following $\mathbf{O}$ denotes the point at infinity on E. If P is a point on E, then $x(P)$ and $y(P)$ denote the rational functions mapping P to its affine x- and y-coordinates.

Let m be a positive integer. We will use the Weil pairing $e_m(\cdot,\cdot)$ definition in [11, p. 107]. To compute $e_m(P,Q)$, given two distinct m-torsion points P and Q on E over an extension field, pick two divisors $\mathcal{A}_P$ and $\mathcal{A}_Q$ which are equivalent to $(P)-(\mathbf{O})$ and $(Q)-(\mathbf{O})$, respectively, and such that $\mathcal{A}_P$ and $\mathcal{A}_Q$ have disjoint support. Let $f_{\mathcal{A}_P}$ be a function on E whose divisor of zeros and poles is $(f_{\mathcal{A}_P}) = m\cdot\mathcal{A}_P$. Similarly, let $f_{\mathcal{A}_Q}$ be a function on E whose divisor of zeros and poles is $(f_{\mathcal{A}_Q}) = m\cdot\mathcal{A}_Q$. Then

$$e_m(P,Q) = \frac{f_{\mathcal{A}_P}(\mathcal{A}_Q)}{f_{\mathcal{A}_Q}(\mathcal{A}_P)}.$$

2.2 Rational Functions Needed in the Evaluation of the Pairing

Fix an integer $m>0$ and an m-torsion point P on an elliptic curve E. Let $\mathcal{A}_P$ be a divisor equivalent to $(P)-(\mathbf{O})$. For a positive integer j, let $f_{j,\mathcal{A}_P}$ be a rational function on E with divisor

$$(f_{j,\mathcal{A}_P}) = j\mathcal{A}_P - (jP) + (\mathbf{O})$$

This means that $f_{j,\mathcal{A}_P}$ has j-fold zeros and poles at the points in $\mathcal{A}_P$, as well as a simple pole at jP and a simple zero at $\mathbf{O}$, and no other zeros or poles. Since $mP=\mathbf{O}$, it follows that $f_{m,\mathcal{A}_P}$ has divisor $m\mathcal{A}_P$, so in fact $f_{\mathcal{A}_P} = f_{m,\mathcal{A}_P}$. Throughout the paper the notation $f_{j,P}$ will be used to denote the function $f_{j,\mathcal{A}_P}$ with $\mathcal{A}_P = (P)-(\mathbf{O})$.

Silverman [11, Cor. 3.5, p. 67] shows that these functions exist. Each $f_{i,\mathcal{A}_P}$ is unique up to a nonzero multiplicative scalar. Miller's algorithm gives an iterative construction of these functions (see for example [1]). The construction of $f_{1,\mathcal{A}_P}$ depends on $\mathcal{A}_P$. Given $f_{i,\mathcal{A}_P}$ and $f_{j,\mathcal{A}_P}$, one constructs $f_{i+j,\mathcal{A}_P}$ as the product

$$f_{i+j,\mathcal{A}_P} = f_{i,\mathcal{A}_P}\cdot f_{j,\mathcal{A}_P}\cdot \frac{g_{iP,jP}}{g_{(i+j)P}}. \tag{1}$$

Here the notation $g_{U,V}$ (two subscripts) denotes the line passing through the points U and V on E. The notation g_U (one subscript) denotes the vertical line through U and $-U$. For more details on efficiently computing $f_{m,\mathcal{A}_P}$, see [6].

2.3 Squared Weil Pairing for Elliptic Curves

The purpose of this section is to construct a new pairing, which we call the 'squared Weil pairing', and which has the advantage of being more efficient to compute than Miller's algorithm for the original Weil pairing. Our algorithm also has the advantage that it is guaranteed to output the correct answer and does not depend on inputting a randomly chosen point. In contrast Miller's algorithm may restart, since the randomly chosen point can cause the algorithm to fail.

2.4 Algorithm for $e_m(P,Q)^2$

Fix a positive integer m and the curve E. Given two m-torsion points P and Q on E, we want to compute $e_m(P,Q)^2$. Start with an addition-subtraction chain for m. That is, after an initial 1, every element in the chain is a sum or difference of two earlier elements, until an m appears. Well-known techniques give a chain of length $O(\log(m))$. For each j in the addition-subtraction chain, form a tuple $t_j = [jP,\ jQ,\ n_j,\ d_j]$ such that

$$\frac{n_j}{d_j} = \frac{f_{j,P}(Q)\ f_{j,Q}(-P)}{f_{j,P}(-Q)\ f_{j,Q}(P)}. \tag{2}$$

Start with $t_1 = [P,\ Q,\ 1,\ 1]$. Given t_j and t_k, this procedure gets t_{j+k}:

1. Form the elliptic curve sums $jP + kP = (j+k)P$ and $jQ + kQ = (j+k)Q$.
2. Find coefficients of the line $g_{jP,kP}(X) = c_0 + c_1 x(X) + c_2 y(X)$.
3. Find coefficients of the line $g_{jQ,kQ}(X) = c_0' + c_1' x(X) + c_2' y(X)$.
4. Set

$$\begin{aligned} n_{j+k} &= n_j n_k (c_0 + c_1 x(Q) + c_2 y(Q))\ (c_0' + c_1' x(P) - c_2' y(P)) \\ d_{j+k} &= d_j d_k (c_0 + c_1 x(Q) - c_2 y(Q))\ (c_0' + c_1' x(P) + c_2' y(P)). \end{aligned}$$

A similar construction gives t_{j-k} from t_j and t_k. The vertical lines through $(j+k)P$ and $(j+k)Q$ do not appear in the formulae for n_{j+k} and d_{j+k}, because the contributions from Q and $-Q$ (or from P and $-P$) are equal. When $j+k = m$, this simplifies to $n_{j+k} = n_j n_k$ and $d_{j+k} = d_j d_k$, since c_2 and c_2' will be zero.

When n_m and d_m are nonzero, then the computation

$$\frac{n_m}{d_m} = \frac{f_{m,P}(Q)\ f_{m,Q}(-P)}{f_{m,P}(-Q)\ f_{m,Q}(P)}.$$

has been successful, and we have the correct output. If, however, n_m or d_m is zero, then some factor such as $c_0 + c_1 x(Q) + c_2 y(Q)$ must have vanished. That line was chosen to pass through jP, kP, and $(-j-k)P$, for some j and k. It does not vanish at any other point on the elliptic curve. Therefore this factor can vanish only if $Q = jP$ or $Q = kP$ or $Q = (-j-k)P$. In all of these cases Q will be a multiple of P, ensuring $e_m(P,Q) = 1$.

2.5 Correctness Proof

Theorem 1 (Squared Weil Pairing Formula). *Let m be a positive integer. Suppose P and Q are m-torsion points on E, with neither being the identity and P not equal to $\pm Q$. Then the squared Weil pairing satisfies*

$$\frac{f_{m,P}(Q)\cdot f_{m,Q}(-P)}{f_{m,P}(-Q)\cdot f_{m,Q}(P)} = (-1)^m e_m(P,Q)^2.$$

Proof. Let R_1, R_2 be points on E such that the divisors $\mathcal{A}_P := (P+R_1)-(R_1)$ and $\mathcal{A}_Q := (Q+R_2)-(R_2)$ have disjoint support. Let $\mathcal{A}_{-Q} := (-Q+R_2)-(R_2)$. Let $f_{\mathcal{A}_P}$ and $f_{\mathcal{A}_Q}$ be as above. Then

$$e_m(P,Q) = \frac{f_{\mathcal{A}_P}((Q+R_2)-(R_2))}{f_{\mathcal{A}_Q}((P+R_1)-(R_1))} = \frac{f_{\mathcal{A}_P}(Q+R_2)}{f_{\mathcal{A}_P}(R_2)} \cdot \frac{f_{\mathcal{A}_Q}(R_1)}{f_{\mathcal{A}_Q}(P+R_1)}.$$

Let $g(X) = f_{m,P}(X-R_1)$. Then $(g) = m(P+R_1) - m(R_1) = m\mathcal{A}_P = (f_{\mathcal{A}_P})$, This implies $g(X)/f_{\mathcal{A}_P}(X)$ is constant and

$$\frac{f_{\mathcal{A}_P}(Q+R_2)}{f_{\mathcal{A}_P}(R_2)} = \frac{g(Q+R_2)}{g(R_2)} = \frac{f_{m,P}(Q+R_2-R_1)}{f_{m,P}(R_2-R_1)}.$$

Similarly

$$\frac{f_{\mathcal{A}_Q}(R_1)}{f_{\mathcal{A}_Q}(P+R_1)} = \frac{f_{m,Q}(R_1-R_2)}{f_{m,Q}(P+R_1-R_2)}.$$

Plugging these into Miller's formula gives

$$e_m(P,Q) = \frac{f_{m,P}(Q+R_2-R_1)}{f_{m,P}(R_2-R_1)} \frac{f_{m,Q}(R_1-R_2)}{f_{m,Q}(P+R_1-R_2)}.$$

Using the same argument for $e_m(P,-Q)$ we obtain

$$\begin{aligned} e_m(P,-Q) &= \frac{f_{m,P}(-Q+R_2-R_1)}{f_{m,P}(R_2-R_1)} \frac{f_{m,-Q}(R_1-R_2)}{f_{m,-Q}(P+R_1-R_2)} \\ &= \frac{f_{m,P}(-Q+R_2-R_1)}{f_{m,P}(R_2-R_1)} \frac{f_{m,Q}(-R_1+R_2)}{f_{m,Q}(-P-R_1+R_2)} \end{aligned}$$

Hence we can simplify $e_m(P,Q)^2$ to

$$\frac{e_m(P,Q)}{e_m(P,-Q)} = \frac{f_{m,P}(Q+R_2-R_1)\, f_{m,Q}(R_1-R_2)\, f_{m,Q}(-P-R_1+R_2)}{f_{m,P}(-Q+R_2-R_1)\, f_{m,Q}(-(R_1-R_2))\, f_{m,Q}(P+R_1-R_2)}.$$

Let $R := R_2 - R_1$. This equation becomes

$$e_m(P,Q)^2 = \frac{f_{m,P}(Q+R)\, f_{m,Q}(-R)\, f_{m,Q}(-P+R)}{f_{m,P}(-Q+R)\, f_{m,Q}(R)\, f_{m,Q}(P-R)}. \tag{3}$$

Fix two linearly independent m-torsion points P and Q. The right side of (3) is a rational function of R; call it $\psi = \psi(R)$. Since $f_{m,P}$ can have zeros and poles only at P and $\mathbf{O}$, and $f_{m,Q}$ can have zeros and poles only at Q and $\mathbf{O}$, this function $\psi(R)$ can have zeros or poles only at $R = -Q$, Q, $P-Q$, $P+Q$, P, and $\mathbf{O}$. By looking at the factors of ψ we can check that at each of these points, the value of $\psi(R)$ is well-defined, because the zeros and poles cancel each other out. Since ψ is a rational function on an elliptic curve which does not have any zeros or poles, ψ must be constant. Since for certain values of R, $\psi(R) = e_m(P,Q)^2$, this must be the case for all values of R. Hence we may in particular choose $R = \mathbf{O}$, or equivalently $R_1 = R_2$. So let $R_1 = R_2$. By Lemma 1 below,

$$\frac{f_{m,Q}(R_1 - R_2)}{f_{m,Q}(-(R_1 - R_2))} = (-1)^m,$$

and by assumption $f_{m,P}$ does not have a zero or pole at Q and $f_{m,Q}$ does not have a zero or pole at P. Hence expression (3) simplifies to

$$e_m(P,Q)^2 = (-1)^m \, \frac{f_{m,P}(Q)\, f_{m,Q}(-P)}{f_{m,P}(-Q)\, f_{m,Q}(P)}. \tag{4}$$

Lemma 1. *Let $f : E \to \mathbb{F}_q$ be a rational function on E with a zero of order m (or a pole of order $-m$) at $\mathbf{O}$. Define $g : E \to \mathbb{F}_q$ by $g(X) = f(X)/f(-X)$. Then $g(\mathbf{O})$ is finite and $g(\mathbf{O}) = (-1)^m$.*

Proof. The rational function $h(X) = x(X)/y(X)$ has a zero of order 1 at $X = \mathbf{O}$. The function $f_1 = f/h^m$ has neither a pole nor a zero at $X = \mathbf{O}$, so $f_1(\mathbf{O})$ is finite and nonzero. We check that the rational function $\phi(X) = h(X)/h(-X)$ has no zeros and poles on E. Hence ϕ is constant. By computing $\phi(X)$ for a finite point $X = (x, y)$ on E with $x, y \neq 0$, we see that ϕ is equal to -1. Hence

$$g(X) = \frac{f(X)}{f(-X)} = \frac{h(X)^m f_1(X)}{h(-X)^m f_1(-X)} = \phi(X)^m \frac{f_1(X)}{f_1(-X)} = (-1)^m \frac{f_1(X)}{f_1(-X)},$$

and $g(\mathbf{O}) = (-1)^m$.

2.6 Estimated Savings

In this section we compare our algorithm for the squared Weil pairing to Miller's algorithm for the Weil pairing. We count operations in the underlying finite field, counting field squarings as field multiplications throughout. This analysis assumes that we use the short Weierstrass form for the elliptic curve E.

In practice, some of these arithmetic operations may be over a base field and others over an extension field. That issue is discussed in more detail in [8]. Without knowing the precise context of the application, we don't distinguish these, although individual costs may differ considerably.

Miller's algorithm. Miller's algorithm chooses two points R_1, R_2 on E, and lets $\mathcal{A}_P = (P+R_1)-(R_1)$ and $\mathcal{A}_Q = (P+R_2)-(R_2)$. Recall that in the notation of Section 2.1, $f_{\mathcal{A}_P}$ is a function whose divisor is $m\mathcal{A}_P$. As in Section 2.2, let $f_{j,\mathcal{A}_P}$ be a function with divisor $(f_{j,\mathcal{A}_P}) = j(P+R_1) - j(R_1) - (jP) + (\mathbf{O})$. This is the function f_j in the notation of [3, p. 611f.]. Then $f_{m,\mathcal{A}_P} = f_{\mathcal{A}_P}$. As pointed out in Equation (B.1) of [3, p. 612], (1) leads to the recurrence

$$f_{i+j,\mathcal{A}_P}(\mathcal{A}_Q) = f_{i,\mathcal{A}_P}(\mathcal{A}_Q) \cdot f_{j,\mathcal{A}_P}(\mathcal{A}_Q) \cdot \frac{g_{iP,jP}(\mathcal{A}_Q)}{g_{(i+j)P}(\mathcal{A}_Q)}. \tag{5}$$

During the computations, each $f_{j,\mathcal{A}_P}(\mathcal{A}_Q)$ is a known field element, unlike the unevaluated functions $f_{j,\mathcal{A}_P}$. Since $\mathcal{A}_Q$ has degree 0, the value of $f_{j,\mathcal{A}_P}(\mathcal{A}_Q)$ is unambiguous, whereas $f_{j,\mathcal{A}_P}$ is defined only up to a multiplicative scalar.

To compute the Weil pairing we need

$$e_m(P,Q) = \frac{f_{\mathcal{A}_P}(Q+R_2)}{f_{\mathcal{A}_P}(R_2)} \frac{f_{\mathcal{A}_Q}(R_1)}{f_{\mathcal{A}_Q}(P+R_1)} = \frac{f_{m,\mathcal{A}_P}(Q+R_2)}{f_{m,\mathcal{A}_P}(R_2)} \frac{f_{m,\mathcal{A}_Q}(R_1)}{f_{m,\mathcal{A}_Q}(P+R_1)}.$$

For integers j in an addition-subtraction chain for m, we will construct a tuple $t_j = [jP, jQ, n_j, d_j]$ where n_j and d_j satisfy

$$\frac{n_j}{d_j} = \frac{f_{j,\mathcal{A}_P}(Q+R_2)}{f_{j,\mathcal{A}_P}(R_2)} \frac{f_{j,\mathcal{A}_Q}(R_1)}{f_{j,\mathcal{A}_Q}(P+R_1)}.$$

To compute t_{i+j} from t_i and t_j, one uses the above recurrence (5) to derive the following expression for n_{i+j}/d_{i+j}:

$$\begin{aligned}\frac{n_{i+j}}{d_{i+j}} = \frac{n_i}{d_i} \cdot \frac{n_j}{d_j} \cdot \frac{g_{iP,jP}(Q+R_2)}{g_{iP,jP}(R_2)} \cdot \frac{g_{(i+j)P}(R_2)}{g_{(i+j)P}(Q+R_2)} \\ \cdot \frac{g_{iQ,jQ}(R_1)}{g_{iQ,jQ}(P+R_1)} \cdot \frac{g_{(i+j)Q}(P+R_1)}{g_{(i+j)Q}(R_1)}.\end{aligned} \tag{6}$$

To evaluate, for example, $g_{iP,jP}(Q+R_2)/g_{iP,jP}(R_2)$, start with the elliptic curve addition $iP+jP = (i+j)P$. This costs 1 field division and 2 field multiplications in the generic case where iP and jP have distinct x-coordinates and neither is **O**. Save the slope λ of the line $g_{iP,jP}(X) = y(X) - y(iP) - \lambda(x(X) - x(iP))$ through iP and jP. Two field multiplications suffice to evaluate $g_{iP,jP}(Q+R_2)$ and $g_{iP,jP}(R_2)$ given $Q+R_2$ and R_2. No more field multiplications or divisions are needed to compute the numerator and denominator of

$$\frac{g_{(i+j)P}(R_2)}{g_{(i+j)P}(Q+R_2)} = \frac{x(R_2) - x((i+j)P)}{x(Q+R_2) - x((i+j)P)}.$$

Repeat this once more to evaluate the last two fractions in (6). Overall these evaluations cost 8 field multiplications and 2 field divisions. We need 10 multiplications to multiply the six fractions, for an overall cost of 18 multiplications and 2 divisions.

Squared pairing. The squared pairing needs n_m/d_m where n_j/d_j is given by (2). The recurrence formula is

$$\frac{n_{i+j}}{d_{i+j}} = \frac{n_i}{d_i} \frac{n_j}{d_j} \frac{g_{iP,jP}(Q)}{g_{iP,jP}(-Q)} \frac{g_{(i+j)P}(-Q)}{g_{(i+j)P}(Q)} \frac{g_{iQ,jQ}(-P)}{g_{iQ,jQ}(P)} \frac{g_{(i+j)Q}(P)}{g_{(i+j)Q}(-P)}. \tag{7}$$

This time the update from $t_i = [iP, iQ, n_i, d_i]$ and t_j to t_{i+j} needs 2 elliptic curve additions. Each elliptic curve addition needs 2 multiplications and 1 division in the generic case. We can evaluate the numerator and denominator of

$$\frac{g_{iP,jP}(Q)}{g_{iP,jP}(-Q)} = \frac{y(Q) - y(iP) - \lambda(x(Q) - x(iP))}{y(-Q) - y(iP) - \lambda(x(-Q) - x(iP))}$$

with only 1 multiplication, since $x(Q) = x(-Q)$.

The fraction $g_{(i+j)P}(-Q)/g_{(i+j)P}(Q)$ simplifies to 1 since $g_{(i+j)P}(X)$ depends only on $x(X)$, not $y(X)$. Overall 6 multiplications and 2 divisions suffice to evaluate the numerators and denominators of the six fractions in (7). We multiply the four non-unit fractions with 6 field multiplications.

Overall, the squared Weil pairing advances from t_i and t_j to t_{i+j} with 12 field multiplications and 2 field divisions in the generic case, compared to 18 field multiplications and 2 field divisions for Miller's method. When $i = j$, each algorithm needs 2 additional field multiplications due to the elliptic curve doublings. Estimating a division as 5 multiplications, this is roughly a 20% savings.

3 Squared Tate Pairing for Elliptic Curves

3.1 Squared Tate Pairing Formula

Let m be a positive integer. Let E be defined over $\mathbb{F}_q$, where m divides $q-1$. Let $E(\mathbb{F}_q)[m]$ denote the m-torsion points on E over $\mathbb{F}_q$. Assume $P \in E(\mathbb{F}_q)[m]$, and $Q \in E(\mathbb{F}_q)$, with neither being the identity and P not equal to a multiple of Q. The Tate pairing $\phi_m(P,Q)$ on $E(\mathbb{F}_q)[m] \times E(\mathbb{F}_q)/mE(\mathbb{F}_q)$ is defined in [8] as

$$\phi_m(P,Q) := (f_{\mathcal{A}_P}(\mathcal{A}_Q))^{(q-1)/m},$$

with the notation and evaluation as for the Weil pairing above. Now we define

$$v_m(P,Q) := \left(\frac{f_{m,P}(Q)}{f_{m,P}(-Q)}\right)^{(q-1)/m},$$

where $f_{m,P}$ is as above, and call v_m the squared Tate pairing. To justify this terminology, we will show below that $v_m(P,Q) = \phi_m(P,Q)^2$.

3.2 Algorithm for $v_m(P,Q)$

Fix a positive integer m and the curve E. Given an m-torsion point P on E and a point Q on E, we want to compute $v_m(P,Q)$. As before, start with an addition-subtraction chain for m. For each j in the chain, form a tuple $t_j = [jP, n_j, d_j]$ such that

$$\frac{n_j}{d_j} = \frac{f_{j,P}(Q)}{f_{j,P}(-Q)}. \tag{8}$$

Start with $t_1 = [P,1,1]$. Given t_j and t_k, this procedure gets t_{j+k}:

1. Form the elliptic curve sum $jP + kP = (j+k)P$.
2. Find the line $g_{jP,kP}(X) = c_0 + c_1 x(X) + c_2 y(X)$.
3. Set

$$\begin{aligned} n_{j+k} &= n_j \cdot n_k \cdot (c_0 + c_1 x(Q) + c_2 y(Q)) \\ d_{j+k} &= d_j \cdot d_k \cdot (c_0 + c_1 x(Q) - c_2 y(Q)). \end{aligned}$$

A similar construction gives t_{j-k} from t_j and t_k. The vertical lines through $(j+k)P$ and $(j+k)Q$ do not appear in the formulae for n_{j+k} and d_{j+k}, because the contributions from Q and $-Q$ are equal. When $j+k=m$, one can further simplify this to $n_{j+k} = n_j \cdot n_k$ and $d_{j+k} = d_j \cdot d_k$, since c_2 will be zero. When n_m and d_m are nonzero, then the computation of (8) with $j=m$ is successful, and after raising to the $(q-1)/m$ power, we have the correct output. If some n_m or d_m were zero, then some factor such as $c_0 + c_1 x(Q) + c_2 y(Q)$ must have vanished. That line was chosen to pass through jP, kP, and $(-j-k)P$, for some j and k. It does not vanish at any other point on the elliptic curve. Therefore this factor can vanish only if $Q = jP$ or $Q = kP$ or $Q = (-j-k)P$ for some j and k. In all of these cases Q would be a multiple of P, contrary to our assumption.

3.3 Correctness Proof

Theorem 2. *Let m be a positive integer. Suppose $P \in E(\mathbb{F}_q)[m]$ and $Q \in E(\mathbb{F}_q)$ with neither being the identity and $P \neq \pm Q$. Then the squared Tate pairing is*

$$\phi_m(P,Q)^2 = \left(\frac{f_{m,P}(Q)}{f_{m,P}(-Q)}\right)^{(q-1)/m}.$$

Proof. Let R_1 and R_2 be as in the proof of Theorem 1. The proof proceeds exactly as the correctness proof for the Weil pairing. The only difference is that the factor of $(-1)^m$ is missing in the Tate pairing and so we have

$$\phi_m(P,Q)^2 = \frac{\phi_m(P,Q)}{\phi_m(P,-Q)} = \left(\frac{f_{m,P}(Q+R_2-R_1)}{f_{m,P}(-Q+R_2-R_1)}\right)^{(q-1)/m}.$$

By the same argument as in the proof for the Weil pairing we may choose $R_2 = R_1$, which gives us the desired formula.

3.4 Estimated Savings

This analysis is almost identical to that for the Weil pairing in Section 2.6. When analyzing Miller's algorithm for the Tate pairing, the main difference from Section 2.6 is that the analog of (6) has 2 fewer fractions to evaluate and combine. An elliptic curve addition costs 1 division and 2 multiplications, while 2 multiplications are needed to evaluate the numerators and denominators of the two fractions. Then 6 multiplications are needed to combine the numerators and denominators of the 4 fractions. Therefore each step of Miller's algorithm performing an addition costs 1 division and 10 multiplications.

For the squared Tate pairing, the analog of (7) also has 2 fewer fractions in it. An elliptic curve addition costs 1 division and 2 multiplications, while only 1 multiplication is needed to evaluate the numerators and denominators of the 2 fractions. Then 4 multiplications are needed to combine the numerators and denominators of the 3 non-unit fractions. Therefore each step of the squared Tate pairing algorithm performing an addition costs 1 division and 7 multiplications.

Overall, the squared Tate pairing advances from t_i and t_j to t_{i+j} with 7 field multiplications and 1 field division in the generic case, compared to 10 field multiplications and 1 field division for Miller's method applied to the usual Tate pairing. When $i = j$, each algorithm needs one additional field multiplication due to the elliptic curve doubling. Estimating a division as 5 multiplications, this is roughly a 20% savings.

Comparing our squared pairing to the algorithm from [2], the algorithms are equally efficient in the case of general base points, where there is no cancellation of denominators in their algorithm. In [2], the authors show that if the security multiplier is even ($k = 2d$) and the x-coordinate of the base point Q lies in a subfield $\mathbb{F}_{q^d}$, then the denominators in the Tate pairing evaluation disappear. This makes their method more efficient, but it is possible that adding this extra structure may weaken the system for cryptographic use. Also, in some situations, restricting to k even may not be desirable.

4 Squared Tate Pairing for Hyperelliptic Curves

Let C be a hyperelliptic curve of genus g given by an affine model $y^2 = f(x)$ with $\deg f = 2g+1$ over a finite field $\mathbb{F}_q$ not of characteristic 2. The curve C has one point at infinity, which we will denote by P_∞. Let $J = J(C)$ be the Jacobian of C. If $P = (x, y)$ is a point on C, then P' will denote the point $P' := (x, -y)$. We denote the identity element of J by **id**.

The Riemann-Roch theorem assures that each element D of J contains a representative of the form $A - gP_\infty$, where A is an effective divisor of degree g. In addition, we will always work with *semi-reduced* representatives, which means that if a point $P = (x, y)$ occurs in A then $P' := (x, -y)$ does not occur elsewhere in A. The effective divisor representing the identity element **id** will be gP_∞. For an element D of J and integer i, a representative for iD will be $A_i - gP_\infty$, where A_i is effective of degree g and semi-reduced.

To a representative $A_i - gP_\infty$ we associate two polynomials (a_i, b_i) which represent the divisor. The first polynomial, $a_i(x)$, is monic and has zeros at the x-coordinates of the points in the support of the divisor A_i. The second polynomial, $b_i(x)$, has degree less than $\deg(a_i(x))$, and the graph of $y = b_i(x)$ passes through the finite points in the support of the divisor A_i.

4.1 Definition of the Tate Pairing

Fix a positive integer m and assume that $\mathbb{F}_q$ contains a primitive mth root of unity ζ_m. The Tate pairing, $\phi_m : J(\mathbb{F}_q)[m] \times J(\mathbb{F}_q)/mJ(\mathbb{F}_q) \to \mathbb{F}_q^*/\mathbb{F}_q^{*m} \cong \langle\zeta_m\rangle$, is defined in [7, p. 871] explicitly as follows. Let $D \in J(\mathbb{F}_q)[m]$ and $E \in J(\mathbb{F}_q)$. Let $h_{m,D}$ be a function on C whose divisor is $(h_{m,D}) = mD$. Then

$$\phi_m(D, E) := h_{m,D}(E)^{\frac{q-1}{m}} \in \langle\zeta_m\rangle.$$

This pairing is known to be well-defined, bilinear, and non-degenerate. The value $h_{m,D}(E)$ is defined only up to mth powers, so we raise the result to the power

$\frac{q-1}{m}$ to eliminate all mth powers. Note that E is a divisor on the curve C, not an elliptic curve. We also assume that the support of E does not contain P_∞ and that E is prime to the A_i's. Actually E needs to be prime to only those representatives which will be used in the addition-subtraction chain for m, so to about $\log m$ divisors.

Frey and Rück [7, pp. 872-873] show how to evaluate the Tate pairing on the Jacobian of a curve assuming an explicit reduction algorithm for divisors on a curve. Cantor [4] gives such an algorithm for hyperelliptic curves when the degree of f is odd. In Section 4.4 below, we use Cantor's algorithm to explicitly compute the necessary intermediate functions. These functions will be used to evaluate the *squared* Tate pairing, but they could just as well be used to evaluate the usual Tate pairing.

4.2 Squared Tate Pairing v_m for Hyperelliptic Curves

Theorem 3. *Given an m-torsion element D of J and an element E of J, with representatives $D = P_1+P_2+\cdots+P_g-gP_\infty$ and $E = Q_1+Q_2+\cdots+Q_g-gP_\infty$ respectively, with P_i not equal to Q_j or Q'_j for any i, j define*

$$v_m(D,E) := \left(h_{m,D}(Q_1 - Q'_1 + Q_2 - Q'_2 + \cdots + Q_g - Q'_g)\right)^{(q-1)/m}.$$

Then $v_m(D,E) = \pm\phi_m(D,E)^2$ where $\phi_m(D,E)$ is the Tate pairing defined above.

Proof. Recall that if $P_1 = (x, y)$ is a point on C, then P'_1 is the point $(x, -y)$. Similarly, if $D = P_1 + P_2 + \cdots + P_g - gP_\infty$, let $D' = P'_1 + P'_2 + \cdots + P_g - gP_\infty$. For the proof, we will compute $\phi_m(2D, 2E)$.

Observe that $E - E' = Q_1 - Q'_1 + Q_2 - Q'_2 + \cdots + Q_g - Q'_g \sim 2E$ in the Jacobian of C, since $E + E' = (Q_1 + Q'_1 - 2P_\infty) + \cdots + (Q_g + Q'_g - 2P_\infty) \sim \mathbf{id}$. Let $h_{m,D}$ denote the rational function on C with divisor $(h_{m,D}) = mP_1 + \cdots + mP_g - 2gmP_\infty$ as above. Then the divisor of $h_{m,D}/h_{m,D'}$ has the form

$$\left(\frac{h_{m,D}}{h_{m,D'}}\right) = mP_1 - mP'_1 + \cdots + mP_g - mP'_g,$$

so $(h_{m,D}/h_{m,D'}) \sim 2mD$ in the Jacobian. That means we can use $h_{m,D}/h_{m,D'}$ to compute the pairing $\phi_m(2D, 2E)$. If Q is any point on C, then we can see by comparing the divisors of the two functions that $h_{m,D}(Q) = c \cdot h_{m,D'}(Q')$, where c is a constant which does not depend on Q.

Hence

$$\begin{aligned}\phi_m(2D, 2E) &= \left(\frac{h_{m,D}(E-E')}{h_{m,D'}(E-E')}\right)^{(q-1)/m} = \left(\frac{h_{m,D}(E-E')}{h_{m,D}(E'-E)}\right)^{(q-1)/m} \\ &= \left(h_{m,D}(E-E')^2\right)^{(q-1)/m}.\end{aligned}$$

Since $\phi_m(2D, 2E) = \phi_m(D, E)^4$, it follows that

$$\phi_m(D, E)^2 = \pm(h_{m,D}(Q_1 - Q'_1 + \cdots + Q_g - Q'_g))^{(q-1)/m}.$$

4.3 Functions Needed in the Evaluation of the Pairings

Let D be an m-torsion element of J. For a positive integer j, let $h_{j,D}$ denote a rational function on C with divisor

$$(h_{j,D}) = jA_1 - A_j - (j-1)gP_\infty.$$

Since D is an m-torsion element, we have that $A_m = gP_\infty$, so the divisor of $h_{m,D}$ is $(h_{m,D}) = mA_1 - m \cdot gP_\infty$. Each $h_{j,D}$ is well-defined up to a multiplicative constant.

Given positive divisors A_i and A_j, we can use Cantor's algorithm to find a positive divisor A_{i+j} and a function $u_{i,j}$ with divisor equal to

$$(u_{i,j}) = A_i + A_j - A_{i+j} - gP_\infty.$$

We construct $h_{j,D}(E)$ iteratively. For $j = 1$, let $h_{1,D}$ be 1. Suppose we have A_i, A_j, $h_{i,D}(E)$ and $h_{j,D}(E)$. Let $u_{i,j}$ be the above function on C. Then

$$h_{i+j,D}(E) = h_{i,D}(E) \cdot h_{j,D}(E) \cdot u_{i,j}(E).$$

4.4 Algorithm to Compute $v_m(D, E)$

Let D and E be as above. Form an addition-subtraction chain for m. For each j in the chain we need to form a tuple $t_j = [A_j, n_j, d_j]$ such that jD has representative $A_j - 2P_\infty$ and

$$\frac{n_j}{d_j} = \frac{h_{j,D}(Q_1)\, h_{j,D}(Q_2)}{h_{j,D}(Q_1')\, h_{j,D}(Q_2')}.$$

Let $t_1 = [A_1, 1, 1]$. Given t_i and t_j, let (a_i, b_i) and (a_j, b_j) be the polynomials corresponding to the divisors A_i and A_j. Do a composition step as in Cantor's algorithm to obtain (a, b) corresponding to $A_i + A_j$, without performing the reduction step. Let $d(x) = \gcd(a_i(x), a_j(x), b_i(x) + b_j(x))$. The output polynomials a, b, and d depend on i and j, but we will omit the subscripts here for ease of notation. If $d(x) = 1$, then $a(x) = a_i(x)a_j(x)$, and $b(x)$ is the polynomial with $\deg(b) < \deg(a)$ such that $y = b(x)$ passes through the distinct finite points in the support of A_i and A_j.

The reduction step described in [4, p. 99] then replaces (a, b) by $(\tilde{a}, \tilde{b})$ where $\tilde{a} = (f - b^2)/a, \tilde{b} \equiv -b \pmod{\tilde{a}}$ and $\deg(\tilde{b}) < \deg(\tilde{a})$. This reduction step is applied repeatedly until $\deg(\tilde{a}) \le g$. In the genus 2 situation, it follows from [4, p. 99] that at most one reduction step is performed.

Case i. If $g = 2$ and $\deg(a(x)) > 2$, a reduction step is performed. If we let

$$v_{i,j}(P) = \frac{a(x(P))}{b(x(P)) + y(P)}, \tag{9}$$

and

$$u_{i,j}(P) := v_{i,j}(P) \cdot d(x(P)),$$

then $(u_{i,j}) = A_i + A_j - A_{i+j} - 2P_\infty$, and

$$\frac{u_{i,j}(P)}{u_{i,j}(P')} = \frac{a(x(P))}{a(x(P'))} \cdot \frac{b(x(P')) + y(P')}{b(x(P)) + y(P)} \cdot \frac{d(x(P))}{d(x(P'))} = \frac{b(x(P')) + y(P')}{b(x(P)) + y(P)}.$$

Let

$$\begin{aligned} n_{i+j} &:= n_i \cdot n_j \cdot (b+y)(Q_1') \cdot (b+y)(Q_2') \\ d_{i+j} &:= d_i \cdot d_j \cdot (b+y)(Q_1) \cdot (b+y)(Q_2). \end{aligned} \tag{10}$$

There is no contribution from a in n_{i+j} and d_{i+j} because the contributions from Q_i and Q_i' are equal. This improves the algorithm for the Tate pairing in [7].

Case ii. If $g = 2$ and $\deg(a(x)) \leq 2$, then $u_{i,j}(P) = d(x(P))$. In this case we let $n_{i+j} := n_i \cdot n_j$ and $d_{i+j} := d_i \cdot d_j$.

Case iii. Suppose $g > 2$. If r reduction steps are needed, then to compute $u_{i,j}$, we obtain intermediate factors $v_{i,j}^{(1)}, \ldots, v_{i,j}^{(r)}$, one factor as in (9) per reduction step. Then $u_{i,j}$ will be the product $u_{i,j} := v_{i,j}^{(1)} \cdot \ldots \cdot v_{i,j}^{(r)} \cdot d(x(P))$.

Note: If we evaluate n_i and d_i at intermediate steps then it is not enough to assume that the divisors D and E are coprime. Instead, E must also be coprime to A_i for all i which occur in the addition chain for m. One way to ensure this condition is to require that E and D be linearly independent and that the polynomial $p(x)$ in the pair $(p(x), q(x))$ representing E be irreducible. There are other ways possible to achieve this, like changing the addition chain for m.

4.5 Estimated Savings for Genus 2

Using a straightforward implementation of Cantor's algorithm, the total costs for doubling and addition on the Jacobian of a hyperelliptic curve of genus 2 in odd characteristic, $C : y^2 = f(x)$, where f has degree 5, are as follows. Doubling an element costs 34 multiplications and 2 inversions. Adding two distinct elements of J costs 26 multiplications and 2 inversions. More efficient implementations of the group law may alter the total impact of our algorithm. Different field multiplication/inversion ratios and field sizes, as well as differing costs in an extension field will also affect the analysis, but these costs are chosen as representative for the purpose of estimating the savings.

Analysis of standard algorithm. Let $D := P_1 + P_2 - 2P_\infty$. Let R_1, R_2, R_3, R_4 be four points on C such that $Q_1 + Q_2 - 2P_\infty \sim R_1 + R_2 - R_3 - R_4$ in J. The algorithm in [7] computes t_{i+j} from t_i and t_j, where $t_i = [A_i, n_j, d_j]$ and

$$\frac{n_j}{d_j} = \frac{h_{j,D}(R_1)\, h_{j,D}(R_2)}{h_{j,D}(R_3)\, h_{j,D}(R_4)}.$$

The expression for n_{i+j}/d_{i+j} becomes

$$\frac{n_{i+j}}{d_{i+j}} = \frac{n_i}{d_i} \frac{n_j}{d_j} \frac{u_{i,j}(R_1)\, u_{i,j}(R_2)}{u_{i,j}(R_3)\, u_{i,j}(R_4)}.$$

To form $u_{i,j}$, we have to perform an addition or doubling step to obtain A_{i+j} from A_i and A_j. This costs 34 multiplications and 2 inversions for a doubling, 26 multiplications and 2 inversions for an addition. Then

$$u_{i,j}(P) = \frac{a(x(P))}{b(x(P)) + y(P)},$$

and to compute (n_{i+j}, d_{i+j}), we need to evaluate $u_{i,j}$ at four different points. Each evaluation of $a(x(P))$ costs 2 multiplications in a doubling step, 3 multiplications in an addition step (square or product of monic quadratics). Evaluation of $b(x(P))$ (cubic) costs 3 multiplications. Finally we multiply the partial numerators and denominators out, using 5 multiplications each, including the multiplications with n_i, n_j, d_i, and d_j. So the total cost for an addition step is 60 multiplications and 2 inversions, and the total cost for a doubling is 64 multiplications and 2 inversions.

Squared Tate pairing. The squared Tate pairing works with the divisor $Q_1 - Q_1' + Q_2 - Q_2' \sim 2Q_1 + 2Q_2 - 4P_\infty$. After adding A_i and A_j to obtain A_{i+j} as above, we need to form

$$\frac{n_{i+j}}{d_{i+j}} = \frac{n_i}{d_i} \frac{n_j}{d_j} \frac{u_{i,j}(Q_1)}{u_{i,j}(Q_2)} \frac{u_{i,j}(Q_1')}{u_{i,j}(Q_2')} .$$

As can be seen from (10) above, no evaluations of $a(x(P))$ are needed. For $i = 1, 2$, we need to evaluate $b(x(Q_i))$ and $b(x(Q_i'))$. This costs only 3 multiplications for each i, since the x-coordinates of Q_i and Q_i' are the same. Finally, we have to multiply the partial numerators and denominators, for a total cost of 12 multiplications for either a doubling or an addition.

So the total cost for an addition step is 38 multiplications and 2 inversions, and the total cost for a doubling is 46 multiplications and 2 inversions. Estimating an inversion as 4 multiplications, this is a 25% improvement in the doubling case and a 33% improvement in the addition case.

5 Example: $g = 2$, $p = 31$, $m = 5$

In this section, we evaluate the squared Tate pairing on 5-torsion on the Jacobian of a hyperelliptic genus 2 curve over a field of 31 elements. Let C be defined by the affine model $y^2 = f(x)$ where $f(x) = x^5 + 13x^4 + 2x^3 + 4x^2 + 11x + 1$. The group of points on the Jacobian of C over $\mathbb{F}_{31}$ has order $N = 1040$. Let D be the 5-torsion element of the Jacobian of C given by the pair of polynomials $D = [x^2 + 23x + 15, 13x + 28]$. Let E be the element of the Jacobian of C of order 260 given by the pair $E = [x^2 + 4x + 2, 29x + 20]$. Then the squared Tate pairing evaluated at D and E is $v_5(D, E) = 4$, where

$$h_{5,D} = \frac{(x+26)^2(x^4 + 19x^3 + 23x^2 + 16x + 19)(x^2 + 23x + 15)}{x^3 + 6x^2 + 9x + 21 + y}.$$

To illustrate the bilinearity of the pairing, look for example at $2D = [x^2 + 25x + 9, 10x + 6]$, $3D = [x^2 + 25x + 9, 21x + 25]$, and $2E = [x^2 + x + 3, 26x + 3]$. Then we compute that indeed $v_5(2D, E) = 16 = v_5(D, E)^2$, with

$$h_{5,2D} = \frac{(x+26)(x^4 + 19x^3 + 23x^2 + 16x + 19)^2(x^2 + 25x + 9)}{(x^3 + 6x^2 + 9x + 21 + y)^2},$$

and $v_5(D, 2E) = 16 = v_5(D, E)^2$, with $h_{5,D}$ as above. Also

$$v_5(3D, E) \equiv 2 \equiv v_5(D, E)^3 \pmod{31},$$

with

$$h_{5,3D} = \frac{(x+26)(x^4 + 19x^3 + 23x^2 + 16x + 19)^2(x^2 + 25x + 9)}{(30x^3 + 25x^2 + 22x + 10 + y)^2}.$$

References

1. Paulo S.L.M. Barreto, Hae Y. Kim, Ben Lynn, and Michael Scott. Efficient algorithms for pairing-based cryptosystems. In M. Yung, editor, *Advances in Cryptology - Crypto 2002*, pages 354–368. LNCS **2442**, Springer-Verlag, 2002.
2. Paulo S.L.M. Barreto, Ben Lynn, and Michael Scott. On the Selection of Pairing-Friendly Groups. In: M. Matsui and R. Zuccherato, editors, *Selected Areas in Cryptography - SAC 2003*, LNCS **3006**, Springer-Verlag, 2004.
3. Dan Boneh and Matthew Franklin. Identity-based encryption from the Weil pairing. *SIAM J. Comput.*, 32(3):586–615 (electronic), 2003.
4. David G. Cantor. Computing in the Jacobian of a hyperelliptic curve. *Math. Comp.*, 48(177):95–101, 1987.
5. Iwan M. Duursma and Hyang-Sook Lee. Tate Pairing Implementation for Hyperelliptic Curves $y^2 = x^p - x + d$. *Advances in Cryptology - Asiacrypt 2003*, pages 111–123, LNCS **2894**, Springer-Verlag, 2003.
6. Kirsten Eisenträger, Kristin Lauter, and Peter L. Montgomery. Fast elliptic curve arithmetic and improved Weil pairing evaluation. In Marc Joye, editor, *Topics in Cryptology - CT-RSA 2003, The Cryptographers' Track at the RSA Conference 2003, San Francisco, CA, USA, April 13-17, 2003, Proceedings*, pages 343–354, LNCS **2612**, Springer-Verlag, 2003.
7. Gerhard Frey and Hans-Georg Rück. A remark concerning m-divisibility and the discrete logarithm in the divisor class group of curves. *Math. Comp.*, 62(206):865–874, 1994.
8. Steven Galbraith, Keith Harrison, and David Soldera. Implementing the Tate Pairing. In Claus Fieker and David R. Kohel, editors, *Algorithmic Number Theory, 5th International Symposium, ANTS-V, Sydney, Australia, July 7-12, 2002*, pages 324–337. LNCS **2369**, Springer-Verlag, 2002.
9. Antoine Joux. The Weil and Tate pairings as building blocks for public key cryptosystems (survey). In Claus Fieker and David R. Kohel, editors, *Algorithmic Number Theory, 5th International Symposium, ANTS-V, Sydney, Australia, July 7-12, 2002*, pages 20–32. LNCS **2369**, Springer-Verlag, 2002.
10. Victor S. Miller. Short programs for functions on curves. Unpublished manuscript, 1986.
11. Joseph Silverman. *The Arithmetic of Elliptic Curves.* GTM **106**, Springer-Verlag, 1986.

Elliptic Curves $x^3 + y^3 = k$ of High Rank

Noam D. Elkies[1] and Nicholas F. Rogers[2]

[1] Department of Mathematics, Harvard University, Cambridge, MA 02140
elkies@math.harvard.edu
Supported in part by NSF grant DMS-0200687
[2] Department of Mathematics, Harvard University, Cambridge, MA 02140
nfrogers@math.harvard.edu

Abstract. We use rational parametrizations of certain cubic surfaces and an explicit formula for descent via 3-isogeny to construct the first examples of elliptic curves $E_k : x^3 + y^3 = k$ of ranks 8, 9, 10, and 11 over $\mathbb{Q}$. As a corollary we produce examples of elliptic curves over $\mathbb{Q}$ with a rational 3-torsion point and rank as high as 11. We also discuss the problem of finding the minimal curve E_k of a given rank, in the sense of both $|k|$ and the conductor of E_k, and we give some new results in this direction. We include descriptions of the relevant algorithms and heuristics, as well as numerical data.

1 Introduction

In the fundamental Diophantine problem of finding rational points on an elliptic curve E, one is naturally led to ask which abelian groups can occur as the group of rational points $E(\mathbb{Q})$. Mordell's theorem guarantees that $E(\mathbb{Q})$ is finitely generated, so we have

$$E(\mathbb{Q}) = E(\mathbb{Q})_{\text{tors}} \oplus \mathbb{Z}^r,$$

where r is the rank of E. Mazur's well-known work [Ma] completely classifies the possibilities for $E(\mathbb{Q})_{\text{tors}}$, but the behavior of the rank remains mysterious. Part of the "folklore" is the conjecture that there exist elliptic curves with arbitrarily large rank over $\mathbb{Q}$. But large rank examples are rare, and the record to date is 24 [MM]. One might further ask about the distribution of ranks in families of twists, or with prescribed Galois structure on the torsion subgroup; there is some evidence to suggest that conditions of this sort do not impose an upper bound on the rank.

A classical question in number theory is to describe the numbers k that can be written as the sum of two rational cubes. This leads one to study the elliptic curves

$$E_k : x^3 + y^3 = k$$

for $k \in \mathbb{Q}^*$. Clearly E_k and $E_{k'}$ are isomorphic if k/k' is a cube, so we can and will restrict our attention to positive cubefree integers k. A Weierstrass equation

D.A. Buell (Ed.): ANTS 2004, LNCS 3076, pp. 184–193, 2004.

for E_k is given by $Y^2 = X^3 - 432k^2$, where

$$X = \frac{12k}{y+x}, \qquad Y = 36k\frac{y-x}{y+x}.$$

As long as $k > 2$, the group $E_k(\mathbb{Q})_{\text{tors}}$ is trivial, so E_k has a nontrivial rational point if and only if its rank is positive. The distribution of ranks in this family is not well understood. Zagier and Kramarz [ZK] used numerical evidence for $k \leq 70000$ to conjecture that a positive proportion of the curves E_k have rank at least 2; however, more recent computations by Mark Watkins [Wa] suggest that, in fact, a curve E_k has rank 0 or 1 with probability 1. Still, the following conjecture is widely believed:

Conjecture 1. *There exist elliptic curves E_k with arbitrarily large rank over $\mathbb{Q}$.*

A proof of this conjecture seems beyond the reach of current techniques. So for now we content ourselves with constructing high-rank examples within this family (thereby adding to the body of supporting evidence), and gathering more data on the distribution of ranks so as to be able to formulate more precise conjectures. The main results of this paper are examples of curves E_k with rank r for each $r \leq 11$. For $r = 8, 9, 10, 11$ these are the first curves known of those ranks; for $r = 6, 7$ our curves have k smaller than previous records, and are proved minimal assuming some standard conjectures. For $r \leq 5$ we recover previously known k, and prove unconditionally that they are minimal.

Throughout, we make use of the fact that the curves E_k are 3-isogenous to the curves

$$E'_k : uv(u+v) = k$$

or, in Weierstrass form, $E'_k : V^2 = U^3 + 16k^2$, where

$$U = \frac{4k}{v}, \qquad V = \frac{8ku + 4kv}{v}$$

The isogeny is given by:

$$\phi : E_k \to E'_k, \qquad (x,y) \mapsto (u,v) = (\frac{y^2}{x}, -\frac{k}{xy}).$$

The dual isogeny, with respect to the Weierstrass equations for E_k and E'_k, is

$$\hat{\phi} : E'_k \to E_k, \qquad (U,V) \mapsto (X,Y) = \left(\frac{U^3 + 64k^2}{U^2}, \frac{V(U^3 - 128k^2)}{U^3}\right).$$

Applying Tate's Algorithm [Ta] to the curves E'_k, we find that a minimal Weierstrass form for E'_k is given by

$$Z^2 = W^3 + \frac{k^2}{4} \qquad (W,Z) = (\frac{U}{4}, \frac{V}{8})$$

in the case that k is even, and

$$Z^2 + Z = W^3 + \frac{k^2-1}{4} \qquad (W,Z) = (\frac{U}{4}, \frac{V-4}{8})$$

in the case that k is odd. The primes of bad reduction for E'_k are the primes dividing k and the prime 3. For a prime factor p of k, the Kodaira type at p is IV* if $p^2 \mid k$ and IV if $p \| k$. If $3 \nmid k$, the Kodaira type at 3 is III if $k \equiv \pm 2 \mod 9$, and II otherwise. It follows that the conductor of E'_k is given by the formula

$$N(E'_k) = \prod_{p|3k} p^{2+\beta_p}$$

where $\beta_p = 0$ if $p \neq 3$, and $\beta_3 = 0$, 1, or 3 for $k \equiv \pm 2 \mod 9$, $k \equiv \pm 1, \pm 4 \mod 9$, or $3 \mid k$ respectively.

The curves E'_k have the rational 3-torsion points $(U,V) = (0, \pm 4k)$. Since the rank is an isogeny invariant, we produce as a corollary to our work examples of elliptic curves E'_k with a rational 3-torsion point and rank as high as 11. Curiously, there are no other known elliptic curves over $\mathbb{Q}$ with a rational 3-torsion point and rank greater than 8 [Du].

In section 2 we describe the geometric underpinnings of our search technique, which heavily uses rational parametrizations of various cubic surfaces, the points of which correspond to pairs of (usually independent) points on the curves E_k. Section 3 gives a formula for an upper bound on the rank of E_k, using descent via 3-isogeny. Section 4 describes some of the specific algorithms we used to produce examples of E_k with high rank. Finally, we give our numerical results in Section 5.

2 Cubic Surfaces

The most naïve approach to constructing curves E_k of high rank is to enumerate small points on the curves E_k, which can be accomplished via the simple observation that a point on some curve E_k corresponds to a pair of whole numbers (x,y) so that the cubefree part of x^3+y^3 (that is, the unique cubefree integer s such that $(x^3+y^3)/s$ is a perfect cube) is k. The second author used essentially this approach to find the first known E_k of rank 7. By incorporating some more sophisticated techniques, such as 3-descent (see below), this approach could yield curves with rank as high as 8. The weakness of this method is that the number of such points up to height H grows as H^2, most of which lie on curves of rank 1 and waste our time and/or memory.

We can reduce this H^2 to $H^{1+\epsilon}$ by considering only curves E_k together with a pair of points, which correspond to points on the cubic surface

$$S_1 : w^3 + x^3 = y^3 + z^3$$

other than the trivial points on the lines $w+x = y+z = 0$, $w+y = x+z = 0$, $w+z = x+y = 0$. (This pairs-of-points idea is also used in [EW] to produce elliptic curves with high rank and smallest conductor known.) The cubic surface whose points correspond to pairs of points on the isogenous curves E'_k,

$$S_2 : wx(w + x) = yz(y + z),$$

and the "mixed" cubic surface

$$S_3 : wx(w + x) = y^3 + z^3,$$

are also fruitful. Each of these cubic surfaces is smooth, and thus rational in the sense that it is birational to $\mathbb{P}^2$ over $\overline{\mathbb{Q}}$; in fact, each has a rational parametrization defined over $\mathbb{Q}$.

A parametrization of S_1 was found by the first author [El], and a parametrization of S_2 follows fairly quickly: there is an obvious isomorphism between S_1 and S_2, defined *a priori* over $\mathbb{Q}(\sqrt{-3})$, but which actually descends to $\mathbb{Q}$. To parametrize S_3, we used the following more general approach, provided by Izzet Coskun [Co].

Let S be a cubic surface defined over $\mathbb{Q}$, and suppose L_1 and L_2 are disjoint lines on S, with L_3 a third line meeting both. Then there is a 3-dimensional space of quadrics that vanish on this set of lines. Use a basis of this space to map S into $\mathbb{P}^2$; the inverse map will be a parametrization of S. The parametrization so obtained is defined over $\mathbb{Q}$ if all of the L_i are rational, or if L_3 is rational and L_1, L_2 are Galois conjugate. This construction realizes S as $\mathbb{P}^2$ blown up at six points; the six blown-down lines are the six lines (other than L_i) that meet exactly two of the three lines L_1, L_2, L_3.

On both S_1 and S_3, the relative paucity of lines defined over $\mathbb{Q}$ means that, up to automorphism of the surface, there is only one choice for the configuration L_1, L_2, L_3. Thus for S_1 the parametrization obtained with this technique,

$$\begin{aligned}(w : x : y : z) = (&t^3 - 2t^2s - 2tsr + ts^2 + r^2s - r^3 - rs^2\\ &: -t^3 - 2t^2s + ts^2 - 2tsr + 2rs^2 - s^3 + r^3 - 2r^2s\\ &: 2t^3 + 3t^2r - 2t^2s - 2tsr + 3tr^2 + ts^2 + 2rs^2 - s^3 + r^3 - 2r^2s\\ &: -2t^3 + t^2s - 3t^2r - 3tr^2 + 4tsr - 2ts^2 - r^3 + r^2s - rs^2),\end{aligned}$$

is equivalent to the one in [El], in the sense that one can be obtained from the other by composing an automorphism of S_1 with a projective linear transformation of $\mathbb{P}^2$. For S_3 we obtain the parametrization

$$(w : x : y : z) = (r^3 - s^3 : s^3 + t^3 : r^2s - s^2t + t^2r : r^2t - s^2r - t^2s).$$

In order to compute the k for which a particular point on S_3 corresponds to a pair of points on E'_k and E_k, we must find the cubefree part of $wx(w+x)$, which we do by factoring this number. It is therefore useful that $wx(w + x)$, which is a polynomial of degree 9 in r, s, and t, decomposes as a product of three linear and three quadratic factors. By contrast, the factorization of $w^3 + x^3$ in the parametrization of S_1 is as a product of one linear, two quadratic, and one quartic factor; the difficulty of factoring the value of this quartic at (r, s, t) severely limits the usefulness of the S_1 parametrization.

On S_2, there are, up to automorphism, seven different configurations of lines L_1, L_2, L_3. Four of them lead to parametrizations where $wx(w+x)$ has five linear and two quadratic factors; the parametrization of S_2 mentioned above is of this type. The other three lead to factorizations into three linear and three quadratic polynomials. There is not a significant computational advantage for one factorization over another, but we do mention here a rather elegant parametrization of S_2 obtained in this way:

$$(w : x : y : z) = (-r^2 s + s^2 t : r^2 s - rt^2 : -r^2 t + st^2 : rs^2 - st^2).$$

3 Descent via 3-isogeny

A powerful tool for obtaining upper bounds for the ranks of the curves E_k is descent, since these curves admit the aforementioned 3-isogeny with E'_k. An analysis of the descent for these curves first appeared in [Se]. What follows is essentially a simplification of the formula given there.

Let $k = \prod p_j^{\epsilon_j}$, where $\epsilon_j = 1$ or 2, and let q_i be the primes dividing k with $q_i \equiv 1 \bmod 3$. Now define a matrix A over $\mathbb{F}_3$ by:

$$\left(\frac{p_j^{\epsilon_j}}{q_i}\right)_3 = \rho^{A_{ij}}$$

if $p_j \neq q_i$, and $A_{ij} = -\sum_{\ell : p_\ell \neq q_i} A_{i\ell}$ if $p_j = q_i$ (equivalently, the rows of A sum to zero). Here $\left(\frac{\cdot}{q}\right)_3$ denotes the cubic residue symbol mod q; note that for each $q \equiv 1 \bmod 3$, there are two choices for this cubic residue symbol, but they lead to proportional rows A_i. If $k \equiv \pm 1$ or $0 \bmod 9$, add an additional row corresponding to cubic characters mod 9 for p_j; if $9|k$, the entry corresponding to $p_j = q_i = 3$ is defined as in general when $p_j = q_i$.

If we let $\phi : E_k \to E'_k$ and $\phi' : E'_k \to E_k$ denote the relevant 3-isogenies, then the row and column null spaces of A correspond, respectively, to the ϕ- and ϕ'-Selmer groups of $E_k/\mathbb{Q}$ and $E'_k/\mathbb{Q}$. We can conclude that, after taking the 3-torsion of E'_k into account,

$$\operatorname{rank}(E_k) \leq \#\text{rows} + \#\text{columns} - 2 \cdot \operatorname{rank}(A) - 1.$$

4 Computational Techniques

An application of the explicit formula for descent via 3-isogeny is another technique for searching for curves of large rank, which tends to be more effective in finding the minimal k such that E_k has a given rank (it was actually this technique that produced the current rank 9 record). The idea is to enumerate all possible k less than some given upper bound which have a sufficiently high 3-Selmer bound, and then to search for points on these curves. To do this, we recursively build up products of primes, and at each stage compute the portion of the matrix A corresponding to the primes chosen so far. Of course, the diagonal entries remain in some doubt, since they depend on the whole row of A; still, at each stage a lower bound for the rank of the matrix A can be computed, and used to give a lower bound on the number of primes still needed. In this way one can vastly reduce the search space. A similar approach can be used to enumerate candidate curves whose conductor is smaller than a given bound, with the formula for the conductor given in the Introduction.

It is also important to have a way to guess which curves are the most promising before committing to a lengthy point search. The most important tool in this regard is provided by a heuristic argument suggesting that high rank curves should have many points on their reductions modulo p, or more specifically,

$$\prod_{p \leq x} \frac{\#E(\mathbb{F}_p)}{p+1} \sim (\log x)^r$$

where r is the rank of E. This formula was conjectured by Birch and Swinnerton-Dyer in [BSD], and the idea of using it to find elliptic curves of high rank is due to Mestre [Me].

Note that for the curves E_k it is only useful to consider primes $p \equiv 1 \bmod 3$, since E_k is supersingular mod p when $p \equiv 2 \bmod 3$. Furthermore, computing $\#E_k(\mathbb{F}_p)$ is quite fast: modulo each prime $p \equiv 1 \bmod 3$, there are only three isomorphism classes of E_k, corresponding to the three cubic residue symbols mod p. We compute the a_p's for each of these isomorphism classes once for all, and then to find the a_p for a given curve E_k, we need only compute the cubic residue symbol of k mod p.

In the end, though, we must still search for points on curves we suspect of having high rank. But here, too, there are improvements over the most naïve approach. As noted above, points on E_k correspond to pairs of whole numbers (x, y) such that $k = d^{-3}(x^3 + y^3)$. Of course, we may further assume that x and y are relatively prime. It follows that $\gcd(x + y, x^2 - xy + y^2)$ is either 1 or 3, whence $x + y$ must be a factor of $3k$ times a perfect cube. For each $x + y$ and d, we must simply decide if the resulting quadratic equation has a rational solution. Furthermore, we can use local conditions to reduce the number of possible $x + y$ we must consider (this is closely related to the descent described in the last section). A similar approach works for a point search on $xy(x + y) = k$.

5 Results

Here we list the minimal known k such that E_k has rank r for each rank $r \leq 11$, as well as the minimal known conductor of a curve E_k of rank r with $r \leq 8$. We include notes where relevant, and r independent points for each of the new record curves. The points are listed on the minimal Weierstrass equation for E'_k, as described in the Introduction. To transfer points back to the curve E_k one may use the dual isogeny $\hat{\phi}$ described there.

rank	k (minimal known with E_k of given rank)
0	1
1	$6 = 2 \cdot 3$
2	19 (prime)
3	$657 = 3 \cdot 3 \cdot 73$
4	$21691 = 109 \cdot 199$
5	$489489 = 3 \cdot 7 \cdot 11 \cdot 13 \cdot 163$
6	$9902523 = 3 \cdot 73 \cdot 103 \cdot 439$
7	$1144421889 = 3 \cdot 13 \cdot 19 \cdot 41 \cdot 139 \cdot 271$
8	$1683200989470 = 2 \cdot 3 \cdot 5 \cdot 7 \cdot 11 \cdot 13 \cdot 17 \cdot 29 \cdot 41 \cdot 47 \cdot 59$
9	$349043376293530 = 2 \cdot 5 \cdot 37 \cdot 41 \cdot 53 \cdot 73 \cdot 1231 \cdot 4831$
10	$137006962414679910 = 2 \cdot 3 \cdot 5 \cdot 7 \cdot 23 \cdot 31 \cdot 37 \cdot 43 \cdot 83 \cdot 109 \cdot 151 \cdot 421$
11	$13293998056584952174157235$
	$= 3 \cdot 5 \cdot 7 \cdot 13 \cdot 19 \cdot 23 \cdot 31 \cdot 43 \cdot 59 \cdot 61 \cdot 73 \cdot 79 \cdot 103 \cdot 109 \cdot 157 \cdot 457$

The following records for ranks up to 5 are known to be minimal (the proof for ranks 4 and 5 seems to be new); for ranks 6 and 7, they are minimal provided the weak Birch and Swinnerton-Dyer Conjecture and the Generalized Riemann Hypothesis are true for all $L(E_{k'}, s)$ with $k' < k$. The records for ranks 8 through 10 are likely to be minimal. In each case, we use the approach described in the last section to enumerate all of the smaller k with a sufficiently large 3-Selmer group. For each of them we compute a partial product of $L(E_k, 1)$ over the first 1000 or so primes; a large partial product should correspond to high rank. In each case of rank 8 through 10, the record k significantly distinguished itself from all smaller k. Note that for large r our record value of k tends to have considerably more prime factors congruent to 1 mod 3 than to 2 mod 3.

rank	k (E_k of given rank and minimal known conductor)
0	1
1	$9 = 3^2$
2	19 (prime)
3	$657 = 3 \cdot 3 \cdot 73$
4	$34706 = 2 \cdot 7 \cdot 37 \cdot 67$
5	$763002 = 2 \cdot 3^2 \cdot 19 \cdot 23 \cdot 97$
6	$24565833 = 3^2 \cdot 17 \cdot 307 \cdot 523$
7	$1144421889 = 3 \cdot 13 \cdot 19 \cdot 41 \cdot 139 \cdot 271$
8	$23381862574950 = 2 \cdot 3^2 \cdot 5^2 \cdot 11 \cdot 19 \cdot 23 \cdot 31 \cdot 83 \cdot 4201$

The k in the following chart are known to correspond to curves E_k of minimal conductor for $k \leq 3$. For ranks 4, 5, and 6, they are minimal provided the weak Birch and Swinnerton-Dyer Conjecture and the Generalized Riemann Hypothesis are true for all $L(E_{k'}, s)$ with $k' < k$. The records for ranks 7 and 8 are likely to be minimal; as above, Mestre's heuristic distinguishes them markedly from all curves of smaller conductor. It is striking that for all ranks less than 8, the k corresponding to minimal conductor are squarefree away from 3. Since divisibility by p^2 as opposed to p in k does not alter the value of the conductor of E_k, one might expect that for high rank, the k corresponding to the curve of minimal conductor would have many square factors.

We conclude by listing independent points on the record curves E'_k for ranks 6 through 11, all of which were newly found using methods described here. A program that implements LLL reduction on the lattice of points of E_k, provided by Randall Rathbun [Ra], was used to reduce their heights where possible.

$$k = 9902523, \ r = 6$$

6 independent points on the Weierstrass minimal curve for E'_k:

$$y^2 + y = x^3 + 24514990441382$$

(100092, 32051170), (−6798, 4919434), (−22338, 3656314),
(43672, 10383069), (−11988, 4774114), (126720, 45380386).

$$k = 24565833, \ r = 6$$

6 independent points on the Weierstrass minimal curve for E'_k:

$$y^2 + y = x^3 + 150870037745972$$

(−37656, 9872932), (86292, 28167835), (187270, 81966083),
(−32058, 10859260), (−39798, 9372019), (236572, 115719195).

$$k = 1144421889, \ r = 7$$

7 independent points on the Weierstrass minimal curve for E'_k:

$$y^2 + y = x^3 + 327425365005582080$$

(267748, 588744383), (1235988, 1488490048), (−333330, 538877944),
(−648774, 233133847), (−422760, 501863680), (5104008, 11545190143),
(−688974, 19483912).

$$k = 1683200989470, \ r = 8$$

8 independent points on the Weierstrass minimal curve for E'_k:

$$y^2 = x^3 + 70829139273819676272025$$

(−88860785, 81396479060), (−87348261, 204569627262),
(−63256830, 674665720365), (−40588401, 800890328532),
(101707060, 1326794024465), (−35705670, 814107128835),
(−44793980, 786391914385), (8308684440, 757353550270065).

In fact, the curve E'_k for $k = 1683200989470$ has the remarkable property that the Diophantine equation $xy(x + y) = k$ actually has 8 *integral* solutions, namely $(11, 391170)$, $(533, 55930)$, $(770, 46371)$, $(1003, 40467)$, $(2639, 23970)$,

(6970, 12441), (7293, 1197), (8555, 10387). These 8 solutions, considered as rational points on E'_k, are independent.

$$k = 23381862574950, \; r = 8$$

8 independent points on the Weierstrass minimal curve for E'_k:

$$y^2 = x^3 + 136677874368461861091875625$$

(1826174700, 78910161016275), (794409100, 25259021976275),
(−259483950, 10918164759975), (499986216, 16176140969139),
(−503804925, 2966885463150), (798185799, 25400836633032),
(165873591, 11884516327764), (215137494, 12109307517153).

$$k = 349043376293530, \; r = 9$$

9 independent points on the Weierstrass minimal curve for E'_k:

$$y^2 = x^3 + 30457819633596695100179965225 :$$

(−734843410, 173381106196815), (5130038900, 406775844709485),
(−2676929565, 106184137901590), (690947990, 175464197227935),
(291207945620, 157146639625792365), (25120488440, 3985280944128435),
(−872639080, 172607374924365), (2918890200, 235215923409485),
(−102315705, 174518619468760).

$$k = 137006962414679910, \; r = 10$$

10 independent points on the Weierstrass minimal curve for E'_k:

$$y^2 = x^3 + 4692726937524378378756566939402025 :$$

(−135797482140, 46781315964225555), (−150436201545, 35891470127810220),
(−42200591214, 67952721291406041), (2327642247924, 3551854243978575507),
(5504535148140, 12914782107290941395), (140506152430, 86409469562070095),
(397507563420, 259814927561209005), (7162660587075, 19169656506442936830),
(73148794740, 71303068454026605), (−102758626586, 60063833881519937).

$$k = 13293998056584952174157235, \; r = 11$$

11 independent points on the Weierstrass minimal curve for E'_k:

$$y^2 + y = x^3 + 441825960821211213171351700256803990465456257113 06 :$$

(−30156002278649820, 4093799681127459731025817),
(11364087102067560, 6756491872572362690626342),
(−20835788771691894, 5927660006237675713476241),
(113426492056989390, 12080316858288251182214780 17),

$(8907565209691176834, 26585114133655761890666064910)$,
$(11184919988612133 4, 37992674604901443769570910)$,
$(11724873521668020, 6767159346634715672034457)$,
$(-138658831412368575/4, 127198194435742683333258 11/8)$,
$(16597106090152224 0, 679417888764028165771389 82)$,
$(994768217796990, 664707307532766224396601 7)$,
$(532896351059436225/16, 5764573107853248832486778 23/64)$.

References

[BSD] Birch, B.J., and Swinnerton-Dyer, H.P.F.: Notes on Elliptic Curves II. *J. reine angew. Math.* **218** (1965) 79–108.

[Co] Coskun, I., personal communication.

[Du] Dujella, A.: High rank elliptic curves with prescribed torsion. `http://www.math.hr/~duje/tors/tors.html` (2000–2003).

[El] Elkies, N.D.: Complete Cubic Parametrization of the Fermat Cubic Surface. 2001 (`http://math.harvard.edu/~elkies/4cubes.html`).

[EW] Elkies, N.D., and Watkins, M.: Elliptic curves of large rank and small conductor. Preprint, 2003 (accepted for ANTS-6).

[Ma] Mazur, B.: Modular curves and the Eisenstein ideal. *IHES Publ. Math.* **47** (1977) 33–186.

[Me] Mestre, J.-F.: Courbes elliptiques de rang ≥ 11 sur $\mathbb{Q}(t)$. (French) [Elliptic curves with rank ≥ 11 over $\mathbb{Q}(t)$.] *C. R. Acad. Sci. Paris Sér. I Math.* **313** (1991) 139–142.

[MM] Martin, R., and McMillen, W.: An elliptic curve over $\mathbb{Q}$ with rank at least 24. Number Theory Listserver, May 2000. Electronically available from `listserv.nodak.edu/archives/nmbrthry.html` and directly from `www.math.harvard.edu/~elkies/rk24_1.html`

[Ra] Rathbun, R., personal communication.

[Se] Selmer, E.S.: The Diophantine Equation $ax^3 + by^3 + cz^3 = 0$. *Acta. Math.* **85** (1951) 203–362.

[Si] Silverman, J.H.: *The Arithmetic of Elliptic Curves.* New York: Springer, 1986.

[Ta] Tate, J.: Algorithm for determining the type of a singular fiber in an elliptic pencil. In *Modular Functions of One Variable IV*, Lect. Notes in Math. 476, B.J. Birch and W. Kuyk, eds. Berlin: Springer-Verlag, 1975, 33–52.

[Wa] Watkins, M.: Rank Distribution in a Family of Cubic Twists. Preprint, 2003 (`http://www.math.psu.edu/watkins/papers/zk.ps`).

[ZK] Zagier, D., and Kramarz, G.: Numerical investigations related to the L-series of certain elliptic curves. *J. Indian Math. Soc.* **52** (1987), 51–69.

Proving the Primality of Very Large Numbers with fastECPP

J. Franke[1], T. Kleinjung[1], F. Morain[2], and T. Wirth[1]

[1] Dept. of Math., Bonn University, Beringstr. 1, 53115 Bonn, Germany
{franke,thor,wirth}@math.uni-bonn.de
[2] Laboratoire d'Informatique de l'École polytechnique (LIX) ***
F-91128 Palaiseau Cedex France
morain@lix.polytechnique.fr

Abstract. The elliptic curve primality proving algorithm is one of the fastest practical algorithms for proving the primality of large numbers. Its fastest version, fastECPP, runs in heuristic time $\tilde{O}((\log N)^4)$. The aim of this article is to describe new ideas used when dealing with very large numbers. We illustrate these with the primality proofs of some numbers with more than 10,000 decimal digits.

1 Introduction

The work by Agrawal, Kayal and Saxena [1] on the existence of a deterministic polynomial time algorithm for deciding primality stimulated the field of primality proving at large. As a result, this caused the study and implementation of a fast version of the elliptic curve primality proving algorithm (ECPP). We refer to [2] for a presentation of the method and [13] for the description of the faster version, originally due to J. O. Shallit. Whereas ECPP has a heuristic running time of $\tilde{O}((\log N)^5)$ for proving the primality of N, the new algorithm has complexity $\tilde{O}((\log N)^4)$. This new approach enabled one of us (FM), to prove the primality of numbers with more than $7,000$ decimal digits.

Independently, three of us (JF, TK and TW) started to write a new implementation of ECPP in November 2002 which was available by December 2002, and this was improved step by step until the team working in Bonn came up with a set of programs used to prove the primality of $10^{9999} + 33603$ on August 19, 2003.

The two teams decided after this to exchange ideas and comparisons, forming the present article that concentrates on issues regarding distributed implementations of fastECPP and its use in the proving of very large numbers. The theory of fastECPP will be described more fully in the final version of [13].

Our article is organized as follows. Section 2 provides a short description of fastECPP. Section 3 gives two strategies for distributing the computations. Section 4 deals with a faster way of looking for small prime factors of a bunch

*** Projet TANC, Pôle Commun de Recherche en Informatique du Plateau de Saclay, CNRS, École polytechnique, INRIA, Université Paris-Sud.

D.A. Buell (Ed.): ANTS 2004, LNCS 3076, pp. 194–207, 2004.

of numbers at the same time. This part has an independent interest and we think that it could be useful in other algorithms. In Section 5, an alternative to the method of [9,5] for the root finding in the proving steps of ECPP is described. Section 6 deals with the use of fast multiplication beyond the GMP level in order to speed up all basic multiplications. In Section 7, we describe an *early abort strategy* for limiting the number of steps in ECPP. We conclude with timings on primality proofs for some very large numbers, obtained with either implementation.

2 The Fast Version of ECPP

Ordinary ECPP is described in [2] and fastECPP in [13]. We sketch the description of the latter, assuming the reader has some familiarity with the algorithm.

We want to prove that N is prime. The algorithm builds a so-called *downrun* that is a sequence of decreasing probable primes $N_0, N_1, \dots, N_k$ such that $N_0 = N$ and the primality of each N_i is sufficient to prove that of N_{i-1}. Theory tells us that we should anticipate a length of $k = O(\log N)$ for the sequence.

If q is an odd prime, put $q^* = (-1)^{(q-1)/2}q$; add to this special primes $-4, -8, 8$ as explained in [13].

The algorithm runs as follows:

[Step 1.]

1.1. Find the r smallest primes q^* such that $\left(\frac{q^*}{N}\right) = 1$, yielding $\mathcal{Q} = \{q_1^*, q_2^*, \dots, q_r^*\}$.

1.2. Compute all $\sqrt{q^*} \bmod N$ for $q^* \in \mathcal{Q}$.

1.3. Try all subsets of distinct elements of $\mathcal{S} = \{q_{i_1}^*, q_{i_2}^*, \dots, q_{i_s}^*\}$ of $\mathcal{Q}$ for which $-D = \prod_{q^* \in \mathcal{S}} q^* < 0$, until a solution of the equation

$$4N = U^2 + DV^2 \tag{1}$$

in rational integers U and V is found, which involves computing $\sqrt{-D} \bmod N$ and use Cornacchia's algorithm. When this is the case, let $\{U_1, \dots U_w\}$ be the different U-values (we have at most $w = 6$ and generally $w = 2$).

[Step 2.] For all U_i's, compute $m_i = N + 1 - U_i$; if none of these numbers can be written as cN' with c a B-smooth number and N' a probable prime, go to Step 1. If there is a good one, call it m.

[Step 3.] Build an elliptic curve $\overline{E}$ over $\overline{\mathbb{Q}}$ having complex multiplication by the ring of integers of $\mathbf{K} = \mathbb{Q}(\sqrt{-D})$.

[Step 4.] Reduce $\overline{E}$ modulo N to get a curve E of cardinality m.

[Step 5.] Find P on E such that $[N']P = O_E$. If this cannot be done, then N is composite, otherwise, it is prime.

[Step 6.] Set $N = N'$ and go back to Step 1.

Note that what differentiates ECPP from its fast version is Step 1. In Step 1.3, we only consider fundamental discriminants, as a curve with CM by a non-principal order has the same cardinality as one with CM by the principal order.

ECPP is a Las Vegas type algorithm. Its running time cannot be proved rigorously, but only in a heuristic way using standard hypotheses in number theory. When given a number N, it can answer one of three things: "N is prime", "N is composite" or "I do not know". In the first two of these cases, the answer is definitely correct and there is an accompanying *proof* that can be verified in polynomial time. The problem is in showing that the third case happens with very low probability.

In real life, programs implementing (fast)ECPP should follow this philosophy and never return something wrong. When the third answer is returned, this corresponds very frequently to the fact that the program ran out of precomputed data (such as discriminants, or class polynomials) or used too small factorisation parameters in Step 2. The programmer has to correct this and start again with the number. We never saw a number resisting indefinitely, though we cannot prove none exists.

All algorithms and tricks [11,12] developed over the years for ECPP apply *mutatis mutandis* to the new version. This includes the invariants developed in [6,4] and the Galois approach for solving the equation $H_D[u](X)$ modulo p (see [9,5]) needed in Step 3, which favors smooth class numbers.

When dealing with very large numbers (10000 decimal digit numbers, say), every part of the algorithm should be scrutinized again, which includes optimizing the basic routines beyond the current level of GMP. In Step 2., B-smooth numbers are to be identified. The number B is important in the actual running time, and its precise value must be set depending on the algorithm used. See Section 4 below. A new strategy (early abort) is described in section 7. Also, Step 3-4 can be merged, as explained in section 5.

3 First Strategies for Distribution

Step 1-2 and Step 3-5 are easy to distribute over clusters of workstations. In this section, we give two distribution strategies.

3.1 Strategy 1

The following is easily implemented when all slaves have the same computing power, making it a parallel implementation.

S1. The master sends to each slave a batch of $\sqrt{q_i^*}$ to compute.

S2. Each slave computes its batch and sends the results back to the master.

S3. The master sends all the squareroots to all the slaves, so that each slave can compute any $\sqrt{-D}$ that is needed.

S4. The master sends batches of D's to all the slaves. Each slave is responsible for the resolution of (1) and the factorization of the m's. If one is found, it is sent back to the master which checks the results and restarts a new phase.

S3 needs synchronization and communications.

In S4, load balancing is not easy, since the results are probabilistic in nature (for which D does N split?). A probabilistic answer is to compute beforehand the average number of splitting D's that can occur. By genus theory, each D with t prime factors has splitting probability $g(-D)/h(-D)$ with $g(-D) = 2^{t-1}$. This suggests to build the whole set of discriminants D in Step 1.3 and to sort them with respect to $(h(-D)/g(-D), D)$. One could also add a criterion describing the difficulty of building the class polynomial $H_D(X)$ later on, maybe using the height of the polynomial (as evaluated in [4]). We send to each slave discriminants $D_{i_1}, D_{i_2}, \ldots, D_{i_k}$ in such a way that

$$\sum_j g(D_{i_j})/h(D_{i_j}) \geq 5$$

(the value of 5 is somewhat arbitrary) which corresponds to the fact that on average, 5 values of D will be splitting values. Of course, this quantity should depend on the power of the slave.

3.2 Strategy 2

Another approach would be to set up a complicated system in which the master keeps trace of the work of each slave and decides what kind of work to do at some point. This is easily implemented on the side of the slaves: they wait for a task from the master, do it and send the result back. We now describe a possible implementation of the master.

There are 6 different tasks which the slaves work on:

T1. Check whether the class number for a discriminant D is good, i.e. is not too big and does not contain a very large prime factor.

T2. Compute a modular square root $\sqrt{q_i^*}$ modulo N.

T3. Try to solve (1) for a given D.

T4. Do trial division for an interval of primes and a batch of m's and return the factored parts. See Section 4.

T5. Do a Fermat test.

T6. Do a Miller-Rabin test.

The master keeps lists of tasks of these six types which at the beginning are all empty. If all task lists are empty the master creates new tasks of type T1. The tasks are sorted according to their priority, T1 having lowest and T6 having highest priority. If a slave requests a new task the master selects from all available tasks one with the highest priority and passes it to this slave. A completed task will create a varying number of new task, e.g. a computed square root (T2) may create many tasks of type T3 whilst a Fermat test (T5) will only on success create a task of type T6. If a certain number of tasks T6 for the same integer are successful one reduction step is finished and pending tasks are cancelled.

4 An Optimized Test for the Divisibility by Small Primes

Let us first analyze the effect of trial division on the number of pseudo-primality tests (the most time consuming part of our implementation). Suppose we do trial division up to B on a number $N = fR$ where f is only divisible by primes $\leq B$ and R only divisible by primes $> B$. Let us assume that $\log B \leq \sqrt[3]{\log N}$, a condition which is almost always satisfied in practice. Let $p_{\max}(n)$ be the largest prime divisor of n. One can combine the prime number theorem with Rankin's trick and Mertens's theorem [14, 9.1.5 and 9.1.8] and related facts to investigate the sums

$$s = \sum_{\substack{f<x \\ p_{\max}(f)<B}} \pi(x/f)$$

$$l = \sum_{\substack{f<x \\ p_{\max}(f)<B}} \log(f)\pi(x/f)$$

where $\pi(x)$ counts the number of primes below x. Assuming $\log B \leq \sqrt[3]{\log x}$, we find them to be

$$s = \frac{x\exp(\gamma)\big(\log(B) + O(1)\big)}{\log(x)}$$

$$l = \frac{x\exp(\gamma)\big(\log(B)^2 + O(\log B)\big)}{\log(x)},$$

where γ is Eulers constant. Since s counts the number of $N < x$ for which R is prime, while l is the sum of $\log(f)$ over such N, one concludes that for a randomly chosen $N \in [(1-\epsilon)x, x]$ with a fixed positive $\epsilon < 1$ and for $x \to \infty$, $z \to \infty$ subject to $\log B \leq \sqrt[3]{\log x}$, the probability that R is prime tends to $\exp(\gamma)\log(B)/\log(N)$ while the expectation value of $\log(l)$ is $\log(B) + O(1)$. By this heuristic argument, one expects the number of pseudo-primality tests for a reduction of fixed size to be proportional to $(\log B)^{-2}$.

We now describe how to perform the trial division in Step 2 more efficiently by doing it on many numbers simultaneously. This is essentially a simplification of the algorithm in [3]. Let N be a (pseudo)prime for which we want to do a reduction step and let $m_1, \ldots, m_t$ be the numbers computed in Step 2 (we may choose t of order $\frac{\log N}{2\log B}$). For simplicity we assume that $t = 2^u$ is a power of two. The following algorithm strips the primes up to B from the numbers m_i:

1. Build the product $P = \prod_{p\leq B, p \text{ prime}} p$ using a binary tree. Unless the bound B is changed this has to be done only once.
2. Compute the product $M = \prod_i m_i$ as follows: Set $m_i^{(0)} = m_i$ and for $a = 1, \ldots, u$ successively compute $m_i^{(a)} = m_{2i-1}^{(a-1)} m_{2i}^{(a-1)}$, $1 \leq i \leq 2^{u-a}$. Set $M = m_1^{(u)}$.
3. Compute $\tilde{M} = P \bmod M$ and set $\tilde{m}_1 = \tilde{M}$.

4. Compute $\tilde{m}_i = P \bmod m_i$ as follows: For $a = u, \ldots, 1$ compute $\tilde{m}_{2i-1}^{(a-1)} = \tilde{m}_i^{(a)} \bmod m_{2i-1}^{(a-1)}$ and $\tilde{m}_{2i}^{(a-1)} = \tilde{m}_i^{(a)} \bmod m_{2i}^{(a-1)}$, $1 \leq i \leq 2^{u-a}$. Set $\tilde{m}_i = \tilde{m}_i^{(0)}$.
5. For $1 \leq i \leq t$ replace repeatedly $\tilde{m}_i$ by $\gcd(m_i, \tilde{m}_i)$ and m_i by $\frac{m_i}{\tilde{m}_i}$ until $\tilde{m}_i = 1$.

Remarks:
Note that for $P < M$ we can save some of the top level computations and the application of the algorithm becomes equivalent to several applications with a smaller u, grouping the m_i appropriately. So we assume that u is chosen such that $P > M$ holds.

If we store the partial products

$$P^{(i)} = \prod_{p \leq \frac{B}{2^i}, p \text{ prime}} p$$

which are computed during the precomputation we can decrease the trial division bound by powers of 2 with no extra effort.

Step 5 can be improved e. g. by replacing $\tilde{m}_i$ by $\gcd(m_i, {\tilde{m}_i}^2)$ in the iteration.

We now analyze the cost of the algorithm. Let $M(n)$ denote the cost of a multiplication of two numbers of size $\exp(n)$. We assume that the FFT is used and set $M(n) = O(n(\log n)^{1+\epsilon})$. The first step is a precomputation whose cost is $O(B(\log B)^2)$. The cost for the second step is

$$\sum_{k=0}^{u-1} 2^{u-1-k} M(2^k \log N) = O(u 2^u \log(N)(\log(2^u \log N))^{1+\epsilon})$$

since all m_i are of size N. In the fourth step the operation count is the same except that an $n \cdot n$-multiplication is replaced by an $2n : n$-division. Since the latter is asymptotically as fast as the former the cost for step 4 is the same as for step 2. Since $P \approx \exp(B)$ the cost of the third step is $O(B(\log(2^u \log N))^{1+\epsilon})$. The last step as described above has complexity $O(2^u (\log N)^2 (\log\log N)^{2+\epsilon})$ since the iteration ends after at most $\log_2 N$ steps each consisting of a division and a gcd. This could be improved by modifying this step but we do not need it here. Note also that in practice this step consists mainly of the first $\gcd(m_i, \tilde{m}_i)$, which with high probability is very small, and the number of iterations also is very small.

Assuming $2^u < \log N$ and neglecting log log-terms the time spent in pseudo-primality tests is $O(\frac{(\log N)^3}{\log B})$ for a reduction of size $\log B$ whereas the time for trial division is $O(B + (\log N)^3)$. So it is optimal to choose $B = O((\log N)^3)$ which also implies that the cost for the precomputation can be neglected.

Some remarks about storage and parallelization:

The algorithm above needs a lot of memory; most of it at the end of the computation of P. To reduce the memory requirement we may compute partial products P_j of the primes below B whose product is P and modify step 3 into computing the residues $P_j \bmod M$ and the modular product of these residues. For this to be efficient the partial products should be larger than M.

For a distributed implementation we propose to split P into as many pieces as slaves are present. Each slave executes steps 2-4 of the algorithm for its own P_j and passes at the end $\gcd(m_i, \tilde{m}_i) = \gcd(m_i, P_j)$ to the master. The master assembles these informations and executes step 5 which in practice is very fast.

For the number $10^{9999} + 33603$, the bound B was set to 2^{32}.

5 Computing Roots of the Hilbert Polynomial Modulo p

The run-down sequence contains, among other things, a list of pairs (p, D), where p is a pseudo-prime and where it is expected that an elliptic curve with complex multiplication by the ring $\mathcal{O}_K$ of integers in $K = \mathbb{Q}(\sqrt{-D})$ can be used to derive the primality of p from the primality of some smaller number. It is necessary to compute an element j_p of $\mathbb{F}_p$ which is the j-invariant of an elliptic curve over $\mathbb{F}_p$ with complex multiplication by $\mathcal{O}_K$.

We outline the method which was used to perform this step for the run-down sequence of $10^{9999} + 33603$. As in [9], the idea is to split this task into several steps, each one involving the determination of the modular solution of an equation of degree ℓ_i, where the ℓ_i are the prime factors of the class number h of K. One difference is that [9] tries to compute a sequence of polynomials which define a sequence of intermediate fields terminating in the Hilbert class field L of K. By contrast, the implementation which was used for $10^{9999} + 33603$ constructs methods to reduce elements x of the intermediate fields modulo an appropriate prime ideal over p, increasing the subfield in each step. The element x is given by floating point approximations to its conjugate algebraic numbers. Thus, the sequence of intermediate fields occurs in both methods but otherwise the language used is somewhat different, making it difficult to explain to what extent the methods are similar. Since the available space is not sufficient for a careful description of the new method, we give a short example explaining how it works in the case $p = 479$, $D = 335$.

The program chooses the modular invariant $x = x_{13}$ from [2, 2.7.1] and a precision of 32 bit is sufficient for the floating point calculations. The class group of K, and therefore $\mathrm{Gal}(L/K)$, is cyclic of order 18. The program has selected a generator σ of the Galois group, and has computed the complex numbers $\sigma^i(x)$. It has then decided to choose the prime ideal $\mathfrak{p}_o \subset \mathcal{O}_K$ such that

$$(a + b\sqrt{-D}) \bmod \mathfrak{p}_o = \frac{1}{2}\Big(([2a] + 12[\frac{2b}{\sqrt{D}}]) \bmod p\Big), \tag{2}$$

where $\sqrt{D} = 18.30300521772312668$ is the positive square root of D, the subexpressions in square brackets are in real life floating point numbers which will be rounded to nearest integers, and the factor 12 in the second summand is of course a square root of $-D \bmod p$.

The Hilbert class field has unique subfields L_1 and L_2 of degrees 2 and 3 over K. The program knows that the genus field L_1, which is the largest subfield of L which is Abelian over $\mathbb{Q}$, is given by $L_1 = K(\sqrt{5})$. It decides to work with the prime ideal $\mathfrak{p}_1 \subseteq \mathcal{O}_{L_1}$ such that, for each element z of L_1 given by a complex floating point approximation to z and $\sigma(z)$, we have

$$z \bmod \mathfrak{p}_1 = \frac{1}{2}\Big(\big((z+\sigma(z)\big) \bmod \mathfrak{p}_o + 196\big((z-\sigma(z))\sqrt{5} \bmod \mathfrak{p}_o\big)\Big), \tag{3}$$

where $\sqrt{5} = 2.23606798$ and the factor 196 in the second summand is of course a modular square root of $1/5 \bmod p$. The reductions modulo $\mathfrak{p}_o$ occuring in this formula are computed by the program using (2).

It is more difficult to describe a prime ideal $\mathfrak{p}_2 \subset \mathcal{O}_{L_2}$ with $\mathfrak{p}_2 \cap \mathcal{O}_K = \mathfrak{p}_o$ in a way which is suitable for calculations. The program considers

$$x_2 = \sum_{i=0}^{5} \sigma^{3i}(x),$$

which is an algebraic integer. We have

$$\begin{aligned} x_2 &= -60.2484307 + 78.0404771\sqrt{-1} \\ \sigma(x_2) &= -14.7805113 - 15.4588718\sqrt{-1} \\ \sigma^2(x_2) &= -10.4710580 + 1.47891293\sqrt{-1}. \end{aligned}$$

Note that $\sigma^i(x_2)$ depends only on $i \bmod 3$. The program computes a complex floating point approximation to the minimal polynomial of x_2 over K and finds, using (2) to reduce polynomial coefficients modulo $\mathfrak{p}_o$, that this polynomial is congruent to $T^3+283T^2+226T+108$ modulo $\mathfrak{p}_o$. It finds that 341, 395 and 418 are the roots of this polynomial modulo p and decides to work with the prime ideal $\mathfrak{p}_2 \subseteq L_2$ such that $x_2 \equiv 341 \pmod{\mathfrak{p}_2}$. It computes coefficients $v_i \in \mathbb{F}_p$, $0 \le i < 3$, such that

$$z \bmod \mathfrak{p}_2 = \sum_{i=0}^{2} v_i \Big(\big(\sum_{j=0}^{2} \sigma^{i+j}(x_2)\sigma^j(z)\big) \bmod \mathfrak{p}_o\Big), \tag{4}$$

where in practice the reduction mod $\mathfrak{p}_o$ is carried out using (2). For this to be possible, it is necessary that x_2 generates a normal basis of $\mathcal{O}_{L_2}$ over $\mathcal{O}_K$ after localisation at $\mathfrak{p}_o$. The program will abort if this assumption fails. This does not happen in this example, nor did it ever happen during the calculations for $10^{9999} + 33603$. But it should be possible to construct examples of failure of the program, although it is very unlikely for this to happen in practice. In order to determine the coefficients of (4), it is also necessary that the modular roots ξ_i of the minimal polynomial have been ordered in such a way that $\xi_i = \sigma^i(x_2) \bmod \mathfrak{p}_2$. The choice of ξ_0 is of course free, since a different ξ_0 only gives a different $\mathfrak{p}_2$. But the order of the other modular zeros is no longer free and the program has to compute them in the correct order. We will describe in a different paper how this is done without a major increase of the required precision, where of course the order of the cyclic extension will often be larger than 3. Once the ξ_i have been computed in the correct order, it is a linear algebra task to determine the v_i such that (4) holds for $z = \sigma^k(x_2)$, $0 \le k < 3$. In the given example, the result is

$$v_0 = 417 \qquad v_1 = 170 \qquad v_2 = 393.$$

The compositum $L_3 = L_1 L_2$ has degree 6 over K, and there is a unique prime ideal $\mathfrak{p}_3 \subset \mathcal{O}_{L_3}$ such that $\mathfrak{p}_3 \cap \mathcal{O}_{L_i} = \mathfrak{p}_i$ for $i \in \{1;2\}$. If an element z of $\mathcal{O}_{L_3}$ is given by complex floating point approximations to $\sigma^i(z)$, where $0 \le i < 6$, then

$$z \bmod \mathfrak{p}_3 = \frac{1}{2}\big(z_o \bmod \mathfrak{p}_2 + 196(z_1 \bmod \mathfrak{p}_o)\big), \tag{5}$$

where the $z_{0,1} \in L_2$ are given by

$$\begin{aligned} \sigma^i(z_0) &= \sigma^i(z) + \sigma^{i+3}(z) \\ \sigma^i(z_1) &= \sqrt{5}\big(\sigma^i(z) - \sigma^{i+3}(z)\big), \end{aligned}$$

and $z_i \bmod \mathfrak{p}_2$ is computed by (4).

The program now computes a complex floating point approximation to $\sigma^i(P)$ for $0 \le i < 6$, where P is the minimal polynomial of x over L_3. Using (5), it finds P to be congruent to $T^3 + 151T^2 + 434T + 346$ modulo $\mathfrak{p}_3$. The largest rounding error was 0.000488281. One modular root of P is 153. This means that there exists a prime ideal $\mathfrak{p}$ dividing p in $\mathcal{O}_L$ such that $x \equiv 153 \bmod \mathfrak{p}$. From x, one can compute the j-invariant of an elliptic curve with complex multiplication by $\mathcal{O}_K$ using the formula

$$j = \frac{(x^4 + 7x^3 + 20x^2 + 19x + 1)^3(x^2 + 5x + 13)}{x}.$$

With $x = 153 \bmod 479$, this gives $j = 307 \bmod 479$.

Calculating the minimal polynomial of x over K and reducing it modulo $\mathfrak{p}_o$, using (2), turns out to be impossible with 32 bit precision. If 48 bits are used, the largest rounding error is 0.0195312. Of course, this increase of the required precision is due to the fact that the theory of the genus field was not used.

The program used for $10^{9999} + 33603$ was a development version, with many possible optimisations not yet implemented. For instance, it is clear from the above example that not all Weber class invariants were implemented.

6 Use of the Fastest Fourier Transform in the West

For most of the calculations for $10^{9999} + 33603$, we used integer multiplication using the Fastest Fourier Transform in the West (see `http://www.fftw.org` and [7,8]). To square a number of size 10^{9999}, it was broken into 1661 digits of size 20 bit. These digits were inserted into an array of 3600 `double` variables, which was then transformed using the functions provided by `libfftw3.a`, the result was squared and transformed back also using `libfftw3.a`. The same thing can be done for a product of two different factors, and if a factor occurs often then its Fourier transform may be precalculated and stored to reduce the time for a multiplication by this factor to the time for a squaring. It is easy to see that this choice of parameters does not guarantee exact results. Therefore, we also calculated a 32-bit checksum. If the checksum test indicates an error, the multiplication is recalculated using the GMP function. In the case of the p10000, these recalculations appear to be rare, if they occur at all. Of course, even the checksum does not make this multiplication method rigorous.

We used this fast multiplication both for primality tests and for the calculation of modular square roots in the calculation of the run down sequence. While a modular square root can be checked, and the calculation repeated if necessary, there appears to be no way to detect a false negative result of a Miller-Rabin test. Therefore, by using this method we accepted a small but probably positive probability of a prime number being declared composite by mistake.

The following benchmark results were obtained on an 800-MHz Athlon, using version 4.1.1 of `libgmp.a`, version 3.0 of `libfftw3.a`, and $10^{9999}+33603$ as the input number:

- One call to the GMP function `mpz_probab_prime_p` with second argument equal to 1, which means that one Fermat and one Rabin-Miller pseudo-primality test is carried out: 317 seconds user time.
- A Rabin-Miller test using 2 as base, and using Montgomery modular multiplication [10] and the GMP functions: 149 seconds.
- A similar program, but using `libfftw3.a` for multiplications: 56 seconds.

The advantage of using `libfftw3.a` could perhaps be reduced if GMP allowed for a way to precalculate Fourier and Toom-Cook transforms of frequently used factors. It is difficult to predict whether this is sufficient to achieve the speed of a multiplication subroutine which is optimized for speed at the expense of producing sometimes (albeit rarely) a false result.

7 The Early Abort Strategy

The idea behind this strategy is to force the new candidate N' in Step 2. to satisfy $N/N' \geq 2^\delta$ for some (small) integer value $\delta = \delta(N)$, with the hope to decrease the number of steps and thereby the number of proving steps. Of course, this might slow down the search for N' a little bit and some optimization is necessary. Yet it appeared critical when used in the primality proof of $10^{9999}+33603$, when it was first implemented and used.

In FM's implementation, the following value for the lower bound on $\delta(N) = b(N) - b(N')$ was used, where $b(x)$ denotes the number of bits of integer x:

$b(N) \leq$	1000	2500	5000	7500	10000	15000	∞
$\delta(N)$	0	5	10	15	20	25	30

The following data were gathered (with the mpi program to be described). The first column contains the time without EAS, the second with. The line H_D stands for the time needed to compute the polynomials representing the class field, as in [9]; CZ is the time needed for the special variant of the Cantor-Zassenhaus algorithm using a trick of Atkin (using roots of unity modulo N); the number of steps in the *downrun* is then given, and the last lines contain the maximal value for h, before the mean value.

Number	$10^{2999} + 1887$		$10^{3999} + 4771$		$10^{4999} + 22669$	
Steps 1-2	81 h	58 h	280 h	164 h	716 h	476 h
Steps 3-5	26 h	26 h	76 h	86 h	209 h	261 h
H_D	1680 s	4880 s	4497 s	7317 s	3 h	5 h
CZ	22 h	22 h	63 h	75 h	179 h	234 h
# steps	436	358	597	437	734	512
$\max h$	1968	2336	2184	2432	3400	4000
$\overline{h}$	86	116	120	164	152	272

The restriction one puts on m and thus on D tends to make D and h larger than in the plain case. This can have an impact on the time for computing H_D, and also on the proving part. In the first phase, fewer N' are ever tested for probable primality, though more must be produced. EAS indeed decreases the number of steps, which tends to decrease the total time for the 1st phase, the 2nd being constant or increasing a little. In any case, a strategy yielding a factor of 2 in the total running time is certainly worthwhile.

8 Some Large Primality Proofs

8.1 The First Records of fastECPP

We begin with some data from FM's implementation that uses MPI on top of his `ecpp` program, and implementing Strategy 1. All computations were done on a cluster of 6 biprocessor Xeon at 2.66 MHz. We took the following numbers from the tables of P. Leyland*. These are numbers of the form $x^y + y^x$. WCT stands for *wall clock time* and includes the time wasted by the distribution process (waiting time of the slaves, typically). The line "Checking" indicates the time needed to check the certificate. Note that the time for this should be $\tilde{O}((\log N)^3)$ and this is compatible with the timings given.

All numbers were proven in 2003. The "when" line indicates the elapsed human dates in big endian notation.

The first number was dealt with an experimental program that turned out to spend too much time in the $\sqrt{-q^*}$ computations. As a matter of fact, a value of $r = 4000$ was used. This led to proceed by chunks of 400 squareroots from a total of 4000, adding 400 more if this was not enough. All discriminants with $D \leq 10^8$, $h \leq 6000$ (later increased to 8000) and the largest prime factor of h

* `http://research.microsoft.com/~pleyland/primes/xyyx.htm`

Table 1. Some large numbers proven with FASTECPP.

x	2177	2589	2551	2438	3571
y	580	218	622	1995	648
#dd	6016	6055	7127	8046	10041
when	0513-0604	0606-0617	0618-0714	0715-0901	1001-1220
# steps	801	736	965	1128	947
Steps 1-2 (CPU)	164 days	103 days	235 days	355 days	531 days
Steps 1-2 (WCT)	164×1.2	103×1.1	235×1.3	355×1.13	531×1.2
$\sqrt{-q^*}$	81 days	30 days	74 days	138 days	204 days
Steps 3-5	28 days	21 days	55 days	77 days	138 days
H_D	2951 sec	1686 sec	18451 sec	22552 sec	20285 sec
CZ	26 days	20 days	50 days	69 days	124 days
Checking	25 hours	22 hours	45 hours	70 hours	85 hours
$\max h$	1980	2080	3312	3640	6176
$\overline{h}$	121	103	190	209	409
$\max D$	7,749,263	19,076,479	52,396,648	87,949,348	95,895,480

not exceeding 200 were decided to be usable. A look at column 3 compared to 4 justifies the claim of complexity of $\tilde{O}((\log N)^4)$. The 8046dd number was done after the annoucement of the proof of $10^{9999}+33603$ (see next section), and EAS was not used for this. The 10041dd number was finished on December 20, 2003, well after the one to be described in the next section. This was the first use of EAS for this implementation.

8.2 A New Frontier

Let us turn our attention to the barrier-breaking number $10^{9999}+33603$, whose primality was verified by JF, TK with the help of TW.

The calculations were done using two programs, a pvm program producing the sequence of discriminants, group orders and their prime number factorization (called a run down sequence in what follows), and the second program calculating the elliptic curves.

The calculation of a run down sequence was started on July 17, 2003 on six 900MHz PIII CPUs. On July 21, the computation was moved to 12 nodes of parnass2, the LINUX cluster built at the Scientific Computing Institute in Bonn. 4 of these nodes had two 800MHz CPUs, the other nodes were double PII/400MHz computers. At 8550 digits (on July 30) and 8286 digits (on July 31) we interrupted these calculations to replace the program by a faster version, using the Fastest Fourier Transform in the West in the way explained above. This improvement resulted by a speedup by a factor of about 2. On August 5, we reached 6574 digits. On August 8, we stopped the program at 3256 digits. The final calculations for the run down sequence took about 8 hours on eight 800MHz CPUs. The total CPU time to produce the run down (i.e., without calculating the elliptic curves) was estimated to 234.5 days on a 1GHz Pentium.

The CPU time spent for the actual certificates is more difficult to estimate, since the program for this step was still under development when the calculation of the run down sequence started, and since these calculations were done in heterogeneous environment. We estimate that it would have taken less than 140 days on a single 800MHz Athlon CPU.

The certificate is available as:
`ftp://ftp.math.uni-bonn.de:pub/people/franke/p10000.cert`

9 Conclusions

We have described some new ideas to speed up practical primality proving of large numbers using fastECPP. These ideas need more testing and improvements. We hope that this will serve as benchmarks and motivations for the study of other primality proving algorithms as well.

Acknowledgments. Thanks to A. Enge for some useful discussions about this work and for his careful reading of the manuscript. The authors of the certificate for $10^{9999}+33603$ would like to thank the Scientific Computing Institute at Bonn University for providing the parallel computer which produced the downrun. One of us (J. F.) would also like to thank G. Zumbusch for pointing out the existence of libfftw3.a to him. Thanks also to D. Bernstein for sending us his remarks on a preliminary version and to the referees that helped clarify some points.

References

1. M. Agrawal, N. Kayal, and N. Saxena. PRIMES is in P. Preprint; available at `http://www.cse.iitk.ac.in/primality.pdf`, August 2002.
2. A. O. L. Atkin and F. Morain. Elliptic curves and primality proving. *Math. Comp.*, 61(203):29–68, July 1993.
3. D. J. Bernstein. How to find small factors of integers. June 2002. Available at `http://cr.yp.to/papers.html`.
4. A. Enge and F. Morain. Comparing invariants for class fields of imaginary quadratic fields. In C. Fieker and D. R. Kohel, editors, *Algorithmic Number Theory*, volume 2369 of *Lecture Notes in Comput. Sci.*, pages 252–266. Springer-Verlag, 2002. 5th International Symposium, ANTS-V, Sydney, Australia, July 2002, Proceedings.
5. A. Enge and F. Morain. Fast decomposition of polynomials with known Galois group. In M. Fossorier, T. Høholdt, and A. Poli, editors, *Applied Algebra, Algebraic Algorithms and Error-Correcting Codes*, volume 2643 of *Lecture Notes in Comput. Sci.*, pages 254–264. Springer-Verlag, 2003. 15th International Symposium, AAECC-15, Toulouse, France, May 2003, Proceedings.
6. A. Enge and R. Schertz. Constructing elliptic curves from modular curves of positive genus. Soumis, 2001.
7. Matteo Frigo and Steven G. Johnson. The fastest Fourier transform in the west. Technical Report MIT-LCS-TR-728, Massachusetts Institute of Technology, September 1997.

8. Matteo Frigo and Steven G. Johnson. FFTW: An adaptive software architecture for the FFT. In *Proc. 1998 IEEE Intl. Conf. Acoustics Speech and Signal Processing*, volume 3, pages 1381–1384. IEEE, 1998.
9. G. Hanrot and F. Morain. Solvability by radicals from an algorithmic point of view. In B. Mourrain, editor, *Symbolic and algebraic computation*, pages 175–182. ACM, 2001. Proceedings ISSAC'2001, London, Ontario.
10. Peter L. Montgomery. Modular multiplication without trial division. *Math. Comp.*, 44:519–521, 1985.
11. F. Morain. Primality proving using elliptic curves: an update. In J. P. Buhler, editor, *Algorithmic Number Theory*, volume 1423 of *Lecture Notes in Comput. Sci.*, pages 111–127. Springer-Verlag, 1998. Third International Symposium, ANTS-III, Portland, Oregon, june 1998, Proceedings.
12. F. Morain. Computing the cardinality of CM elliptic curves using torsion points. Submitted, October 2002.
13. F. Morain. Implementing the asymptotically fast version of the elliptic curve primality proving algorithm. June 2003. Available at `http://www.lix.polytechnique.fr/Labo/Francois.Morain/`.
14. M. Ram Murty. *Problems in Analytic Number Theory*, volume 206 of *Graduate Texts in Mathematics*. Springer-Verlag, 2001.

A Low-Memory Parallel Version of Matsuo, Chao, and Tsujii's Algorithm

Pierrick Gaudry[1] and Éric Schost[2]

[1] Laboratoire LIX, École polytechnique, 91128 Palaiseau, France
gaudry@lix.polytechnique.fr
[2] Laboratoire STIX, École polytechnique, 91128 Palaiseau, France
Eric.Schost@polytechnique.fr

Abstract. We present an algorithm based on the birthday paradox, which is a low-memory parallel counterpart to the algorithm of Matsuo, Chao and Tsujii. This algorithm computes the group order of the Jacobian of a genus 2 curve over a finite field for which the characteristic polynomial of the Frobenius endomorphism is known modulo some integer. The main tool is a 2-dimensional pseudo-random walk that allows to heuristically choose random elements in a 2-dimensional space. We analyze the expected running time based on heuristics that we validate by computer experiments. Compared with the original algorithm by Matsuo, Chao and Tsujii, we lose a factor of about 3 in running time, but the memory requirement drops from several GB to almost nothing. Our method is general and can be applied in other contexts to transform a baby-step giant-step approach into a low memory algorithm.

1 Introduction

Jacobians of small genus curves have now become an important tool for public-key cryptography; computing the Zeta function of such curves remains one of the central problems in this area.

For elliptic curves, the question found a first answer with Schoof's algorithm and the subsequent improvements (see [4] and the references therein). Furthermore, if the characteristic of the definition field is small, p-adic methods initiated with Satoh's algorithm [18] give a tremendous speed-up.

For higher genus curves, the p-adic methods (based on either Mestre's [15] or Kedlaya's [12] algorithms) also yield very satisfactory solutions in small characteristic, both in theory and in practice. However, in large characteristic, the question remains delicate. From the theoretical point of view, Pila's algorithm [16] and subsequent improvements [2,10] give polynomial time solutions, following Schoof's strategy. However, these have been turned into practical algorithms for genus 2 curves only [8], and it was only very recently that a Jacobian of cryptographic size was counted that way [9].

In this paper, we concentrate on the last part of a Schoof-like genus 2 algorithm. We first recall the basics of such algorithms.

D.A. Buell (Ed.): ANTS 2004, LNCS 3076, pp. 208–222, 2004.

Let us denote by $\chi \in \mathbb{Z}[T]$ the characteristic polynomial of the Frobenius endomorphism, so that $\chi(1)$ equals the Jacobian order. A basic task is to compute the reduction of χ modulo ℓ, for sufficiently many small primes or prime powers ℓ. Such information is obtained through the study of the torsion in the Jacobian; if the characteristic is medium, the Cartier-Manin operator [8,6] can also be used.

Thanks to Weil's bounds, one can then collect modular information until χ is known. However, it is far better to switch to a baby-step giant-step (BSGS) algorithm before the end, since collecting modular information is costly. For instance, in [9], the highest prime ℓ was 19, whereas without the BSGS phase, it would have been 59.

If $\chi(1)$ is known modulo m, then using standard BSGS techniques, the time necessary to compute the Jacobian order varies like $1/\sqrt{m}$. In 2002, Matsuo, Chao and Tsujii [14] showed that if we use not only the value $\chi(1)$ modulo m, but rather all coefficients of χ modulo m, it is possible to speed-up the BSGS computation. The runtime of their method (called MCT in the following) varies like $1/m$, which is an important improvement. The main drawback is the space complexity: the largest example shown in [14] used 12 GB of central memory, whereas the runtime (5 days on a single processor) was quite reasonable.

Standard BSGS techniques have low-memory, parallelizable, probabilistic counterparts, based on the rho or lambda (kangaroo) methods of Pollard's [17]: such techniques were presented in [24,22,23,8,21]. In this paper, we apply the ideas of Matsuo *et al.* to improve the variant of [8]; we obtain a probabilistic algorithm, with a heuristic complexity analysis, but which requires almost no memory and is immediately parallelizable. The expected running time is about 3 times the running time of the MCT algorithm.

In order to simplify the exposition, we concentrate on genus 2 curves. However, our idea works for more general curves as soon as the characteristic polynomial χ is known modulo some integer m. Another application of our method is the algorithm of [3] for Picard curves, which is a BSGS type algorithm.

The paper is organiz ed as follows. Section 2 defines the necessary notation and recalls the BSGS algorithm of [14]. Then, to introduce methods based on the birthday paradox, we start in Section 3 by a special case, where the modular information is complete enough, so we can use the method of [8]. In Section 4, we deal with the general case, and introduce bidimensional analogues of these techniques. Section 5 finally presents our experimental results.

2 MCT Algorithm for Genus 2 Hyperelliptic Curves

2.1 Characteristic Polynomial of Frobenius Endomorphism

Let $\mathcal{C}$ be a genus 2 curve, defined over the finite field $\mathbb{F}_q$ with q elements. The Jacobian group of $\mathcal{C}$ is denoted by $\mathbf{J}(\mathcal{C})$ and the characteristic polynomial of the Frobenius endomorphism is denoted by $\chi(T)$. This polynomial has the form

$$\chi(T) = T^4 - s_1T^3 + s_2T^2 - qs_1T + q^2,$$

where s_1 and s_2 are integers. The group order of the Jacobian is then given by

$$\#\mathbf{J}(\mathcal{C}) = \chi(1) = q^2 + 1 - s_1(q+1) + s_2.$$

By Weil's theorem, the roots of χ have absolute value $\sqrt{q}$ and bounds on s_1 and s_2 follow directly. Better bounds can be found using the fact that the roots of χ come in pairs of complex conjugates [13, Proposition 7.1]:

$$|s_1| \le 4\sqrt{q}, \qquad 2|s_1|\sqrt{q} - 2q \le s_2 \le \frac{s_1^2}{4} + 2q. \tag{1}$$

The values of (s_1, s_2) satisfying these bounds form the hatched zone in Figure 1.

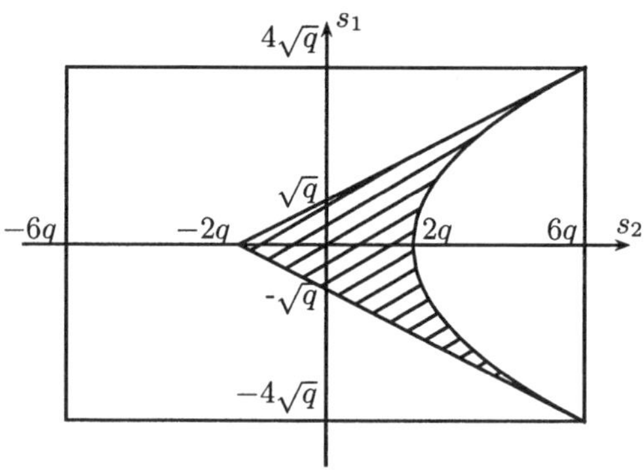

Fig. 1. Bounds on s_1 and s_2

2.2 Review of the MCT Algorithm

In large characteristic, point counting algorithms work by collecting modular information; this information is then recombined using the Chinese Remainder Theorem and the work is finished using some BSGS strategy. We now describe the MCT algorithm for this last step.

From now on, we assume that the characteristic polynomial χ is known modulo some positive integer m, *i.e.* s_1 and s_2 are known modulo m. Therefore we introduce new variables for the known and unknown parts of s_1 and s_2:

$$s_1 = \overline{s_1} + m\widetilde{s_1} \quad \text{and} \quad s_2 = \overline{s_2} + m\widetilde{s_2},$$

so that our goal is now to find $\widetilde{s_1}$ and $\widetilde{s_2}$. To this effect, a random divisor D is first picked in $\mathbf{J}(\mathcal{C})$: the strategy is to compute the order of D, hoping that it is large enough to be able to conclude (the case when $\mathbf{J}(\mathcal{C})$ is highly non-cyclic is rare in practice and easily tackled).

The order of D divides the group order $\chi(1)$, therefore we have the equality

$$(q^2 + 1 - \overline{s_1}(q+1) + \overline{s_2}) \cdot D + (-\widetilde{s_1}(q+1) + \widetilde{s_2}) \cdot m \cdot D = 0.$$

To obtain a BSGS algorithm, it is usual to separate the two unknowns, one on each side of the equation. Here the unknowns $\widetilde{s_1}, \widetilde{s_2}$ lie in intervals that are of different sizes, therefore it is necessary to split again $\widetilde{s_2}$ into two pieces. Let n be a parameter to be fixed later, and let us write $\widetilde{s_2} = t_2 + nu_2$ with $0 \leq t_2 < n$. Then $\chi(1) \cdot D = 0$ rewrites

$$\left(q^2 + 1 - \overline{s_1}(q+1) + \overline{s_2} + m(-\widetilde{s_1}(q+1) + nu_2)\right) \cdot D = -t_2 m \cdot D.$$

The algorithm proceeds as follows: first, all possible values for the right-hand side are computed and stored in a data-structure in which searching is fast. Then the left-hand side is computed for all possible values of $\widetilde{s_1}$ and u_2, until a match is found with a value of the right-hand side. For each value of $\widetilde{s_1}$, the bounds of Equation (1) are used to find the range of the possible values for u_2; a precise study of the area of the space of search leads to an optimal value $n \approx q^{3/4}/m$, yielding a running time of about $4q^{3/4}/m$ operations in $\mathbf{J}(\mathcal{C})$.

3 Preliminaries: The Case $m \geq 8\sqrt{q}$

The main drawback of the previous method is the storage requirement. We now introduce low-memory variants and start by describing a special case, for which most features of the general treatment are already present.

Suppose that χ is known modulo m, with $m \geq 8\sqrt{q}$. Because $|s_1| \leq 4\sqrt{q} \leq m/2$, the coefficient s_1 is known exactly. Corresponding to this value of s_1, we have bounds for s_2 which yield bounds on $\widetilde{s_2}$, which is the only remaining unknown. We are thus in the setting of the search of a group order in a bounded arithmetic progression. Among the many variants inspired by Pollard's lambda (or kangaroos) method that yield a low-memory solution to this question, we chose the one based on the birthday paradox described in [8]. We recall it here briefly; detailed descriptions and analyses of several other variants can be found in [17,24,22,23,21].

3.1 Random Search in Two Intersecting Intervals

Let $K = q^2 + 1 - s_1(q+1) + \overline{s_2}$. Then the group order can be written $\#\mathbf{J}(\mathcal{C}) = K + m\widetilde{s_2}$, where K is known and $\widetilde{s_2}$ is unknown. Performing a suitable shift, it is possible to adjust K by some multiple of m so that the bounds on $\widetilde{s_2}$ become symmetric, *i.e.* such that $|\widetilde{s_2}| \leq B$ for some integer B.

As before, we pick at random an element D in $\mathbf{J}(\mathcal{C})$, of presumably large order. We then define two sets of divisors that we denote W and T:

$$W = \{(K + m\sigma_2) \cdot D \ ; \ \sigma_2 \in [-B, B]\}, \quad T = \{m\sigma_2 \cdot D \ ; \ \sigma_2 \in [-B, B]\}.$$

By definition W and T intersect, since the zero divisor is in both of them, by taking $\sigma_2 = \widetilde{s_2}$ for W and $\sigma_2 = 0$ for T. More precisely, the size of the intersection is $\#(W \cap T) \in [B+1, 2B+1]$, depending on how far $\widetilde{s_2}$ is from 0. The algorithm then proceeds as follows: we pick random elements uniformly alternatively in W

and T. If the same divisor is obtained as an element of both W and T then (a multiple of) the order of D can be deduced.

Assuming that picking a random element in W or T has unit cost, that all elements are stored in table, and that a search in the table has unit cost, then by the birthday paradox, a collision can be obtained in expected time and space $O(\sqrt{B})$. The constant in the $O(\ )$ can actually be made more explicit. First, for fixed $\widetilde{s_2}$, the expected running time grows like $\sqrt{(2B+1)\pi/\gamma}$, where $\gamma = \#(W \cap T)/(2B+1) = (2B+1-|\widetilde{s_2}|)/(2B+1)$. Thus, assuming that $\widetilde{s_2}$ is uniformly distributed in $[-B, B]$, the expected number of operations is asymptotic to

$$\frac{1}{2B+1}\int_{-B}^{B}\sqrt{\frac{(2B+1)\pi}{\gamma(\widetilde{s_2})}}\,d\widetilde{s_2} = \int_{-B}^{B}\sqrt{\frac{\pi}{2B+1-|\widetilde{s_2}|}}\,d\widetilde{s_2},$$

which itself is asymptotic to $2(2-\sqrt{2})\sqrt{2\pi B} \approx 2.07\sqrt{2B}$. Though, one should note that even if $\mathcal{C}$ is randomly chosen uniformly among all curves, the number of points of the Jacobian is not uniformly distributed, as the concentration is slightly higher in the middle of the Hasse-Weil interval; therefore $\widetilde{s_2}$ is not uniformly distributed.

3.2 Pseudo-Random Walk and Distinguished Points

We now address the main issues raised in the above algorithm: the generation of random elements in W and T and their storage. The key to the answer is to replace randomness by a deterministic pseudo-random walk.

Let $r \geq 0$ and ℓ be parameters to be fixed later. The pseudo-random walk is initialized as follows: for all k in $[1, r]$, an offset $\mathcal{O}_k \in \mathbf{J}(\mathcal{C})$ is precomputed as $\mathcal{O}_k = \alpha_k m \cdot D$, where α_k is a random integer in $[0, 2\ell]$. We also need a hash function $\mathcal{H}$ that maps elements of $\mathbf{J}(\mathcal{C})$ to $[1, r]$; this hash function should have good statistical properties, but no cryptographically strong property, like one-wayness, is required. Typically, $\mathcal{H}$ is obtained by taking a few bits in the internal representation of the elements and the integer obtained this way is taken modulo r. Then, starting with an element P in W (resp. in T) for which we know the corresponding σ_2, we define another point Q by $Q = P + \mathcal{O}_{\mathcal{H}(P)}$.

Assuming that ℓ is not too large compared to B, with high probability the point Q is still an element of W (resp. of T). Furthermore, the value of σ_2 corresponding to Q is obtained by adding $\alpha_{\mathcal{H}(P)}$ to the value of σ_2 for P.

Iterating this process yields a chain of pseudo-randomly chosen elements in W (resp. in T). However, the chain should not be too long, to keep the probability to go out of the domain moderate. Thus the average length ℓ of an offset must be adapted according to the average number of steps we expect to do in one chain (see below). With this device it is then possible to produce each new pseudo-random element in W or T for one operation in $\mathbf{J}(\mathcal{C})$.

We now deal with storage requirements. To this effect, we introduce the concept of distinguished points, originally appeared in [7]. We say that an element of $\mathbf{J}(\mathcal{C})$ is a distinguished point if its image by a second hash function is 0. Again, we do not ask much of this hash function, except that its behavior looks independent of any arithmetic property of the element, and that the probability of being distinguished can be effectively estimated and tuned to a prescribed value. As usual, looking at some bits in the internal representation of the elements is a good way of doing. We denote by $p_{\mathcal{D}}$ the probability for an element of being distinguished.

The algorithm then proceeds as follows: starting with a random point alternatively in W and T, we produce pseudo-random elements using the pseudo-random walk, until a distinguished point is hit. Then this point is stored and another chain is started. The length of the chains is $1/p_{\mathcal{D}}$ on average, and the parameter ℓ should be tuned accordingly. If all the parameters are well chosen, then after (say) 1000 chains on average, we have produced enough points to expect a collision. If this occurs at a point that is not distinguished, then the two chains continue on the same track because the pseudo-random walk is deterministic; the two chains will end at the same distinguished point, thus allowing the detection and the solution of the problem.

Many practical experiments were made to test this strategy against the idealized algorithm described above. They are very satisfactory and in [22] it was suggested that taking $r = 20$ is enough to simulate a random walk in this context.

As a conclusion, in the case when $m \geq 8\sqrt{q}$, the number of points can be computed using a parallel low-memory algorithm that requires on average $O(\sqrt{q/m})$ operations in $\mathbf{J}(\mathcal{C})$. This is worse than the $O(q^{3/4}/m)$ complexity announced in [14], but their complexity analysis implicitly excludes that case.

4 The General Case

In the case when $m < 8\sqrt{q}$, there are several choices for s_1 and still many more for s_2. We could loop over all the possible values of s_1 and look for a corresponding s_2 using the algorithm of the previous section, but this approach has worse complexity than the MCT algorithm. The workaround is to take into account the bidimensional nature of the problem.

Recall that we have $s_1 = \overline{s_1} + m\widetilde{s_1}$ and $s_2 = \overline{s_2} + m\widetilde{s_2}$, where $\overline{s_1}$ and $\overline{s_2}$ are known integers in $[0, m-1]$, the goal being to find $\widetilde{s_1}$ and $\widetilde{s_2}$. From the bounds (1) on s_1 and s_2, we deduce similar bounds for $\widetilde{s_1}$ and $\widetilde{s_2}$. In order to make the description more generic, we write these bounds in the following form:

$$B_{1,\min} \leq \widetilde{s_1} \leq B_{1,\max}, \qquad B_{2,\min} \leq \widetilde{s_2} \leq B_{2,\max}.$$

4.1 Random Search in Two Intersecting Rectangles

Let D be a random element of $\mathbf{J}(\mathcal{C})$; as previously, we assume that the order of D is large enough compared to the group order. Since $\chi(1) \cdot D = 0$, we have

$$\left(q^2 + 1 - \overline{s_1}(q+1) + \overline{s_2}\right) \cdot D + \left(-\widetilde{s_1}(q+1) + \widetilde{s_2}\right) \cdot m \cdot D = 0.$$

Let K be $q^2+1-\overline{s_1}(q+1)+\overline{s_2}$, which is a known integer; then we have to find $\widetilde{s_1}$ and $\widetilde{s_2}$ such that $K \cdot D + (-\widetilde{s_1}(q+1)+\widetilde{s_2})\, m \cdot D = 0$. Let R be a rectangle containing the possible values for the pair $(\widetilde{s_1}, \widetilde{s_2})$:

$$R = \{(\sigma_1, \sigma_2);\ \sigma_1 \in [B_{1,\min}, B_{1,\max}],\ \sigma_2 \in [B_{2,\min}, B_{2,\max}]\}\,.$$

Since $B_{1,\min}$ (resp. $B_{2,\min}$) is not necessarily the opposite of $B_{1,\max}$ (resp. $B_{2,\max}$), we have to normalize the situation. We define a value K' to be used in place of K, so as to center our search:

$$K' = K + m\left(-\left\lfloor \frac{B_{1,\min}+B_{1,\max}}{2} \right\rfloor (q+1) + \left\lfloor \frac{B_{2,\min}+B_{2,\max}}{2} \right\rfloor\right).$$

We can now define two sets of points:

$$W = \{K' \cdot D + (-\sigma_1(q+1)+\sigma_2)\cdot m \cdot D;\ (\sigma_1,\sigma_2) \in R\}\,,$$

$$T = \{(-\sigma_1(q+1)+\sigma_2)\cdot m \cdot D;\ (\sigma_1,\sigma_2) \in R\}\,.$$

We assume that these two sets have cardinality exactly $\#R$. This may not hold in general, but is true with the further assumptions that D is of large order and that m is larger than 8.

By construction, the sets W and T have a non-trivial intersection. Let D_W be in W and D_T in T. We write $(\sigma_{1W}, \sigma_{2W})$ the values corresponding to D_W and $(\sigma_{1T}, \sigma_{2T})$ the values corresponding to D_T. Then, assuming again that the order of D is large enough, $D_W = D_T$ if and only if

$$\begin{aligned} \sigma_{1W} - \sigma_{1T} &= \widetilde{s_1} - \lfloor (B_{1,\min}+B_{1,\max})/2 \rfloor\,, \\ \sigma_{2W} - \sigma_{2T} &= \widetilde{s_2} - \lfloor (B_{2,\min}+B_{2,\max})/2 \rfloor\,. \end{aligned} \tag{2}$$

Hence it is easily checked that

$$N = \#(W \cap T) \in \left[\frac{\#R}{4}, \#R\right].$$

When $B_{1,\min} = -B_{1,\max}$ and $B_{2,\min} = -B_{2,\max}$, we get the picture of Figure 2.

A first version of our algorithm now proceeds as follows: random elements of W and T are constructed by picking random elements in R. These elements are stored in a data structure in which it is possible to detect quickly collisions between an element of W and an element of T. Together with these elements, we also store the corresponding pair (σ_1, σ_2). On average, due to the birthday paradox, a collision occurs after having constructed $O(\sqrt{N})$ elements of W and T; taking the difference of the pairs σ_T and σ_W then gives the result by Equations (2).

Since the bounds on $\widetilde{s_1}$ and $\widetilde{s_2}$ yield $|\widetilde{s_1}| = O(\sqrt{q}/m)$ and $|\widetilde{s_2}| = O(q/m)$, we have $N = O(q^{3/2}/m^2)$, and therefore the expected number of points to construct is in $O(q^{3/4}/m)$. Just as in the unidimensional case, we now give an estimate of the constant hidden in the $O(\,)$, using simplifying assumptions.

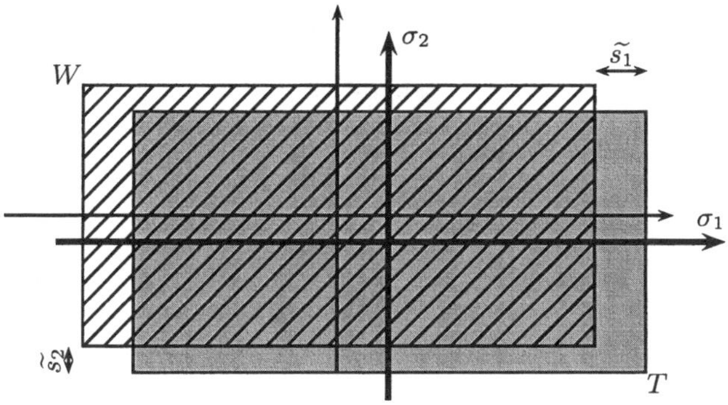

Fig. 2. Intersection of W and T

For such estimates, we can assume that the problem is centered, and therefore we put $B_1 = B_{1,\max} = -B_{1,\min}$ and $B_2 = B_{2,\max} = -B_{2,\min}$; hence $\#R = (2B_1+1)(2B_2+1)$. We denote by $\gamma \in [\frac{1}{4}, 1]$ the ratio $\#(W \cap T)/\#R$: this parameter is easily computed as:

$$\gamma(\widetilde{s_1}, \widetilde{s_2}) = \frac{(2B_1 + 1 - |\widetilde{s_1}|)(2B_2 + 1 - |\widetilde{s_2}|)}{\#R}.$$

From the birthday paradox, we see that the expected number of elements that have to be created before a collision between an element of W and an element of T occurs is asymptotic to $\sqrt{\pi\,\#R/\gamma}$.

We now assume that $\widetilde{s_1}$ and $\widetilde{s_2}$ are uniformly distributed. Then the average number of points to construct grows like

$$\frac{1}{\#R}\int_{\widetilde{s_1}=-B_1}^{B_1}\int_{\widetilde{s_2}=-B_2}^{B_2}\sqrt{\frac{\pi\,\#R}{\gamma(\widetilde{s_1},\widetilde{s_2})}}\,d\widetilde{s_1}\,d\widetilde{s_2}.$$

This integral is easily computed and shows that the expected number of points to construct is asymptotic to $8\sqrt{\pi}(3-2\sqrt{2})\sqrt{\#R} \approx 2.43\sqrt{\#R}$. Now, bounds (1) on s_1 and s_2 yield approximate bounds for $\widetilde{s_1}$ and $\widetilde{s_2}$:

$$B_{1,\max} - B_{1,\min} = 8\sqrt{q}/m,\ B_{2,\max} - B_{2,\min} = 8q/m,$$

hence $\#R = 64\,q^{3/2}/m^2$. The approximate value of 2.43 then yields a running time of about $19.5\,q^{3/4}/m$ operations in $\mathbf{J}(\mathcal{C})$. Hence, we see that a constant factor of about 5 is lost compared to the original (memory consuming) MCT algorithm. This difference is partly due to the fact that in the MCT algorithm, the BSGS approach allows to search only in the area described in Figure 1, whereas ours does not take this specificity into account.

Note that this analysis is idealized in several places: first, the pseudo-random walk that will be used below is not purely random and can cause some discrepancies. Next, the assumption that $\widetilde{s_1}$ and $\widetilde{s_2}$ are uniformly distributed in a rectangle is actually wrong. However, we expect that this should still give a good estimate.

4.2 A Bidimensional Pseudo-Random Walk

The main questions are now the generation of random elements in W and T, and how to avoid storing them. These are the same issues as in Section 3 and similar techniques will be applied to solve them. The definition of distinguished points still makes sense in this new context, and will be used with no modification. However, the previous pseudo-random walk needs to be converted into a bidimensional one.

Let r, ℓ_1 and ℓ_2 be parameters to be fixed later: r controls the number of offsets that we are going to precompute, and ℓ_1 and ℓ_2 are the average lengths of the horizontal and vertical offsets. For each k and k' in $[1, r]$, we select a random non-negative integer $\alpha_{k,k'}$ uniformly in $[0, 2\ell_1]$ and a random non-negative integer $\beta_{k,k'}$ uniformly in $[0, 2\ell_2]$. Then for each k and k' in $[1, r]$ and b in $\{0, 1\}$, we compute and store the offsets $\mathcal{O}_{k,k',b} = (-1)^b \alpha_{k,k'}(q+1)mD + \beta_{k,k'}mD \in \mathbf{J}(\mathcal{C})$, where D is the base point whose order is to be computed.

Starting with a point P in W (resp. in T) for which we know the corresponding pair (σ_1, σ_2), we define another point as follows. We compute k, k', and b as pseudo-random deterministic functions of P, by using some hash functions. Then we define $Q = P + \mathcal{O}_{k,k',b}$. If ℓ_1 and ℓ_2 are small enough, with high probability Q is still in W (resp. in T) and the corresponding pair is given by $(\sigma_1 - (-1)^b \alpha_{k,k'}, \sigma_2 + \beta_{k,k'})$. Iterating this process allows us to produce chains of pseudo-randomly chosen elements in W (resp. in T). As before, the chains can not be too long, otherwise they go out of W (resp. of T). The cost of producing one element is one group operation.

Note that we have only used positive values for the offset in the second direction: this is intended to reduce the chance of creating cycles. However, experiments with alternative strategies turned out to yield similar results.

4.3 Setting the Parameters

Let λ be such that the expected number of points to construct is $\lambda\sqrt{\#R}$ and let C be the number of chains we expect to construct. Note that C is fixed by the user; it should be large enough so that averaging considerations make sense (say $C \geq 1000$), and small enough so that the cost of initializing a chain is negligible compared to the cost of the steps that are done in that chain. Also the number of chains is essentially the number of distinguished points that have to be stored and therefore should be small enough. Knowing an approximation of λ and having fixed C, the probability of being distinguished follows from

$$p_{\mathcal{D}} = \frac{C}{\lambda\sqrt{\#R}}.$$

Now we fix ℓ_1 and ℓ_2; to control the probability of going out of W (or T), we first evaluate the average length of a chain. There are about $1/p_{\mathcal{D}}$ steps and each step goes on average ℓ_1 in the first direction and ℓ_2 in the second direction. In the second direction, all the offsets are positive and therefore the length of a chain is $\ell_2/p_{\mathcal{D}}$ on average. We want this to be small enough compared to the size of R in that direction, say one tenth of $B_{2,\max} - B_{2,\min}$:

$$\ell_2 = \frac{(B_{2,\max} - B_{2,\min})\, p_{\mathcal{D}}}{10}.$$

In the first direction, the situation is different since the offsets can be positive or negative, but still with an average absolute value of ℓ_1. The central limit theorem applied to the 1-dimensional random walks gives that the average distance to the origin after $1/p_{\mathcal{D}}$ steps is about $2\sqrt{2/3\pi}\,\ell_1/\sqrt{p_{\mathcal{D}}}$: the factor $2/\sqrt{3}$ corresponds to the standard deviation of the lengths of the offsets. For convenience, we approximate $2\sqrt{2/3\pi}$ by $9/10$. Again, imposing that this value is one tenth of the size of R in that direction yields:

$$\ell_1 = \frac{(B_{1,\max} - B_{1,\min})\, \sqrt{p_{\mathcal{D}}}}{9}.$$

It may happen that ℓ_1 or ℓ_2 is very small, and even smaller than 1. This is especially the case when the bounds on $\widetilde{s_1}$ and $\widetilde{s_2}$ have a different order of magnitude. Unfortunately this is the case in the genus 2 point counting case, where $\widetilde{s_2}$ is on average about $\sqrt{q}$ times larger than $\widetilde{s_1}$. A solution would be to enlarge C, but this is not satisfactory since it implies more storage.

A better choice is to modify the random walk as follows. Assume that ℓ_1 is small (ℓ_1 and ℓ_2 can not be simultaneously small). Then with probability p, we add either the same offset $\mathcal{O}_{k,k',b}$ as before or a modified offset $\widetilde{\mathcal{O}}_{k,k'} = \beta_{k,k'} mD$ which does not include a progression in the first direction. The probability p is fixed so that the apparent mean value of ℓ_1 is the one we wanted. Note that the decision of adding $\mathcal{O}_{k,k',b}$ or $\widetilde{\mathcal{O}}_{k,k'}$ is a deterministic choice that depends on D.

4.4 Reducing the Search Space

In order to have a better understanding of the distribution of (s_1, s_2), we ran some statistics. Note that similar statistics, supported by theoretical and heuristic considerations were done in [20], but with the purpose of finding the mean value of the class number.

Here, for $p = 10^6 + 3$, we randomly selected 10,000 monic squarefree polynomials of degree 5 over $\mathbb{F}_p$ and computed the (s_1, s_2) values for the corresponding curves. As expected, the pairs (s_1, s_2) tend to be not too close to the borders of the domain. In Figure 3, we represented the domain where (s_1, s_2) are allowed to stay according to bounds (1), and inside the domain, the darkness of a point means that the density of pairs (s_1, s_2) is high.

On the picture, we see that there are very few pairs for which s_2 is large, because the "wings" are very thin. In fact, these points correspond to curves which are close to maximal curves, and it is no surprise that they are rare. More precisely, in our tests, the proportion of curves for which s_2 is larger than $3q$ is about 2.2% and the proportion of curves for which s_2 is larger than $4q$ is about 0.23% (remember that in theory, s_2 can be as large as $6q$).

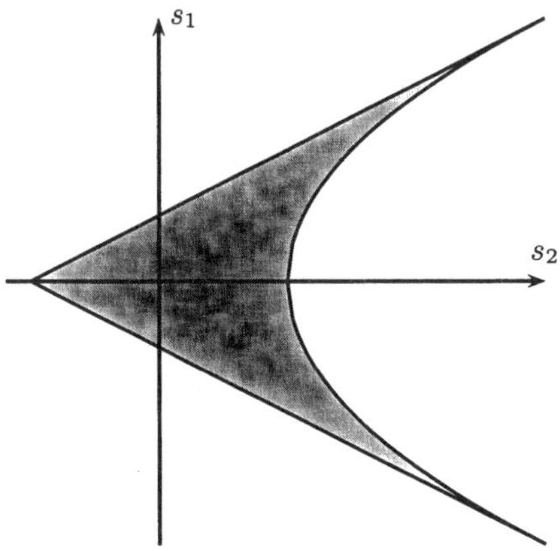

Fig. 3. Statistics on (s_1, s_2)

In view of these results, in order to improve our constant 19.5, and remarking that there is no point in spending too much time for large values of s_2, we will restrict the rectangle R to the following bounds:

$$R = \{(\sigma_1, \sigma_2);\ \sigma_1 \in [-2.5\sqrt{q}/m, 2.5\sqrt{q}/m],\ \sigma_2 \in [-2q/m, 3q/m]\}.$$

In our statistics, for more than 97% of the curves, $\widetilde{s_2} < 3q/m$, and therefore the overlapping factor of W and T is at least 1/4; and for about 99.7% of the curves, $\widetilde{s_2} < 4q/m$ and therefore the overlapping factor is at least $\frac{3}{25} = 0.12$. The overlapping decreases as the point (s_1, s_2) gets closer to the end of the "wings" of the arrow on the picture and if $s_2 > 5.5\,q$, the sets W and T do not overlap.

With this strategy, the area of R is reduced to $25\,q^{3/2}/m^2$ so that the expected runtime is about $12\,q^{3/4}/m$ group operations. Therefore we lose "only" a factor of 3 compared to the original MCT algorithm; this strategy is used in the experiments presented below.

We expect that this is very unlikely to get curves with s_1, s_2 outside of the above rectangle R by random constructions. In case the algorithm does not find an answer after say 10 times the above expected time it could pay to start a classical MCT algorithm to search deterministically in the ends of the wings if we have enough memory for this much smaller subproblem. Otherwise, another method is to start a chain corresponding to these wings in the area outside the rectangle with a small probability, so that we do not perturb the average runtime but we can guarantee that the program finishes.

5 Practical Experiments

We did two kinds of experiments: first with a high level implementation using the Magma computer algebra system [5], we ran some simulations to check the validity of our approach in various situations. Then, we wrote a parallel C++ implementation of our algorithm using the NTL [19] and the MPICH [1] libraries, in order to run tests with real sized curves. In our experiments, we used the reduced search space given in Section 4.4, with $\#R \approx 25\, q^{3/2}/m^2$.

5.1 Simulations

In order to test the validity of our heuristics on sufficiently many examples of reasonable size, we used the following simulated algorithm. Let p be an integer. We pick two integers s_1 and s_2 at random uniformly in the area of Figure 1. Then we form the integer $N = p^2 + 1 - s_1(p+1) + s_2$ and work in the group $\mathbb{Z}/N\mathbb{Z}$ instead of $\mathbf{J}(\mathcal{C})$. An integer m is chosen, and using only p and s_1, s_2 modulo m, our goal is to recover N. Thus, we can construct appropriate examples with different values for the parameters at almost no cost.

We chose several pairs (p, m) yielding always $\#R \approx 5 \times 10^{10}$, so that $p_\mathcal{D}$ can be taken to be 2^{-8}, and with ℓ_1 varying from 0.01 to 21. For each such pair (p, m), 100 random pairs (s_1, s_2) where tested and the average number of steps is measured and compared to $p^{3/4}/m$. The results are reported in Table 1.

Table 1. Simulations with cyclic groups $\mathbb{Z}/N\mathbb{Z}$. Each line corresponds to 100 runs.

p	m	ℓ_1	ℓ_2	Avg ratio nb jumps / $(p^{3/4}/m)$
5.7×10^7	14	21	$8,040$	12.7
9.2×10^8	118	10	$15,250$	12.7
1.5×10^{10}	949	5	$30,350$	12.5
1.5×10^{14}	953674	0.5	$304,000$	12.6
1.2×10^{17}	144675925	0.1	$1,621,271$	14.3
1.9×10^{18}	1157407407	0.05	$3,242,542$	14.5
2.5×10^{21}	250000000000	0.01	$19,455,252$	22.2

Our conclusion is that even if the rectangle R is very thin ($i.e.$ ℓ_1 is tiny), then the measured running time is quite close to the predictions. In the extreme, the last case $\ell = 0.01$ corresponds to a case where $m \approx 4\sqrt{q}$, therefore $B_{1,\max} - B_{1,\min}$ is just 2. In that case, the heuristics we used hardly make sense, but this is a good surprise that the average running time is still within a factor 2 of the heuristic analysis.

5.2 Several Runs on the Same Curve

We ran our software implementation many times on a given curve, in order to check that the average measured running time is close to the heuristic estimate, for a medium sized problem.

Let $p = 5 \times 10^{24} + 8503491$ and f a random monic squarefree polynomial of degree 5 over $\mathbb{F}_p$. Using the Schoof-like method described in [9] we have deduced the values of s_1 and s_2 modulo $m = 44696171520$. The curve is such that $s_1/\sqrt{p} \approx -0.84$ and $s_2/p \approx 0.38$, therefore it is not close to any border of the bounds.

With the MCT algorithm, the computation is feasible and requires about 1.5 GB of memory. We ran our software 100 times on that input, and for each run we measured the number of operations in the Jacobian. The average value is $11.27\,p^{3/4}/m$, which is in accordance with our estimates. The minimal value is $2.69\,p^{3/4}/m$ and the maximal value is $31.55\,p^{3/4}/m$. In Figure 4 we give the histogram for the number of runs whose running time is in a given range.

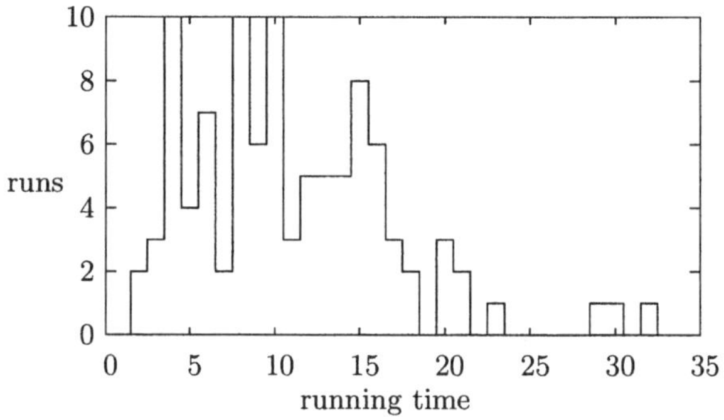

Fig. 4. Histogram showing the number of runs whose running time is in the given range (divided by $p^{3/4}/m$).

5.3 A Larger Example

In order to test the scalability of our method, we ran a larger example. Let $p = 5 \times 10^{24} + 8503491$ as above, and f a random monic squarefree polynomial of degree 5 over $\mathbb{F}_p$. We now suppose that s_1 and s_2 are known modulo only $m = 1655413760$ (note that the modulus 44696171520 mentioned above equals 27×1655413760).

With this more restricted information, we cannot conclude using the original MCT algorithm: the computation would require to store about 5×10^9 points together with their indices. Even using hash tables, it seems difficult to use less than 8 bytes for each entry, which means at least 40 GB of memory.

We ran our algorithm on that input. The probability of being distinguished was set to $p_{\mathcal{D}} = 2^{-24}$. The program was run in parallel on a cluster of 24 Pentium IV at 3 GHz. After 4 hours of computation, 544 distinguished points were computed and a useful collision occurred. About 9×10^9 steps were performed: this is a "lucky" run, since this is about $4.3\,p^{3/4}/m$. The memory requirement was about 1.5 MB on each node, mostly for the executable code, not the data. The amount of communication is reduced to a few KB between the nodes and a "master" node.

For that curve, $s_1/\sqrt{p} \approx -1.16$ and $s_2/p \approx 1.25$, so the curve was not exceptional, in the sense that it was not too close to the border of the area of Figure 1.

6 Curves of Higher Genus

The MCT algorithm can be extended to higher genus curves for which the characteristic polynomial of the Frobenius endomorphism is known modulo some integer m. In [11], different variants of this extension are studied and compared. Our extension of the original genus 2 MCT algorithm also applies to all these variants with a few modifications. Indeed, the number k of unknown coefficients in the characteristic polynomial can be larger than 2 and then a k-dimensional random walk should be designed in order to search for a collision in the intersection of two k-dimensional boxes. It is a tedious task to fill in the details, in particular the expected constants hidden in the $O(\ )$, but for fixed genus, we certainly lose only a constant factor compared to the memory-costly algorithms.

Another range of application of our method is the BSGS algorithm developed in [3] for counting points of the Jacobians of Picard curves. Without giving many details, let us mention that their algorithm ends by the search for a collision in two arithmetic progressions; therefore our algorithm applies almost directly to that case and should dramatically reduce the space complexity, which is the main drawback of their method.

7 Conclusion

We have presented a low-memory, parallelizable analogue of the algorithm by Matsuo, Chao and Tsujii for genus 2 hyperelliptic curves; the method works in other cases, such as the BSGS algorithm of [3] for Picard curves.

The main tool we used is a bidimensional pseudo-random walk. As usual with this kind of algorithms, it is impossible to make a rigorous analysis and heuristics and computer experiments are necessary to validate the approach. Our numerical data are positive in that sense, therefore our algorithm can be used when the memory constraint becomes problematic.

Acknowledgements. We thank Gérard Guillerm for giving us access to his PC cluster, Gilles Schaeffer for sharing his knowledge of the central limit theorem, and Edlyn Teske for many helpful comments on a first version of this paper. Part of the computations were done at LIX, on a cluster paid by the *ACI Cryptologie* funding.

References

1. MPICH: A portable implementation of MPI.
   `http://www-unix.mcs.anl.gov/mpi/mpich/`.
2. L. Adleman and M.-D. Huang. Counting points on curves and abelian varieties over finite fields. *J. Symbolic Comput.*, 32:171–189, 2001.
3. M. Bauer, E. Teske, and A. Weng. Point counting on Picard curves in large characteristic, 2003. Preprint.

4. I. Blake, G. Seroussi, and N. Smart. *Elliptic curves in cryptography*, volume 265 of *London Math. Soc. Lecture Note Ser.* Cambridge University Press, 1999.
5. W. Bosma and J. Cannon. *Handbook of Magma functions*, 1997. `http://www.maths.usyd.edu.au:8000/u/magma/`.
6. A. Bostan, P. Gaudry, and É. Schost. Linear recurrences with polynomial coefficients and computation of the Cartier-Manin operator on hyperelliptic curves. 2003. To appear in Proceedings Fq'7.
7. J.-P. Delescaille and J.-J. Quisquater. How easy is collision search? Application to DES. In J.-J. Quisquater, editor, *Advances in Cryptology – EUROCRYPT'89*, volume 434 of *Lecture Notes in Comput. Sci.*, pages 429–434, 1990.
8. P. Gaudry and R. Harley. Counting points on hyperelliptic curves over finite fields. In W. Bosma, editor, *ANTS-IV*, volume 1838 of *Lecture Notes in Comput. Sci.*, pages 313–332. Springer–Verlag, 2000.
9. P. Gaudry and É. Schost. Construction of secure random curves of genus 2 over prime fields. In *Eurocrypt 2004*, To appear.
10. M.-D. Huang and D. Ierardi. Counting points on curves over finite fields. *J. Symbolic Comput.*, 25:1–21, 1998.
11. F. Izadi and K. Murty. Counting points on an abelian variety over a finite field. Preprint, 2003.
12. K. S. Kedlaya. Counting points on hyperelliptic curves using Monsky-Washnitzer cohomology. *J. Ramanujan Math. Soc.*, 16(4):323–338, 2001.
13. H. W. Lenstra, Jr., J. Pila, and C. Pomerance. A hyperelliptic smoothness test, II. *Proc. London Math. Soc.*, 84:105–146, 2002.
14. K. Matsuo, J. Chao, and S. Tsujii. An improved baby step giant step algorithm for point counting of hyperelliptic curves over finite fields. In C. Fiecker and D. Kohel, editors, *ANTS-V*, volume 2369 of *Lecture Notes in Comput. Sci.*, pages 461–474. Springer-Verlag, 2002.
15. J.-F. Mestre. Utilisation de l'AGM pour le calcul de $E(\mathbb{F}_{2^n})$. Letter to Gaudry and Harley, December 2000.
16. J. Pila. Frobenius maps of abelian varieties and finding roots of unity in finite fields. *Math. Comp.*, 55(192):745–763, October 1990.
17. J. M. Pollard. Monte Carlo methods for index computation mod p. *Math. Comp.*, 32(143):918–924, July 1978.
18. T. Satoh. The canonical lift of an ordinary elliptic curve over a finite field and its point counting. *J. Ramanujan Math. Soc.*, 15:247–270, 2000.
19. V. Shoup. *NTL: A library for doing number theory.* `http://www.shoup.net/ntl/`.
20. A. Stein and E. Teske. Explicit bounds and heuristics on class numbers in hyperelliptic function fields. *Math. Comp.*, 71:837–861, 2002.
21. A. Stein and E. Teske. The parallelized Pollard kangaroo method in real quadratic function fields. *Math. Comp.*, 71:793–814, 2002.
22. E. Teske. Speeding up Pollard's rho method for computing discrete logarithms. In J. P. Buhler, editor, *ANTS-III*, volume 1423 of *Lecture Notes in Comput. Sci.*, pages 541–554. Springer–Verlag, 1998.
23. E. Teske. Computing discrete logarithms with the parallelized kangaroo method. *Discrete Appl. Math.*, 130:61–82, 2003.
24. P. C. van Oorschot and M. J. Wiener. Parallel collision search with cryptanalytic applications. *J. of Cryptology*, 12:1–28, 1999.

Function Field Sieve in Characteristic Three

R. Granger[1], A.J. Holt[2], D. Page[1], N.P. Smart[1], and F. Vercauteren[1]

[1] Dept. Computer Science,
University of Bristol,
Merchant Venturers Building,
Woodland Road,
Bristol, BS8 1UB,
United Kingdom.
{granger,page,nigel,frederik}@cs.bris.ac.uk
[2] Dept. Computer Science,
University of Bath,
Bath, BA2 7AY,
United Kingdom.
ajh@cs.bath.ac.uk

Abstract. In this paper we investigate the efficiency of the function field sieve to compute discrete logarithms in the finite fields $\mathbb{F}_{3^n}$. Motivated by attacks on identity based encryption systems using supersingular elliptic curves, we pay special attention to the case where n is composite. This allows us to represent the function field over different base fields. Practical experiments appear to show that a function field over $\mathbb{F}_3$ gives the best results.

1 Introduction

Research into the discrete logarithm problem (DLP) in finite fields has typically only focused on characteristic two and large prime fields, because these offer the greatest efficiency for implementations. However, since the proposal of the first fully functional identity based encryption scheme (IBE) by Boneh and Franklin [4], characteristic three fields have gained a significant cryptographic interest. These fields allow optimal security parameters with reduced bandwidth in the context of supersingular elliptic curves and pairings [9], used by concrete IBE implementations.

First suggested by Shamir [22] in 1984, the concept of identity based cryptography has been an attractive target for researchers because it has the potential to massively reduce the complexity of current PKI systems. By using the notion of identity as a users public key, the amount of infrastructure required is greatly reduced since a message sender implicitly knows the public key of the recipient. An often used example is that of an email address. Within this context, an identity for Alice might be the string

`alice@hotmail.com.`

D.A. Buell (Ed.): ANTS 2004, LNCS 3076, pp. 223–234, 2004.

If Bob wants to send Alice secure email, he implicitly knows her email address and hence her identity and public key. Therefore he can encrypt the email to her without the same level of involvement from, for example, certificate and trust authorities.

Since the paper of Boneh and Franklin [4] a large body of work on efficient implementation of the underlying algorithms [5,10] and arithmetic in both software [13] and hardware [6,19], has made such applications viable. Almost all these implementations have made use of elliptic curves defined over finite fields of characteristic three. The security of these IBE implementations is therefore dependent on the difficulty of solving the discrete logarithm problem both on an elliptic curve over a field of characteristic three and in a finite field of characteristic three.

Algorithms for discrete logarithm problems on elliptic curves do not depend, essentially, on the underlying finite field. Hence, their behaviour is well understood. The same cannot be said for discrete logarithm algorithms over finite fields. Such algorithms depend quite closely on the characteristic. Indeed different algorithms are chosen for different size characteristics.

The best known algorithm for solving the DLP in fields $\mathbb{F}_{p^n}$ of small characteristic is currently the Function Field Sieve (FFS). First developed by Adleman [1] and Adleman and Huang [2] in 1994, the FFS may be regarded as an extension of Coppersmith's earlier discrete logarithm algorithm for characteristic two fields [7,8]. Initially the algorithm applied to small fields subject to the condition $p^6 \leq n$ as $p^n \to \infty$, where p^n is the size of the finite field. Schirokauer [21] then extended the method to relax the condition to $p \leq n^{o\sqrt{n}}$. Asymptotically it has complexity

$$L_{p^n}[1/3,(32/9)^{\frac{1}{3}}] = \exp\left(((32/9)^{\frac{1}{3}} + o(1))\log(p^n)^{\frac{1}{3}}\log(\log(p^n))^{\frac{2}{3}}\right).$$

More recently, Joux and Lercier [14] provided a slightly more practical algorithm, which chooses the initial function field and resulting polynomials more efficiently. It is this latter method that we adopt for our implementation making minor changes to facilitate the use of superelliptic curves which provide some efficiency benefits.

In the past discrete logarithm computations in finite fields of small characteristic have been carried out only in characteristic two, see for example [11,24,14]. In each of these works the extension degree was prime, since the underlying protocols being attacked were assumed to come from traditional discrete logarithm based systems in which one almost always selects prime extension degrees.

The fields of characteristic three arising in identity based encryption systems are all of composite extension degree. The so called MOV embedding is usually chosen to be six, see [18,9]. We then have the option to apply the FFS method relative to any subfield. A theoretical analysis indicates that in this case, some representations may be more amenable to attack than others [12], and we investigate these possibilities also.

The remainder of this paper is organized as follows: in Section 2 we recall the mathematical background of the function field sieve and in Section 3 we provide a detailed description of our implementation. In Section 4 we report on several practical experiments which investigate the efficiency of the function field sieve in characteristic three and in Section 5 we present some conclusions.

2 The Function Field Sieve

The finite fields of interest to pairing based cryptography in characteristic three are given by $\mathbb{K} = \mathbb{F}_{3^n}$ where $n = 6 \cdot m$. In what follows we set $q = 3^m$. In practice, we limit ourselves to finite fields which could arise as the MOV embedding of a supersingular elliptic curve over the field $\mathbb{F}_q$ whose group order is divisible by a prime l of size comparable to that of q. Such elliptic curves have group orders given by

$$q \pm \sqrt{3q} + 1.$$

We wish to investigate not only the practicality of the function field sieve for the field extension $\mathbb{F}_{3^n}/\mathbb{F}_3$, but also the effect of taking different base fields, i.e. looking at the extension $\mathbb{F}_{3^n}/\mathbb{F}_{3^e}$ where $e = 1, 2, 3$ or 6. To this end we let $k = \mathbb{F}_{3^e}$, $N = n/e$ and $p = 3^e$.

We assume we are given $\alpha, \beta \in \mathbb{K}$, both of order l, such that

$$\beta = \alpha^x$$

for some unknown $x \in \{1, \ldots, l-1\}$. The discrete logarithm problem then is to compute x given α and β.

We will use a function field $F = k(X)[Y]/(H)$ defined by the polynomial $H(X,Y) \in k[X,Y]$ over the rational function field $k(X)$. Note, we shall abuse notation slightly and refer to the polynomial $H(X,Y)$ as the curve H, by which we mean the curve defined by $H(X,Y) = 0$. For practical reasons [17], one usually selects a C_{ab}-curve for H. However, in our examples we used a superelliptic curve of the form

$$H(X,Y) = Y^d + R(X)$$

where $R(X)$ is a polynomial in $k[X]$ of degree b. This enables more efficient calculation of the functions on the algebraic side, at the expense of a little less generality of our implementation. The class number of the function field F defined by H, or equivalently the number of points on the Jacobian of the curve J_H defined by H over k, we shall denote by $h = \#J_H(k)$.

We assume that the field $\mathbb{K}$ is defined by a polynomial basis with respect to the polynomial $f \in k[X]$ of degree N, i.e.

$$\mathbb{K} \cong k[X]/(f).$$

To define the function field sieve algorithm we need to specify two polynomials $u_1, u_2 \in k[X]$ such that the norm of the function $u_1 + u_2 Y$, given by the resultant

$$u_2^d H(X, -u_1/u_2) = (-u_1)^d + u_2^d R$$

is divisible by f. In such a situation we have a surjective homomorphism

$$\phi : \begin{cases} k[X,Y]/(H) \longrightarrow \mathbb{K} \cong k[X]/(f) \\ \quad Y \quad \mapsto \quad -u_1/u_2. \end{cases}$$

We select a rational factorbase $\mathcal{R}$ of small degree irreducible polynomials in $k[X]$ and an algebraic factorbase $\mathcal{A}$ of small prime divisors in the divisor group $\mathrm{Div}(F)$. Hence, $\mathcal{R}$ and $\mathcal{A}$ are therefore defined by

$$\mathcal{R} = \{\mathfrak{p} : \deg \mathfrak{p} \leq B, \mathfrak{p} \text{ irreducible }\},$$
$$\mathcal{A} = \{\langle \mathfrak{p}, Y - \mathfrak{r}\rangle : \deg \mathfrak{p} \leq B, \mathfrak{p} \text{ irreducible }, \mathfrak{r} \equiv -u_1/u_2 \pmod{\mathfrak{p}}\},$$

for some smoothness bound B.

The goal of the function field sieve is to find relatively prime pairs of polynomials (r, s), with $\deg r, \deg s \leq l$, such that the polynomial $(su_2 - ru_1)$ and the divisor $\langle s + rY\rangle$ simultaneously factor over the respective factorbases, i.e.

$$su_2 - ru_1 = \prod_{\mathfrak{p}_i \in \mathcal{R}} \mathfrak{p}_i^{a_i}$$
$$\langle s + rY\rangle = \sum_{\langle \mathfrak{p}_j, \mathfrak{r}_j\rangle \in \mathcal{A}} b_j \langle \mathfrak{p}_j, Y - \mathfrak{r}_j\rangle.$$

Determining the factorization of $\langle s + rY\rangle$ over $\mathcal{A}$ is easily done by examining the factorization of the norm

$$N(\langle s + rY\rangle) = (-s)^d + r^d R.$$

Since $h \langle \mathfrak{p}_j, Y - \mathfrak{r}_j\rangle$ is a principal divisor, for each $\langle \mathfrak{p}_j, Y - \mathfrak{r}_j\rangle \in \mathcal{A}$ there exists a function $\lambda_j \in F^*$ with $h \langle \mathfrak{p}_j, Y - \mathfrak{r}_j\rangle = \langle \lambda_j\rangle$ and such a function is unique up to multiplication by an element in k^*. We then have that our algebraic relation is given by

$$(s + rY)^h = \mu \prod_{\lambda_j} \lambda_j^{b_j},$$

with μ an element in k^*. We then apply the homomorphism ϕ above to obtain

$$(s - r\frac{u_1}{u_2})^h \equiv \prod_{\lambda_j} \phi(\lambda_j)^{b_j},$$

where $\equiv$ denotes equality modulo a possible factor in k^*. Assuming h is coprime to $(p^N - 1)/(p - 1)$, we can take h-th roots of both sides of this equation and if we write $\kappa_j = \phi(\lambda_j)^{1/h}$ we obtain

$$(s - r\frac{u_1}{u_2}) \equiv \prod_{\lambda_j} \kappa_j^{b_j}.$$

Combining the relation on the rational side with the above equation we find

$$\frac{1}{u_2}\prod_{\mathfrak{p}_i\in\mathcal{R}} \mathfrak{p}_i^{a_i} \equiv \prod_{\lambda_j} \kappa_j^{b_j}.$$

Hence, we obtain the relation between discrete logarithms given by

$$\sum_{\mathfrak{p}_i} a_i \log_g \mathfrak{p}_i - \log_g(u_2) = \sum_{\gamma_j} b_j \log_g \kappa_j, \qquad (1)$$

where g is a multiplicative generator of the field $\mathbb{K}$. Note that we do not need at any point to compute the values of the κ_j, or for that matter the values of the λ_j, all that we require is that they exist. This is guaranteed by the condition that h should be coprime to $(p^N - 1)/(p-1)$.

If sufficiently many independent relations of the form (1) have been obtained, we can solve for the discrete logarithms themselves using structured Gaussian elimination combined with the Lanczos method. Determining the discrete logarithm of β with respect to α is then performed using a standard recursive sieving strategy as explained in [21].

3 Choice of Parameters and Implementation Details

The various parameters of the function field sieve, namely the size d of the function field extension, the size B of the largest factorbase element and the size l of the polynomials r and s, are approximated by a heuristic analysis [14] of the function field sieve as

$$\begin{aligned} l &= B,\\ B &= \left\lceil \left(\frac{4}{9}\right)^{1/3} N^{1/3}\log_p(N)^{2/3}\right\rceil,\\ d &= \left\lceil \sqrt{\frac{N}{B+1}}\right\rceil. \end{aligned}$$

Since we have restricted ourselves to superelliptic curves we need to ignore values of d which are divisible by three. However, in the range of our experiments this is not an issue and extending our results to the case where $d \equiv 0 \pmod 3$ can be accomplished using C_{ab}-curves [17], at the expense of more complicated formulae on the algebraic side.

3.1 Selection of f

Following Joux and Lercier [14] we first select u_1 and u_2 and then find a suitable value of f. We set $N' = N \pmod d$. However, since the degree $N = n/e$ may not be prime we split into three possible sub-cases:

- Case (i) : $\gcd(N', d) = 1$.
  We choose a curve H such that $R(X)$ is of degree $b = N'$ and such that $\#J_H(k)$ is coprime to $(p^N - 1)/(p - 1)$. If this is not possible we go to Case (iii). We then select u_1 and u_2 of exact degrees $m - 1$ and m respectively where $m = (N - b)/d$. The degree of the polynomial $f(x) = u_2^d H(X, -u_1/u_2)$ is then given by $\max\{d \cdot (m - 1), d \cdot m + b\} = N$. Hence, we keep selecting u_1 and u_2 until f is irreducible.
- Case (ii) : $N' = 0$.
  We select $R(X)$ of smallest possible degree b such that $\#J_H(k)$ is coprime to $(p^N - 1)/(p - 1)$. We then select u_1 and u_2 of exact degrees m and $m - 1$ where $m = N/d$. The degree of the polynomial $f(x) = u_2^d H(X, -u_1/u_2)$ is then given by $\max\{d \cdot m, d \cdot (m - 1) + b\} = N$, assuming $\deg R < d$. Hence, we keep selecting u_1 and u_2 until f is irreducible.
- Case (iii) : $\gcd(N', d) \neq 1, d$, or no suitable curve found above.
  Here we select $R(X)$ of degree $b < d$, with $\gcd(b, d) = 1$, such that $\#J_H(k)$ is coprime to $(p^N - 1)/(p - 1)$ and for which one of t_1 and t_2 is minimal where

$$t_1 = \max\{N, d \cdot m, d \cdot (m - 1) + b\},$$
$$t_2 = \max\{N, d \cdot (m - 1), d \cdot m + b\}.$$

  In the case of t_1 being minimal we then select u_1 and u_2 of degree m and $m - 1$ until $u_2^d H(X, -u_1/u_2)$ if divisible by an irreducible polynomial f of degree N. In the case of t_2 being minimal we select u_1 and u_2 to be of degree $m - 1$ and m, until we obtain an irreducible polynomial f of degree N which divides $u_2^d H(X, -u_1/u_2)$.

Note that Joux and Lercier [14] only considered Case (i) above, since they used arbitrary C_{ab}-curves (and not simply superelliptic ones), and because the extension degree N was always prime. Clearly, Case (iii) will lead to marginally less efficient relation collection, since the degree of the algebraic side will be slightly higher than if one was in Case (i) or (ii). However, if one is to deal with non-prime values of N and superelliptic curves one is led to such a case.

3.2 Lattice Sieving

The finding of (r, s) is performed using a lattice sieve. In the lattice sieve one selects a prime polynomial $\mathfrak{p} \in \mathcal{R}$ (resp. prime divisor $\langle \mathfrak{p}, Y - \mathfrak{r} \rangle \in \mathcal{A}$). Then one looks at the sub-lattice of the (r, s) plane such that $su_2 - ru_1$ (resp. $\langle s + rY \rangle$) is divisible by the chosen polynomial (resp. divisor). We found it more efficient to then sieve in the sub-lattice on a line-by-line basis, rather than sieve in a two-dimensional manner in the sub-lattice.

The use of lattice sieving has a number of advantages over sieving in the (r, s) plane. Firstly, it is better at yielding a large number of relations. Secondly, one can target factorbase elements for which one does not yet have a relation. This enables one to obtain a matrix involving all elements in the factorbase reasonably

efficiently. Thirdly, the use of lattice sieving is crucial in the final stage where one wishes to target individual discrete logarithms using the recursive sieving strategy mentioned earlier.

3.3 Factorbase Size

A major consideration is the balancing of the effort required for the sieving and the matrix step. In theory one balances the two halves of the computation and so derives the complexity estimates such as those above. The size of both factorbases is approximated by p^B/B, however one should treat this size as a continuous function of a real variable B, as opposed to the discrete B above. See [12] for further analysis of this point.

However, sieving can be performed in a highly scalable manner. After all, to move from using a single computer performing the sieving to around 100-200 computers is relatively easy given the resources of most organizations these days. However, our code for the matrix step required the use of a single machine and hence does not scale.

Hence, in practice we can devote less total time to the matrix step compared to the sieving step. Since the matrix step has complexity approximately $O(T^2)$ where $T = \#\mathcal{R} + \#\mathcal{A} \approx 2p^B/B$, we see that we have a physical constraint on the size of the factorbases we can accommodate. In our experiments we assumed that solving a matrix with over half a million rows and columns was infeasible given our resources and matrix code. This sometimes led us to choose non-optimal, from a theoretical perspective, values for the other parameters.

3.4 Linear Algebra Step

Several studies involving use of index calculus-type methods discuss the linear algebra step, since it is a major practical bottleneck in the procedure. This is because parallelising existing algorithms is rather difficult. Various authors [15,20], have identified several effective techniques for the solution of sparse linear systems modulo a prime. These include iterative schemes such as the Lanczos, Conjugate Gradient and Wiedemann algorithms, with or without a pre-processing step involving structured Gaussian elimination. Recently attention has moved onto attempting to perform this step in parallel, see for example [25] for the case of Lanczos over the field $\mathbb{F}_2$.

In our implementation we followed [15] and used a basic structured Gaussian elimination routine, such as that described in [3,15,20], so as to reduce the size of the linear system whilst maintaining a degree of sparsity. This submatrix is subsequently solved by the Lanczos algorithm [16], and a full result set is then recovered via back-substitution. Our implementation for this stage made use of Victor Shoup's NTL C++ library for multiprecision integers [23].

4 Experiments

As mentioned in the introduction, we are not only interested in how efficient the function field sieve method is on fields of relevance to IBE based systems. We are also interested in the effect of choosing a different base field $k = \mathbb{F}_{3^e}$ in the function field sieve. Usually for traditional discrete logarithm systems this would not be an issue since the extension degree is often prime. One should note that a precise theoretical analysis [12] of the function field sieve method shows that for a fixed field size the effect of the size of the base field is not as one would expect. This strange effect of the base field size 3^e on the overall performance for a fixed field size 3^n can be explained due to the non-continuous behaviour of various parameters, in particular the function field extension degree d.

Note, the expected run time would be T^2 operations in $k[x]$. Each field operation requires at most $O(e^2)$ operations in $\mathbb{F}_3$ and the polynomials involved are of degree around

$$\mathfrak{d} = \max\{N/d, dB\}.$$

Hence, one would expect the run time to be proportional to

$$(\mathfrak{d}eT)^2.$$

In the experiments below we selected field sizes $q = 3^n$ which arise from supersingular elliptic curves with group orders which are "almost" prime.

4.1 Field Size 3^{186}

This corresponds to a field size of approximately 295 bits. The rough theoretical estimates, given above, for the various values of $e = 1, 2, 4, 6$ are in the table below. A more careful analysis as in [12] reveals the following estimates, where for the factorbases we take the first $T/2$ primes (resp. $T/2$ prime places) on the rational (resp. algebraic side), in other words the value of B is not used directly.

	Rough Analysis			Analysis of [12]	
e	d	B	$T \approx$	d	$T \approx$
1	4	12	89000	5	85000
2	4	6	180000	4	75000
3	4	4	270000	4	150000
6	3	2	530000	4	330000

We compared these results to the yields provided by our implementation and found that the best possible values seemed to be given by

e	d	$T \approx$
1	5	70000
2	4	90000
3	4	190000

This latter table was produced by comparing the yields of the implementation over a fixed time period for various parameter sizes and then selecting the one which would produce a full matrix in the shortest time period. We were however unable to generate suitable experimental data for the case $e = 6$ since this field is really too large to apply the FFS method in practice for such a value of n.

4.2 Field Size 3^{222}

This corresponds to a field size of approximately 352 bits. The rough theoretical estimates, given above, for the various values of $e = 1, 2, 4, 6$ along with the more precise estimates of [12], are in the following table

	Rough Analysis			Analysis of [12]	
e	d	B	$T \approx$	d	$T \approx$
1	4	13	250000	5	130000
2	4	6	180000	4	190000
3	4	4	270000	4	270000
6	4	2	530000	4	460000

We compared these results to the yields provided by our implementation and found that the best possible values seemed to be given by

e	d	$T \approx$
1	5	100000
2	4	190000
3	4	320000

Again the values for $e = 6$ are not given since n is still too small for this base field size to apply the FFS method successfully.

4.3 Field Size 3^{582}

We present this field since it is the first one which is usable in pairing based systems and which "could" be secure against current computing power on both the elliptic curve and the finite field sides. It corresponds to a bit length of 923 bits. The rough theoretical estimates are given below which should be compared to the the analysis in [12]

	Rough Analysis			Analysis of [12]	
e	d	B	$T \approx$	d	$T \approx$
1	5	21	$9.0 \cdot 10^8$	6	$3.0 \cdot 10^9$
2	5	10	$7.0 \cdot 10^8$	7	$3.0 \cdot 10^9$
3	5	6	$1.0 \cdot 10^8$	7	$2.8 \cdot 10^9$
6	5	3	$2.0 \cdot 10^8$	5	$4.0 \cdot 10^8$

Note that these parameters would imply that such key sizes are currently out of range of the FFS method.

To completely confirm both our theoretical estimates and our partial experiments we ran a few experiments through to the final matrix stage and computation of individual logarithms:

Our experiments were run on a network of around 150 Unix based workstations. The network contained a number of older Sparc 5s and 10s running Solaris, plus a large number of Linux based machines with AMD1600 or Pentium 4 processors. We only used idle cycles of the machines whilst other people were using them, hence during the day we had the equivalent of 50 Pentium 4 machines working flat out. At night this increased to around 100 Pentium 4 machines. The matrix step was run on a Sun Blade 1000 workstation.

In the following table we present the wall clock time t_1 needed to produce the relations, the time t_2 needed to solve the matrix step (divided into the time t_2' needed to perform the structured Gaussian elimination and the time t_2'' to solve the reduced system). We ran the sieving clients for time t_1 producing a total of R relations on T_a elements of the factorbase. For many examples we did not try to find relations on all the factorbase elements, since finding the relations on the last few elements can take a disproportionate amount of time. The line m denotes the approximate row-size of the resulting (approximately square) matrix after the application of structured Gaussian elimination. Our times for the linear algebra step can be considerably improved we believe, and work is ongoing in this area.

n	186	186	186	222	222
N	186	93	62	222	111
e	1	2	3	1	2
d	5	4	4	5	4
$R(X)$	X	X	X	$X^3 + 2X + 1$	$X^3 + X$
T	70000	80000	140000	100000	160000
R	80045	96365	139376	96956	181675
T_a	69345	76425	137750	95169	158952
m	12634	13218	24746	20719	32148
t_1	8h	7h	30h	48h	50h
t_2'	43m	1h	2h 24m	3h	4h 20m
t_2''	13h	16h	1d 18h	2d 21h	4d 14h

5 Conclusion

We have reported on the first implementation of the function field sieve in characteristic three. We have paid particular attention to the case of finite fields which arise in pairing based cryptosystems. In particular such fields are of a composite nature and we have seen that this provides at best a marginal benefit in allowing one to apply the function field sieve over either $\mathbb{F}_3$ or $\mathbb{F}_{3^2}$. We have shown how the exact analysis of [12] is more able to predict the behaviour, and thereby parameter choices, than the naive simple analysis. We have also shown how the key sizes one would use in a simple pairing based system are likely to be secure against current algorithms and computing power.

Finally we hope that our work will encourage others to investigate discrete logarithm algorithms in composite fields of characteristic three, and thereby allow the community to have greater faith in the security of pairing based systems which are based over such fields.

References

1. L. M. Adleman. The function field sieve. In *Algorithmic Number Theory Symposium – ANTS-I*, Springer-Verlag LNCS 877, 108–121, 1994.
2. L. M. Adleman and M. A. Huang. Function field sieve method for discrete logarithms over finite fields. *Inform. and Comput.*, **151**, 5–16, 1999.
3. E. A. Bender and E. R. Canfield. An approximate probabilistic model for structured Gaussian elimination. *J. Algorithms*, **31**, 271–290, 1999.
4. D. Boneh and M. Franklin. Identity based encryption from the Weil pairing. In *Advanced in Cryptology – CRYPTO 2001*, Springer-Verlag LNCS 2139, 213–229, 2001.
5. P.S.L.M. Barreto, H.Y. Kim, B. Lynn and M. Scott. Efficient algorithms for pairing-based cryptosystems. In *Advances in Cryptology – CRYPTO 2002*, Springer-Verlag LNCS 2442, 354–368, 2002.
6. G. Bertoni, J. Guajardo, S. Kumar, G. Orlando, C. Paar and T. Wollinger. Efficient GF(p^m) arithmetic architectures for cryptographic applications. In *Topics in Cryptology – CT-RSA* Springer-Verlag LNCS 2612, 158–175, 2003.
7. D. Coppersmith. Evaluating logarithms in GF(2^n). In *16th ACM Symp. Theory of Computing*, 201–107, 1984.
8. D. Coppersmith. Fast evaluation of logarithms in fields of characteristic two. *IEEE Transactions in Information Theory*, **30**, 587–594, July 1984.
9. S.D. Galbraith. Supersingular curves in cryptography. In *Advances in Cryptography – AsiaCrypt - 2001*, Springer-Verlag LNCS 2248, 495–513, 2001.
10. S.D. Galbraith, K. Harrison and D. Soldera. Implementing the Tate pairing. In *Algorithmic Number Theory Symposium – ANTS-V*, Springer-Verlag LNCS 2369, 324–337, 2002.
11. D. M. Gordon and K. S. McCurley. Massively parallel computation of discrete logarithms. In *Advances in Cryptology – CRYPTO'92*, Springer-Verlag LNCS 740, 312–323, 1993.
12. R. Granger. Estimates for discrete logarithm computations in finite fields of small characteristic. In *Cryptography and Coding*, Springer-Verlag LNCS 2898, 190–206, 2003.
13. K. Harrison, D. Page and N.P. Smart,. Software implementation of finite fields of characteristic three. *LMS Journal of Computation and Mathematics*, **5**, 181–193, 2002.
14. A. Joux and R. Lercier. The function field sieve is quite special. In *Algorithmic Number Theory Symposium – ANTS-V*, Springer-Verlag LNCS 2369, 431–445, 2002.
15. B. A. LaMacchia and A. M. Odlyzko. Solving large sparse linear systems over finite fields. In *Advances in Cryptology – CRYPTO '90* Springer-Verlag LNCS 537, 109–133, 1991.
16. C. Lanczos. Solution of systems of linear equations by minimized iterations. *J. of Research of the National Bureau of Standards* , **49**, 33–53, 1952.

17. R. Matsumoto. Using C_{ab} Curves in the Function Field Sieve. *IEICE Trans. Fundamentals*, **82**, March 1999.
18. A.J. Menezes, T. Okamoto and S. Vanstone. Reducing elliptic curve logarithms to logarithms in a finite field. *IEEE Transactions in Information Theory*, **39**, 1639–1646, 1993.
19. D. Page and N.P. Smart. Hardware implementation of finite fields of characteristic three. In *Cryptographic Hardware and Embedded Systems – CHES 2002*, Springer-Verlag LNCS 2523, 529–539, 2002.
20. C. Pomerance and J. W. Smith. Reduction of huge, sparse linear systems over finite fields via created catastrophes. *Exper. Math.*, **1**, 89–94, 1992.
21. O. Schirokauer. The special function field sieve. *SIAM Journal on Discrete Mathematics*, **16**, 81–98, 2002.
22. A. Shamir. Identity-based cryptosystems and signature schemes. In *Advanced in Cryptology – CRYPTO - 1984*, Springer-Verlag LNCS 196, 47–53, 1984.
23. V. Shoup. NTL – A Library for Number Theory. `http://www.shoup.net/ntl/`.
24. E. Thomé. Computation of discrete logarithms in $\mathbb{F}_{2^{607}}$. In *Advances in Cryptology – AsiaCrypt 2001*, Springer-Verlag LNCS 2248, 107–124, 2001.
25. L.T. Yang and R.P. Brent. The parallel improved Lanczos method for integer factorization over finite fields for public key cryptosystems. In *International Conference on Parallel Programming Workshops (ICPPW)*, IEEE Press, 106-114, 2001.

A Comparison of CEILIDH and XTR

R. Granger, D. Page, and M. Stam

University of Bristol, Department of Computer Science,
Merchant Venturers Building, Woodland Road,
Bristol, BS8 1UB, UK.
{granger,page,stam}@cs.bris.ac.uk

Abstract. We give a comparison of the performance of the recently proposed torus-based public key cryptosystem CEILIDH, and XTR. Underpinning both systems is the mathematics of the two dimensional algebraic torus $T_6(\mathbb{F}_p)$. However, while they both attain the same discrete logarithm security and each achieve a compression factor of three for all data transmissions, the arithmetic performed in each is fundamentally different. In its inception, the designers of CEILIDH were reluctant to claim it offers any particular advantages over XTR other than its exact compression and decompression technique. From both an algorithmic and arithmetic perspective, we develop an efficient version of CEILIDH and show that while it seems bound to be inherently slower than XTR, the difference in performance is much smaller than what one might infer from the original description. Also, thanks to CEILIDH's simple group law, it provides a greater flexibility for applications, and may thus be considered a worthwhile alternative to XTR.

1 Introduction

From a representation perspective, basing cryptography in the multiplicative group of a finite field is inefficient. The reason for this is that current index calculus algorithms can solve discrete logarithms in finite fields approximately six times the size of groups to which generic algorithms apply [1,6,8,20,26]. As a consequence this representation is about six times less efficient in terms of memory and bandwidth, than the information-theoretic limit. In the future this ratio will continue to increase as recommended key sizes become larger, since algorithms to solve finite field discrete logarithms are subexponential.

So what can one do if bandwidth, power consumption and speed are essential factors in the design of a cryptosystem? The most obvious answer is to use a system for which no known subexponential algorithm exists, such as a secure elliptic curve [2] or perhaps a genus two hyperelliptic curve [27]. Assuming the discrete logarithm problem (DLP) in these groups is as hard as is currently believed, these systems provide an effective solution to the issues highlighted above, and have been thoroughly researched since introduced in 1985 [10,15], and 1989 [11] respectively.

In recent years systems exploiting the algebraic structure of finite field extensions have been studied. Whereas in elliptic curve cryptography so called

D.A. Buell (Ed.): ANTS 2004, LNCS 3076, pp. 235–249, 2004.

'composite' finite fields can introduce potential weaknesses [5], for torus-based cryptography they present the possibility for highly compressed data transmission, compact implementation, and significantly improved performance.

Torus-based cryptography may be regarded as a natural extension of classical Diffie-Hellman and El-Gamal in a finite field $\mathbb{F}_p$, where key agreement, encryption and signature schemes are performed in the multiplicative group $\mathbb{F}_p^*$. For any positive integer n, one can define an algebraic torus T_n over $\mathbb{F}_p$ such that over $\mathbb{F}_{p^n}$, this variety is isomorphic to $\phi(n)$ copies of $\mathbb{F}_p^*$, where $\phi(n)$ is the dimension of T_n. In fact T_n is nothing other than the cyclotomic subgroup of $\mathbb{F}_{p^n}^*$ [19]. When T_n is 'rational', it is possible to embed T_n in $\phi(n)$-dimensional affine space, and thus represent every element by just $\phi(n)$ elements of $\mathbb{F}_p$.

Only recently was the connection between algebraic tori and the existing trace-based systems LUC [25] and XTR [13] made explicit [19]. Of particular interest is a current conjecture about algebraic tori that if true, implies the existence of practical cryptosystems based in $\mathbb{F}_{p^n}^*$ for suitable n with higher data compression ratios than both CEILIDH and XTR. An analogous statement has been proven for Abelian varieties over finite fields [18], though whether these are at all practical is debatable. It has also been demonstrated that attempts to extend the trace-methods of LUC and XTR to more general symmetric function-based compression maps can not work [19]. Thus if one requires smaller representations of elements than those afforded by XTR, torus-based cryptography probably holds the greatest potential for doing so.

In this paper we content ourselves with a comparison of the torus-based public key system CEILIDH, introduced at CRYPTO 2003 by Rubin and Silverberg [19], and its closest relative, XTR, which was introduced at CRYPTO 2000 by Lenstra and Verheul [13]. Based on current understanding, both systems attain the security of $\mathbb{F}_{p^6}^*$ while transmitting only two elements of $\mathbb{F}_p$ in communications. The compression factor of three is possible since the underlying sets of each possess a birational embedding into $\mathbb{A}^2(\mathbb{F}_p)$, the 2-dimensional affine space. For implementation the basic difference between the two is that while XTR essentially performs all arithmetic in $\mathbb{F}_{p^2}$ using a third order recurrence, the group operation in CEILIDH is performed in $\mathbb{F}_{p^6}^*$, with compression and decompression effected by the birational map to $\mathbb{A}^2$. In this sense CEILIDH may be viewed purely as a compression mechanism attached to the usual finite field arithmetic. Using a rather simplistic analysis, it would seem that since the elements of CEILIDH needed for general group operations are three times the size of those in XTR, its performance is bound to be inferior.

Our main contribution is a new representation of the cyclotomic subgroup of $\mathbb{F}_{p^6}^*$ which permits near-optimal multiplication efficiency. Together with another squaring-efficient representation of the same subgroup [22], and some arithmetic refinements of the birational maps consistuting the CEILIDH compression mechanism, we considerably develop the basic method of CEILIDH. In contrast to what one might naively expect from the original exposition, our implementation performs nearly as efficiently as the fastest known version of XTR. We hope the authors of [19] do not object if for simplicity we refer throughout the paper to our improvements to CEILIDH, by the same name as well.

The authors are grateful to F. Vercauteren for many fruitful discussions and for his insightful comments, and would like to thank an anonymous referee for their helpful suggestions.

2 Overview

In this section we give a brief description of algebraic tori and the mathematics underlying CEILIDH and XTR. Further details can be found in [19], and a good reference for algebraic tori is [28]. In the following, we take as our base field $\mathbb{F}_p$, which we use in our implementation, though any finite field would suffice.

2.1 The Torus $T_n(\mathbb{F}_p)$

Throughout, let $\mathbb{F}_p$ be the prime field consisting of p elements. Let ϕ be the Euler ϕ-function, and let Φ_n be the n-th cyclotomic polynomial. We write $G_{p,n}$ for the subgroup of $\mathbb{F}_{p^n}^*$ of order $\Phi_n(p)$, and let $\mathbb{A}^n(\mathbb{F}_p)$ denote n-dimensional affine space over $\mathbb{F}_p$, i.e., the variety whose points lie in $\mathbb{F}_p^n$.

For cryptographic purposes, when using finite extensions of prime fields, the cyclotomic subgroup plays a prominent role. Given current index calculus algorithms, to attain the full security of $\mathbb{F}_{p^n}^*$ one should make sure that the chosen subgroup G can not be embedded in any proper subfield of $\mathbb{F}_{p^n}$. One should also ensure that no information can be obtained regarding the discrete logarithm of an element by a simple application of the norm function, i.e., the norm of a generator should be one with respect to every intermediate subfield. These two conditions are in fact equivalent, and define the cyclotomic subgroup $G_{p,n}$ [12, 19]. Fortunately for cryptography, the second condition coincides with the definition of the algebraic torus $T_n(\mathbb{F}_p)$, which we now introduce. We do not give the complete algebraic definition, but for our purposes this is sufficient.

Definition 1. *Let $k = \mathbb{F}_p$ and $L = \mathbb{F}_{p^n}$. Define the torus T_n to be the intersection of the kernels of the norm maps $N_{L/F}$, for all subfields $k \subset F \subsetneq L$:*

$$T_n(k) := \bigcap_{k \subset F \subsetneq L} \mathrm{Ker}[N_{L/F}]. \tag{1}$$

The dimension of T_n is $\phi(n)$. Since $T_n(\mathbb{F}_p)$ is a subgroup of $\mathbb{F}_{p^n}^*$, the group operation is just ordinary multiplication in the larger field. The following lemma provides some essential properties of T_n [19].

Lemma 1. *1. $T_n(\mathbb{F}_p) \cong G_{p,n}$.*
2. $\#T_n(\mathbb{F}_p) = \Phi_n(p)$.
3. If $h \in T_n(\mathbb{F}_p)$ is an element of prime order not dividing n, then h does not lie in a proper subfield of $\mathbb{F}_{p^n}/\mathbb{F}_p$.

When T_n is rational, elements can be represented by just $\phi(n)$ elements of $\mathbb{F}_p$, and hence a compression factor of $n/\phi(n)$ is achieved over the usual representation. T_n is known to be rational when n is either a prime power, or is a product of two prime powers, and it is conjectured to be rational for all n. For current key size recommendations, this would have interesting cryptographic applications for $n = 30$ and $n = 210$, which would give compression ratios of $3\frac{3}{4}$ and $4\frac{3}{8}$ respectively, assuming good parameters can be found efficiently. This means for instance that one could achieve 1024-bit security while transmitting 274 or 235 bits respectively in communications. Both CEILIDH and XTR currently require 342 bits in transmissions for the same security, which provides a significant improvement over ordinary RSA for example in terms of bandwidth.

2.2 CEILIDH

Much of this section is a slightly abridged version of the original description of CEILIDH [19]. We point out that the construction given there for the rationality of T_6 also provides an explicit rational parametrisation for the torus T_2 *en passant.* We take advantage of this in our algorithms in Section 3.3.

Fix $x \in \mathbb{F}_{p^2} \backslash \mathbb{F}_p$, so $\mathbb{F}_{p^2} = \mathbb{F}_p(x)$, and let $\{\alpha_1, \alpha_2, \alpha_3\}$ be a basis for $\mathbb{F}_{p^3}$ over $\mathbb{F}_p$. Then $\{\alpha_1, \alpha_2, \alpha_3, x\alpha_1, x\alpha_2, x\alpha_3\}$ is a basis for $\mathbb{F}_{p^6}$ over $\mathbb{F}_p$. Let $\sigma \in \mathrm{Gal}(\mathbb{F}_{p^6}/\mathbb{F}_p)$ be the element of order two. Define $\psi_0 : \mathbb{A}^3(\mathbb{F}_p) \hookrightarrow \mathbb{F}_{p^6}$ by

$$\psi_0(u_1, u_2, u_3) = \frac{\gamma + x}{\gamma + \sigma(x)},$$

where $\gamma = u_1\alpha_1 + u_2\alpha_2 + u_3\alpha_3$. Then $N_{\mathbb{F}_{p^6}/\mathbb{F}_{p^3}}(\psi_0(\mathbf{u})) = \mathbf{1}$ for every $\mathbf{u} = (\mathbf{u_1}, \mathbf{u_2}, \mathbf{u_3})$. Let $U = \{\mathbf{u} \in \mathbb{A}^\mathbf{3} : \mathbf{N}_{\mathbb{F}_{\mathbf{p}^6}/\mathbb{F}_{\mathbf{p}^2}}(\psi_\mathbf{0}(\mathbf{u})) = \mathbf{1}\}$. By (1), $\psi_\mathbf{0}(\mathbf{u}) \in \mathbf{T_6}(\mathbb{F}_\mathbf{p})$ if and only if $\mathbf{u} \in \mathbf{U}$, so restricting ψ_0 to U gives a morphism $\psi_0 : U \longrightarrow T_6$. It follows from Hilbert's Theorem 90 that every element of $T_6(\mathbb{F}_p) \setminus \{1\}$ is in the image of ψ_0, and so ψ_0 defines an isomorphism

$$\psi_0 : U \xrightarrow{\sim} T_6 \setminus \{1\}.$$

The equation defining U is a quadratic hypersurface in u_1, u_2, u_3. Fix a point $\mathbf{a} = (\mathbf{a_1}, \mathbf{a_2}, \mathbf{a_3}) \in \mathbf{U}(\mathbb{F}_\mathbf{p})$. By adjusting the basis $\{\alpha_1, \alpha_2, \alpha_3\}$ of $\mathbb{F}_{p^6}$ if necessary, one can assume without loss of generality that the tangent plane at $\mathbf{a}$ to the surface U is just the plane $u_1 = a_1$. If $(v_1, v_2) \in \mathbb{F}_p \times \mathbb{F}_p$, then the intersection of U with the line $\mathbf{a} + t(1, v_1, v_2)$ consists of two points, namely $\mathbf{a}$ and a point of the form $\mathbf{a} + \frac{1}{f(v_1,v_2)}(1, v_1, v_2)$ where $f(v_1, v_2) \in \mathbb{F}_p[v_1, v_2]$ is an explicit polynomial independent of p. The map that takes (v_1, v_2) to the latter point is a birational isomorphism

$$g : \mathbb{A}^2 \setminus V(f) \xrightarrow{\sim} U \setminus \{\mathbf{a}\},$$

where $V(f)$ denotes the subvariety of $\mathbb{A}^2$ defined by $f(v_1, v_2) = 0$. Thus $\psi_0 \circ g$ defines an isomorphism

$$\psi : \mathbb{A}^2 \setminus V(f) \xrightarrow{\sim} T_6 \setminus \{1, \psi_0(\mathbf{a})\}.$$

For the inverse isomorphism, suppose that $\beta = \beta_1 + \beta_2 x \in T_6(\mathbb{F}_p) \setminus \{1, \psi_0(\mathbf{a})\}$ with $\beta_1, \beta_2 \in \mathbb{F}_{p^3}$. One can check that $\beta_2 \neq 0$, and if $\gamma = (1 + \beta_1)/\beta_2$, then $\gamma/\sigma(\gamma) = \beta$. Write $(1 + \beta_1)/\beta_2 = u_1\alpha_1 + u_2\alpha_2 + u_3\alpha_3$ with $u_i \in \mathbb{F}_p$, and define

$$\rho(\beta) = \left(\frac{u_2 - a_2}{u_1 - a_1}, \frac{u_3 - a_3}{u_1 - a_1} \right).$$

Then $\rho : T_6(\mathbb{F}_p) \setminus \{1, \psi_0(\mathbf{a})\} \xrightarrow{\sim} \mathbb{A}^2(\mathbb{F}_p) \setminus V(f)$ is the inverse isomorphism of ψ, and hence we have an efficient compression and decompression mechanism for all (bar two) elements of $T_6(\mathbb{F}_p)$.

2.3 XTR

In common with CEILIDH, XTR is based on the cyclotomic subgroup $G_{p,6}$. Let g be a generator for this subgroup. In XTR elements of $\langle g \rangle$ are represented by their trace over $\mathbb{F}_{p^2}$

$$Tr_{\mathbb{F}_{p^6}/\mathbb{F}_{p^2}}(g) = g + g^{p^2} + g^{p^4} \in \mathbb{F}_{p^2},$$

and hence need only two elements of $\mathbb{F}_p$ to specify. The set of traces constitute the XTR 'group'. Clearly,

$$Tr_{\mathbb{F}_{p^6}/\mathbb{F}_{p^2}}(g) = Tr_{\mathbb{F}_{p^6}/\mathbb{F}_{p^2}}(g^{p^2}) = Tr_{\mathbb{F}_{p^6}/\mathbb{F}_{p^2}}(g^{p^4}),$$

and so given an element in the XTR group one can not distinguish between g and its conjugates g^{p^2} and g^{p^4}. Hence decompression to $\mathbb{F}_{p^6}$ is not unique, though this can easily be resolved. The analogue of the DLP is to compute n given $Tr_{\mathbb{F}_{p^6}/\mathbb{F}_{p^2}}(g)$ and $Tr_{\mathbb{F}_{p^6}/\mathbb{F}_{p^2}}(g^n)$. One can convert this to an ordinary DLP by mapping both back to $\mathbb{F}_{p^6}^*$ by finding the correct root of

$$X^3 - Tr_{\mathbb{F}_{p^6}/\mathbb{F}_{p^2}}(g)X^2 + Tr_{\mathbb{F}_{p^6}/\mathbb{F}_{p^2}}(g)^p X - 1 = (X - g)(X - g^{p^2})(X - g^{p^4}),$$

and similarly for $Tr_{\mathbb{F}_{p^6}/\mathbb{F}_{p^2}}(g^n)$. The real benefit of the XTR representation is the ease with which arithmetic can be performed. Let $c_n = Tr_{\mathbb{F}_{p^6}/\mathbb{F}_{p^2}}(g^n)$. To compute c_n given c_1, one uses some properties of third order addition chains over $\mathbb{F}_{p^2}$ applied to the recurrence

$$c_{u+v} = c_u c_v - c_u^p c_{u-v} + c_{u-2v}.$$

As a consequence, exponentiations can be performed faster than with an optimal representation of $\mathbb{F}_{p^6}$ [23,22]. The main drawback of XTR is that one can not perform straightforward multiplication since the set of traces is not a group. By keeping track of the correct conjugate however, this can be accomplished. Rather than being based on the torus T_6, XTR may be viewed algebraically as a quotient of T_6 by an action of the symmetric group S_3 [19]. What makes XTR possible is that the trace map from this quotient variety to $\mathbb{F}_{p^2}$ provides an explicit rational parametrisation. For various algorithmic improvements and implementation tricks for XTR we refer the reader to [23].

3 Efficient Representations and Algorithms for $T_6(\mathbb{F}_p)$

In this section we develop suitable field representations and algorithms for the efficient implementation of the CEILIDH cryptosystem. We base our implementation on Example 11 of [19], since this allows key generation to be done exactly as in XTR, and so a fair comparison can be made. In addition to our new multiplication-efficient representation, we make some basic observations that result in a considerable improvement over the original algorithms.

Depending on the particular protocol one wishes to implement, and even on which part of the protocol, some representations of the underlying field $\mathbb{F}_{p^6}$ may perform better than others. For example, in the first stage of a Diffie-Hellman key agreement, both Alice and Bob exponentiate a fixed public base g, so here one can precompute some powers of g, and one should use a field representation that permits fast multiplication. In the second stage, both parties exponentiate a random element of $\mathbb{F}_{p^6}^*$, and so here one should use a representation which permits fast squaring. As we show, optimising each consideration leads to different field representations, whilst switching between them is a simple matter.

For the two operations of exponentiating a fixed and a random base, two field representations suffice. The first is a degree six extension of $\mathbb{F}_p$ and allows us to use many implementation tricks [22], which include very fast squaring in the cyclotomic subgroup $G_{p,6}$. We refer to this representation as F_1.

The second representation F_2 fulfills two functions: built as a quadratic extension of a cubic extension of $\mathbb{F}_p$, it firstly permits the efficient use of the birational maps ψ and ρ which are essential to CEILIDH, given our cheap inversion method; but primarily, it provides the basis for arithmetic in F_3. What we refer to as F_3 is a semi-compressed fractional form of the torus T_6, which is actually just the torus T_2. Its utility is that one can perform the group operation, but with much better multiplication efficiency. Together, F_1, F_2, F_3, $\mathbb{A}^2$ and the isomorphisms between them constitute our implementation of CEILIDH, which may be depicted as follows:

$$F_1 \underset{\sigma^{-1}}{\overset{\sigma}{\rightleftarrows}} F_2 \underset{\tau^{-1}}{\overset{\tau}{\rightleftarrows}} F_3 \underset{\psi}{\overset{\rho}{\rightleftarrows}} \mathbb{A}^2(\mathbb{F}_p). \qquad (2)$$

For F_1 we give a brief description of the arithmetic, and for F_2 and F_3 we give full details of all operations. In Lemma 3 we provide a simple cost analysis where M,A, and I represent the cost of an $\mathbb{F}_p$ multiplication, addition, and inversion respectively. In $\mathbb{F}_p$ we assume that a subtraction amounts to the same as an addition, and also that squaring costs the same as a multiplication, since the former operation is seldom used.

For $(n, p) = 1$, let ζ_n denote a primitive n-th root of unity mod p, and as in XTR let $p \equiv 2 \bmod 9$ throughout ($p \equiv 5 \bmod 9$ is equally valid).

3.1 The Representation F_1

A full derivation of the results of this section can be found in [22]. Let $z = \zeta_9$, so that $\mathbb{F}_{p^6} = \mathbb{F}_p(z)$, and let our basis for $\mathbb{F}_{p^6}$ be $\{z, z^2, z^3, z^4, z^5, z^6\}$. Using a

Karatsuba-type method for multiplication and squaring, these can be performed in $18M+53A$ and $12M+42A$ respectively. However, for the cyclotomic subgroup $G_{p,6}$ in which we work, some improvements can be made. For example, if $g \in G_{p,6}$ inversion is just the cube of the Frobenius endomorphism, since $\Phi_6(p) = (p^2 - p + 1)|(p^3+1)$, and so $g^{-1} = g^{p^3}$. Also the condition $g^{\Phi_6(p)} = 1$ gives a set of six equations on the six coefficients of g which enable squaring to be performed with a cost of $6M + 21A$.

3.2 The Representation F_2

Let $x = \zeta_3$ and $y = \zeta_9 + \zeta_9^{-1}$. Then $\mathbb{F}_{p^3} = \mathbb{F}_p(y)$, and $\mathbb{F}_{p^6} = \mathbb{F}_{p^3}(x)$. The bases we use are $\{1, y, y^2-2\}$ for $\mathbb{F}_{p^3}$, and $\{1, x\}$ for the degree two extension. We now describe the basic arithmetic in each of these extensions.

$\mathbb{F}_{p^3}$ Frobenius :
For our basis, since $p \equiv 2 \bmod 9$, the Frobenius map gives $y^p = y^2 - 2$, and $(y^2-2)^p = -y - (y^2-2)$. Hence for $a = a_0 + a_1 y + a_2(y^2-2)$, $a^p = a_0 - a_2 y + (a_1 - a_2)(y^2-2)$.

$\mathbb{F}_{p^3}$ Multiplication :
Let $a = a_0 + a_1 y + a_2(y^2-2)$, $b = b_0 + b_1 y + b_2(y^2-2)$. Then $ab = (a_0b_0 + 2a_1b_1 + 2a_2b_2 - a_1b_2 - a_2b_1) + (a_0b_1 + a_1b_0 + a_1b_2 + a_2b_1 - a_2b_2)y + (a_0b_2 + a_2b_0 + a_1b_1 - a_2b_2)(y^2-2)$. Precompute $t_{00} = a_0b_0$, $t_{11} = a_1b_1$, $t_{22} = a_2b_2$, and $t_{01} = (a_0+a_1)(b_0+b_1)$, $t_{12} = (a_1-a_2)(b_2-b_1)$, $t_{20} = (a_2-a_0)(b_0-b_2)$. Then $ab = (t_{00} + t_{11} + t_{22} - t_{12}) + (t_{01} + t_{12} - t_{00})y + (t_{20} + t_{00} + t_{11})(y^2-2)$.

$\mathbb{F}_{p^3}$ Inversion :
Usually, to invert an element in an extension of a prime field one must either use a basic GCD algorithm, or one of many suggestions based on exponentiating to a power one less than the group order [7]. However, since the extension degree is small, we can perform inversion directly: one uses the multiplication formula and sets the result to the identity, i.e., one solves

$$\begin{pmatrix} a_0 & 2a_1 - a_2 & 2a_2 - a_1 \\ a_1 & a_0 + a_2 & a_1 - a_2 \\ a_2 & a_1 & a_0 - a_2 \end{pmatrix} \begin{pmatrix} b_0 \\ b_1 \\ b_2 \end{pmatrix} = \begin{pmatrix} 1 \\ 0 \\ 0 \end{pmatrix}$$

for b. This gives

$$\begin{pmatrix} b_0 \\ b_1 \\ b_2 \end{pmatrix} = \frac{1}{\Delta} \begin{pmatrix} -a_0^2 + a_1^2 + a_2^2 - a_1a_2 \\ a_2^2 + a_0a_1 - 2a_1a_2 \\ -a_1^2 + a_2^2 + a_0a_2 \end{pmatrix},$$

where $\Delta = -a_0^3 + a_1^3 + a_2^3 + 3a_0a_2^2 + 3a_0a_1^2 + 3a_1a_2^2 - 6a_1^2a_2 - 3a_0a_1a_2$. Computing $t_{00} = a_0^2$, $t_{11} = a_1^2$, $t_{22} = a_2^2$, $t_{01} = a_0a_1$, $t_{12} = a_1a_2$, $t_{20} = a_2a_0$, $t_{012} = a_0 + a_1 + a_2$ and $t = t_{12}(a_0 + a_1)$, then $\Delta = t_{012}^3 - t_{00}(3t_{012} - a_0) - 9t$. To finish, we perform one $\mathbb{F}_p$ inversion and obtain a^{-1} equals

$$\Delta^{-1}((t_{11} + t_{22} - t_{00} - t_{12}) + (t_{01} - 2t_{12} + t_{22})y + (t_{20} + t_{22} - t_{11})(y^2-2)).$$

$\mathbb{F}_{\mathbf{p}^6}$ **Frobenius** :
Let $c = c_0 + c_1x$. Then $c^p = (c_0 + c_1x)^p = (c_0^p - c_1^p) - c_1^p x$.

$\mathbb{F}_{\mathbf{p}^6}$ **Multiplication** :
For $c = c_0 + c_1x$, $d = d_0 + d_1x$, we have $cd = (c_0d_0 - c_1d_1) + (c_0d_1 + c_1d_0 - c_1d_1)x$. If we compute $t_{00} = c_0d_0$, $t_{11} = c_1d_1$, and $t_{01} = (c_0 + c_1)(d_0 + d_1)$, then $cd = (t_{00} - t_{11}) + (t_{01} - t_{00} - 2t_{11})x$.

$\mathbb{F}_{\mathbf{p}^6}$ **Squaring** :
$c^2 = (c_0 + c_1x)^2 = (c_0^2 - c_1^2) + c_1(2c_0 - c_1)x$. We compute $t_{01} = (c_0 + c_1)(c_0 - c_1)$, giving $c^2 = t_{01} + c_1(2c_0 - c_1)x$.

$\mathbb{F}_{\mathbf{p}^6}$ **Inversion** :
Performing again a direct inversion as in $\mathbb{F}_{p^3}$, we find

$$d_0 + d_1x = (c_0 + c_1x)^{-1} = \frac{1}{c_0^2 - c_0c_1 + c_1^2}\begin{pmatrix} c_1 - c_0 \\ -c_1 \end{pmatrix},$$

so that we still only require one $\mathbb{F}_p$ inversion. Precomputing $t_0 = (c_1 - c_0)$, $t_{01} = c_0c_1$, and $\Delta = t_0^2 + t_{01}$, the coefficients of the inverse d are given by $d_0 = \Delta^{-1}t_0$, $d_1 = -\Delta^{-1}c_1$.

If we are working in $G_{p,6}$, then as in F_1 inversions are essentially free thanks to the cheap Frobenius endomorphism.

$\sigma : \mathbf{F_1} \to \mathbf{F_2}$:
In addition to the individual arithmetic of F_1 and F_2 we need to specify an efficiently computable isomorphism between them. Writing x and y in terms of z we find that $x = z^3$, and $y = z - z^2 - z^5$, and so $\sigma^{-1} : F_2 \to F_1$ can be evaluated with just a few additions:

$$\sigma^{-1} = \begin{pmatrix} 0 & 1 & -1 & 0 & 0 & 1 \\ 0 & -1 & 1 & 0 & 1 & 0 \\ -1 & 0 & 0 & 1 & 0 & 0 \\ 0 & 0 & -1 & 0 & 1 & 0 \\ 0 & -1 & 0 & 0 & 0 & 1 \\ -1 & 0 & 0 & 0 & 0 & 0 \end{pmatrix}.$$

Since σ^{-1} has determinant three, a naive evaluation of σ necessitates four divisions by three. It is not possible to eliminate all of these since for our F_1, writing $\mathbb{F}_{p^6}$ as a quadratic extension of a cubic extension, all bases have determinant divisible by three. We can reduce this to just one division by three however, (or a multiplication by its precomputed inverse) by writing

$$\sigma = \begin{pmatrix} 1 & 0 & 0 & 0 & 0 & 0 \\ 0 & 1 & -1 & 0 & -1 & 1 \\ 0 & 0 & 1 & 0 & 1 & -1 \\ 0 & 0 & 0 & 1 & 0 & 0 \\ 0 & 0 & 0 & 0 & 1 & -1 \\ 0 & 0 & 0 & 0 & 0 & 1 \end{pmatrix} \begin{pmatrix} 0 & 0 & 0 & 0 & 0 & -1 \\ 1 & 1 & 0 & -1 & -1 & 0 \\ 0 & 0 & 0 & -1 & 0 & 0 \\ 0 & 0 & 1 & 0 & 0 & -1 \\ 1 & 1 & 0 & 0 & 0 & 0 \\ \frac{2}{3} & \frac{1}{3} & 0 & -\frac{1}{3} & \frac{1}{3} & 0 \end{pmatrix}.$$

3.3 The Representation F_3

In our notation, the group operation in CEILIDH as originally described is performed in F_2 [19], and the inverse birational maps $\psi : \mathbb{A}^2(\mathbb{F}_p) \setminus V(f) \xrightarrow{\sim} T_6(\mathbb{F}_p) \setminus \{1, x^2\}$, and $\rho : T_6(\mathbb{F}_p) \setminus \{1, x^2\} \xrightarrow{\sim} \mathbb{A}^2(\mathbb{F}_p) \setminus V(f)$, are given by

$$\psi(\alpha_1, \alpha_2) = \frac{1 + \alpha_1 y + \alpha_2(y^2 - 2) + (1 - \alpha_1^2 - \alpha_2^2 + \alpha_1\alpha_2)x}{1 + \alpha_1 y + \alpha_2(y^2 - 2) + (1 - \alpha_1^2 - \alpha_2^2 + \alpha_1\alpha_2)x^2}, \tag{3}$$

where $V(f)$ is the set of zeros of $f(\alpha_1, \alpha_2) = 1 - \alpha_1^2 - \alpha_2^2 + \alpha_1\alpha_2 = 0$ in $\mathbb{A}^2(\mathbb{F}_p)$; and for $\beta = \beta_1 + \beta_2 x$, with $\beta_1, \beta_2 \in \mathbb{F}_{p^3}$, let $(1 + \beta_1)/\beta_2 = u_1 + u_2 y + u_3(y^2 - 2)$. Then $\rho(\beta) = (u_2/u_1, u_3/u_1)$.

Since the torus T_6 is two-dimensional, given compressed points $P_1 = (\alpha_1, \alpha_2)$ and $P_2 = (\beta_1, \beta_2)$, it would be aesthetically appealing to compute their composition directly without having to map the affine representation back to $\mathbb{F}_{p^6}$, ie., to find the $(\gamma_1, \gamma_2) \in \mathbb{A}_2$ such that $(\alpha_1, \alpha_2) \circ (\beta_1, \beta_2) = (\gamma_1, \gamma_2)$, where $\circ$ refers to an operation equivalent to multiplication in $\mathbb{F}_{p^6}$. The drawback with this approach is that the group law on T_6 as we have presented it is defined only in terms of the arithmetic of $\mathbb{F}_{p^6}$, and so the decompression of P_1 and P_2 before multiplication seems essential.

If one insists on representing intermediate results in their compressed form in an exponentiation for example, we found that this costs $24M + 43A + I$ for a multiplication, and $21M + 38A + I$ for a squaring. Note that these are even more costly than both a general elliptic curve add, or double for example. The reason these operations are so expensive is that the rational representation of T_6 does not lend itself favourably to performing the group law operation: this statement does not hold for the XTR representation, which is why the corresponding arithmetic is so much faster (cf. Figure 1).

The arithmetic developed in the previous section shows that if one alternatively performs a one-time decompression at the start of, and a re-compression at the end of an exponentiation, then the cost of a general multiplication and squaring is only $18M + 54A$ and $12M + 33A$ respectively in F_2, and $18M + 53A$ and $6M + 21A$ respectively in F_1. Clearly decompressing to $\mathbb{F}_{p^6}$ seems to be the better method.

This is not the whole story though. Since $T_6(\mathbb{F}_p)$ is by definition the intersection of the kernels of the two norm maps $N_{\mathbb{F}_{p^6}/\mathbb{F}_{p^3}}$ and $N_{\mathbb{F}_{p^6}/\mathbb{F}_{p^2}}$, if we have a good representation for either of these, we can save some work. The following lemma emphasises the relevance to CEILIDH of the parametrisation of elements in the kernel of the former norm map, which is implicit in the construction of ψ.

Lemma 2. *There is an isomorphism*

$$\tau : \mathrm{Ker}[N_{\mathbb{F}_{p^6}/\mathbb{F}_{p^3}}] \xrightarrow{\sim} \left\{ \frac{b + x}{b + x^2}, b \in \mathbb{F}_{p^3} \right\} \cup \{1\},$$

where for $a = a_0 + a_1 x \in \mathrm{Ker}[N_{\mathbb{F}_{p^6}/\mathbb{F}_{p^3}}] \setminus \{1\}$,

$$\tau(a) = \left(\frac{b+x}{b+x^2}\right),$$

with $b = (1+a_0)/a_1$ *if* $a_1 \neq 0$ *and* $b = a_0/(a_0 - 1)$ *otherwise, and*

$$\tau^{-1}(b) = \frac{b^2-1}{b^2-b+1} + \frac{2b-1}{b^2-b+1}x.$$

Proof. For $a \in \mathbb{F}_{p^6}$, $N_{\mathbb{F}_{p^6}/\mathbb{F}_{p^3}}(a) = a^{1+p^3}$, and so the kernel of this map has at most $1 + p^3$ solutions. For each $(b+x)/(b+x^2), b \in \mathbb{F}_{p^3}$, the stated norm is one, and counting also the identity, we have all $1 + p^3$ solutions. Solving $a_0 + a_1 x = (b+x)/(b+x^2)$ for b gives the second part, while for the third, we solve the same equation for a_0, a_1, where we have used the condition that $a_0^2 + a_1^2 - a_0 a_1 = 1 \Leftrightarrow a_0 + a_1 x \in \mathrm{Ker}[N_{\mathbb{F}_{p^6}/\mathbb{F}_{p^3}}]$.

□

With this we can introduce the following:

Definition 2. F_3 *is the set of elements*

$$\left\{\frac{a_0 + a_1 x}{a_0 + a_1 x^2}, a_i \in \mathbb{F}_{p^3}\right\}.$$

When the coefficient a_1 *of this representation equals* 1, *we say the element is reduced.*

Note that if we do not need the reduced form of an element in F_3, then evaluating τ simplifies to $(1+a_0+a_1x)/(1+a_0+a_1x^2)$, so no inversion is necessary. Mapping an unreduced element back to F_2, we obtain

$$\tau^{-1}\left(\frac{a_0 + a_1 x}{a_0 + a_1 x^2}\right) = \frac{a_0^2 - a_1^2}{a_0^2 - a_0 a_1 + a_1^2} + \frac{2a_0 a_1 - a_1^2}{a_0^2 - a_0 a_1 + a_1^2}x.$$

As we have pointed out, this construction is no more than the rational parametrisation of $T_2(\mathbb{F}_{p^3})$. As a result of the symmetry of x and x^2 in the irreducible polynomial defining the quadratic extension of F_2, for all arithmetic operations between elements of F_3, the coefficients of the numerator and those of the denominator correspond exactly. Hence we need only work with the numerator, which will in general have the form $a = a_1 + a_2 x$. For expositional purposes, in the following we still write elements of T_6 as fractions but for our implementation this is of course unnecessary.

Using this fractional form alone does not seem to offer any advantage over the F_2 representation. However, considering the exponentiation of a reduced element $(g+x)/(g+x^2)$ of T_6, we already save one $\mathbb{F}_{p^3}$ multiplication for every multiplication by this element, since

$$\left(\frac{g+x}{g+x^2}\right) \times \left(\frac{a_0 + a_1 x}{a_0 + a_1 x^2}\right) = \left(\frac{(ga_0 - a_1) + (ga_1 + a_0 - a_1)x}{(ga_0 - a_1) + (ga_1 + a_0 - a_1)x^2}\right), \quad (4)$$

so we only need to compute ga_0 and ga_1, and a few additions, when our multiplier is in this form.

Furthermore, we can exploit the fractional form of elements of F_3 for squaring and inversion, and if additions are sufficiently cheap compared to multiplications, we can actually reduce the cost of a basic $\mathbb{F}_{p^3}$ multiplication to the theoretical minimum measured in the number of $\mathbb{F}_p$ multiplications, using a Toom-Cook-style interpolation [9]. We do not have space to detail this last point here unfortunately.

The reason this all works is that the rational representation of elements of T_2 can be embedded efficiently as a fraction in the field extension. Noting that we need only work with the numerator, the group law can be performed directly on this compressed element.

The reduced form of an element may be viewed as the affine representation of T_2, with the non-compressible identity element being the point at infinity, with the non-reduced form of an element corresponding to the projective representation, with identity λ for any $\lambda \in \mathbb{F}_{p^3}^*$. This point was essentially made in [19], but without reference to its applicability to CEILIDH as well. There however, it was suggested the group law be performed entirely in $\mathbb{F}_{p^3}$, which would require an $\mathbb{F}_{p^3}^*$ inversion for every multiplication: this would be very inefficient.

In fact the idea of using the torus T_2, as applied to cryptography, is not new. It has been suggested by other authors that elements of $T_2(\mathbb{F}_p)$ be represented in the (group-theoretically) isomorphic set $\mathbb{F}_{p^2}^*/\mathbb{F}_p^*$. If $x = \zeta_3$, then elements are represented as $a_0 + a_1 x$, and $a_0 + a_1 x \equiv b_0 + b_1 x$ iff $a_0/a_1 = b_0/b_1$. As a consequence, elements for which $a_1 \neq 0$ can be represented by a_0/a_1, and all the arithmetic above follows *mutatis mutandis.*

F_3 Frobenius :
Let $a \in F_3$. Then

$$\left(\frac{a_0 + a_1 x}{a_0 + a_1 x^2}\right)^p = \left(\frac{a_0^p + a_1^p x^2}{a_0^p + a_1^p x}\right) = \left(\frac{(a_0^p - a_1^p) - a_1^p x}{(a_0^p - a_1^p) - a_1^p x^2}\right) = \left(\frac{(a_1^p - a_0^p) + a_1^p x}{(a_1^p - a_0^p) + a_1^p x^2}\right).$$

F_3 Multiplication :
Multiplication by a reduced element is performed as in (4), or if by a non-reduced element, exactly as in F_2.

F_3 Squaring :
This is performed as in F_2.

F_3 Inversion :
This is straightforward, since elements are represented as fractions.

$$\left(\frac{a_0 + a_1 x}{a_0 + a_1 x^2}\right)^{-1} = \left(\frac{a_0 + a_1 x^2}{a_0 + a_1 x}\right) = \left(\frac{(a_1 - a_0) + a_1 x}{(a_1 - a_0) + a_1 x^2}\right).$$

Also, since we use the intermediate representation F_3 between $\mathbb{A}^2(\mathbb{F}_p)$ and F_2, we must adjust the map $\rho : F_3 \setminus \{1, x^2\} \xrightarrow{\sim} \mathbb{A}^2(\mathbb{F}_p) \setminus V(f)$. Let $\beta =$

$(\beta_1 + \beta_2 x)/(\beta_1 + \beta_2 x^2) \in F_3$, with $\beta_1/\beta_2 = u_1 + u_2 y + u_3(y^2 - 2)$: then $\rho(\beta) = (u_2/u_1, u_3/u_1)$. We summarise the results regarding arithmetic in F_1, F_2 and F_3 in the following:

Lemma 3. *The cost of arithmetical operations in F_1, F_2 and F_3 are:*

Operation	F_1	F_2	F_3
Multiply	$18M + 53A$	$18M + 54A$	$18M + 54A$
Square	$6M + 21A$	$12M + 33A$	$12M + 33A$
Inverse	$2A$	$6A$	$3A$
Frobenius	$1A$	$10A$	$10A$
Reduce	n/a	n/a	$19M + 35A + I$
Mixed Mul.	n/a	n/a	$12M + 33A$

Map	Cost
$F_1 \to F_2$	$1M + 11A$
$F_2 \to F_3$	$1A$
$F_3 \to \mathbb{A}^2$	$14M + 19A + I$
$\mathbb{A}^2 \to F_3$	$2M + 3A$
$F_3 \to F_2$	$25M + 41A + I$
$F_2 \to F_1$	$8A$

Here the operation *Reduce* refers to obtaining the reduced form of an element of F_3, and a *Mixed Mul.* refers to multiplying a non-reduced element by a reduced one. The cost of the map $\rho : F_3 \to \mathbb{A}_2$ assumes the element being compressed is in non-reduced form, as this is the case after an exponentiation in both F_1, or F_3. Also, for the map $\tau^{-1} : F_3 \to F_2$ we assume that the x-coefficient is in $\mathbb{F}_p$ only as in (3), and not $\mathbb{F}_{p^3}$, as in practice one would only perform this operation when decompressing from $\mathbb{A}_2$ to F_1, and not from a non-reduced element.

3.4 Exponentiation

For F_1, F_2 and F_3 one can use the Frobenius map to obtain fast exponentiation. In a subgroup of order l where $l|(p^2 - p + 1)$, we write an exponent m as $m \equiv m_1 + m_2 p \mod l$, where m_1, m_2 are approximately half the bitlength of m [22]. One can find m_1 and m_2 very quickly having performed a one-time Gaussian two dimensional lattice basis reduction, and using this basis to find the closest vector to $(m, 0)^T$. To compute a^m for a random a, we perform a double exponentiation $a^{m_1}(a^p)^{m_2}$ using the Joint Sparse Form (JSF) of the integers m_1, m_2 [21] and Shamir's trick, originally due to Straus [24], which on average halves the number of pairs of non-zero bits in their paired binary expansion. The use of the JSF is possible since we have virtually free inversion.

When the base of the exponentiation is fixed we also use the JSF and Shamir's trick but perform some precomputation as well. Fixed elements are important since we can spend some time and space to precompute values that might accelerate an operation yet be reused often enough to make the cost of doing so acceptable. We store $(a^i(a^p)^j)^{2^k}$ for $i, j \in \{0, \pm 1\}$, and k from 1 to half the bitlength of l, where l is the size of the subgroup we work with. For a 1024-bit field, and with l approximately of length 160 bits, for the price of storing $4 \times 80 = 320$ fields elements, we eliminate all squarings from the exponentiation routine. In F_3 the storage of 320 reduced elements requires about 22.5Kb, whereas for F_1 and F_2 this amount is doubled to 45Kb, since elements can not be reduced. In XTR only one element is precomputed (it is unclear how to exploit more precomputation). This provides nearly as good a speed up as for CEILIDH, but here the cost of the precomputation is much cheaper in both time and space.

For cryptosystems where the group law is just ordinary multiplication in a finite field though, one can always exchange space for time, so CEILIDH has a slight advantage here over XTR, if these resources are available. We concede that more efficient precomputation methods can be applied given our chosen level of storage [14], but are confident our method provides an accurate reflection of the possible gains resulting from this approach.

For double exponentiation we assume that both bases are random, so that no precomputation can be employed. Using the JSF for each exponent, we combined the squarings for both exponentiations while performing the multiplications separately. It is possible to make this slightly more efficient [17], and to use some precomputation, but again we are satisfied that our results are indicative of the general performance of the algorithms.

4 Implementation Results

To demonstrate the different performance characteristics of the three representations of CEILIDH, we constructed an implementation of the entire system, based on the algorithms described in the previous section. We also implemented the fastest algorithms for the equivalent XTR protocols [23], so that our comparison was made between the best possible implementations of both systems.

We based this implementation on a special purpose library for arithmetic in $\mathbb{F}_p$ that represents and manipulates field elements using Montgomery reduction [16]. Montgomery arithmetic facilitates fast field operations given some modulus specific precomputation [3]. This is ideal for our purposes since after key-generation the field $\mathbb{F}_p$ remains the same for all subsequent operations.

We used a GCC 3.3 compiler suite to build our implementation and ran timing experiments on a Linux based PC incorporating a 2.80 GHz Intel Pentium 4 processor. The entire system was constructed in C++ except for small assembly language fragments to accelerate operations in $\mathbb{F}_p$. We accept that further performance improvements could be made through aggressive profiling and optimisation but are confident our results are representative of the undorlying algorithms and allow a comparison between them.

For our experiments we randomly chose 500 key pairs (p, l) with the field characterisitc p of length 176 bits and subgroup size l 160 bits. These parameters heuristically provide the equivalent of 1024 bit RSA security. With the same key pairs for both CEILIDH and XTR we performed 500 instances of each operation listed in Figure 4. For exponentiations, exponents were chosen randomly modulo l in all cases.

The left table shows timings for operations pertinent to use in real cryptosystems. We use x_R to denote a random element in a given representation and x_F to represent a fixed element. Thus, x_R^n represents a single exponentiation with a random base and exponent while $x_R^n.y_R^m$ represents a double exponentiation without precomputation.

These timings are useful since they form the basis of all public key cryptographic protocols, and so we have a good idea of the comparative performance

Operation	Time			
	F_1	F_2	F_3	XTR
$x_R \cdot y_R$	37.8 μs	41.2 μs	41.8 μs	8.7 μs
x_R^2	16.8 μs	26.9 μs	27.7 μs	4.9 μs
Reduce	n/a	n/a	170.6 μs	n/a
Mixed Mul.	n/a	n/a	28.4 μs	n/a
x_R^n	2.99 ms	3.94 ms	3.94 ms	2.57 ms
$x_R^n \cdot y_R^m$	4.71 ms	5.75 ms	5.79 ms	2.98 ms
x_F^n	1.56 ms	1.74 ms	1.21 ms	1.49 ms
Precomp.	5.89 ms	9.33 ms	63.65 ms	1.43 ms

Mapping	Time
$F_1 \to F_2$	7.1 μs
$F_2 \to F_3$	1.9 μs
$F_3 \to \mathbb{A}^2$	161.7 μs
$F_1 \leftarrow F_2$	2.4 μs
$F_2 \leftarrow F_3$	222.8 μs
$F_3 \leftarrow \mathbb{A}^2$	5.1 μs

Fig. 1. Timing Results

of the different representations of $T_6(\mathbb{F}_p)$, and XTR. One can see that the results are in general agreement with our arithmetic cost analysis. Indeed F_1 provides the most efficient representation when exponentiating a random element, and with precomputation, F_3 offers a slight improvement over XTR, allowing for the cost of $\rho : F_3 \to \mathbb{A}^2$ as well.

The right table in Figure 4 demonstrates the cost of applying mapping operations on elements to transform them between our different representations. These results represent the time taken to map a random element in the source representation and transform it into the corresponding element in the target representation. One notes from these that it is unfortunately not advantageous to 'mix' representations, so that in a single exponentiation squaring is performed in F_1, and multiplication in F_3. This would be analogous to the use of mixed coordinate systems in elliptic curve cryptography [4].

The authors note that it is possible to negate the necessity of F_1 altogether by basing all arithmetic in a field of even characteristic, since then squarings are just as inexpensive in F_3. With suitable keys, this would simplify implementations of CEILIDH and far more efficient precomputation strategies could be brought to bare.

References

1. L.M. Adleman and M.A. Huang. Function Field Sieve Method for Discrete Logarithms over Finite Fields. *Info. and Comp.*, **151(1–2)**, 5–16, 1999.
2. I.F. Blake, G. Seroussi and N.P. Smart. Elliptic Curves in Cryptography. Cambridge University Press, 1999.
3. A. Bosselaers, R. Govaerts and J. Vandewalle. Comparison of Three Modular Reduction Functions. In *Advances in Cryptology (CRYPTO)* Springer LNCS 773, 175–186, 1994.
4. H. Cohen, A. Miyaji and T. Ono. Efficient Elliptic Curve Exponentiation Using Mixed Coordinates. In *Advances in Cryptology (ASIACRYPT)*, Springer LNCS 1514, 51–65, 1998.
5. P. Gaudry, F. Hess and N.P. Smart. Constructive And Destructive Facets Of Weil Descent On Elliptic Curves. In *J. of Cryptology*, **15**(1), 19–46, 2002.

6. D.M. Gordon. Discrete Logarithms in $GF(p)$ Using the Number Field Sieve. In *SIAM J. of Disc. Math.*, **6**, 124–138, 1993.
7. T. Itoh and S. Tsujii. A Fast Algorithm for Computing Multiplicative Inverses in $GF(2^m)$ Using Normal Bases. *Info. and Comp.*, **78**(3), 171–177, 1988.
8. A. Joux and R. Lercier. The Function Field Sieve is Quite Special. In *Alg. Numb. Th. Symp. (ANTS-V)*, Springer LNCS 2369, 431–445, 2002.
9. D. E. Knuth. The Art of Computer Programming. Vol. 2, Addison-Wesley, 1981.
10. N. Koblitz. Elliptic Curve Cryptosystems. In *Math. Comp.*, **48**, 203–209, 1987.
11. N. Koblitz. Hyperelliptic Cryptosystems. In *J. of Cryptology*, **1**, 139–150, 1989.
12. A.K. Lenstra. Using Cyclotomic Polynomials to Construct Efficient Discrete Logarithm Cryptosystems over Finite Fields. In *Proc. ACISP97*, Springer LNCS 1270, 127–138, 1997.
13. A.K. Lenstra and E.R. Verheul. The XTR Public Key System. In *Advances in Cryptology (CRYPTO)*, Springer LNCS 1880, 1–19, 2000.
14. C.H. Lim and P.J. Lee. More flexible exponentiation with precomputation. In *Advances in Cryptology (CRYPTO)*, Springer LNCS 839, 95–107, 1994.
15. V. Miller. Uses of Elliptic Curves in Cryptography. In *Advances in Cryptology (CRYPTO)*, Springer LNCS 218, 417–426, 1985.
16. P.L. Montgomery. Modular Multiplication Without Trial Division. *Math. Comp.*, **44**, 519–521, 1985.
17. J. Proos. Joint Sparse Forms and Generating Zero Columns when Combing. University of Waterloo, Technical Report CORR 2003-23,
18. K. Rubin and A. Silverberg. Supersingular abelian varieties in Cryptology. In *Advances in Cryptology (CRYPTO)*, Springer LNCS 2442, 336–353, 2002.
19. K. Rubin and A. Silverberg. Torus-Based Cryptography. In *Advances in Cryptology (CRYPTO)*, Springer LNCS 2729, 349–365, 2003.
20. V. Shoup. Lower Bounds for Discrete Logarithms and Related Problems. In *Advances in Cryptology (EUROCRYPT)*, Springer LNCS 1233, 256–266, 1997.
21. J.A. Solinas. Low-Weight Binary Representations for Pairs of Integers. University of Waterloo, Technical Report CORR 2001-41,
22. M. Stam and A. Lenstra. Efficient Subgroup Exponentiation in Quadratic and Sixth Degree Extensions. In *Cryptographic Hardware and Embedded Systems (CHES)*, Springer LNCS 2523, 318–332, 2002.
23. M. Stam and A. Lenstra. Speeding Up XTR. In *Advances in Cryptology (ASIACRYPT)*, Springer LNCS 2248, 125–143, 2001.
24. E.G. Straus. Problems and Solutions: (5125) Addition Chains of Vectors. In *American Mathematical Monthly*, **71**, 806–808, 1964.
25. P. Smith and C. Skinner. A public-key cryptosystem and a digital signature system based on the Lucas function analogue to discrete logarithms. In *Advances in Cryptology (ASIACRYPT)* Springer LNCS 917, 357–364, 1995.
26. E. Teske. Speeding up Pollard's Rho Method for Computing discrete Logarithms. In *Alg. Numb. Th. Symp. (ANTS-III)*, Springer LNCS 1423, 541–554, 1998.
27. N. Thériault. Index Calculus Attack for Hyperelliptic Curves of Small Genus. In *Advances in Cryptology (ASIACRYPT)*, Springer LNCS 2894, 75–92, 2003.
28. V.E. Voskresenskiĭ. Algebraic Groups and Their Birational Invariants. In *Translations of Mathematical Monographs*, **179**, American Mathematical Society, 1998.

Stable Models of Elliptic Curves, Ring Class Fields, and Complex Multiplication

Jordi Guàrdia[1⋆], Eugenia Torres[1⋆⋆], and Montserrat Vela[2⋆⋆⋆]

[1] Departament de Matemàtica Aplicada IV. Escola Universitària Politècnica de Vilanova i la Geltrú, Avinguda Víctor Balaguer s/n. E-08800 Vilanova i la Geltrú
{guardia,eugenia}@mat.upc.es

[2] Departament de Matemàtica Aplicada II. Facultat de Matemàtiques i Estadística, c/ Pau Gargallo, 5, E-08028 Barcelona
Montse.Vela@upc.es

Abstract. We introduce a new model for elliptic curves over rings of odd characteristic, and study its properties and its utility in numerical computations. They turn to be particularly interesting for elliptic curves with complex multiplication, for which they provide very simple stable equations. The invariants associated to these models allow an easy construction of ring class fields of certain imaginary quadratic orders, with interesting theoretical consequences and practical utility in numerical computations.

1 Introduction

We introduce a new model for elliptic curves over rings of odd characteristic. The symmetry of this model gives good reduction properties to them: they are stable curves at any odd prime. As a consequence, the discriminants of these models are quite small, since they are only divisible by a relatively low power of 2. The aim of this paper is the study of these models, both from the theoretical and the practical viewpoint.

We show the reduction properties of our model in section 2, which ends up with a table of elliptic curves with complex multiplication by imaginary quadratic orders with class number 1 and 2 and good reduction outside 2. The invariants related to our model can be seen as certain modular functions, which are described in section 3. In section 4 we describe an easy construction of certain ring class fields based on the use of the discriminant of these models. The class polynomials originated by these discriminants are then analyzed in section 5. Finally, in section 6 we discuss the possibility of using the new invariants for the computation of Hilbert class polynomials.

⋆ The first author wa partially supported by MCYT BFM2003-06768-C02-02, 2002SGR00148.

⋆⋆ The second author was partially supported by MCYT BFM2003-06768-C02-02.

⋆⋆⋆ The third author was partially supported by MCYT BFM2003-06768-C02-01, HPRNC-CT-2000-00114, 2002SGR00148.

D.A. Buell (Ed.): ANTS 2004, LNCS 3076, pp. 250–262, 2004.

2 Symmetric Models of Elliptic Curves

Let A be an integral domain, with field of fractions K of odd characteristic. For $G \in A$ we consider the elliptic curve

$$E_G : Y^2 = X^3 + GX^2 + X. \tag{1}$$

We shall call such an equation for an elliptic curve a *symmetric model*; the word symmetry refers both to the coefficients of the polynomial $X^3 + GX^2 + X$ and to the role of its non-zero roots ℓ_1, ℓ_2 which satisfy $\ell_1\ell_2 = 1$.

The discriminant and the j-invariant of E_G are:

$$\mathfrak{D} = 16(G^2 - 4), \qquad j = \frac{256(G^2-3)^3}{G^2-4} = \frac{(\mathfrak{D}+16)^3}{\mathfrak{D}}. \tag{2}$$

The first expression of j in terms of G shows that G is integral over $A[\frac{1}{2}][j]$, since it satisfies the sextic equation

$$G^6 - 9G^4 - \left(-27 + \frac{j}{256}\right)G^2 + \frac{j}{64} - 27 = 0. \tag{3}$$

Moreover, this expression shows that every elliptic curve E over a ring A of odd characteristic admits a symmetric model E_G over an algebraic extension of A: it can be obtained from the j-invariant of the curve, solving equation (3). In general we will find six symmetric models E_G not defined over A, but over a triquadratic extension of A. If one needs a model defined over a smaller extension of A, the following quadratic twist of E_G can be used:

$$Y^2 = X^3 + G^2X + G^2X = X^3 + \frac{\mathfrak{D}+64}{16}X^2 + \frac{\mathfrak{D}+64}{16}X.$$

The discriminant of this model is $\mathfrak{D}' = 256G^6(G^2-4) = \mathfrak{D}(\mathfrak{D}+64)^3/256$.

The expression of j in terms of $\mathfrak{D}$ provides the relation:

$$\mathfrak{D}^3 + 48\mathfrak{D}^2 + (768 - j)\mathfrak{D} + 2^{12} = 0, \tag{4}$$

which shows that $\mathfrak{D}$ is a unit in $A\left[\frac{1}{2}\right][j]$, and suggest that the curve E_G should have very nice reduction properties outside 2. The following result confirms this point:

Proposition 1.
a) E_G is an stable curve over $A[1/2][G]$.
b) The odd factor of the conductor of E_G is $N_{odd} = \prod_{\mathfrak{p}|G\pm 2}\mathfrak{p}$.
c) For an odd prime $\mathfrak{p} \mid G+2$, the curve has E_G split multiplicative reduction.
d) For an odd prime $\mathfrak{p} \mid G-2$ the curve has E_G multiplicative reduction. The reduction is split if $-1 \in \mathbb{F}_\mathfrak{p}^{*2}$ and non-split otherwise.

Proof. For an odd prime $\mathfrak{p}$ dividing $G \pm 2$ the reduction of E_G mod $\mathfrak{p}$ is

$$Y^2 = X^3 \mp 2X^2 + X = X(X \mp 1)^2,$$

which has an evident node. The tangent lines of this node are the linear factors of $Y^2 \pm (X \mp 1)^2$. ■

In fact, we can say more about the reduction properties of the symmetric models.

Theorem 1. *Let E be an elliptic curve over A. Let L be the field of fractions of $A[G]$ and O_L its integer ring.*

a) If an odd prime $\mathfrak{p} \in \operatorname{Spec} O_L$ satisfies $v_\mathfrak{p}(G) < 0$ (or equivalently $v_\mathfrak{p}(\mathfrak{D}) < 0$) then E has potentially multiplicative reduction at $\mathfrak{p}$.
b) If $j(E) \in A$, the symmetric models E_G are integral minimal models of E over $A[\frac{1}{2}][G]$.

Proof. If $v_p(\mathfrak{D}) < 0$, then also $v_p(j(E)) = v_p(\frac{(\mathfrak{D}+16)^3}{\mathfrak{D}}) < 0$ so that E cannot have potentially good reduction. Hence, when $j(E) \in A$ we will have $v_\mathfrak{p}(G) \geq 0$ for every odd prime $\mathfrak{p} \in \operatorname{Spec} A$. In fact, equation (4) shows that the norm $N_{L/K}(\mathfrak{D})$ is a power of two, and hence $v_\mathfrak{p}(\mathfrak{D}) = 0$. ■

Similar results can be proved for the Legendre models $Y^2 = X(X-1)(X-\lambda)$. The advantage of symmetric models is that their invariants are simpler than the invariants associated to the Legendre models. We have already seen that G satisfies a triquadratic equation over $A[j]$, while the equation relating λ and j is a general sextic. The discriminant $\Delta(\lambda)$ of the Legendre model satisfies

$$2^{12}\Delta(\lambda)^3 + (-196608 + 1536\,j - j^2)\ \Delta(\lambda)^2 + 2^{13}\ (384 + j)\ \Delta(\lambda) - 2^{24} = 0,$$

which is clearly worse than (4).

We finish this section with a table of symmetric models of the elliptic curves E_G with complex multiplication by an imaginary quadratic order of class number 1. We also include a number of examples for class number 2 quadratic orders. All these models have good reduction outside the primes dividing 2 in its field of definition.

D	j	$\mathfrak{D}$	$E_G : Y^2 = f(X)$
		$h = 1$	
-3	0	-2^4	$X^3 + \sqrt{3}X^2 + X$
-4	$2^6 3^3$	-64	$X^3 + X$
-7	$-3^3 5^3$	-1	$X^3 + \frac{3\sqrt{7}}{4}X^2 + X$
-8	$2^6 5^3$	2^6	$X^3 + 2\sqrt{2}X^2 + X$

D	j	$\mathfrak{D}$	$E_G : Y^2 = f(X)$
-11	-2^{15}	$\mathfrak{D} = 16\,\frac{4\,\alpha^2 - 96 - 3\,\alpha}{3\alpha}$ $\alpha = \sqrt[3]{27 + 21\,\sqrt{33}}$	$X^3 + \frac{\sqrt{3}\sqrt{4\,\alpha^2-96+9\,\alpha}}{3\sqrt{\alpha}} X^2 + X$
-12	$2^4 3^3 5^3$	-2^8	$X^3 + 2\sqrt{-3}X^2 + X$
-16	$2^3 3^3 11^3$	2^9	$X^3 + 6X^2 + X$
-19	$-2^{15} 3^3$	$\mathfrak{D} = -64 + 48\frac{4\alpha^2+\alpha-32}{\alpha}$ $\alpha = \sqrt[3]{1 + 3\sqrt{57}}$	$X^3 + \sqrt{3\frac{4\alpha^2+\alpha-32}{\alpha}}\,X^2 + X$
-27	$2^{15} 3 \cdot 5^3$	$16(-1 - 100\sqrt[3]{2} + 80\sqrt[3]{4})$	$X^3 + \sqrt{-3 - 100\sqrt[3]{2} + 80\sqrt[3]{4}}\,X^2 + X$
-28	$3^3 5^3 17^3$	-2^{12}	$X^3 + 6\sqrt{-7}X^2 + X$
-43	$-2^{18} 3^3 5^3$	$\mathfrak{D} = -64 + 48\frac{40\alpha^2+\alpha-3200}{\alpha}$ $\alpha = \sqrt[3]{1 + 63\sqrt{129}}$	$X^3 + \sqrt{3\frac{40\alpha^2+\alpha-3200}{\alpha}}\,X^2 + X$
-67	$-2^{15} 3^3 5^3 11^3$	$\mathfrak{D} = -64 + 48\frac{220\alpha^2+\alpha-96800}{\alpha}$ $\alpha = \sqrt[3]{1 + 651\sqrt{201}}$	$X^3 + \sqrt{3\frac{220\alpha^2+\alpha-96800}{\alpha}}\,X^2 + X$
-163	$-2^{18} 3^3 5^3 23^3 29^3$	$\mathfrak{D} = -64 + 48\frac{26680\alpha^2+\alpha-1423644800}{\alpha}$ $\alpha = \sqrt[3]{1 + 557403\sqrt{489}}$	$X^3 + \sqrt{3\frac{26680\alpha^2+\alpha-1423644800}{\alpha}}\,X^2 + X$

$h = 2$			
-15	$-\frac{135}{2}(1415 + 637\sqrt{5})$	$-\left(\frac{\sqrt{5}-1}{2}\right)^8$	$X^3 + \sqrt{\frac{3}{32}(27 + 7\sqrt{5})}\,X^2 + X$
-20	$320(1975 + 884\sqrt{5})$	$-2^6\left(\frac{1+\sqrt{5}}{2}\right)^6$	$X^3 - 4\sqrt{-2 - \sqrt{5}}\,X^2 + X$
-24	$1728(1399 + 988\sqrt{2})$	$2^6(1 + \sqrt{2})^4$	$X^3 + 2\sqrt{6(3 + 2\sqrt{2})}\,X^2 + X$
-32	$1000(26125 + 18473\,\sqrt{2})$	$2^9(1 + \sqrt{2})^3$	$X^3 + 2(5 + 4\sqrt{2})\,X^2 + X$
-36	$192(399849 + 230888\,\sqrt{3})$	$-2^6(2 + \sqrt{3})^4$	$X^3 + 4\sqrt{-2(12 + 17\sqrt{3})}\,X^2 + X$
-40	$8640(24635 + 11016\,\sqrt{5})$	$2^6\left(\frac{1+\sqrt{5}}{2}\right)^{12}$	$X^3 + 6\sqrt{2(9 + 4\sqrt{5})}\,X^2 + X$
-48	$40500(35010 + 20213\,\sqrt{3})$	$2^{10}(2 + \sqrt{3})^3$	$X^3 + 2(15 + 8\sqrt{3})\,X^2 + X$
-52	$216000(15965 + 4428\,\sqrt{13})$	$-2^6\left(\frac{3+\sqrt{13}}{2}\right)^6$	$X^3 + 12\sqrt{-(18 + 5\sqrt{13})}\,X^2 + X$
-60	$\frac{135}{2}(274207975 + 122629507\,\sqrt{5})$	$-2^{12}\left(\frac{1+\sqrt{5}}{2}\right)^8$	$X^3 + 2\sqrt{-3(501 + 224\sqrt{5})}\,X^2 + X$
-64	$54(761354780 + 538359129\,\sqrt{2})$	$2^{10}\sqrt{2}(1 + \sqrt{2})^6$	$X^3 + 6(11 + 8\sqrt{2})\,X^2 + X$

3 Modular Interpretation

We now assume that E_G is defined over $\mathbb{C}$ so that it is analytically isomorphic to a complex torus $\mathbb{C}/\langle 1, \tau\rangle$. To establish the relation between τ and G we use the classical Thetanullwerte (cf. [9]):

$$\theta_2(\tau) = \sum_{n\in\mathbb{Z}} q^{(\frac{n-1}{2})^2/2}, \qquad \theta_3(\tau) = \sum_{n\in\mathbb{Z}} q^{n^2/2}, \qquad \theta_4(\tau) = \sum_{n\in\mathbb{Z}} (-1)^n q^{n^2/2}.$$

We can substitute $j(\tau) = \frac{32\left(\theta_2(\tau)^8+\theta_3(\tau)^8+\theta_4(\tau)^8\right)^3}{\theta_2(\tau)^8\,\theta_3(\tau)^8\,\theta_4(\tau)^8}$ in equation (3) and solve it for G. We find six solutions:

$$G_{\pm 2}(\tau) := \pm\frac{\theta_3(\tau)^4+\theta_4(\tau)^4}{\theta_3(\tau)^2\theta_4(\tau)^2},$$

$$G_{\pm 3}(\tau) := \pm\frac{i(\theta_2(\tau)^4-\theta_4(\tau)^4)}{\theta_2(\tau)^2\theta_4(\tau)^2},$$

$$G_{\pm 4}(\tau) := \pm\frac{\theta_2(\tau)^4+\theta_3(\tau)^4}{\theta_2(\tau)^2\theta_3(\tau)^2}.$$

Concerning the discriminant $\mathfrak{D}$, we find three possible expressions for it:

$$\mathfrak{D}_2(\tau) = \frac{16\theta_2(\tau)^8}{\theta_3(\tau)^4\theta_4(\tau)^4},\quad \mathfrak{D}_3(\tau) = -\frac{16\theta_3(\tau)^8}{\theta_2(\tau)^4\theta_4(\tau)^4},\quad \mathfrak{D}_4(\tau) = \frac{16\theta_4(\tau)^8}{\theta_2(\tau)^4\theta_3(\tau)^4}. \tag{5}$$

It is worth noting that

$$\begin{aligned}\mathfrak{D}_2(\tau)+\mathfrak{D}_3(\tau)+\mathfrak{D}_4(\tau) &= -48,\\ \mathfrak{D}_2(\tau)\mathfrak{D}_3(\tau)\mathfrak{D}_4(\tau) &= -2^{12},\end{aligned} \tag{6}$$

so that the irreducible polynomial of $\mathfrak{D}(\tau)$ over $\mathbb{Q}(j(\tau))$ is

$$\mathfrak{D}_r(\tau)^3 + 48\mathfrak{D}_r(\tau)^2 + (768 - j(\tau))\mathfrak{D}_r(\tau) + 2^{12} = 0. \tag{7}$$

Proposition 2. *Let us denote by* $S = \begin{pmatrix}1&1\\0&1\end{pmatrix}$, $T = \begin{pmatrix}0&1\\-1&0\end{pmatrix}$ *the usual generators of the modular group* $\Gamma = SL_2(\mathbb{Z})$, *and consider the groups* $\Gamma_0(2) = \langle S, TS^2T\rangle$, $\Gamma_\theta = S^{-1}\Gamma^0(2)S = \langle S^2, T\rangle$, $\Gamma^0(2) = T\Gamma_0(2)T^{-1} = \langle S^2, TST\rangle$ *and the associated modular curves* $X_0(2)$, X_θ, $X^0(2)$.

a) $\mathfrak{D}_2(\tau)$ *is a* Hauptmodul *for the modular curve* $X_0(2)$.
b) $\mathfrak{D}_3(\tau)$ *is a* Hauptmodul *for the modular curve* X_θ.
c) $\mathfrak{D}_4(\tau)$ *is a* Hauptmodul *for the modular curve* $X^0(2)$.

Proof. The action of S, T on the set of Thetanullwerte is well-known:

$$\begin{aligned}\theta_2(S\tau) &= \zeta_8\theta_2(\tau), & \theta_2(T\tau) &= \zeta_8^{-1}\sqrt{\tau}\theta_4(\tau),\\ \theta_3(S\tau) &= \theta_4(\tau), & \theta_3(T\tau) &= \zeta_8^{-1}\sqrt{\tau}, \theta_3(\tau),\\ \theta_4(S\tau) &= \theta_3(\tau), & \theta_4(T\tau) &= \zeta_8^{-1}\sqrt{\tau}, \theta_2(\tau),\end{aligned}$$

where $\zeta_8 = \exp(2\pi i/8)$. From these relations we derive the equalities

$$\begin{aligned}\mathfrak{D}_2(S\tau) &= \mathfrak{D}_2(\tau), & \mathfrak{D}_2(T\tau) &= \mathfrak{D}_4(\tau),\\ \mathfrak{D}_3(S\tau) &= \mathfrak{D}_4(\tau), & \mathfrak{D}_3(T\tau) &= \mathfrak{D}_3(\tau),\\ \mathfrak{D}_4(S\tau) &= \mathfrak{D}_3(\tau), & \mathfrak{D}_4(T\tau) &= \mathfrak{D}_2(\tau),\end{aligned}$$

which imply that the $\mathfrak{D}_r(\tau)$ are modular functions for the given groups. The index of these groups in Γ is 3, and this is precisely the degree of the irreducible equation of $\mathfrak{D}_r(\tau)$ over $\mathbb{Q}(j(\tau))$ ■.

Consequently, equation (7) is a rational model for the modular curves $X_0(2)$, X_θ, $X^0(2)$.

In a similar way we can prove:

Proposition 3. *The symmetric invariants $G_{\pm 2}(\tau)$, $G_{\pm 3}(\tau)$ and $G_{\pm 4}(\tau)$ are* Hauptmoduln *for the modular curves associated to the congruence groups $\Gamma_0(4)$, $S^{-1}\Gamma^0(4)S$ and $\Gamma^0(4)$ respectively.*

3.1 Fourier Expansions

The Fourier expansions of the modular functions $\mathfrak{D}_r(\tau)$ begin with:

$$\mathfrak{D}_2(\tau) = \sum_{n\geq 0} d_n q^n = 2^{12}(q + 24q^2 + 300q^3 + 2624q^4 + 18126q^5 + \ldots),$$

$$\mathfrak{D}_3(\tau) = \frac{-1}{\sqrt{q}} - 24 - 276\sqrt{q} - 2048\,q - 11202\,q^{\frac{3}{2}} - 49152\,q^2 - 184024\,q^{\frac{5}{2}} + O(q)^3,$$

$$\mathfrak{D}_4(\tau) = \frac{1}{\sqrt{q}} - 24 + 276\sqrt{q} - 2048\,q + 11202\,q^{\frac{3}{2}} - 49152\,q^2 + 184024\,q^{\frac{5}{2}} + O(q)^3. \tag{8}$$

In figure 1 we have plotted the logarithmic height of the coefficients d_n and c_n of the Fourier series of $\mathfrak{D}_2(\tau)$ and $j(\tau)$ respectively. It is apparent that the d_n are really smaller than the c_n:

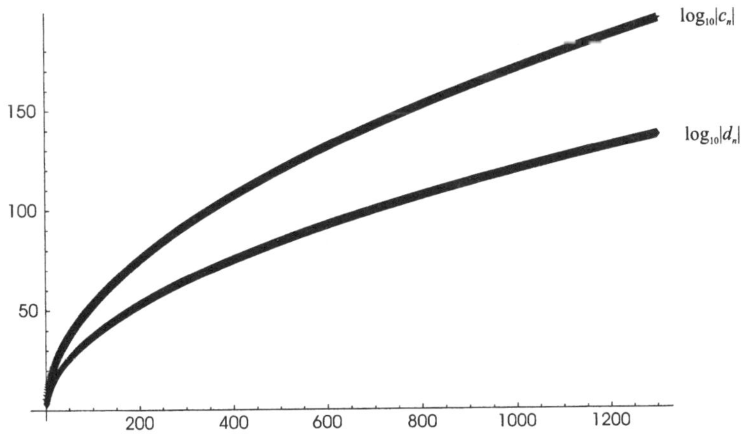

Fig. 1.

The series expansions of $\mathfrak{D}_r(\tau)$ show that these invariants will be smaller in absolute value than the j-invariant, but in order to compute the values of these functions one should use their expressions (7) in terms of the Thetanullwerte, or use the following relations

$$\mathfrak{D}_1(\tau) = f_2(\tau)^{24} = 2^{12}\frac{\Delta(2\tau)}{\Delta(\tau)},$$

$$\mathfrak{D}_2(\tau) = -f(\tau)^{24} = \frac{\Delta(\frac{\tau+1}{2})}{\Delta(\tau)},$$

$$\mathfrak{D}_3(\tau) = f_1(\tau)^{24} = \frac{\Delta(\frac{\tau}{2})}{\Delta(\tau)},$$

where

$$f(\tau) = e^{-\pi i/24}\frac{\eta\left(\frac{\tau+1}{2}\right)}{\eta(\tau)}, \qquad f_1(\tau) = \frac{\eta\left(\frac{\tau}{2}\right)}{\eta(\tau)}, \qquad f_2(\tau) = \sqrt{2}\frac{\eta(2\tau)}{\eta(\tau)},$$

are the Schläfli functions, $\eta(\tau) = e^{2\pi i\tau/24}(1 + \sum_{n\geq 1}(-1)^n(q^{n(3n-1)/2} + q^{n(3n+1)/2}))$ is Dedekind's eta function, and $\Delta(\tau) = \eta(\tau)^{24}$ is the classical discriminant modular form.

4 Ring Class Fields

We recall the construction of ray class fields of quadratic imaginary extensions of $\mathbb{Q}$. For the remainder of the paper, $K = \mathbb{Q}(\sqrt{D})$ will be a quadratic imaginary number field of discriminant $D < 0$, with ring of integers $\mathcal{O}_K = \mathbb{Z}[\tau]$ and class number $h = h(K)$. We take an elliptic curve

$$E : Y^2 = X^3 + AX + B$$

in Weierstrass form with complex multiplication by $\mathcal{O}_K$. We assume that $A, B \in K(j)$, where j is the j-invariant of E, which generates the Hilbert class field of K. The Weber function on E is defined as:

$$h((x,y)) = \begin{cases} x & \text{if } j(\tau) \neq 0, 1728, \\ x^2 & \text{if } j(\tau) = 1728, \\ x^3 & \text{if } j(\tau) = 0. \end{cases}$$

Theorem 2. *([13], [12]) Given an integral ideal $\mathfrak{a}$ of $\mathcal{O}_K$, the ray class modulo $\mathfrak{a}$ is $K(j(E), h(E[\mathfrak{a}]))$.*

We shall use this result to express the ring class field $K_{\mathcal{O}_2}$ of the order $\mathcal{O}_2 = \mathbb{Z}[2\tau]$ of conductor 2 in K in terms of our invariants $\mathfrak{D}_r(\tau)$.

First of all, we note that

$$Y^2 = X^3 - \frac{(\mathfrak{D}_r(\tau)+16)^3}{48\,(\mathfrak{D}_r(\tau)-8)^2\,(\mathfrak{D}_r(\tau)+64)}X + \frac{(\mathfrak{D}_r(\tau)+16)^3}{864\,(\mathfrak{D}_r(\tau)-8)^2\,(\mathfrak{D}_r(\tau)+64)}$$

is an elliptic curve with j-invariant $j(\tau) = \frac{(\mathfrak{D}_r(\tau)+16)^3}{\mathfrak{D}_r(\tau)}$ defined over $\mathbb{Q}(j(\tau))$. The X-coordinates of the non-trivial 2-torsion points of this curve are

$$\frac{\mathfrak{D}_r(\tau)+16}{12\,(\mathfrak{D}_r(\tau)-8)}, \frac{-(\mathfrak{D}_r(\tau)+16)\left(\mathfrak{D}_r(\tau)+64 \pm 3\sqrt{\mathfrak{D}_r(\tau)(\mathfrak{D}_r(\tau)+64)}\right)}{24\,(\mathfrak{D}_r(\tau)-8)\,(\mathfrak{D}_r(\tau)+64)}.$$

For $D=-3$, $j=0$ the three invariants $\mathfrak{D}_r(\tau)$ take the value -16, and we get $K_{\mathcal{O}_2} = K$. For $D=-4$, $j=1728$ the values of the $\mathfrak{D}_r(\tau)$ are 8 and -64, and again $K_{\mathcal{O}_2} = K = K(\mathfrak{D}_r(\tau))$. In the general case, we have:

$$\begin{aligned} K_{\mathcal{O}_2} &= K\left(j(\tau), \tfrac{\mathfrak{D}_r(\tau)+16}{\mathfrak{D}_r(\tau)-8}, \sqrt{\tfrac{\mathfrak{D}_r(\tau)+64}{\mathfrak{D}_r(\tau)}}\right) \\ &= K\left(j(\tau)-1728, \tfrac{\mathfrak{D}_r(\tau)+16}{\mathfrak{D}_r(\tau)-8}, \sqrt{\tfrac{\mathfrak{D}_r(\tau)+64}{\mathfrak{D}_r(\tau)}}\right) \\ &= K\left(\tfrac{(\mathfrak{D}_r(\tau)-8)^2(\mathfrak{D}_r(\tau)+64)}{\mathfrak{D}_r(\tau)}, \tfrac{\mathfrak{D}_r(\tau)+16}{\mathfrak{D}_r(\tau)-8}, \sqrt{\tfrac{\mathfrak{D}_r(\tau)+64}{\mathfrak{D}_r(\tau)}}\right) \\ &= K\left((\mathfrak{D}_r(\tau)-8)^2, (\mathfrak{D}_r(\tau)+16)(\mathfrak{D}_r(\tau)-8), \sqrt{\tfrac{\mathfrak{D}_r(\tau)+64}{\mathfrak{D}_r(\tau)}}\right) \\ &= K\left(\mathfrak{D}_r(\tau), \sqrt{\tfrac{\mathfrak{D}_r(\tau)+64}{\mathfrak{D}_r(\tau)}}\right). \end{aligned}$$

Since the discriminant of the polynomial $F_{\mathfrak{D},j}(X) = X^3+48X^2+(768-j(\tau))X+2^{12}$ is precisely

$$4j(\tau)^2(j(\tau)-1728) = \frac{4(\mathfrak{D}_r(\tau)-8)^2(\mathfrak{D}_r(\tau)+16)^6(\mathfrak{D}_r(\tau+64))}{\mathfrak{D}_r(\tau)^3},$$

we see that $K_{\mathcal{O}_2}$ is the splitting field of $F_{\mathfrak{D},j}(X)$ and we have proved:

Theorem 3. *For any fundamental discriminant* $D < 0$, $K_{\mathcal{O}_2} = K(\mathfrak{D}_2(\tau), \mathfrak{D}_3(\tau))$.

Corollary 1. *$K(\mathfrak{D}_r(\tau))$ is an abelian extension of $K(j(\tau))$.*

Lemma 1. *Let $\mathcal{O}_D$ and $\mathcal{O}_{4D}$ be the imaginary quadratic orders of discriminant D and $4D$ respectively, with $D < -4$. Then:*

$$h(\mathcal{O}_{4D}) = \begin{cases} h(\mathcal{O}_D) & \text{if } D \equiv \pm 1 \pmod 8, \\ 2h(\mathcal{O}_D) & \text{if } D \equiv 0 \pmod 4, \\ 3h(\mathcal{O}_D) & \text{if } D \equiv \pm 5 \pmod 8. \end{cases}$$

Proof. The formula follows from the comparison of both class numbers with the class number of the maximal order $\mathcal{O}_K$ of the common field of fractions K of $\mathcal{O}_D$ and $\mathcal{O}_{4D}$, by means of the well-known formula relating the class number of the order $\mathcal{O}$ of conductor t with the class number of $\mathcal{O}_K$ (cf. [4, p. 146]):

$$h(\mathcal{O}) = \frac{h(\mathcal{O}_K)t}{[\mathcal{O}_K^* : \mathcal{O}^*]} \prod_{p|t} \left(1 - \left(\frac{D}{p}\right)\frac{1}{p}\right). \blacksquare$$

Combining this lemma with theorem 3 and the decomposition of the discriminant of $F_{\mathfrak{D},j}(X)$ we have a very simple proof of the following well-known result ([6]):

Corollary 2. *Let* $\tau = \frac{b+\sqrt{D}}{2}$ *a quadratic number of discriminant* $0 < D \equiv \pm 1 \pmod 4$. *The singular value* $j(\tau) - 1728$ *is a square in* $K(j(\tau))$.

5 The Polynomial $H_D[\mathfrak{D}](X)$

We fix a system of representatives of the classes of quadratic forms of discriminant D, and compute the corresponding quadratic numbers $\tau_1, \dots, \tau_h \in K$, so that the Hilbert class polynomial is

$$H_D(X) = (X - j(\tau_1)) \cdots (X - j(\tau_h)) = X^h + c_{h-1}X^{h-1} + \cdots + c_0 \in \mathbb{Z}[X].$$

We want now to study the polynomial:

$$H_D[\mathfrak{D}](X) = \prod_{r,k}(X - \mathfrak{D}_r(\tau_k)) = \prod_{k=1}^{h}(X^3 + 48X^2 + (768 - j(\tau_k))X + 2^{12}) = \sum_{k=0}^{3h} d_k X^k.$$

The second expression shows that $H_D[\mathfrak{D}](X) \in \mathbb{Z}[X]$ since its coefficients are symmetric functions of $j(\tau_1), \dots, j(\tau_h)$. It is not a class polynomial in a strict sense, since it is not irreducible in many cases. But it has a number of properties which make it interesting. For instance, some of their coefficients are independent of D and $j(\tau_k)$:

$$H_D[\mathfrak{D}](X) = X^{3h} + 48hX^{3h-1} + \cdots + 2^{12h}.$$

The a-priori knowledge of these coefficients of $H_D[\mathfrak{D}](X)$ is very useful for the numerical computation of the remaining coefficients: one can use them either to shorten computations or to check the validity of the results.

The remaining coefficients of $H_D[\mathfrak{D}](X)$ can be related to the coefficients of the Hilbert class polynomial. For instance:

$$\begin{aligned} d_1 &= 2^{20}3h + 2^{12}c_{h-1}, \\ d_2 &= 2^{12(h-2)}\left(2^{16}3h + 2^{16}3^2\frac{h(h-1)}{2} + 2^8 3(h-1)c_{h-1} + c_{h-2}\right), \\ &\vdots \\ d_{3h-2} &= 2^8 3^2 \frac{h(h-1)}{2} + 2^8 3h + c_{h-1}. \end{aligned} \tag{9}$$

All these relations have integral coefficients, an important fact that since it implies that they remain valid over any finite field of odd characteristic.

We now proceed to study the factorization of $H_D[\mathfrak{D}](X)$ over $\mathbb{Q}$.

Lemma 2. *Let* $\tau_0 = \frac{D+\sqrt{D}}{2}$ *and* $j_0 = j(\tau_0)$. *If* $D \equiv 0 \pmod 4$ *then* $\mathfrak{D}_2(\tau_0)$, $\mathfrak{D}_3(\tau_0)$, $\mathfrak{D}_4(\tau_0) \in \mathbb{R}$. *If* $D \equiv 1 \pmod 4$ *only one of the three* $\mathfrak{D}_r(\tau_0)$ *is real.*

Proof. The polynomial $F_{\mathfrak{D},j}(X)$ with $j \in \mathbb{R}$ has three real roots if and only if $j > 0$. It is well known that $j(-1/2 + it) < 0$ for $t > 1/2$, $j\left(\frac{-1+\sqrt{(-3)}}{2}\right) = 0$ and $j(it) > 0$ for $t \geq 1$. ■

Lemma 3. *If* $\mathfrak{D}_r(\tau_0) \in K(j_0) \cap \mathbb{R}$ *then the irreducible polynomial of* $\mathfrak{D}_r(\tau_0)$ *over* $\mathbb{Q}$ *has degree* h.

Corollary 3.

a) For $D \equiv 1 \pmod 8$, $H_D[\mathfrak{D}](X)$ *factors over* $\mathbb{Q}[X]$ *as a product* $f_h(X)f_{2h}(X)$, *with* $f_k(X) \in \mathbb{Z}[X]$ *irreducible polynomial of degree* k. *Both* $f_h(X)$ *and* $f_{2h}(X)$ *split linearly in* $K(j_0)$.

b) For $D \equiv 0 \pmod 4$, $H_D[\mathfrak{D}](X) = f_h(X)f_{2h}(X)$, *with* $f_k(X) \in \mathbb{Z}[X]$ *irreducible polynomial of degree* k. *The polynomial* $f_h(X)$ *splits linearly in* $K(j_0)$ *and* $f_{2h}(X)$ *splits as a product of* h *irreducible polynomials of degree 2.*

c) For $D \equiv 5 \pmod 8$, $H_D[\mathfrak{D}](X)$ *is irreducible in* $\mathbb{Z}[X]$. *It splits as a product of* h *irreducible polynomials of degree 3 over* $K(j_0)$.

6 Computation of Class Polynomials

The Hilbert class polynomial $H_D(X)$ has been widely studied both from the theoretical and numerical viewpoint ([2], [6], [8], [14], [7]). The construction of class polynomials with smaller coefficients for defining Hilbert class fields has also been worked a lot ([3], [5], [10]). However, for practical applications (cryptography, primality proving) it is necessary to find the roots modulo p of the Hilbert class polynomial in order to build elliptic curves with complex multiplication. The available methods ([5]) derive these roots from the reduction modulo p of a simpler class polynomial. We don't pretend to improve these algorithms, but to introduce the use of the invariants $\mathfrak{D}_r(\tau)$ in this problem. While the corresponding class polynomials are not the smallest possible, their properties make them really manageable.

Expressions (9) show that one can derive easily the Hilbert class polynomial $H_D(X)$ from the polynomial $H_D[\mathfrak{D}](X)$. Unfortunately, these expressions also suggest that we can expect the coefficients d_h to be of the same order of magnitude as the c_h, so that it is not a good strategy to use the $H_D[\mathfrak{D}](X)$ directly to compute the Hilbert class polynomials. On the other hand, the series expansions (8) provide very good bounds of the size of the roots $\mathfrak{D}_r(\tau_k)$ of $H_D[\mathfrak{D}](X)$. For instance, the highest absolute value of the roots of $H_D[\mathfrak{D}](X)$ is approximately $\exp(\pi\sqrt{|D|}/2)$, which is more or less the square root of the largest root of $H_D(X)$. Whenever $D \equiv 1 \pmod 8$ or $D \equiv 0 \pmod 4$ the polynomial $H_D[\mathfrak{D}](X)$ has an irreducible factor $f_h(X) \in \mathbb{Q}[X]$ of degree h. The size of the coefficients of these factors is at most the square root of the coefficients of the Hilbert class polynomial (in general they are quite smaller), so that they provide nice class polynomials. We will show with a couple of examples that these factors allow an efficient computation of Hilbert class polynomials. Moreover, the use of the $\mathfrak{D}_r(\tau)$ for the computation of the Hilbert class polynomial produces simultaneously a polynomial defining the ring class field $K_{\mathcal{O}_2}$.

6.1 Computation of the Hilbert Class Polynomial for the Discriminant $D = -60$

We shall find the irreducible factors of the polynomial $H_{-15}[\mathfrak{D}](X)$ and derive from them the Hilbert class polynomial $H_{-15}(X)$. By corollary 3, the decomposition of $H_{-15}[\mathfrak{D}](X)$ in this case must be $H_{-15}[\mathfrak{D}](X) = f_2(X)f_4(x)$ with $f_k(X) \in \mathbb{Q}[X]$ a monic polynomial of degree k. A system of representatives of quadratic forms of discriminant -15 is $g_1 = X^2+XY+4Y^2$, $g_2 = 2X^2+XY+2$, with roots $\tau_1 = \frac{-1+\sqrt{-15}}{2}$ and $\tau_2 = \frac{-1+\sqrt{-15}}{4}$. We know that $j(\tau_1) < 0$ and $j(\tau_2) \notin \mathbb{R}$ so that only two of the $\mathfrak{D}_r(\tau_j)$ are real. Moreover, it can be seen that $\mathfrak{D}_2(\tau)$ takes real values on the half line $\tau = -\frac{1}{2} + it$, so that we can assume that $\mathfrak{D}_2(\tau_1) = -0.021286\ldots$ is a root of $f_2(X)$. We write

$$f_2(X) = (X - \mathfrak{D}_2(\tau_1))(X - 2^r/\mathfrak{D}_2(\tau_1))$$

and look for the value of r which produces integral coefficients for this polynomial. We find $r = 0$ and

$$f_2(X) = X^2 + 47X + 1.$$

From the equality

$$f_2(X)f_4(X) = H_{-15}[\mathfrak{D}](X)$$

we obtain a linear system in the coefficients of $f_4(X)$ and $H_{-15}[\mathfrak{D}](X)$, whose solution gives:

$$f_4(X) = X^4 + 49X^3 + 192561X^2 + 200704\,X + 2^{24},$$

$$H_{-15}(X) = X^2 + 191025X - 121287375.$$

The ring class field of the quadratic order of discriminant -60 is generated over $\mathbb{Q}(\sqrt{D})$ by any root of the polynomial $f_4(X)$. Note that we have needed only 6 significant figures in $\mathfrak{D}_2(\tau_1)$ for the whole computation, while 17 figures of $j(\tau_1)$ and $j(\tau_2)$ are necessary to compute the Hilbert class polynomial directly.

Similar tricks can be applied to compute the irreducible factors of the polynomial $H_D[\mathfrak{D}](X)$ for any discriminant $D \equiv 1 \pmod 8$ of $D \equiv 0 \pmod 4$ with class number 2.

6.2 Computation of the Hilbert Class Polynomial for the Discriminant $D = -1588$

We know in this case that $H_{-1588}[\mathfrak{D}](X)$ splits as the product of two irreducible factors $f_6(X), f_{12}(X) \in \mathbb{Z}[X]$. We could study the action of the Galois group of the extension $K_{\mathcal{O}_2}/K$ on the $\mathfrak{D}_r(\tau)$, and then precise the roots of the polynomials $f_6(X), f_{12}(X)$. Anyway, it is not strictly necessary, at least for low class numbers.

We begin finding the six reduced forms of discriminant -1588 and their corresponding roots τ_k:

$$\tau_1 = \frac{\sqrt{-397}}{2}, \quad \tau_2 = \frac{-1+\sqrt{-397}}{2}, \; \tau_3 = \frac{-3+\sqrt{-397}}{7},$$

$$\tau_4 = \frac{3+\sqrt{-397}}{7}, \; \tau_4 = \frac{-3+\sqrt{-397}}{14}, \; \tau_6 = \frac{-3+\sqrt{-397}}{14}.$$

We now compute the six values $\mathfrak{D}_2(\tau_k)$ with $\exp(\pi\sqrt{397}) + 10 \approx 40$ significant digits. (We may take into account that $\mathfrak{D}_2\left(\frac{b+\sqrt{-D}}{2a}\right) = \overline{\mathfrak{D}_2\left(\frac{-b+\sqrt{-D}}{2a}\right)}$ to save some time). The remaining values $\mathfrak{D}_3(\tau_k)$ and $\mathfrak{D}_4(\tau_k)$ are computed solving the system

$$\mathfrak{D}_3(\tau_k) + \mathfrak{D}_4(\tau_k) = -48 - \mathfrak{D}_2(\tau_k),$$
$$\mathfrak{D}_3(\tau_k)\mathfrak{D}_4(\tau_k) = -2^{12}/\mathfrak{D}_2(\tau_k).$$

We knew a priori that only four of the $\mathfrak{D}_r(\tau_k)$ should be real: $\mathfrak{D}_2(\tau_1)$, $\mathfrak{D}_3(\tau_1)$, $\mathfrak{D}_4(\tau_1)$ and $\mathfrak{D}_2(\tau_2)$. The last one and one of the $\mathfrak{D}_r(\tau_1)$ must be roots of the polynomial $f_6(X)$. We form the products $T_r = \mathfrak{D}_r(\tau_1)\mathfrak{D}_2(\tau_2)$. We now look for the product $\mathfrak{D}_{r_3}(\tau_3)\mathfrak{D}_{r_3}(\tau_4)\mathfrak{D}_{r_5}(\tau_5)\mathfrak{D}_{r_5}(\tau_6)$ of pairs of complex conjugates which multiplied by the T_r gives a power of 2. We find that

$$\mathfrak{D}_3(\tau_1)\mathfrak{D}_2(\tau_2)\mathfrak{D}_2(\tau_3)\mathfrak{D}_2(\tau_4)\mathfrak{D}_4(\tau_5)\mathfrak{D}_4(\tau_6) = 2^{36},$$

and we deduce that these are the roots of $f_6(X)$. Hence:

$$\begin{aligned} f_6(X) = {} & X^6 + 153110953822469070362689816X^5 - 5138592259654383232415544053760X^4 + \\ & 8944384305675598321768767469125632\,0X^3 - 21047673895544353719974068444200960X^2 + \\ & 25687755442455892467940464846176256X + 2^{36}. \end{aligned}$$

As in the previous example, we could now find the polynomials $f_{12}(X)$ and $H_{-1588}(X)$ solving a linear system. The direct computation of $H_{-1588}(X)$ from its roots $j(\tau_k)$ requires 90 significant digits for every root. The largest coefficient of $H_{-1588}(X)$ has 102 decimal figures, while the largest coefficient of f_6 has 35 digits. It is worth noting that if we would like to construct an elliptic curve over $\mathbb{F}_p$ ($p > 2$) with complex multiplication by the quadratic order of discriminant -1588, it is unnecessary to compute the Hilbert polynomial, since the roots of the polynomial $f_6(X)$ modulo p provide directly equations for such curves.

References

1. Atkin, O.,Morain, F., 'Elliptic curves and primality proving', Math. Comp. 61, 29-68, 1993.
2. Berwick, W.E.H, 'Modular invariants expressible in terms of quadratic and cubic irrationalities', Proc. London Math. Soc. 28, 53-69, 1928.
3. Cohen, H. *Advanced Topics in Computational Number Theory*, Graduate Texts in Math. 193, Springer-Verlag, 2000.
4. Cox, D. *Primes of the form $x^2 + ny^2$: Fermat, Class Field Theory and Complex Multiplication*, John Wiley & Sons, 1989.
5. Enge, A., Morain, F., 'Comparing Invariants for Class Fields of Imaginary Quadratic Fields', C. Fieker, D.R. Kohel, eds., *Algorithmic Number Theory*, Lecture Notes in Comput. Sci., 2369, 252-266, Springer-Verlag, 2002. 5th. International Symposium, ANTS-V, Sydney, Australia, July 2002.
6. Gross, B., Zagier, D., 'On singular moduli', J. Reine Angew. Math. 355, 191-220, 1985.
7. Gee, A., Stevenhagen, P., 'Generating class invariant using Shimura reciprocity', *Algorithmic number theory (Portland, 1998)*, Lecture Notes in in Comput Sci. 1423, pp.441-453, Springer, Berlin 1998.
8. Kaltofen, E., Yui, N., 'Explicit construction of the Hilbert class fields of imaginary quadratic fields by integer lattice reduction', *Number Theory, New York Seminars 1989-90*, Springer-Verlag, 1991, 149-202.
9. McKean, H., and Moll, V., *Elliptic curves: function theory, geometry, arithmetic*, Cambridge Univ. Press, 1999.
10. Schertz, R., 'Die singulären Werte der Weberschen Funktionen $f, f_1, f_2, \gamma_2, \gamma_3$', J. Reine Angew. Math. 286/287, 46-74, 1976.
11. Schertz, R., 'Weber's class invariants revisited', J. Théorie des Nombres de Bordeaux, no. 14, 325-343, 2002.
12. Shimura, G., *Introduction to the Arithmetic Theory of Automorphic Functions*, Publications of the Mathematical Society of Japan, 11, Princeton University Press, 1971.
13. Silverman, J. H., *Advanced Topics in the Arithmetic of Elliptic Curves*, Graduate Texts in Math. 151, Springer-Verlag, 1994.
14. Yui, N., Zagier, D., 'On the singular values of Weber modular functions', Math. of Computation 66, no. 220, 1645-1662, 1997.

An Algorithm for Computing Isomorphisms of Algebraic Function Fields

F. Hess

Technical University of Berlin, Faculty II,
Institute of Mathematics, Secr. MA8-1,
Straße des 17. Juni 136, 10623 Berlin, Germany

Abstract. We develop an algorithm for computing isomorphisms and automorphisms of algebraic function fields of transcendence degree one in characteristic zero and positive characteristic.

1 Introduction

Let F_1/k and F_2/k denote algebraic function fields of transcendence degree one. An isomorphism ϕ of F_1/k and F_2/k is an isomorphism of fields $\phi : F_1 \to F_2$ whose restriction to k is the identity map. The main objective of this paper is to develop an algorithm to compute one or all isomorphisms of F_1/k and F_2/k if these algebraic function fields are isomorphic and have genus greater than one. We think of F_1/k and F_2/k as the function fields of some suitable, explicitly given and not necessarily non-singular irreducible algebraic curves such as plane curves defined over k, and describe ϕ by its action on the corresponding coordinates or field generators. For the special case $F = F_1 = F_2$ the algorithm computes the elements of the automorphism group $\mathrm{Aut}_k(F)$ of F/k. We restrict to genus greater than one since otherwise the number of isomorphisms or the automorphism groups may be infinite and the task is related to finding unique models by computing rational points on curves over k, a hard and from the techniques of this paper quite different problem.

Assume that F_1 and F_2 are given as finite and separable extensions of an algebraic function field K/k of transcendence degree one. Since F_1 and F_2 can be obtained by adjoining roots of monic, irreducible and separable polynomials $f_1, f_2 \in K[t]$ to K respectively the computation of isomorphisms ψ of F_1/k and F_2/k which restrict to the identity on K can essentially be reduced to the computation of the roots of f_1 in F_2 or of f_2 in F_1. This latter task can in turn be solved by special Hensel lifting and reconstruction techniques, or by general polynomial factorization algorithms. A concrete algorithm for computing such isomorphisms of F_1 and F_2 and for computing the elements of $\mathrm{Aut}_K(F)$ for finite extensions F_1/K, F_2/K, F/K and $K = \mathbb{Q}(x)$ is given in [6]. The computation of $\mathrm{Aut}_k(F)$ for example cannot be done this way since the fixed field of $\mathrm{Aut}_k(F)$, which would be a candidate for K, is a priori not known.

In the following we need to define and compare various entities for the function fields F_1 and F_2. To facilitate this we use the subscripts (1), (2) for either

D.A. Buell (Ed.): ANTS 2004, LNCS 3076, pp. 263–271, 2004.

of the two cases and (α) for both (hence $\alpha \in \{1, 2\}$). Entities with no subscript are related to both or completely independent of the two fields. We are thus considering the function fields $F_{(1)}$, $F_{(2)}$ and the isomorphisms ϕ.

Throughout the paper $F_{(\alpha)}/k$ denotes an algebraic function fields of transcendence degree one and genus greater than one. Unless otherwise stated, k is assumed to be the exact constant field of $F_{(\alpha)}/k$ (ie. algebraically closed in $F_{(\alpha)}$) and perfect. Our algorithms rely on the ability to compute in $F_{(\alpha)}/k$ as a field and $k(x_{(\alpha)})$-vector space for $x_{(\alpha)}$ a separating element of $F_{(\alpha)}/k$, and to compute with places, divisors and Riemann-Roch spaces $\mathcal{L}(D_{(\alpha)}) = \{a \in F_{(\alpha)}^{\times} \mid (a) + D_{(\alpha)} \geq 0\} \cup \{0\}$ for divisors $D_{(\alpha)}$. In particular an algorithm for computing Weierstrass places is required. We refer to [3,4] for algorithmic aspects and to [8] for the theoretical background. Implementations of these algorithms are available in Kash [5] and Magma [1,2].

2 The Basic Idea

Assume that ϕ is an isomorphism of $F_{(1)}/k$ and $F_{(2)}/k$. We derive a number of necessary conditions for the existence of ϕ, which altogether will also be sufficient. This also leads to a test for $F_{(1)}/k$ and $F_{(2)}/k$ being not isomorphic.

First, the genus of $F_{(1)}/k$ and $F_{(2)}/k$ is equal and ϕ maps places $P_{(1)}$ of $F_{(1)}/k$ to places $P_{(2)}$ of $F_{(2)}/k$, preserving the degree and gap numbers and inducing k-linear isomorphisms of the vector spaces $\mathcal{L}(nP_{(1)})$ and $\mathcal{L}(nP_{(2)})$. In particular, places of $F_{(1)}/k$ of smallest degree are mapped to places of $F_{(2)}/k$ of smallest degree and Weierstrass places of $F_{(1)}/k$ are mapped to Weierstrass places of $F_{(2)}/k$. For global function fields (that is k a finite field) there are only finitely many places of any fixed degree, and in general there are only finitely many Weierstrass places. This restricts the possibilities $P_{(2)} = \phi(P_{(1)})$ to a finite number. We remark that both sets of places (smallest degree, Weierstrass) can be computed efficiently.

Let $P_{(1)}$ and $P_{(2)}$ be places of degree one with $P_{(2)} = \phi(P_{(1)})$. Let g be the genus of $F_{(1)}$ and $F_{(2)}$ and let $n_i \in \mathbb{Z}^{>0}$ for $1 \leq i \leq r \leq g+1$ be the first successive pole numbers greater than zero at $P_{(1)}$, equal to those at $P_{(2)}$, such that $\gcd\{n_i \mid 1 \leq i \leq r\} = 1$. Let $x_{(\alpha),i} \in \mathcal{L}(n_i P_{(\alpha)})$ be elements with $v_{P_{(\alpha)}}(x_{(\alpha),i}) = -n_i$. We define $x_{(\alpha),0} = 1$ and $n_0 = 0$. Obviously, $x_{(\alpha),i}$ is uniquely defined up to multiplication by elements from $k^{\times}$ and addition by k-linear combinations of the $x_{(\alpha),j}$ for $j < i$. As a result, $\phi(x_{(1),i}) = \sum_{j=0}^{i} \mu_{i,j} x_{(2),j}$ for suitable $\mu_{i,j} \in k$ and $\mu_{i,i} \neq 0$.

These observations are interesting because $F_{(\alpha)} = k(x_{(\alpha),1}, \ldots, x_{(\alpha),n_1})$, so ϕ is completely defined by the $\mu_{i,j}$. This equality holds because $P_{(\alpha)}$ is fully ramified and unramified over $k(x_{(\alpha),1}, \ldots, x_{(\alpha),n_1})$ due to the gcd condition. We let $I_{(\alpha)}$ be the kernel of the substitution homomorphism $k[t_1, \ldots, t_{n_1}] \to F_{(\alpha)}$, $t_i \mapsto x_{(\alpha),i}$. Thus generators of $I_{(1)}$ and $I_{(2)}$ define irreducible affine curves whose function fields are equal to $F_{(1)}$ and $F_{(2)}$ respectively. Since ϕ is an isomorphism, substituting $\sum_{j=0}^{i} \mu_{i,j} t_j$ for t_i yields an automorphism of $k[t_1, \ldots, t_{n_1}]$ which accordingly maps $I_{(1)}$ to $I_{(2)}$.

Turning things around assume now that the $\mu_{i,j}$ are indeterminates and $R = k[\{\mu_{i,j} | j \leq i\}]$. Then $I_{(1)}$ is also an ideal of $R[t_1, \ldots, t_{n_1}]$ and we define $I'_{(2)}$ to be the ideal of $R[t_1, \ldots, t_{n_1}]$ obtained from $I_{(1)}$ by substituting $\sum_{j=0}^{i} \mu_{i,j} t_j$ for t_i. Any solution $\mu_{i,j} \in k$ with $\mu_{i,i} \neq 0$ such that $I'_{(2)} = I_{(2)}$ yields an isomorphism of the coordinate rings $k[t_1, \ldots, t_{n_1}]/I_{(1)} \to k[t_1, \ldots, t_{n_1}]/I_{(2)}$ and thus an isomorphism ϕ of $F_{(1)}/k$ and $F_{(2)}/k$. It is actually sufficient to require only $I'_{(2)} \subseteq I_{(2)}$.

Summing up, our algorithm for computing one or all isomorphisms of $F_{(1)}/k$ and $F_{(2)}/k$ proceeds as follows. We first choose a suitable place of smallest degree or Weierstrass place $P_{(1)}$ of $F_{(1)}/k$ and compute all places $P_{(2)}$ which could correspond to $P_{(1)}$ under an isomorphism, respecting the condition of smallest degrees or Weierstrass with same gap sequences. Then we compute $I_{(1)}$, $I_{(2)}$ and $I'_{(2)}$ and solve for the $\mu_{i,j}$ with $\mu_{i,i} \neq 0$ such that $I'_{(2)} \subseteq I_{(2)}$. This gives all possible isomorphisms ϕ of $F_{(1)}/k$ and $F_{(2)}/k$.

This concludes the description of the basic idea. Of course, there are various problems to overcome. For one we need to find a convenient way to compute $I_{(1)}$ and $I_{(2)}$ and to check $I'_{(2)} \subseteq I_{(2)}$ in terms of the $\mu_{i,j}$. Another problem is that the number of the $\mu_{i,j}$ is roughly $g^2/2$ which makes the computation of $I'_{(2)}$ and finding a solution to $I'_{(2)} \subseteq I_{(2)}$ hard. To this end we essentially reduce the $\mu_{i,j}$ to two parameters. These issues are addressed in the following sections.

We finally remark that another way of computing isomorphisms is by using canonical curves. One advantage of this would be that the auxiliary places $P_{(1)}$ and $P_{(2)}$ could be avoided. Since these curves are uniquely determined up to linear transformation we could compare a canonical curve for $F_{(1)}/k$ with a generically linearly transformed canonical curve of $F_{(2)}/k$ similar as above. As is this would involve roughly g^2 indeterminates, and attempts to reduce this even larger number may again require the use of auxiliary places. Also, the computation of canonical curves is in general not very easy. Moreover, for our above strategy the number of the required $x_{(\alpha),i}$ can be considerably smaller than g, resulting in easier to handle affine curves, and it also works for hyperelliptic function fields. Therefore, we do not pursue the use of canonical curves in this paper.

3 Relating Affine Models

We now do not just choose the first r pole numbers n_i but let the n_i be special generators of the Weierstrass semigroup at $P_{(\alpha)}$ satisfying the following condition. We require that $n_j \not\equiv n_i \bmod n_1$ for $1 \leq i < j$ and $1 \leq j \leq r$. We observe that the n_i are uniquely determined, because n_1 is the smallest pole number greater than zero at $P_{(\alpha)}$. The $x_{(\alpha),i}$ are again chosen in $\mathcal{L}(n_i P_{(\alpha)})$ such that $v_P(x_{(\alpha),i}) = -n_i$.

Given the congruence inequality it is not difficult to see that the elements $1, x_{(\alpha),2}, \ldots, x_{(\alpha),r}$ are a $k[x_{(\alpha),1}]$-basis of the integral closure $\mathrm{Cl}(k[x_{(\alpha),1}], F)$. Because $[F : k(x_{(\alpha),1})] = n_1$ this implies $r = n_1$. Since $\mathrm{Cl}(k[x_{(\alpha),1}], F)$ is a free

$k[x_{(\alpha),1}]$-module we know that there are uniquely determined $\lambda_{(\alpha),i,j,\nu} \in k[t]$ such that $x_{(\alpha),i}x_{(\alpha),j} = \lambda_{(\alpha),i,j,1}(x_{(\alpha),1}) + \sum_{\nu=2}^{n_1} \lambda_{(\alpha),i,j,\nu}(x_{(\alpha),1})x_{(\alpha),\nu}$ for $2 \leq i,j \leq n_1$. Looking at the valuations $v_{P_{(\alpha)}}$ we obtain $\deg(\lambda_{(\alpha),i,j,\nu}) \leq (n_i+n_j-n_\nu)/n_1$, and equality must hold for at least one ν because of the incongruence relations of the n_i. An algorithm to compute these $\lambda_{(\alpha),i,j,\nu}$ is given in [3].

Let $I_{(\alpha)}$ be the kernel of the substitution homomorphism $k[t_1,\ldots,t_{n_1}] \to F_{(\alpha)}$, $t_i \mapsto x_{(\alpha),i}$. From the previous discussion we see that $I_{(\alpha)}$ has a nice set of generators of the form $t_it_j - \lambda_{(\alpha),i,j,1}(t_1) - \sum_{\nu=2}^{n_1} \lambda_{(\alpha),i,j,\nu}(t_1)t_\nu$ for $2 \leq i,j \leq n_1$. In particular, given any polynomial $f \in k[t_1,\ldots,t_{n_1}]$ we can reduce it modulo $I_{(\alpha)}$ to a polynomial of degree at most one in $t_2,\ldots,t_{n_1}$ by substituting terms t_it_j with $\lambda_{(\alpha),i,j,1}(t_1) + \sum_{\nu=2}^{n_1} \lambda_{(\alpha),i,j,\nu}(t_1)t_\nu$.

Since we are working with a possibly smaller number of $x_{(\alpha),i}$ than in the previous section these elements do not necessarily realize all pole numbers anymore. Because of the congruence conditions on the n_i this can now be done using elements of the form $x_{(\alpha),1}^j x_{(\alpha),i}$, and a basis of $\mathcal{L}(n_rP)$ is given by $\{x_{(\alpha),1}^j x_{(\alpha),i} \,|\, jn_1 + n_i \leq n_r\}$. It follows that the transformation of Section 2 takes the form $\phi(x_{(1),i}) = \sum_{j=0}^{i} \mu_{i,j}(x_{(2),1})x_{(2),j}$ for suitable $\mu_{i,j} \in k[t]$ with $\deg(\mu_{i,j}) \leq (n_i - n_j)/n_1$ and $\mu_{i,i} \neq 0$. Reducing the generators of $I'_{(2)}$ modulo $I_{(2)}$ in $R[t_1,\ldots,t_{n_1}]$ and equating the coefficients of the t_i and 1 with zero finally yields equations for the coefficients of the $\mu_{i,j}$.

4 Relating Expansions at a Place Depending on Two Parameters

We now discuss how to relate p-adic expansions at $P_{(1)}$ and $P_{(2)}$ in a meaningful way, depending on only two parameters.

We let $\pi_{(\alpha)}$ denote a local uniformizer at $P_{(\alpha)}$. Computing expansions in terms of $\pi_{(\alpha)}$ we may embed $F_{(\alpha)} \subseteq k((\pi_{(\alpha)}))$ and the isomorphism ϕ extends to an isomorphism of $k((\pi_{(1)}))$ and $k((\pi_{(2)}))$. Local uniformizers are not uniquely determined and we cannot hope for $\phi(\pi_{(1)}) = \pi_{(2)}$ if $\pi_{(1)}$ and $\pi_{(2)}$ are chosen independently. However, $\phi(\pi_{(1)})$ is a local uniformizer at $P_{(2)}$ and there are $c_i \in k$, $c_1 \neq 0$ such that $\phi(\pi_{(1)}) = \sum_{i=1}^{\infty} c_i\pi_{(2)}^i$. Of course, the c_i are unkown to us. Looking at $x_{(1),1}$ and $x_{(2),1}$ we see that $\phi(x_{(1),1}) = ax_{(2),1}+b$ for some $a,b \in k$, $a \neq 0$, which are also unknown to us. Furthermore, $x_{(\alpha),1} = \sum_{i=0}^{\infty} d_{(\alpha),i}\pi_{(\alpha)}^{i-n_1}$ for $d_{(\alpha),i} \in k$, $d_{(\alpha),0} \neq 0$. These coeffcients can be computed since $x_{(\alpha),1}, \pi_{(\alpha)}$ are known and lie in the same field respectively. Putting things together relates the c_i to a, b and gives the equation

$$\sum_{i=0}^{\infty} d_{(1),i}\phi(\pi_{(1)})^i = a\phi(\pi_{(1)})^{n_1}/\pi_{(2)}^{n_1} \sum_{i=0}^{\infty} d_{(2),i}\pi_{(2)}^i + b\phi(\pi_{(1)})^{n_1}. \tag{1}$$

We want to solve this equation for $\phi(\pi_{(1)})$ and recursively for c_1, c_2, ... Let $n = n_1$. Equating the coeffcients of $\pi_{(2)}^i$ for $i = 0$ gives $d_{(1),0} = ac_1^n d_{(2),0}$, hence $c_1^n = d_{(1),0}(d_{(2),0}a)^{-1}$. Write $n = p^{r_n}n'$ with $n' \neq 0 \bmod p$, where p denotes the characteristic of k. Let $s \geq 1$ be such that $d_{(1),s} \neq 0$, and write $s = p^{r_s} s'$ with $s' \neq 0 \bmod p$. For $p = 0$ we let $n = n'$, $s = s'$ and $r_n = r_s = 0$. The terms of the smallest degree in $\pi_{(2)}$ involving c_j are of the form $d_{(1),s}(s' c_1^{s'-1} c_j)^{p^{r_s}} \pi_{(2)}^{s+(j-1)p^{r_s}}$, $a(n' c_1^{n'-1} c_j)^{p^{r_n}} \pi_{(2)}^{(j-1)p^{r_n}}$ and $b(n' c_1^{n'-1} c_j)^{p^{r_n}} \pi_{(2)}^{n+(j-1)p^{r_n}}$ for the three main expressions in equation (1) respectively, where $s+(j-1)p^{r_s}$ is minimal with $d_{(1),s} \neq 0$. As a result, using $a = d_{(1),0}(c_1^n d_{(2),0})^{-1}$, there are monic $f_j \in k[c_1, c_1^{-1}][t]$ consisting only of p-power terms in t and $g_j \in k[b, c_1, c_1^{-1}, c_2, c_3, \ldots, c_{j-1}]$ for every $j \geq 2$ such that equation (1) implies $f_j(c_j) = g_j$. Regarding a, b and the c_i as indeterminates we define $R_{a,b} = k[a, b, c_1, c_1^{-1}, c_2, c_3, \ldots]/I_{a,b}$ where $I_{a,b}$ is the ideal generated by $d_{(1),0} - a c_1^n d_{(2),0}$ and $f_j(c_j) - g_j$ for $j \geq 2$ and obtain the generic embedding $\phi_{a,b} : k((\pi_{(1)})) \to R_{a,b}((\pi_{(2)}))$ since the image of $\pi_{(1)}$ under $\phi_{a,b}$ is invertible in $R_{a,b}$. Now $\phi_{a,b}$ specializes to ϕ if we substitute the correct elements of k for a, b and the c_i corresponding to ϕ.

For $n \neq 0 \bmod p$ or $p = 0$ the terms of the smallest degree in $\pi_{(2)}$ involving c_j are $a(n' c_1^{n'-1} c_j)^{p^{r_n}} \pi_{(2)}^{(j-1)p^{r_n}} = a n c_1^{n-1} c_j \pi_{(2)}^{j-1}$, and the f_j are hence all linear. This means that c_1 is an n-th root depending on a and that all other c_i are uniquely determined by c_1 and b. For $n = 0 \bmod p$ a more detailed analysis of the f_j shows that there can be at most n solution vectors $(c_i)_{i \geq 1}$ satisfying equation (1). The $(c_i)_{i \geq 1}$ thus depend also only on a, b up to finitely many possibilities. Looking at higher powers $\pi_{(2)}$ may give (usually gives) additional linear conditions on the c_j if the corresponding $d_{(1),i}$ are not zero. We remark that this strategy is not efficient if the series expansions of $x_{(1),1}$ and $x_{(2),1}$ are such that a larger number of the f_j do not have degree one ($n = 0 \bmod p$ necessarily).

A particularly easy form of the powers of $\phi(\pi_{(1)})$ in terms of $\pi_{(2)}$ can be achieved for $n \neq 0 \bmod p$ or $p = 0$, if we choose $\pi_{(1)}$ and $\pi_{(2)}$ in a special way. Let $h_{(\alpha)} = t^n - 1/x_{(\alpha),1}$. Since $v_{P_{(\alpha)}}(1/x_{(\alpha),1}) = n_1$ it follows that $h_{(\alpha)}$ has a root $\tilde{\pi}_{(\alpha)} \in k[[\pi_{(\alpha)}]]$, and $\tilde{\pi}_{(\alpha)}$ is a local uniformizer. If we require $\tilde{\pi}_{(\alpha)} = \pi_{(\alpha)} + O(\pi_{(\alpha)})$ then $\tilde{\pi}_{(\alpha)}$ is uniquely determined by $x_{(\alpha),1}$. Inverting the representation of $\tilde{\pi}_{(\alpha)}$ in terms of $\pi_{(\alpha)}$ gives a representation of $\pi_{(\alpha)}$ in terms of $\tilde{\pi}_{(\alpha)}$, and we may hence write everything in terms of $\tilde{\pi}_{(\alpha)}$ or equivalently embed $F_{(\alpha)} \subseteq k((\tilde{\pi}_{(\alpha)}))$. Also, ϕ extends to an isomorphism of $k((\tilde{\pi}_{(1)}))$ and $k((\tilde{\pi}_{(2)}))$ but again $\phi(\tilde{\pi}_{(1)}) = \tilde{\pi}_{(2)}$ is not necessarily true. Equation 1 then simplifies to the following equation,

$$1 = a\phi(\tilde{\pi}_{(1)})^n / \tilde{\pi}_{(2)}^n + b\phi(\tilde{\pi}_{(1)})^n. \tag{2}$$

With $\phi(\tilde{\pi}_{(1)}) = \tilde{c}_1 \tilde{\pi}_{(2)} + O(\tilde{\pi}_{(2)}^2)$ and $\tilde{c}_1^n = 1/a$ equation (2) yields

$$\phi(\tilde{\pi}_{(1)})^r = \sum_{i=0}^{\infty} \rho_i b^i \tilde{c}_1^{r+in} \tilde{\pi}_{(2)}^{r+in} \tag{3}$$

where $r \neq 0$ and $\rho_i \in k$. This explicit form will be used in the next section.

5 Relating Affine Models at a Place Depending on Two Parameters

We now want to reduce the number of indeterminate coefficients in the polynomials $\mu_{i,j}$ to basically two indeterminates.

Using the notation of the previous section we have $\mu_{1,1} = a$, $\mu_{1,0} = b$ and $\mu_{0,0} = 1$. A local uniformizer at $P_{(\alpha)}$ is given by $\pi_{(\alpha)}$, and for $n_1 \neq 0 \bmod p$ or $p = 0$ we may assume that $\pi_{(\alpha)}$ is an n_1-th root of $1/x_{(\alpha),1}$.

For every pole number d_i of $P_{(\alpha)}$ let $w_{(\alpha),i} = x_{(\alpha),1}^{\nu} x_{(\alpha),j}$ where $d_i = \nu n_1 + n_j$. As mentioned the $w_{(\alpha),0}, \ldots, w_{(\alpha),i}$ form a basis of $\mathcal{L}(d_i P_{(\alpha)})$ because of the congruence relations of the n_j. We can compute the expansions $w_{(\alpha),j} = \sum_{i=-d_j}^{\infty} \rho_{(\alpha),i,j} \pi_{(\alpha)}^i$ up to any precision. However, precision $O(\pi_{(\alpha)})$ will be sufficient. Using a Gaussian elimination procedure we may assume that $\rho_{(\alpha),i,\nu} = 0$ for all $i = -d_j$ and $\nu > j \geq 0$ and $\rho_{(\alpha),-d_j,j} = 1$ for all $0 \leq j \leq d_1$: For decreasing j we simply eliminate the $-d_j$-th coefficients in the expansions of $w_{(\alpha),\nu}$ by replacing $w_{(\alpha),\nu}$ with $w_{(\alpha),\nu} - (\rho_{(\alpha),-d_j,\nu}/\rho_{(\alpha),-d_j,j}) w_{(\alpha),\nu}$ for all $\nu > j$. Finally, we replace $w_{(\alpha),j}$ by $(1/\rho_{(\alpha),-d_j,j}) w_{(\alpha),j}$. We now define $x_{(\alpha),j} = w_{(\alpha),i}$ where $n_j = d_i$. The resulting $x_{(\alpha),j}$ are then uniquely determined depending on the chosen local uniformizer.

In the two fields situation we assume that the $x_{(2),i}$ have been transformed using this Gaussian elimination procedure. We now know that the Gaussian elimination for $\phi(x_{(1),i})$ would yield the $x_{(2),i}$, but the $\phi(x_{(1),i})$ are unknown to us. We can however perform this Gaussian elimination for the $\phi_{a,b}(x_{(1),i})$ over $R_{a,b}$ in a generic way. Note that the "leading" coefficients of $\phi_{a,b}(x_{(1),1}^{\nu} x_{(1),i})$ are basically powers of c_1 and are hence invertible in $R_{a,b}$. The Gaussian elimination procedure thus precisely yields the $\mu_{i,j}$ as elements of $R_{a,b}[t]$. Since we can work with precision $O(\pi_{(2)})$, we have only to deal and to compute with elements of $R_{a,b}$ involving c_i for $i \leq n_r + 1$.

Taking up the strategy of Section 3 (see in particular its end) and clearing c_1 from the denominators in the coefficients of the $\mu_{i,j}$ and in $I_{a,b}$ yields equations in a, b and the c_i with $i \leq n_1 + 1$. The corresponding ideal in $k[a, b, c_1, \ldots, c_{n_1+1}]$ is zero dimensional, since there can only be finitely many solutions for a, b and the c_i can only assume finitely many different values given a, b. Thus the intersection with $k[a]$ and $k[a, b]$ also results in zero dimensional ideals, and the possible values of a and b and then c_i can be computed from this.

For $n_1 \neq 0 \bmod p$ or $p = 0$ things are much more explicit. Using the special local uniformizer we perform the Gaussian elimination on the $x_{(1),i}$ and the $x_{(2),i}$. Because of equation 2 this special form is almost preserved by ϕ, namely we have that $\phi(x_{(\alpha),i}) = c_1^{n_i} x_{(2),i}$ with $c_1^n = 1/a$ for $i \geq 2$ and $\phi(x_{(1),1}) = a x_{(2),1} + b$. Then

$$\phi(x_{(1),i} x_{(1),j}) = \lambda_{(1),i,j,1}(a x_{(2),1} + b) + \sum_{\nu=2}^{n_1} \lambda_{(1),i,j,\nu}(a x_{(2),1} + b) c_1^{n_\nu} x_{(2),\nu},$$

$$\phi(x_{(1),i}) \phi(x_{(1),j}) = c_1^{n_i+n_j} \lambda_{(2),i,j,1}(x_{(2),1}) + \sum_{\nu=2}^{n_1} c_1^{n_i+n_j} \lambda_{(2),i,j,\nu}(x_{(2),1}) x_{(2),\nu}.$$

Combining the right hand sides via $\phi(x_{(1),i}x_{(1),j}) = \phi(x_{(1),i})\phi(x_{(1),j})$ yields easy equations for a, b and c_1.

6 Computing the Isomorphisms

The choice of the sets of places of $F_{(1)}/k$ and $F_{(2)}/k$ which must be mapped to each other under any isomorphism ϕ depends mainly on the constant field and the genus. For constant fields other than finite fields we always consider subsets of the set of Weierstrass places of $F_{(1)}/k$ and $F_{(2)}/k$ of smallest cardinality, where the places have a particular common gap sequence. The number of such places can be quite high, a lower and upper bound in characteristic zero are given by $2(g+1)$ and $(g-1)g(g+1)$ and in general by $O(g^3)$. This leads to roughly g^2 up to g^6 comparisons of places $P_{(1)}$ and $P_{(2)}$. For finite fields we check whether the estimated number $q \pm 2gq^{1/2}$ of places of degree one is (considerably) smaller than an approximate expected number of Weierstrass places. This would lead to roughly q^2 comparisons of places and is in $O(g^6)$, but can also be much smaller. Note that places of a prescribed small degree can be computed efficiently for global function fields. A bound for the number of isomorphisms is given by the Hurwitz bound $84(g-1)$ in characteristic zero and by roughly $16g^4$ in positive characteristic (details can be found in [7]). Such large automorphism groups are only obtained for very special function fields.

It may happen that we cannot find suitable places of degree one, but this was essential for the strategy described above. A solution to this problem is to consider constant field extensions, over which we would obtain suitable places of degree one. For example, if $P_{(\alpha)}$ is a place of degree two over k which splits into Weierstrass places over an algebraic closure it is sufficient to consider the constant field extension $F_{(\alpha)}k(P_{(\alpha)})/k(P_{(\alpha)})$ by the residue class field of $P_{(\alpha)}$, since $P_{(\alpha)}$ already splits in $F_{(\alpha)}k(P_{(\alpha)})/k(P_{(\alpha)})$. We would then compute an isomorphism of $k(P_{(1)})$ and $k(P_{(2)})$ and isomorphisms ϕ of $F_{(1)}k(P_{(1)})/k(P_{(1)})$ and $F_{(2)}k(P_{(2)})/k(P_{(2)})$ which possibly do not come from isomorphisms of $F_{(1)}/k$ and $F_{(2)}/k$. This can be checked as follows. A generating system of $F_{(\alpha)}/k$ is also a generating system of $F_{(\alpha)}k(P_{(\alpha)})/k(P_{(\alpha)})$. If we compute the effect of ϕ in terms of these generating systems it is easy to check whether ϕ restricts to an isomorphism of $F_{(1)}/k$ and $F_{(2)}/k$. This operation requires the inversion of isomorphisms or inverting the representation of one set of generators in terms of another set of generators. A way of doing this computation is described in [3] and can also be achieved for our special elements $x_{(\alpha),i}$ by linear algebra over k involving algebraic functions of bounded degree. Since every isomorphism of $F_{(1)}/k$ and $F_{(2)}/k$ extends to an isomorphism of $F_{(1)}k(P_{(1)})/k(P_{(1)})$ and $F_{(2)}k(P_{(2)})/k(P_{(2)})$ this method yields all isomorphisms of $F_{(1)}/k$ and $F_{(2)}/k$.

We summarize the single steps for computing the isomorphisms between $F_{(1)}/k$ and $F_{(2)}/k$.

Algorithm 4 *(Isomorphisms)*

Input: Function fields $F_{(1)}/k$ and $F_{(2)}/k$.
Output: A set L of all isomorphisms of $F_{(1)}/k$ and $F_{(2)}/k$.

1. *Check whether* $F_{(1)}/k$ *and* $F_{(2)}/k$ *have the same genus* g. *If not then return the empty list* L.
2. *Let* $p = \mathrm{char}(k)$, $q = \#k$ *and* $d = 1$.
3. *If* $g^3 < q^d$ *compute lists* $S_{(1)}$ *and* $S_{(2)}$ *of all Weierstrass places of* $F_{(1)}/k$ *and of* $F_{(2)}/k$ *(degrees greater one allowed). Check whether numbers, degrees and gap sequences coincide. Determine suitable small subsets* $S'_{(1)}$ *and* $S'_{(2)}$ *as discussed above.*
4. *If* $g^3 \geq q^d$ *compute lists* $S_{(1)}$ *and* $S_{(2)}$ *of all places of degree* d. *If there are no such places let* $d \leftarrow d+1$ *and go to step 3.*
5. *Choose a fixed* $P_{(1)} \in S'_{(1)}$. *If* $\deg(P_{(1)}) > 1$ *work with* $F_{(1)}k(P_{(1)})/k(P_{(1)})$ *in the following. Compute the* n_i, $x_{(1),i}$ *and* d_j, $w_{(1),j}$ *as in Section 3 and Section 5. If* $n_1 \neq 0 \bmod p$ *or* $p = 0$ *set* $c = 1$, *otherwise set* $c = 0$. *Choose a local uniformizer* $\pi_{(1)}$ *at* $P_{(1)}$. *If* $c = 1$ *then choose the special* $\pi_{(1)}$ *of Section 4. Apply the Gaussian elimination procedure of Section 5 to the* $x_{(1),i}$.
6. *The following three steps are done for every* $P_{(2)} \in S'_{(2)}$.
7. *Check that* $k(P_{(2)})/k$ *is isomorphic to* $k(P_{(1)})/k$. *If not, take the next* $P_{(2)}$. *If yes work with* $F_{(2)}k(P_{(2)})/k(P_{(2)})$ *in the following and identify* $k(P_{(2)})$ *and* $k(P_{(2)})$.
8. *Compute the* $x_{(2),i}$ *and* $w_{(2),j}$ *for* $P_{(2)}$ *as in Section 3 and Section 5. If* $c = 1$ *then choose the special* $\pi_{(2)}$ *of Section 4. Apply the Gaussian elimination procedure of Section 5 to the* $x_{(2),i}$.
9. *If* $c = 1$ *then solve for* a, b, c_1 *as described at the end of section 5. Otherwise solve for* $a, b, c_1, c_2, \ldots$ *as described in section 5. For any solution recover the* $\mu_{i,j}$ *and* ϕ. *If* ϕ *restricts to an isomorphism* ϕ' *of* $F_{(1)}/k$ *and* $F_{(2)}/k$, *then* $L \leftarrow L \cup \{\phi'\}$.
10. *Return* L.

We remark that we have implemented a prototype of Algorithm 4 in Magma [1,2] which shows that the algorithm is quite practical in the global function field case.

References

1. W. Bosma, J. Cannon, and C. Playoust. The Magma algebra system I: The user language. *J. Symbolic Comp.*, 24, 3/4:235–265, 1997.
2. Comp. algebra group. Magma. `http://www.maths.usyd.edu.au:8000/u/magma/`, 2004.
3. F. Hess. An algorithm for computing Weierstrass points. In C. Fieker and D. R. Kohel, editors, *Proceedings of the Fifth Symposium on Algorithmic Number Theory, ANTS-V*, LNCS 2369, pages 357–371, Sydney, Australia, 2002. Springer-Verlag, Berlin-Heidelberg-New York.

4. F. Hess. Computing Riemann-Roch spaces in algebraic function fields and related topics. *J. Symbolic Comp.*, 33(4):425–445, 2002.
5. Kant group. Kash. `http://www.math.tu-berlin.de/~kant`, 2004.
6. J. Klüners. Algorithms for function fields. *Exp. Math.*, 11:171–181, 2002.
7. H. Stichtenoth. Über die Automorphismengruppe eines algebraischen Funktionenkörpers von Primzahlcharakteristik. *Arch. Math.*, 24:527–544, 1973.
8. H. Stichtenoth. *Algebraic Function Fields and Codes.* Springer-Verlag, Berlin-Heidelberg-New York, 1993.

A Method to Solve Cyclotomic Norm Equations $f * \bar{f}$

Nick Howgrave-Graham[1] and Mike Szydlo[2]

[1] NTRU Cryptosystems, Burlington, MA, USA
nhowgravegraham@ntru.com
[2] RSA Laboratories, Burlington, MA, USA
mszydlo@rsasecurity.com

Abstract. We present a technique to recover $f \in \mathbb{Q}(\zeta_p)$ where ζ_p is a primitive p_{th} root of unity for a prime p, given its norm $g = f * \bar{f}$ in the totally real field $\mathbb{Q}(\zeta_p + \zeta_p^{-1})$. The classical method of solving this problem involves finding generators of principal ideals by enumerating the whole class group associated with $\mathbb{Q}(\zeta_p)$, but this approach quickly becomes infeasible as p increases. The apparent hardness of this problem has led several authors to suggest the problem as one suitable for cryptography. We describe a technique which avoids enumerating the class group, and instead recovers f by factoring N_f, the absolute norm of f, (for example with a subexponential sieve algorithm), and then running the Gentry-Szydlo polynomial time algorithm for a number of candidates. The algorithm has been tested with an implementation in PARI.

1 Introduction

We present an algorithm to solve degree two norm equations corresponding to the field extention $\mathbb{Q}(\zeta_p)$ / $\mathbb{Q}(\zeta_p + \zeta_p^{-1})$, where ζ_p is a p_{th} primitive root of unity.

The previously best known technique for solving this problem involves finding a principal ideal generator of the ideal of (f) by enumerating representatives of the whole class group, and then applying index calculus techniques to obtain a generator of (f). This approach is explained in a little more detail in section 2, but we note it becomes very expensive as p increases (even the class group is on the order of p^p).

In [6,12,13] the authors explicitly assume the hardness of the cyclotomic norm equation problem to build cryptographic applications. It has been observed by several sources [5,8,10] that their constructions can easily be modified to ones which are not reliant on this particular assumption, for example by adding some kind of perturbation or noise factor. Such enhancements are interesting avenues for further cryptographic research, but in this paper we concentrate on the purely mathematical problem of solving the norm equation.

Our work is motivated by a concrete question concerning polynomial arithmetic in the ring $R = \mathbb{Z}[X]/(X^p - 1)$. To describe our problem, we define, for each polynomial $f \in R$, its *reversal*, f_{rev}, to be the polynomial $f(X^{-1})$. An

D.A. Buell (Ed.): ANTS 2004, LNCS 3076, pp. 272–279, 2004.

element in R which is equal to its own reversal is called a *palindrome*. Given a palindromic element in R of the form $g = f * f_{rev}$, the task is to find such an f.

Our algorithm builds on an earlier work [9] which is able to recover an element $f \in \mathbb{Z}[X]/(X^p - 1)$, up to multiplication by $\pm X^k$ from the principal ideal (f), and from the quantity $f * f_{rev}$.

This problem of "factoring" $f * f_{rev}$ is easily seen to be related to a norm equation as follows. Elements of R naturally map to the quotient $\mathbb{Z}[X]/(1+X+\ldots X^{p-1})$, the ring of integers in the p_{th} cyclotomic field $\mathbb{Q}(\zeta_p)$. Under this map, a polynomial and its reversal map to Galois conjugates in the field extension $\mathbb{Q}(\zeta_p)$ / $\mathbb{Q}(\zeta_p + \zeta_p^{-1})$, and the product of the conjugates is the norm in $\mathbb{Q}(\zeta_p + \zeta_p^{-1})$. Solving such a norm equation yields a cyclotomic integer, and knowledge of $f(1)$ is sufficient to determine f as an element of $\mathbb{Z}[X]/(X^p - 1)$. The integer $f(1)$, in turn, is known up to sign since $f * f_{rev}(1) = f^2(1)$.

Some authors call g, the *autocorrelation* of f, and work more generally in the ring $\mathbb{R}[X]/(X^p - 1)$. In [6,7] instances of the problem are considered where the coefficients of f are taken to all be in $\{0, 1\}$, and the problem of recovering an f from the product $f * f_{rev}$ is called *bit retrieval.*

In this paper, unless otherwise noted, polynomials are elements of the ring R. When it is clear, we use the same symbol for the projection to the ring of cyclotomic integers.

2 The Classical Approach to Solving This Problem

The general principal ideal problem is, given an ideal $\mathcal{I}$ which is known to be principal, to find a generator of $\mathcal{I}$.

The standard technique for solving the principal ideal problem involves constructing a huge "factor base" consisting of ideals with small norm, representing every element of the class group.

The collection of a similarly large number of relations (which are "smooth" over the factor base), will yield a generator for the target ideal. See [3], section 6.5.5 for more details of this approach.

The class number of cyclotomic fields is on the order of p^p, see [4] for a more detailed analysis of this distribution. As a result of this, the factor base approach becomes infeasible very quickly, and had lead various people to suggest it as a basis for cryptographic schemes, see [2,6,12,13].

3 Statement of the Problem

Problem 1. *Let p be prime and let $f \in R = \mathbb{Z}[X]/(X^p - 1)$. Let $g = f * f_{rev}$. Given g, determine such an f.*

Clearly, the problem does not have a unique solution; if f is a solution, then so is $\pm X^k f$. To understand precisely the ambiguity of f, let us consider several ways in which we might enlarge the set of acceptable solutions. Below, f is considered as a cyclotomic integer, and $\bar{f}$ denotes complex conjugation.

Variant 1. *Find fractional ideals F such that $F * \bar{F} = (g)$.*

For an ideal F which is a solution, the ideal FA will also be a solution if $A\bar{A} = (1)$. Each ideal of the form $B/\bar{B}$ has this property, so there will be infinitely many solutions.

Variant 2. *Find integral ideals F such that $F * \bar{F} = (g)$.*

To restrict solutions to integral ideals, let $F_1^{a_1} \ldots\ldots F_k^{a_k}$ be the prime factorization of (f). Then for any $0 \leq b_i \leq a_i$, the ideal

$$F_1^{b_1} \bar{F_1}^{(a_1-b_1)} \cdots F_k^{b_k} \bar{F_k}^{(a_k-b_k)}$$

will be a solution. So there will be exactly $\prod(a_i + 1)$ solutions in total.

Variant 3. *Find cyclotomic field elements f such that $f * \bar{f} = g$.*

To restrict the solutions of variant 1, which are ideals, to those which are principally generated. If a field element f is a solution, we see that the element fa will also be a solution, provided that $a\bar{a} = 1$. Such a, are simply the elements of the form $a = b/\bar{b}$, and so if there is one solution, there will also be infinitely many solutions.

Variant 4. *Find cyclotomic integers f such that $f * \bar{f} = g$.*

This is in fact the flavor of the problem we are most interested in. Restricting solutions to cyclotomic integers, we see that the generator f of each solution ideal of variant 2 will yield a solution for this fourth variant. For a solution f, the multiples fu which are also solutions correspond to the values of u which are $2p_{th}$ roots of unity.

Thus, there are $2p$ times the number of principal ideal solutions. In particular we obtain $2p\prod(a_i + 1)$ as a bound on the total. In the case that (f) is a prime ideal, there are exactly $2p$ solutions.

4 Algorithm Overview

At a high level, there are two components to the algorithm. The first step computes candidate ideals F for the desired ideal (f). The second step consists of recovering f from the ideal (f), and the element $g = f * f_{rev}$. The second step can be accomplished with the result of [9], reviewed below in section 5.1, or in some cases, by a simplified version of it.

We illustrate the algorithm first in a somewhat special case, then proceed to show how it can be extended. In this first case, we are going to assume that the prime p is congruent to 3 (mod 4), and additionally that the norm of f, down to $\mathbb{Z}$ does not have any repeated factors. Our first assumption implies that $\mathbb{Q}(\sqrt{-p})$ is a subfield of the p_{th} cyclotomic field, so that there is a diagram of field extentions.

$$\begin{array}{ccc} \mathbb{Q}(\zeta) & \leftarrow & \mathbb{Q}(\zeta+\zeta^{-1}) \\ \uparrow & & \uparrow \\ \mathbb{Q}(\sqrt{-p}) & \leftarrow & \mathbb{Q} \end{array}$$

The essence of the first step of the algorithm is the determination of a set of potential ideals for (f). As in the classical approach to our problem, this may be accomplished by factoring $f * f_{rev}$ into prime ideals. The candidate ideals will be the ideals of R which have the same norm as f, and also contain $(f * f_{rev})$.

To factor an ideal in the ring cyclotomic integers, one appeals to the classical prime decomposition theory [1], using the prime factorization of the absolute norm. Algorithmically, beginning with $f*f_{rev}$ in $\mathbb{Q}(\zeta+\zeta^{-1})$, it is easy to compute the norm N_f down to $\mathbb{Q}$. Our assumption that each rational prime q divides this norm with multiplicity at most 1, implies that q splits completely in $\mathbb{Q}(\zeta)$. Thus, it makes the use of arithmetic in the quadratic subfield even simpler.

There are two primes above each such q in $\mathbb{Q}(\sqrt{-p})$. Among the various possible products, $\mathcal{I}$, of these ideals in $\mathbb{Q}(\sqrt{-p})$ are the norms of the solutions f. By lifting each such $\mathcal{I}$ to $\mathbb{Q}(\zeta)$, and computing the greatest common divisor with $(f * f_{rev})$, one obtains an ideal F, with the property that $F * F^{\sigma} = (f * f_{rev})$. This can be considered a solution to the problem variant 2, above, and will be deemed a candidate ideal.

The main algorithmic difficultly of determining the potential ideals in $\mathbb{Q}(\sqrt{-p})$ is the integer factorization of N_f in $\mathbb{Z}$. Once this has been accomplished the above procedure yields a short list of candidate ideals.

Once one is in possession of both (f) and $f * f_{rev}$, the algorithm [9] can be used to find f up to a root of 1. We provide further details in the next section.

5 Algorithm Details for $p = 3 \bmod 4$

In this section we provide further details of each step.

Norm Calculation. The initial step consists in calculating the norm of f down to $\mathbb{Q}$. This norm is the product $\prod f(\zeta_i)$ over all p_{th} primitive roots of 1. Grouping the factors into conjugate pairs, we see that this can be efficiently computed from the $(p-1)/2$ conjugates of $(f * f_{rev})(\zeta_i)$. Equivalently, the norm of f is the square root of the norm of $f * f_{rev}$ in $\mathbb{Z}$. Let N_f denote this integer.

Prime Factorization. Next, the prime factors of N_f must be determined. Unless N_f turns out to be prime, or the factorization easy, this step is the most computationally difficult part of the algorithm, and prohibitive if p is large. Depending on size, a standard elliptic curve or sieving algorithm may be successful. Let $q_1^{a_1} \dots q_k^{a_k}$ denote the prime factors of N_f.

Ideals in $\mathbb{Q}(\sqrt{-p})$. Each prime in $\mathbb{Z}$ either splits or is inert in $\mathbb{Q}(\sqrt{-p})$. By making the assumption that N_f has no repeated factors, we ensure that each prime q_i, splits. The two factors are then $(q) = (q, r+\frac{1+\sqrt{-p}}{2})$, and $(q, r-\frac{1+\sqrt{-p}}{2})$.

A candidate ideal $\mathcal{I}$ in $\mathbb{Q}(\sqrt{-p})$ is computed by multiplying together exactly one ideal above each q_i. The ideal arithmetic in this step is particularly efficient, given that we are only working in a quadratic field.

Testing Principality

A necessary condition that the ideal $\mathcal{I}$ be the norm of a solution f, is that $\mathcal{I}$ is a principal ideal. If the class number of $\mathbb{Q}(\sqrt{-p})$ is not large, the ideal $\mathcal{I}$ can be efficiently tested for principality as in [3]. This may be feasible, since the class numbers of quadratic imaginary fields seem to grow at a reasonable rate. If it is concluded that $\mathcal{I}$ is not principal, the subsequent steps of our algorithm can not succeed, so $\mathcal{I}$ may be discarded. We remark that this step is optional. If all of the potential ideals $\mathcal{I}$ were to be processed in parallel, eventually a solution would be found.

GCD Computation. In this step we determine the candidate ideals F of $\mathbb{Q}(\zeta)$. Another consequence of assuming that N_f has no repeated factors is the fact that f and f_{rev} are relatively prime. Additionally, f_{rev} is prime to the other $(p-3)/2$ Galois conjugates of f dividing $\mathcal{I}$.

Now consider both $\mathcal{I}$ and $(f * f_{rev})$ as ideals of $\mathbb{Q}(\zeta)$, and compute the greatest common divisor ideal. This is efficiently accomplished by forming the $\mathbb{Z}$ span of generators of the two $\mathbb{Z}$-modules [3]. The resulting ideal is an ideal F, with the property that $F * \bar{F} = (g)$, and is thus a solution to variant 2 of our problem. Only if F is principal will it correspond to an element f such that $f * f_{rev} = g$.

Recovering f

Given the candidate ideal $\mathcal{I}$ and the product $f * f_{rev}$, apply the algorithm, reviewed below in section 5.1. Note that if $\mathcal{I}$ does not correspond to a solution, this process must fail. One method of dealing with this detail is to process all possible candidate ideals F in parallel.

5.1 Review of Gentry-Szydlo Algorithm

In this section we review the algorithm of [9], which recovers an element from $Z[X]/(X^p - 1)$ given the ideal it generates, and the product $f * f_{rev}$.

A simplified version of the algorithm is as follows. We will express ideals as $\mathbb{Z}$-modules over the power basis $\{1, X, X^2, \ldots\}$. For a p element vector f, the *circulant* matrix $Cir(f)$ is the matrix of all rotations of f, and the *columns* of $Cir(f)$ generate the principal ideal f in $\mathbb{Q}(\zeta)$.

First, note that if F denotes $Cir(f)$, then for any other basis H of (f), $H = FU$ for some unimodular matrix. Next, we combine this information with $D = Cir(f f_{rev})$ by forming the product

$$\begin{aligned} G &= H^t D^{-1} H \\ &= U^t F^t (F f^t)^{-1} F U \\ &= U^t U. \end{aligned}$$

This last matrix can be viewed as the Gram matrix of an auxiliary lattice. The auxiliary lattice might be called a *hypercubic* lattice [19] since it is a rotation of the trivial lattice $\mathbb{Z}^p$. Moreover the paper [19] suggests that it may be easier to reduce hypercubic lattices than general lattices of a similar dimension.

This lattice, expressed via the Gram matrix G, may be reduced with LLL, or one of its variants, producing a unimodular matrix V such that $V^t G V$ is the more reduced lattice. If very successful (perhaps because of the easier problem the hypercubic lattices pose) the lattice reduction might reduce the Gram matrix right down to the Identity matrix (e.g $V = U^{-1}$ would). In such a case, $V^t U^t U V = Id$, implies that $W = UV$ is a signed permutation matrix, and multiplying the matrix FU by V yields FW, whose columns are all signed rotations of the sought vector, f.

Thus, provided that the lattice reduction algorithm manages to fully reduce G to the identity, f can be recovered. However, the variants of LLL, which are guaranteed to produce a shortest vector, are exponential time algorithms. In practice, LLL, and its higher block-size variants [17] often produce much shorter vectors than the available bounds.

The approach taken in the more complicated algorithm in [9] provides a strategy that will guarantee that we find the shortest vector in polynomial time, even though LLL only returns a *multiple* of the shortest vector in polynomial time. The technical trick is to attempt to reduce the ideal F^{R-1}, for a prime R congruent to 1 mod p, in such a way that an element $f^{R-1}\alpha$ is produced where α has small L_2 norm. The congruence $f^{R-1}\alpha \cong \alpha \bmod R$ can be used to find α when it is small compared to R. A final calculation finds f from f^{R-1}.

6 General Case

The assumptions made in the previous section, were convenient but not essential. We now explain how to extend the algorithm to the general case, where we do not assume that $p = 3 \bmod 4$ or that the norm does not have repeated prime factors.

We suppose that the cyclotomic norm, N_f, factors as $q_1^{a_1} \ldots . q_k^{a_k}$, and wish to determine a list of candidate ideals F for (f).

Fix a prime q. Then the factorization of q into prime ideals in $\mathbb{Q}(\zeta_p)$ may be efficiently computed by factoring the polynomial $1 + X + X^2 + \ldots + X^{p-1}$ over $\mathbb{F}_q$ using either the Berlekamp or Cantor-Zassenhaus algorithms (see [3], section 3.4 for more details). (The number of prime ideals above q depends on the order of q mod p.). For each prime ideal Q above q, the exact exponent of Q dividing $(f * \bar{f})$ may be computed via ideal divisibility tests. Thus, given the factorization of the norm N_f, the factorization of $(f * \bar{f})$ into ideals may be efficiently computed.

Now consider a prime ideal Q, and let r be the largest power of Q dividing q. For each Q, let $\bar{Q}$ be the complex conjugate ideal. In the case that $\bar{Q} = Q$, the power of Q dividing (f), is clearly $r/2$. When $\bar{Q} \neq Q$, then we know that the maximal power of Q, a_Q say, that divides (f) must be in $\{0, \ldots, r\}$, and moreover $a_{\bar{Q}} + a_Q = r$. This leaves $r + 1$ possible candidates to enumerate over.

Considering all prime ideal pairs $Q_i \neq \bar{Q}_i$ such that $Q^{r_i} | (f * \bar{f})$, we see that the total number of candidate ideals for (F) is $\prod(r_i + 1)$.

Once the list of candidate ideals F has been established, the algorithm may proceed as described above. The only difference is that one does not have the simplicity and efficiency gain obtained by working in the quadratic imaginary subfield.

7 Experiments

The norm, N_f, of f down to $\mathbb{Z}$, can be calculated as $1/f(1)$ times the determinant of an associated cyclic matrix, and thus can be bounded by $|f|^p$, where $|\cdot|$

denotes the Euclidean norm of the entries of f. In experiments both the number of factors and the multiplicity of these factors followed what one would expect from the Prime Number Theorem, which leads us to conjecture that the prime factorizations of the norms are distributed in a similar fashion to randomly sampled numbers with a similar bit length.

In obtaining practical results, we implemented just the $p = 3 \bmod 4$ case in PARI.

7.1 Recovering f When $p = 3 \bmod 4$

For small values of p, congruent to 3 mod 4, the entire algorithm, as described above was implemented in PARI [11]. This symbolic and computational calculator has the advantage of being able to apply the LLL algorithm to a lattice expressed via a Gram matrix, which the (otherwise flexible) NTL [18] LLL implementation does not allow for.

The arithmetic in $\mathbb{Q}(\sqrt{-p})$ was used to create candidate ideals. Our experiment also implemented the simplified version of the Gentry-Szydlo algorithm described above, which was sufficient for the primes tested. That is, the original LLL algorithm was able to reduce the Gram matrix to the identity, in the cases that yielded a solution.

Result were (quickly) obtained with p varying from 19 to 71, with the fortunate outcome that for these smallish primes, the simplified Gentry-Szydlo approach was always sufficient to completely reduce the lattice[1]. As expected, the results reflected the fact that not all candidate ideals must yield a solution for f, e.g. there were many instances where the norm had exactly two prime factors, and only two of the four candidates produced a valid f.

8 Conclusions

We have shown how the difficulty of factoring $f * \bar{f}$ into elements reduces to the problem of factoring an integer N_f, and the problem of applying the (polynomial time) Gentry-Szydlo algorithm to a number of candidates. Under the assumption that the factorization of the norms N_f are distributed in a similar fashion to average numbers of a similar size, then the bottleneck of the algorithm occurs in the integer factorization stage (since the number of candidates is exponential in the number of prime divisors of N_f but this is logarithmic in p).

In particular this paper shows that f such that $|f|$ is small, where $|\cdot|$ denotes the Euclidean norm of the entries of f, are particularly weak, since the integer N_f is particularly small in these cases.

We remark that if one is trying to build a cryptosystem on the basis of the hardness of factoring $f * \bar{f}$, then p should be chosen such that it is highly unlikely that the norm N_f can be factored in a reasonable time.

---

[1] This is consistent with the conjecture that hypercubic lattice may indeed be an easier class of lattice to reduce.

Acknowledgments. The authors wish to thank Veit Elser, Craig Gentry and Jeff Hoffstein for useful correspondence.

References

1. Z. I. Borevich and I. R. Shafarevich *Number Theory*, Academic Press, 1966.
2. J. Buchmann, M. Maurer and B. Möller, *Cryptography based on number fields with large regulator*. Journal de Théorie des Nombres de Bordeaux, 2000, pp. 293–307
3. H. Cohen, *A Course in Computational Algebraic Number Theory*, Graduate Texts in Mathematics, 138. Springer, 1993.
4. H. Cohen and H. Lenstra, *Heuristics on class groups of number fields*, Number Theory, Lecture Notes in Mathematics, vol. 1068, Springer-Verlag, 1983, pp. 33–62.
5. V. Elser, *Private Communication.*
6. V. Elser, *Bit retrieval: intractability and application to digital watermarking*, `http://arxiv.org/abs/math.NT/0309387`
7. V. Elser, *Phase retrieval challenges*, `http://www.cecm.sfu.ca/~veit/`
8. C. Gentry, *Private Communication.*
9. C. Gentry, M. Szydlo, *Cryptanalysis of the Revised NTRU signature scheme*, in Proc. of Eurocrypt '02, LNCS 2332, pages 299–320. Springer-Verlag, 2002.
10. J. Hoffstein, *Private Communication.*
11. PARI, `http://pari.math.`ubordeaux.fr/-
12. J. Hoffstein, D. Lieman, J.H. Silverman, *Polynomial Rings and Efficient Public Key Authentication*, in Proc. International Workshop on Cryptographic Techniques and E-Commerce (CrypTEC '99), Hong Kong, (M. Blum and C.H. Lee, eds.), City University of Hong Kong Press.
13. J. Hoffstein, J.H. Silverman, *Polynomial Rings and Efficient Public Key Authentication II*, in Proceedings of a Conference on Cryptography and Number Theory (CCNT '99), Birkhauser.
14. A.K. Lenstra, H.W. Lenstra Jr., L. Lovász, *Factoring Polynomials with Rational Coefficients*, Mathematische Ann. 261 (1982), 513–534.
15. D. Micciancio, *The Shortest Vector in a Lattice is Hard to Approximate to within Some Constant*, in Proc. 39th Symposium on Foundations of Computer Science, 1998, 92–98.
16. P. Nguyen and J. Stern, *Lattice Reduction in Cryptology: An Update*, in Proc. of Algorithm Number Theory (ANTS IV), LNCS 1838, pages 85–112. Springer-Verlag, 2000.
17. C.-P. Schnorr, *A Hierarchy of Polynomial Time Lattice Basis Reduction Algorithms*, Theoretical Computer Science 53 (1987), 201–224.
18. Shoup, V., NTL: A Library for Doing Number Theory. Available at `http://www.shoup.net/ntl/`.
19. Szydlo, Michael, *Hypercubic Lattice Reduction*, Eurocrypt '03
20. L. Washington, *Introduction to Cyclotomic Fields*, Graduate Texts in Mathematics 83, 1982.

Imaginary Cyclic Quartic Fields with Large Minus Class Numbers

M.J. Jacobson, Jr., H.C. Williams, and K. Wooding

Centre for Information Security and Cryptography, University of Calgary, 2500 University Drive NW, Calgary, Alberta, Canada, T2N 1N4
jacobs@cpsc.ucalgary.ca
{williams,kjell}@math.ucalgary.ca

Abstract. It is well-known that the minus class number h_p^- of an imaginary cyclic quartic field of prime conductor p can grow arbitrarily large, but until now no one has been able to exhibit an example for which $h_p^- > p$. In an attempt to find such an example, we have tabulated h_p^- for all primes $p \equiv 5 \pmod 8$ with $p < 10^{10}$ and for primes $p < 10^{14}$ satisfying certain quartic character restrictions. An analysis of these data yields unconditional numerical evidence in support of the Cohen-Martinet heuristics, but as we did not find a value of p for which $h_p^- > p$ by these methods, we constructed a 77-digit value of p for which one can prove $h_p^- > p$ assuming the Extended Riemann Hypothesis.

1 Introduction

Let p be a prime and let ζ_p be a primitive pth root of unity. Let N denote the maximal imaginary subfield of degree d a power of 2 of the cyclotomic field $\mathbb{Q}(\zeta_p)$ and let N^+ denote the real quadratic subfield of degree $d/2$ of N. The *minus (relative) class number* h_N^- of N is given by

$$h_N^- = h_N / h_{N^+}$$

where h_N and h_{N^+} are the class numbers of N and N^+, respectively. For example, if $p \equiv -1 \pmod 4$, then $N_p := N = \mathbb{Q}(\sqrt{-p})$, $N^+ = \mathbb{Q}$, and $h_p^- := h_N^-$ is the class number of $\mathbb{Q}(\sqrt{-p})$. In this case we always have $h_p^- < p$.

If $p \equiv 5 \pmod 8$, there exist integers a and b such that $p = a^2 + b^2$, $a \equiv -1 \pmod 4$, $b \equiv 2 \pmod 4$, and $ab \equiv 2 \pmod 8$. (These conditions suffice to determine a and b uniquely.) In this case

$$N_p := N = \mathbb{Q}\left(\sqrt{-(p + b\sqrt{p})}\right), \quad N_p^+ := N^+ = \mathbb{Q}(\sqrt{p}) \ ,$$

and if h_p^- is the minus class number of N_p, then h_p^- is not necessarily less than p. In fact, for any $c > 0$ it can be shown that there exists an infinitude of values of p such that $h_p^- > cp$. In the sequel we will confine our attention to h_p^- when $p \equiv 5 \pmod 8$.

D.A. Buell (Ed.): ANTS 2004, LNCS 3076, pp. 280–292, 2004.

Let χ_p denote the only quartic Dirichlet character of conductor p such that $\chi_p(2) = i$ ($i^2 + 1 = 0$). If, as usual, we denote the Dirichlet L-function as

$$L(s, \chi_p) = \sum_{n=1}^{\infty} \frac{\chi_p(n)}{n^s} ,$$

then

$$h_p^- = \frac{p}{2\pi^2} |L(1, \chi_p)|^2 = \frac{1}{2} |L(0, \chi_p)|^2 . \tag{1}$$

Louboutin [5] has pointed out that we may write $L(0, \chi_p)$ as

$$L(0, \chi_p) = \frac{W(\chi_p)}{\pi} \sqrt{p} \sum_{n=1}^{\infty} \left(\frac{\overline{\chi_p(n)}}{n} e^{-\pi n^2/p} \right) + \frac{2}{\pi} \sum_{n=1}^{\infty} \chi_p(n) \int_{n\sqrt{\pi/p}}^{\infty} e^{-x^2} dx , \tag{2}$$

where

$$W(\chi_p) = \frac{\tau(\chi_p)}{i\sqrt{p}}$$

and $\tau(\chi_p)$ is the Gauss sum

$$\tau(\chi_p) = \sum_{j=1}^{p-1} \chi_p(j) e^{2j\pi i/p} .$$

Also, $\tau(\chi_p)$ can be evaluated as $\epsilon_p \alpha_p$ where

$$\alpha_p = \sqrt{\frac{p + a\sqrt{p}}{2}} + i \frac{b}{|b|} \sqrt{\frac{p - a\sqrt{p}}{2}}$$

and ϵ_p is either 1 or -1.

Louboutin used the fact the $L(0, \chi_p) \in \mathbb{Z}[i]$ to develop a method of using only a certain number of the earliest terms of (2) to approximate $L(0, \chi_p)$ sufficiently accurately to be able to deduce its exact value. By using this technique he was able to compute unconditionally all the values of h_p^- for $p < 10^7$. Furthermore, by restricting the value of p such that $\chi_p(q) = 1$ for $q \in \{3, 5, 7, 11, 13, 17, 19, 23, 29\}$, he was able to use his method to discover that if $p = 1679516029$, then $h_p^- = 904595821 > p/2$ ($h_p^-/p \approx 0.5386$), but he was unable to say whether this was the least such p for which $h_p^- > p/2$. Later in [6], he showed how his technique for computing h_p^- could be improved, but did not describe any implementations of his new ideas.

The purpose of this paper is to describe and use a further modification of Louboutin's later procedure to find all the values of h_p^- for $p < 10^{10}$. On implementing this technique, we were able to establish that the least p such that $h_p^-/p > 1/2$ is $p = 599630509$. We also discuss our attempt to find values of

p for which we get larger values of the ratio h_p^-/p by prescribing the quartic character $\chi_p(q)$ for the first several odd primes q. By doing this we found that if $p = 34131117018541$, then $h_p^- = 23754887413613$ and $h_p^-/p \approx 0.6960$, the largest value of this ratio that we were able to find unconditionally. Finally, as we had been challenged by Louboutin to find a p for which $h_p^-/p > 1$, we exhibit such a prime and describe our procedure for constructing it and verifying that $h_p^-/p > 1$. However, we must emphasize that this result is contingent on the truth of the extended Riemann hypothesis (ERH).

2 Tabulation of h_p^- for $p < 10^{10}$

As indicated earlier, Louboutin provided an improved version of his earlier algorithm for evaluating h_p^- in [5]. This method makes use of the following special case of his more general Theorem 4.

Theorem 1. *Let $M \geq 1$ be given and let $m = \left\lceil \sqrt{\frac{p}{2\pi}\left(\log \frac{Mp}{\pi}\right)} \right\rceil$. If we define $L_M(0,\chi_p)$ by*

$$L_M(0,\chi_p) = \frac{W(\chi_p)}{\pi}\sqrt{p}\sum_{n=1}^{m}\frac{\overline{\chi_p(n)}}{n}e_n + \frac{1}{\sqrt{p}}\sum_{n=1}^{m}(e_n + e_{n+1})S_n(\chi_p) \ , \quad (3)$$

where

$$S_n(\chi_p) = \sum_{k=1}^{n}\chi_p(k) \quad \text{and} \quad e_n = \exp(-n^2\pi/p) \ ,$$

then

$$|L\,(0,\chi_p) - L_M(0,\chi_p)| \leq \frac{3}{2\sqrt{\pi M}} + \frac{3}{8\sqrt{p}} \ .$$

Since $|L\,(0,\chi_p)| \in \mathbb{Z}$, we see that if $p > 5$ and M is sufficiently large that

$$\frac{3}{2\sqrt{\pi M}} + \frac{3}{8\sqrt{p}} < \frac{1}{2} \quad \text{(i.e., } M > 4.57) \ ,$$

then we can compute h_p^- by using only $O(p^{1/2+o(1)})$ terms in $L_M(0,\chi_p)$. This, of course, is predicated on our being able to evaluate $W(\chi_p)$ or, equivalently, determining ϵ_p. Louboutin (unpublished) developed a method for doing this, but it is quite slow and unwieldy. Indeed, as noted by Berndt and Evans [2], there is no known efficient technique for computing ϵ_p. However, as there are only two possible values for ϵ_p, we simply evaluated $L_M(0,\chi_p)$ for both and selected that one which is closest to a Gaussian integer, after having checked that the other one is not close to any Gaussian integer, i.e. does not satisfy $|L_M(0,\chi_p) - \mathbb{Z}[i]| \leq 3/2\sqrt{\pi M} + 3/8\sqrt{p}$. Any possible ambiguity can be resolved by increasing M. For example, if $p = 17333$, then selecting M such that

$3/2\sqrt{\pi M} + 3/8\sqrt{p} < 0.01$ yields an approximation for which the two possible values of $L_M(0,\chi_p)$, $-26.0021066 - 35.9987971i$ and $32.9999231 + 38.9996724i$, have distances 0.002425 and 0.0003365 from the nearest Gaussian integer, both of which are less than 0.01. Using 0.001 for the error bound separates the two cases successfully, as we get $-26.0021379 - 35.9985175i$ and $33.0000346 + 38.9999419i$, and only the latter has distance 0.00006760 less than 0.001 from the nearest Gaussian integer (the former has distance 0.002601). Computationally, this strategy worked very well — an error bound of 0.01 worked in almost all cases.

Louboutin also mentioned that we can lessen the work needed to evaluate e_n by using the recursive formulas $e_{n+1} = f_n e_n$, $f_{n+1} = h f_n$, where $h = \exp(-2\pi/p)$, $f_0 = \exp(-\pi/p)$, $e_0 = 1$. Thus, any e_n can be computed by performing two multiplications. Note that $f_n = \exp(-\pi(2n+1)/p)$.

There remains the problem of evaluating $\chi_p(n)$. It is easy to see that

$$\chi_p(a) = i^{\lambda(a)} , \tag{4}$$

where

$$\lambda(a) = \min\left\{\lambda \geq 0 \mid n^{(p-1)/4} \equiv t_p^\lambda \pmod{p}\right\} \in \{0,1,2,3\}$$

and t_p is defined as $t_p \equiv 2^{(p-1)/4} \pmod{p}$. However, for large values of n ($n > \sqrt{p}$), the determination of all the characters needed in (3) can be very time-consuming. Thus, we found it convenient to make use of an idea in Stein and Williams [8]. We put $r_{-2} = n$, $r_{-1} = p$ and execute the Euclidean algorithm to obtain the continued fraction expansion of n/p. On doing this we get

$$r_{j-2} = q_j r_{j-1} + r_j \quad (0 \leq r_j < r_{j-1}, \quad j = 0,1,2,\dots) .$$

Putting $A_{-2} = 0$, $A_{-1} = 1$, $B_{-2} = 1$, $B_{-1} = 0$ and

$$A_{j+1} = q_{j+1}A_j + A_{j-1}, \quad B_{j+1} = q_{j+1}B_j + B_{j-1} \quad (j = -1,0,1,2,\dots) ,$$

we get

$$nB_j - pA_j = (-1)^j r_j .$$

If we find that k such that

$$B_k \leq \lfloor\sqrt{p}\rfloor < B_{k+1} ,$$

then

$$r_k - |nB_k - pA_k| < \sqrt{p} .$$

Also, since

$$B_k n \equiv (-1)^k r_k \pmod{p} ,$$

Table 1. h^-/p hichamps

p	h^-	h^-/p	p	h^-	h^-/p
13	1	0.07692	1908421	667025	0.34952
109	17	0.15596	2405701	891545	0.37060
757	125	0.16513	3521389	1315845	0.37367
829	145	0.17491	4975909	1860241	0.37385
1621	333	0.20543	5020261	1936701	0.38578
7669	1585	0.20668	6056989	2342125	0.38668
12301	2825	0.22966	6389629	2765413	0.43280
24469	6029	0.24639	14283229	6447985	0.45144
29989	8325	0.27760	30903469	13975937	0.45224
101581	29753	0.29290	89599381	41338865	0.46137
126949	39593	0.31188	125105821	57930237	0.46305
199021	62145	0.31225	169614589	78807397	0.46463
410029	135533	0.33054	182338381	88134569	0.48336
578029	198725	0.34380	599630509	304391965	0.50763
661621	227869	0.34441	1679516029	904595821	0.53861
1039021	362389	0.34878	6033132109	3381836985	0.56054

we can evaluate $\chi_p(n)$ as

$$\chi_p(n) = \chi_p(-1)^k \chi_p(r_k)\chi_p(B_k)^{-1} = (-1)^k \chi_p(r_k)\chi_p(B_k)^{-1} \ .$$

Thus, if we tabulate $\chi_p(x)$ for all $1 \le x \le \lfloor\sqrt{p}\rfloor$, we can compute $\chi_p(n)$ in no more than two table look-ups after evaluating r_k and B_k via a partial Euclidean algorithm. This is faster than using (4) to evaluate $\chi_p(n)$ when $n > \sqrt{p}$, and since $m > \sqrt{p}$, increases the speed with which we can evaluate $L_M(0, \chi_p)$. In our implementation, the modified character evaluation was from 1.4 to 1.9 times faster, depending upon the size of p.

We used this method, implemented using the NTL C++ number theory library [7], to tabulate values of h_p^- for the 113764515 primes $p < 10^{10}$, $p \equiv 5 \pmod 8$. The computation took 1564 days of CPU time on 269 2.4 GHz Xeon processors running Linux, i.e., about 6 days of real time or approximately 1.2 seconds per field. We used 104 bits of precision for the floating-point approximations, and found that this sufficed in all cases.

The h_p^-/p hichamps for these primes are listed in Table 1 and the lochamps in Table 2. Our computation shows that $p = 599630509$ is the smallest prime such that $h_p^-/p > 1/2$. The largest ratio h_p^-/p found by this method, 0.56054, is larger than that found by Louboutin [5] (0.5386), but still unfortunately quite far from 1. The slow growth-rate of h_p^-/p leads us to believe that this approach is unlikely to produce such a p without further significant algorithmic advances.

2.1 The Cohen-Martinet Heuristics

Although our tabulation was unsuccessful in finding p with $h_p^- > p$, the data are still useful for providing evidence in support of the Cohen-Martinet heuristics

Table 2. h^-/p lochamps

p	h^-	h^-/p	p	h^-	h^-/p
13	1	0.0769231	39829	289	0.0072560
29	1	0.0344828	226669	1429	0.0063043
37	1	0.0270270	986749	5365	0.0054370
53	1	0.0188679	13136509	70669	0.0053796
61	1	0.0163934	27733861	142225	0.0051282
349	5	0.0143266	63449149	313909	0.0049474
373	5	0.0134048	181584901	814333	0.0044846
1789	13	0.0072666	3090045781	13284769	0.0042992

[3]. Among other things, these heuristics provide predictions on the probability that an odd integer l divides h_p^-. For an integer $p \geq 2$ and α an integer or ∞, set $(p)_\alpha = \prod_{1\leq k\leq\alpha}(1-p^{-k})$, $(p)_0 = 1$. Then the probability that an odd integer l divides h_p^- is estimated to be $P_1(l)P_3(l)$, where

$$P_1(l) = \prod_{\substack{p^\alpha \,\|\, l \\ p\equiv 1 \pmod 4}} \left(1 - (p)_\infty^2 \sum_{0\leq\beta+\gamma<\alpha} \left(p^{\beta+\gamma}(p)_\beta(p)_\gamma\right)^{-1}\right)$$

$$P_3(l) = \prod_{\substack{p^\alpha \,\|\, l \\ p\equiv 3 \pmod 4}} \left(1 - (p^2)_\infty/(p^2)_{[(\alpha-1)/2]}\right)$$

($[x]$ denotes the nearest integer to x). For example, $\mathrm{prob}(3\,|\,h_p^-) \approx 0.123440$, $\mathrm{prob}(5\,|\,h_p^-) \approx 0.421894$, and $\mathrm{prob}(7\,|\,h_p^-) \approx 0.020825$.

If we assume that the subset of imaginary cyclic quartic fields of prime conductor behaves the same as all such fields, then our data are especially well-suited for this purpose, as the minus class numbers computed with Louboutin's method are unconditionally correct. A similar assumption, restricting to prime discriminants, has proved to be reasonable in other contexts, in particular real quadratic fields [9].

Let $r_l(x)$ denote the fraction of $p \leq x$ for which $l\,|\,h_p^-$, i.e.,

$$r_l(x) = \frac{|\{p \leq x \;:\; l\,|\,h_p^-,\ p \text{ prime}, p \equiv 5 \pmod 8\}|}{|\{p \leq x \;:\; p \text{ prime}, p \equiv 5 \pmod 8\}|}\ ,$$

and let $q_l(x)$ be given by

$$q_l(x) = \frac{r_l(x)}{P_1(l)P_3(l)}\ ,$$

the ratio between $r_l(x)$ and the predicted probability that $l\,|\,h_p^-$. If the Cohen-Martinet heuristics are accurate, we would expect $q_l(x)$ to approach 1 as x increases.

Table 3. $q_l(x)$ values

	l							
x	3	5	7	9	11	13	15	27
10000000	1.00239	0.99408	0.99245	1.00239	0.99066	1.00492	0.99462	0.99038
20000000	0.99904	0.99466	0.98476	0.99904	1.01804	0.99902	0.98544	0.97838
30000000	0.99994	0.99614	0.98937	0.99994	1.00453	0.99843	0.99157	0.98537
40000000	1.00196	0.99538	0.99411	1.00196	1.00864	0.99963	0.99403	0.99860
50000000	1.00214	0.99643	1.00067	1.00214	1.01580	0.99970	0.99510	0.99172
60000000	1.00346	0.99650	1.00024	1.00346	1.01078	1.00002	0.99721	0.99857
70000000	1.00516	0.99719	1.00089	1.00516	1.00961	1.00021	1.00026	0.99993
80000000	1.00360	0.99682	1.00028	1.00360	1.00739	0.99948	0.99881	0.99829
90000000	1.00317	0.99687	1.00309	1.00317	1.00497	0.99915	0.99832	1.00153
100000000	1.00231	0.99649	1.00085	1.00231	1.00120	0.99872	0.99673	0.99742
200000000	1.00167	0.99785	0.99999	1.00167	1.00099	0.99876	0.99979	1.00068
300000000	1.00058	0.99832	0.99866	1.00058	0.99966	0.99923	0.99896	1.00051
400000000	0.99919	0.99894	0.99847	0.99919	1.00174	1.00020	0.99884	0.99754
500000000	0.99903	0.99892	0.99901	0.99903	0.99808	1.00083	0.99884	0.99611
600000000	0.99905	0.99900	0.99877	0.99905	0.99782	1.00067	0.99832	0.99512
700000000	0.99883	0.99905	0.99880	0.99883	0.99840	1.00060	0.99841	0.99430
800000000	0.99870	0.99918	0.99988	0.99870	0.99931	1.00031	0.99802	0.99578
900000000	0.99912	0.99913	0.99990	0.99912	0.99758	1.00018	0.99871	0.99515
1000000000	0.99884	0.99919	0.99970	0.99884	0.99769	1.00024	0.99832	0.99642
2000000000	0.99868	0.99962	1.00136	0.99868	1.00041	1.00035	0.99799	0.99782
3000000000	0.99929	0.99966	1.00136	0.99929	1.00010	1.00036	0.99925	0.99810
4000000000	0.99969	0.99968	1.00094	0.99969	0.99968	1.00031	0.99960	0.99906
5000000000	0.99942	0.99966	0.99985	0.99942	0.99777	1.00022	0.99902	0.99834
6000000000	0.99962	0.99969	0.99987	0.99962	0.99782	1.00037	0.99933	0.99871
7000000000	0.99964	0.99966	0.99974	0.99964	0.99767	1.00034	0.99949	0.99873
8000000000	0.99973	0.99969	0.99991	0.99973	0.99772	1.00025	0.99954	0.99880
9000000000	0.99976	0.99971	1.00006	0.99976	0.99743	1.00004	0.99964	0.99897
10000000000	0.99980	0.99983	1.00015	0.99980	0.99768	1.00004	0.99983	0.99932

In Table 3 we give values of $q_l(x)$ for several small values of l, and in Figure 1 and Figure 2 we plot the $q_3(x)$ and $q_5(x)$ values, respectively. As these values do appear to approach 1 (this is especially clear in the graphs), our data strongly suggest that the Cohen-Martinet heuristics are indeed accurate in this particular case.

3 Further Tabulation

As it was our intention to find some p for which $h_p^- > p$, the results of the computer run described in the previous section were somewhat disappointing. We therefore decided to modify and extend Louboutin's earlier idea of prescribing the quartic character of the primes for which we could evaluate h_p^-. Instead of using only the first 9 primes, we prescribed p as follows:

- $\chi_p(q) = 1$ for $3 \le q \le 31$ (the first 10 odd primes),

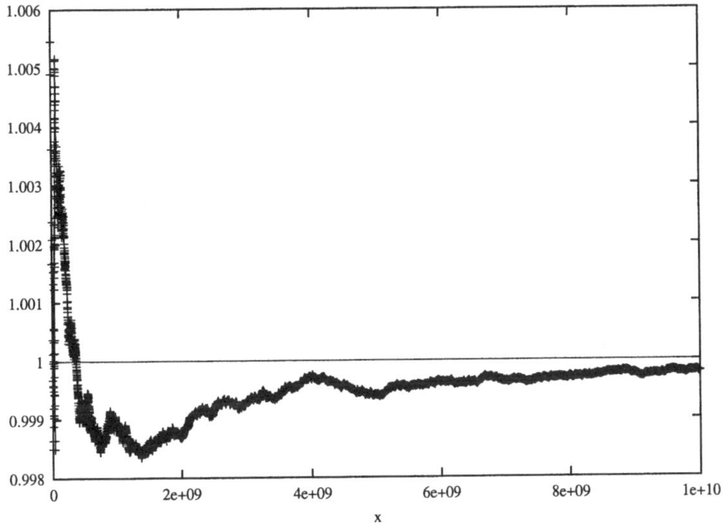

Fig. 1. x vs. $q_3(x)$

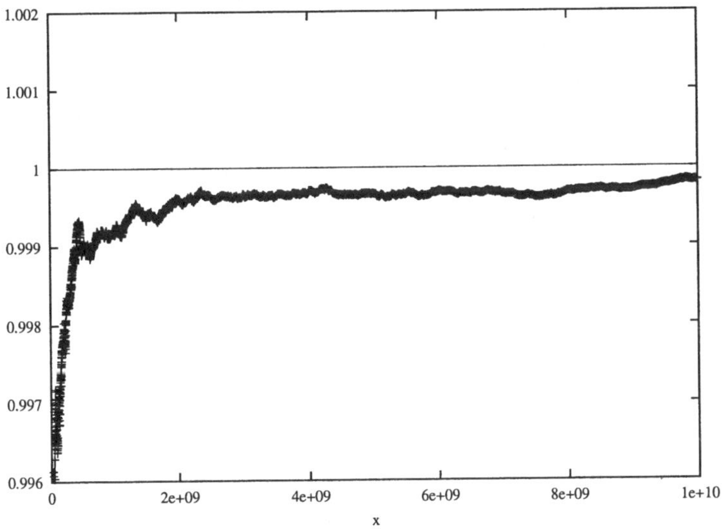

Fig. 2. x vs. $q_5(x)$

- $\chi_p(q) = 1$ for 5 of the next 10 primes.

Since

$$L(1, \chi_p) = \prod_q \left(\frac{q}{q - \chi_p(q)} \right) ,$$

this maximizes the first several terms of the Euler product for $|L(1,\chi_p)|$.

In order to filter out possible prime candidates quickly by recognizing whether or not this characterization holds, we made use of a result of Lehmer [4]. If $p = a^2 + b^2$ as mentioned in Section 1 and $(q/p) = 1$ (clearly if $\chi_p(q) = 1$, then $(q/p) = 1$), then

- $\chi_p(q) = (-1/q)$ if $q \mid b$
- $\chi_p(q) = (-2/q)$ if $q \mid a$
- $\chi_p(q) = (-2\lambda(\lambda+1)/q)$ if $q \not| ab$ and $a \equiv \mu b \pmod{q}$ for any μ such that $\mu^2 \equiv (\lambda^2-1)^{-1} \pmod{q}$ ($\lambda \not\equiv 0, \pm 1 \pmod{q}$).

She also showed that there are exactly $(q - 4 - 3(-1/q) - 2(-2/q))/4$ such values of μ mod q.

For our run we

1. precomputed a table of all possible values of μ for the first 20 primes;
2. precomputed a table of inverses modulo q for each of the first 20 primes.

For $p = a^2 + b^2$, we tested whether $ab^{-1} \equiv \mu \pmod{q}$ for some μ for each q. This required two table look-ups and one multiplication modulo q for each of the 20 values of q.

Once a value of p had survived this process, we computed h_p^- by the techniques of Section 2. This process was used on all primes less than 10^{14}. Out of the 801235712049 primes $p \equiv 5 \pmod{8}$ less than 10^{14}, 59814 had the required quartic character. This computation took 3172 days of CPU time on 269 2.4 GHz Xeon processors running Linux, i.e., about 12 days of real time.

The h_p^-/p hichamps from these primes are listed in Table 4. The largest h_p^-/p found by this method is considerably larger than that of the previous section, but still quite far from 1. Given the slow growth-rate of h_p^-/p, extending the computation to 10^{15}, which would require about 4 months with our resources, seems unlikely to produce a p with $h_p^-/p > 1$.

Another possible strategy would be to strengthen the quartic residuacity requirements for the search in order to reduce the number of candidates for which h_p^- has to be computed. Indeed, searching for p values with more restricted quartic character would likely yield better results, as our best $p = 34131117018541$, has $\chi_p(q) = 1$ for $q \leq 67$, the first 18 odd primes. Unfortunately, such an approach is unlikely to result in a significantly faster search. The strategy we employed filtered out all but 59814 primes (indeed, our filtering parameters were selected so that only a small number of primes passed), and each h_p^- can be evaluated in under 2 minutes. Thus, the vast majority of computing time was spent on the filtering process, and a further reduction in the number of primes p considered will not greatly impact the speed of the search.

We also attempted to use CASSIE, our most recently completed number sieve, to find values for μ modulo the product P of all odd primes up to 43 to 73 and test prime values of $9 + b^2$ and $81 + b^2$, where $b \equiv \mu^{-1}a \pmod{P}$, $b \equiv 2 \pmod{4}$. However, the results were not as good as those found by our previous search.

Table 4. h^-/p hichamps for prescribed $p < 10^{14}$

p	a	b	h^-	h^-/p
926179549	29643	-6890	441494505	0.47668
6033132109	68547	-36530	3381836985	0.56054
17513700229	128223	32750	10341416269	0.59048
178488307141	269679	-325210	106156720801	0.59475
611097213301	78951	-777730	366037283093	0.59898
977833136149	965943	211630	595915703469	0.60942
1636256708629	779223	1014430	1025566748785	0.62678
2749263025909	-212553	-1644410	1723720314185	0.62698
4233467000701	1173051	1690390	2710099480069	0.64016
6514175296381	2512059	-451370	4236111745733	0.65029
6725488386061	2584131	-218530	4401865345149	0.65450
18989849070949	4332207	470990	12549857707309	0.66087
32956902209221	2230239	-5289890	21797169170461	0.66138
34131117018541	-4811229	-3314090	23754887413613	0.69599

4 Constructing a Suitable p

As our search to this point did not reveal any p such that $h_p^- > p$, we attempted to construct such a value of p by using a process similar to that employed in a different context by Teske and Williams [10]. We first computed an approximation to $|L(1,\chi_p)|^2$ by computing

$$\prod_{q \le Q} \left(\frac{q}{q-1}\right)^2 .$$

(We assume $\chi_p(q) = 1$ for all primes $q \le Q$.) We found that if $Q = 257$, then this quantity exceeds $2\pi^2$; thus, since the tail of the modulus of the Euler product is likely to be near 1, by (1) it is reasonable to hope that $h_p^- > p$ for p such that $\chi_p(q) = 1$ for all odd primes $q \le 257$.

To find such values of p we first computed

1. Q_1, the product of all the primes in the set

$$\mathcal{Q}_1 = \{5, 37, 53, 61, 101, 109, 113, 137, 149, 157, 173, 181, 193, 197, 229, 233, 241\} .$$

Note that each of the primes in $\mathcal{Q}_1$ is congruent to 1 mod 4.

2. Q_2, the product of all the primes in the set

$$\mathcal{Q}_2 = \{3, 107, 131, 139, 163, 179, 211, 227, 251\} .$$

Each of the primes in $\mathcal{Q}_2$ is congruent to 3 mod 8.

Now if p is a prime and $p = Q_2^2X^2 + 4Q_1^2$, then by Lehmer's results in Section 3 we have $\chi_p(q) = 1$ for $q \mid Q_1Q_2$. For the remaining primes $q \leq 251$ in the set

$$\mathcal{Q}_3 = \{7, 11, 13, 17, 19, 23, 29, 31, 41, 43, 47, 59, 67, 71, 73, 79, 83, 89, 97, 103, \\ 127, 151, 167, 191, 199, 223, 239\},$$

we found X such that $X \equiv 1 \pmod 2$ and $X \equiv 2Q_1Q_2^{-1}\mu \pmod q$, where $\mu^2 \equiv (\lambda^2 - 1)^{-1} \pmod q$ and $(-2\lambda(\lambda+1)/q) = 1$. Again, we used CASSIE to produce suitable values of X modulo $2Q_3$ where Q_3 is the product of all primes in $\mathcal{Q}_3$. When a prime value of $Q_2^2X^2 + 4Q_1^2$ is produced, we must have $\chi_p(q) = 1$ for all odd primes $q \leq 251$.

By selecting $\mathcal{Q}_1$, $\mathcal{Q}_2$, and $\mathcal{Q}_3$ in this manner, $Q_2^2X^2$ and $4Q_1^2$ will be approximately the same size, and the values of p produced will be almost as small as possible using this method. Although we're only forcing $\chi_p(q) = 1$ for odd primes $q \leq 251$ as opposed to 257, we expect that some solutions p will still have $\chi_p(q) = 1$ for a few values of $q > 251$ as well, and will thus have $h_p^- > p$.

The difficulty with this approach is that the resulting values for p are much too large for h_p^- to be evaluated by our techniques in Section 2. However, we can estimate h_p^-/p by using Bach's [1] technique. We made use of the slightly modified version of this process described in te Riele and Williams [9].

If $S(x) = \sum_{j=0}^{x-1}(x+j)\log(x+j)$, $B(x, \chi_p) = \sum_{q<x} q/(q - \chi_p(q))$, $a_j = (x+j)\log(x+j)/S(x)$, then under the ERH

$$\left|\log L(1, \chi_p) - \sum_{j=0}^{T-1} a_j \log B(T+j, \chi_p)\right| < A(T, p) \ ,$$

where

$$A(T, p) = c(p)G(T) + H(T) \ ,$$

$c(p) = 2/3(\log p + 5/3)$ and $G(x)$, $H(x)$ are defined in [9].

If we put

$$S(T, p) = \sum_{i=0}^{T-1} a_i \mathrm{Re}(\log B(T+j, \chi_p)) = \sum_{q \leq 2T-1} w(q) \log \left|\frac{q}{q - \chi_p(q)}\right| \ ,$$

where

$$w(q) = \begin{cases} 1 & \text{for } q < T \\ \sum_{j=q-T+1}^{T-1} a_j & \text{for } T \leq q < 2T-1 \ , \end{cases}$$

then

$$\left|\log|L(1, \chi_p)| - S(T, p)\right| < A(T, p) \ .$$

Hence

$$\log |L(1, \chi_p)| > S(T,p) - A(T,p) \ . \tag{5}$$

To search for a suitable p, we used CASSIE in conjunction with the Linux cluster described in Section 2 to produce prime values of $p = Q_2^2 X^2 + 4Q_1^2$ as described above. After about 33 hours of computing time, 174 values of X satisfying all the congruence conditions for odd primes $q \leq 239$ were produced, of which 4 yielded prime values of $p = Q_2^2 X^2 + 4Q_1^2$. We could not prove that $h_p^- > p$ for the first three primes, but the fourth prime, the 77-digit

$$p = 1677669146857353273867054733983333681557877980629891674053125487587280304 1429 \ .$$

obtained using $X = 63174133262220373797$ has $\chi_p(q) = 1$ for all odd primes $q \leq 251$ and did yield the desired result. Using Bach's method with $T = 11030000$, we get

$$S(T,p) > 1.50976730, \quad A(T,p) < 0.01499840 \ .$$

This computation took 2 minutes, 13 seconds carrying 53 bits of precision. In order to ensure that our $S(T,p)$ values have sufficient numerical accuracy, we repeated the computation with 300 bits of precision (about 1 hour). These results agreed with the first computation to 12 decimal digits. Thus

$$\begin{aligned} \log |L(1, \chi_p)| &> S(T,p) - A(T,p) \\ &> 1.49476890 \\ &> \log(\sqrt{2\pi^2}) \approx 1.491303476 \ , \end{aligned}$$

which together with (1) and (5) implies that $h_p^- > p$ under the ERH.

5 Conclusions

Finding a smaller such value of p for which $h_p^- > p$, ideally one for which we can prove $h_p^- > p$ unconditionally, remains an open problem. Although tabulating values of h_p^- for $p > 10^{10}$ can be done, the growth rate of h_p^-/p is so slow that it appears to be infeasible to find a suitable p with this method unless a significantly faster method for evaluating h_p^- is developed. Tabulating h_p^- for primes with restricted quartic character and $p > 10^{14}$ also seems unlikely to yield a suitable p value for reasons outlined above. Constructing special values of p does work, but we do not know how to modify this method to find sufficiently small values of p for which Louboutin's algorithm is computationally feasible. One possibility would be to sieve for values of p directly rather than via the special construction we have employed, but it is unknown how to sieve for pseudoquartics in this manner efficiently. Thus, without some new results on either evaluating h_p^- or rapidly generating pseudoquartics, finding the smallest p with $h_p^- > p$, or even any p for which $h_p^- > p$ can be proved unconditionally, is currently beyond our capabilities.

Acknowledgment. The authors wish to thank Stéphane Louboutin for bringing this topic to their attention and for encouraging them to use their techniques to try to gain more information about it.

References

1. E. Bach, *Improved approximations for Euler products*, Number Theory: CMS Proc., vol. 15, Amer. Math. Soc., Providence, RI, 1995, pp. 13–28.
2. B.C. Berndt and R.J. Evans, *The determination of Gauss sums*, Bulletin (New Series) of the AMS **5** (1981), no. 1, 107–129.
3. H. Cohen and J. Martinet, *Class groups of number fields: numerical heuristics*, Math. Comp. **48** (1987), no. 177, 123–137.
4. E. Lehmer, *Criteria for cubic and quadratic residuacity*, Mathematika **5** (1958), 20–29.
5. S. Louboutin, *Computation of relative class numbers of imaginary abelian number fields*, Experimental Mathematics **7** (1998), no. 4, 293–303.
6. ———, *Computation of class numbers of quadratic number fields*, Math. Comp. **71** (2002), no. 240, 1735–1743.
7. V. Shoup, *NTL: A library for doing number theory*, Software, 2001, See http://www.shoup.net/ntl.
8. A. Stein and H.C. Williams, *An improved method for evaluating the regulator of a real quadratic function field*, Algorithmic Number Theory - ANTS-III (Portland, Oregon), Lecture Notes in Computer Science, vol. 1423, Springer-Verlag, Berlin, 1998, pp. 607–620.
9. H.J. te Riele and H.C. Williams, *New computations concerning the Cohen-Lenstra heuristics*, Experimental Mathematics **12** (2003), no. 1, 99–113.
10. E. Teske and H.C. Williams, *A problem concerning a character sum*, Experimental Mathematics **8** (1999), no. 1, 63–72.

Nonic 3-adic Fields

John W. Jones[1] and David P. Roberts[2]

[1] Department of Mathematics and Statistics, Arizona State University, Box 871804, Tempe, AZ 85287
jj@asu.edu

[2] Division of Science and Mathematics, University of Minnesota–Morris, Morris, MN 56267
roberts@mrs.umn.edu

Abstract. We compute all nonic extensions of $\mathbf{Q}_3$ and find that there are 795 of them up to isomorphism. We describe how to compute the associated Galois group of such a field, and also the slopes measuring wild ramification. We present summarizing tables and a sample application to number fields.

1 Introduction

This paper is one of three accompanying our online database of low degree p-adic fields, located at `http://math.la.asu.edu/~jj/localfields/`. The first paper, [10], describes the database in general. There are two cases enormously more complicated than all the others in the range considered, octic 2-adic fields and nonic 3-adic fields. The paper [9] describes the 1823 octic 2-adic fields and this paper describes the 795 nonic 3-adic fields.

Our online database has an interactive feature which allows one to enter an irreducible polynomial $f(x) \in \mathbf{Z}[x]$ and obtain a thorough analysis of the ramification in the corresponding number field $K = \mathbf{Q}[x]/f(x)$. The inclusion of octic 2-adic fields and nonic 3-adic fields in the database greatly extends the number fields K that can be analyzed mechanically by our programs. Certainly, the degree of K can be very much larger than 9.

Section 2 discusses a standard resolvent construction and then three more specialized resolvent constructions for nonic fields with a cubic subfield. Section 3 centers on the Galois theory of nonic fields over general ground fields, describing the 34 possibilities for the Galois group associated to a nonic field. Also this section gives further information useful for our particular ground field $\mathbf{Q}_3$. For example, 11 of the 34 possible Galois groups can be immediately ruled out over $\mathbf{Q}_3$, because they don't have appropriate filtration subgroups. Somewhat coincidentally, the 23 groups that remain are exactly those with a normal Sylow 3-subgroup.

The nonic 3-adic field section of our database would run to some twenty printed pages, so here we give only summarizing tables. Section 4 centers on Table 4.1 which sorts the 795 fields we find according to discriminant and Galois

D.A. Buell (Ed.): ANTS 2004, LNCS 3076, pp. 293–308, 2004.

group. Of the 23 eligible groups, 22 actually appear. Section 5 describes ramification in nonic 3-adic fields in terms of slopes, with Tables 5.1 and 5.2 summarizing our results. Finally, Section 6 gives an application to number fields.

2 Resolvent Polynomials

Resolvents play a major role in the computation of Galois groups over $\mathbf{Q}$. Some resolvents can be computed quickly using exact arithmetic with resultants. However more often one computes resolvents via approximations to complex roots, knowing *a priori* that the resolvents in question have integer coefficients.

The fields studied here are represented by monic polynomials $f \in \mathbf{Z}[x]$. The computation of absolute resolvents can then follow standard methods. However, most applications of resolvents to computing Galois groups in high degree utilize relative resolvents. In the relative case, one somehow gives structure to the roots of f to reflect the fact that $\mathrm{Gal}(f)$ is known to lie in some proper subgroup G^u of S_n. One speaks of resolvents relative to the upper bound group G^u.

A complication for us is that for a given nonic polynomial $\tilde{f} \in \mathbf{Z}[x]$, we may have $\mathrm{Gal}_{\mathbf{Q}_p}(\tilde{f}) \leq G^u$ but $\mathrm{Gal}_{\mathbf{Q}}(\tilde{f}) \not\leq G^u$. In this case, the resolvent will have coefficients which are p-adic integers, but generally not rational integers, and so the method of computing via complex approximations does not work directly. The method [6] of using p-adic approximations rather than complex approximations does not help here. It involves choosing a prime p unramified for the given extension of $\mathbf{Q}$, whereas we are starting with p-adic extensions which are highly ramified.

For our three relative resolvents, the upper bound group G^u is the wreath product $S_3 \wr S_3$, which is the generic Galois group of nonic fields with a cubic subfield. Given $\tilde{f} \in \mathbf{Z}[x]$ which defines a nonic extension of $\mathbf{Q}_3$ with a unique cubic subfield, we work around the problem described in the preceding paragraph by computing $f \in \mathbf{Z}[x]$ which defines the same nonic extension of $\mathbf{Q}_3$, but where f has a corresponding cubic subfield over $\mathbf{Q}$. Then we use complex roots in the resolvent construction applied to f. The computation of f from $\tilde{f}$ is described further in the last paragraph of this section.

We now describe the four resolvent constructions which we will use systematically in the sequel. The first is a standard absolute resolvent. One starts with a degree n polynomial $f(x)$ with complex roots $\alpha_1, \ldots, \alpha_n$. The resolvent corresponds to the subgroup $S_2 \times S_{n-2}$ of S_n and is given by

$$f_{\mathrm{disc}}(x) = \prod_{i<j}(x - (\alpha_i - \alpha_j)^2) \in \mathbf{Z}[x]. \tag{1}$$

It can be computed quickly without approximations to roots via the formula

$$f_{\mathrm{disc}}(x^2) = \mathrm{Resultant}_y(f(y), f(x+y))/x^n. \tag{2}$$

In one case, we will also make use of the variant $f_{\mathrm{Disc}}(x) = f_{\mathrm{disc}}(x^2)$, which is itself an absolute resolvent for $S_1 \times S_1 \times S_{n-2} < S_n$. In general, we will systematically denote polynomial resolvent constructions by $f \mapsto f_*$ for some symbol $*$. We will use similar notation for the same resolvent constructions on the level of fields starting in the next section.

For the remaining three resolvents, we start with an irreducible monic degree nine polynomial $f \in \mathbf{Z}[x]$ such that $\mathbf{Q}[x]/f(x)$ has a unique cubic subfield. We choose $g(y) \in \mathbf{Z}[y]$ so that $\mathbf{Q}[y]/g(y)$ is isomorphic to this subfield. Let

$$h(y,x) = x^3 + \sum_{k=0}^{2}\sum_{\ell=0}^{2} c_{k\ell}y^k x^\ell \tag{3}$$

be a cubic factor of $f(x)$ over $\mathbf{Q}[y]/g(y)$. Let $\beta_1, \beta_2, \beta_3$ be the complex roots of $g(y)$. For $i = 1, 2, 3$, let $\alpha_{i,1}$, $\alpha_{i,2}$ and $\alpha_{i,3}$ be the complex roots of $h(\beta_i, x)$. Then we can recover

$$f(x) = \prod_{i=1}^{3}\prod_{j=1}^{3}(x - \alpha_{i,j})\,. \tag{4}$$

A formula bypassing the $\alpha_{i,j}$ is

$$f(x) = \mathrm{Resultant}_y(g(y), h(y,x)). \tag{5}$$

The three resolvents are

$$f_{18}(x) = \prod_{i=1}^{3}\prod_{\sigma\in S_3}\left(x - \sum_{j=1}^{3}\alpha_{i,j}\alpha_{i+1,\sigma(j)}\right), \tag{6}$$

$$f_{27}(x) = \prod_{i=1}^{3}\prod_{j=1}^{3}\prod_{k=1}^{3}(x - (\alpha_{1,i} + \alpha_{2,j} + \alpha_{3,k}))\,, \tag{7}$$

$$f_{36}(x) = \prod_{\sigma\in S_3}\prod_{\tau\in S_3}\left(x - \sum_{i=1}^{3}\alpha_{1,i}\alpha_{2,\sigma(i)}\alpha_{3,\tau(i)}\right). \tag{8}$$

Now we return to describing how we adjust a given polynomial $\tilde{f}$ to a better one f. We first find a global model $\mathbf{Q}[y]/g(y)$ for the cubic subfield of the extension $\mathbf{Q}_3[x]/\tilde{f}(x)$ of $\mathbf{Q}_3$; this is easy using the database [10]. Then, we use Algorithm 3.6.4 from [2] to reduce the factorization of $\tilde{f}(x)$ over $\mathbf{Q}_3[y]/g(y)$ to a factorization problem over $\mathbf{Q}_3$. We loosely approximate a cubic factor $\tilde{h}(y,x) \in \mathbf{Z}_p[x]$ that we obtain by a polynomial $h(y,x) \in \mathbf{Z}[y,x]$, and use Equation (5) to compute a candidate for f. Finally, we test if our candidate f defines the same nonic field over $\mathbf{Q}_3$ as $\tilde{f}$ by using Panayi's p-adic root finding algorithm [13]. If it doesn't, we repeat with a better approximation to $\tilde{h}(y,x)$. Since we ultimately use complex roots in computing resolvents f_{18}, f_{27}, and f_{36}, we aim throughout to keep the coefficients of f relatively small.

3 Galois Theory of Nonics

We will set things up over a general base field F, and specialize when necessary to our case of interest, $F = \mathbf{Q}_3$. We work in the context of abstract separable fields of finite degree over F. We say a nonic field K is multiply imprimitive, uniquely imprimitive, or primitive, iff it has ≥ 2, 1, or 0 cubic subfields respectively.

To bring in Galois theory, we imagine that a separable closure $\overline{F}$ of F is given. The number of subfields of $\overline{F}$ isomorphic to a separable degree n extension K/F is $n/|\operatorname{Aut}(K)|$. We associate to K is the n-element set X of homomorphisms $\sigma : K \to \overline{F}$. Let K^{gal} be the subfield of $\overline{F}$ generated by the images of these n homomorphisms. Then we call $G = \operatorname{Gal}(K^{\text{gal}}/F)$ the Galois group of K with respect to the fixed separable closure of $\overline{F}$. So G is a transitive subgroup of the symmetric group S_X and a quotient group of the absolute Galois group $\operatorname{Gal}(\overline{F}/F)$. Occasionally we will use this notation when K is a separable algebra which is only a product of fields. Then G is no longer a transitive subgroup of S_X as indeed its minimal orbits correspond to the factor fields of K.

In the case $F = \mathbf{Q}$, one can take $\overline{\mathbf{Q}} \subset \mathbf{C}$ as a separable closure. However in other cases, like our case $F = \mathbf{Q}_3$, there is no simple choice of $\overline{F}$ and in practice one must work with objects which are independent of the choice of $\overline{F}$. We will therefore consider the Galois group of K to be a subgroup of S_n which is only defined up to conjugation.

There are 34 transitive subgroups of S_9 up to conjugation, 4 corresponding to multiply primitive fields, 19 to uniquely imprimitive fields, and 11 to primitive fields. So, given a nonic field K, one wants first to identify its Galois group among the 34 possibilities. The literature contains several accounts of computing Galois groups, with [8] being a recent survey. Some of these accounts pay particular attention to nonics [7,4]. The approach presented here is tailored to 3-adic fields, where it is easier to compute subfields and automorphism groups than it is to work with many different relative resolvents and/or large degree absolute resolvents.

Twenty-three of the thirty-four groups have just one Sylow 3-subgroup, while the remaining seven solvable groups have four Sylow 3-subgroups. The twenty-three groups will be particularly important for us and a partial inclusion diagram for just these groups is given in Figure 1. We use the T-notation of [1] to indicate the possible Galois groups, with T standing for transitive. This T-notation will be our main notation in the sequel as well. However in Tables 3.1, 3.2, and 3.3 we will also present a more descriptive notation based on [5].

In Figure 1, a line from T_i down to T_j means there are subgroups $G_i, G_j \subset S_n$ of type T_i, T_j respectively, with $G_i \subset G_j$ and $|G_i| = |G_j|/2$. The text in Figure 1 briefly indicates how some of the phenomena presented in Tables 3.1, 3.2, and 3.3 relate to index two inclusions.

Over $F = \mathbf{Q}_3$, some of the groups can be easily ruled out. In general, let F be a p-adic field, meaning a finite extension of either $\mathbf{Q}_p$ of $\mathbf{F}_p((t))$. Then a Galois extension K^{gal}/F has a filtration by Galois subfields,

$$F \subseteq K^{\text{gal},u} \subseteq K^{\text{gal},t} \subseteq K^{\text{gal}}, \tag{9}$$

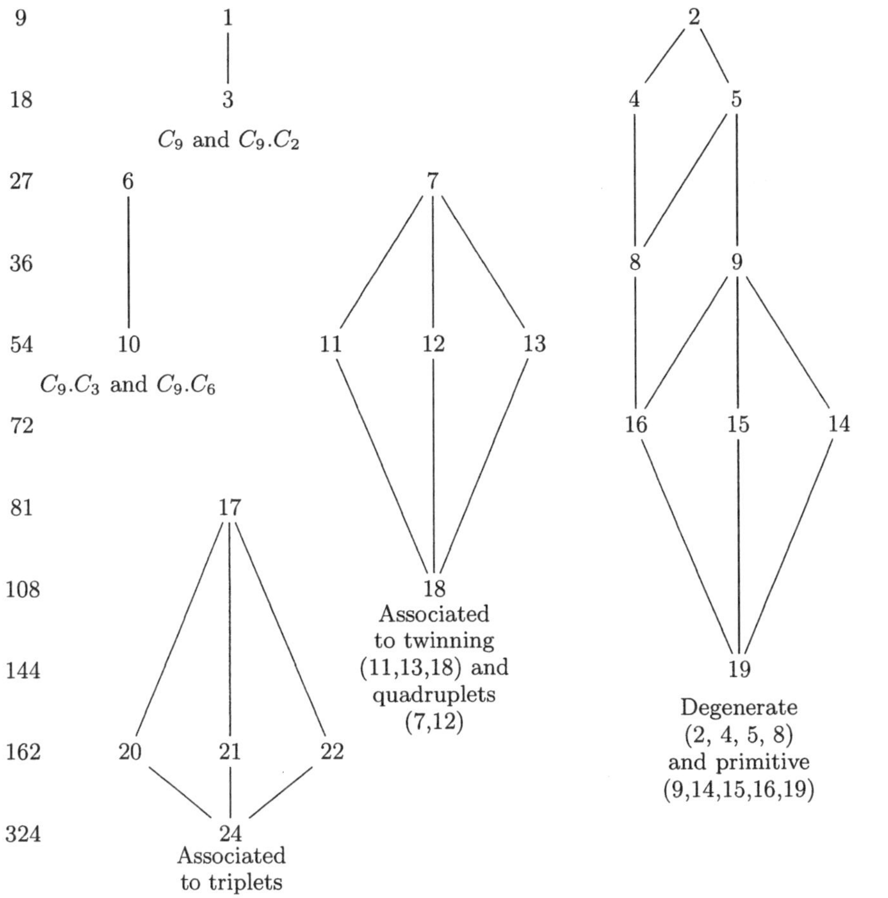

Fig. 1. Nonic groups having a normal Sylow 3-subgroup and their index two inclusions

with $K^{\text{gal},u}$ being the maximal unramified subextension and $K^{\text{gal},t}$ being the maximal tamely ramified subextension. The group $\text{Gal}(K^{\text{gal},u}/F)$ is necessarily cyclic. Similarly, $\text{Gal}(K^{\text{gal},t}/K^{\text{gal},u})$ is cyclic; moreover, this subquotient has order prime to p. Finally $\text{Gal}(K^{\text{gal}}/K^{\text{gal},t})$ is a p-group. Eleven of the thirty-four candidate Galois groups fail to have a corresponding chain of normal subgroups. Five of these excluded groups are uniquely imprimitive and six are primitive. They are presented with a dash in the last column in Table 3.2 and Table 3.3 respectively. The twenty-three groups which do have a corresponding chain are exactly those in Figure 1.

There are different approaches for identifying $\text{Gal}(K^{\text{gal}}/F)$. We begin by computing three quantities directly associated to K, the cubic subfields of K, the automorphism group of K, and the parity of K. There can be one, two, or four cubic subfields, and the possible Galois groups for a cubic subfield are C_3

and S_3. The automorphism group Aut(K) has nine elements exactly when K is itself Galois, in which case $\mathrm{Aut}(K) \cong \mathrm{Gal}(K^{\mathrm{gal}}/F)$. Otherwise $|\mathrm{Aut}(K)|$ is 1 or 3 while $|\mathrm{Gal}(K^{\mathrm{gal}}/F)| > 9$. In our case of $F = \mathbf{Q}_3$, we used Panayi's root finding algorithm [13] for the computation of both subfields and automorphism groups. Parities are easier. By definition $K = \mathbf{Q}[x]/f(x)$ has parity $\epsilon = +$ if the polynomial discriminant of f is a square in F and $\epsilon = -$ otherwise. In the former case $\mathrm{Gal}(K^{\mathrm{gal}}/F)$ is in A_9 and in the latter case it is not.

Tables 3.1, 3.2, and 3.3 have columns corresponding to the objects just discussed. A blank for $|\mathrm{Aut}(K)|$ signifies that K has only the identity automorphism. Tables 3.2 and 3.3 also present information related to resolvents. A resolvent construction $f \mapsto f_*$ on the level of polynomials induces a well-defined resolvent construction $K \mapsto K_*$ on the level of algebras, where $K = F[x]/f(x)$ and $f(x)$ is chosen so that $f_*(x)$ is separable, in which case $K_* = F[x]/f_*(x)$. Finally, the column headed by # previews the next section by giving the number of nonic 3-adic fields with the given Galois group. If the Galois group is ruled out by the lack of an unramified/tame/wild filtration, we print a dash rather than a 0, as mentioned before.

3.1 Multiply Imprimitive Fields

Suppose K is a nonic field with more than one cubic subfield. Then if K_1 and K_2 are any two distinct cubic subfields, $K = K_1 \otimes K_2$. Table 3.1 gives the four possible Galois groups. In two cases, there are two cubic subfields, and in two cases there are four cubic subfields, as indicated. In this category, no resolvents are necessary for distinguishing the Galois groups

Table 3.1. Nonic groups corresponding to nonic fields with more than one cubic subfield

G	**Name**	$\lvert G\rvert$	$\lvert A\rvert$	**Subs**	ϵ	#
2	$E(9)$	3^2	9	$CCCC$	+	1
4	$S_3 \times 3$	$2^1 3^2$	3	SC	−	24
5	$3^2 : 2$	$2^1 3^2$		$SSSS$	+	1
8	$S_3 \times S_3$	$2^2 3^2$		SS	−	9

3.2 Uniquely Imprimitive Fields

It is here that we will use the three specialized resolvent constructions of the previous section. They let us canonically construct algebras K_{18}, K_{27}, and K_{36} of the indicated degree from a nonic field K with a unique cubic subfield. Table 3.2 gives the nineteen possibilities for the Galois group of a nonic field with a unique cubic subfield.

Table 3.2. Nonic groups corresponding to nonic fields with exactly one cubic subfield. The first fourteen groups have one Sylow 3-subgroup and the last five groups have four Sylow 3-subgroups.

G	**Name**	$\|G\|$	$\|A\|$	**Sub**	ϵ	K_{9b}	K_{9x}	K_{27}	K_{27b}	K_{9y}	m	#
1	$C(9)$	3^2	9	C	+	T_1	C_3^3	T_1^3	T_1^3	C_3^3		12
3	$D(9)$	$2^1 3^2$		S	+	T_3	S_3^3	T_3^3	T_3^3	S_3^3		5
6	$\frac{1}{3}[3^3]3$	3^3	3	C	+	T_6	T_2	27_+	27_+	T_2		8
7	$[3^2]3$	3^3	3	C	+	T_7	T_2	$T_7'T_7''T_7'''$	27_+	$C_3'C_3''C_3'''$	4	4
10	$[3^2]S_{3;6}$	$2^1 3^3$		S	+	T_{10}	T_4	27_+	27_+	T_4		49
11	$E(9):6$	$2^1 3^3$		S	+	T_{11}	T_4	$T_{13}18_-$	27_+	$G_{6,18-}C_3$	2^*	20
12	$[3^2]S_3$	$2^1 3^3$	3	S	−	T_{12}	T_5	$T_{12}'T_{12}''T_{12}'''$	27_+	$S_3'S_3''S_3'''$	4	36
13	$E(9):D_6$	$2^1 3^3$		C	−	T_{13}	T_4	$T_{11}18_-$	27_+	$G_{6,18-}C_3$	2^*	20
17	$3 \wr 3$	3^4	3	C	+	T_{17}'	T_7	27_+	27_+	T_7	3	36
18	$E(9):D_{12}$	$2^2 3^3$		S	−	T_{18}	T_8	$T_{18}'18_-$	27_+	$G_{6,36-}S_3$	2	48
20	$3 \wr S_3$	$2^1 3^4$	3	S	−	T_{20}'	T_{11}	27_-	27_+	T_{13}	3	180
21	$\frac{1}{2}[3^2:2]S_3$	$2^1 3^4$		S	+	T_{21}'	T_{12}	27_+	27_+	T_{12}	3	108
22	$[3^3:2]3$	$2^1 3^4$		C	−	T_{22}'	T_{13}	27_-	27_+	T_{11}	3	60
24	$[3^3:2]S_3$	$2^2 3^4$		S	−	T_{24}'	T_{18}	27_-	27_+	T_{18}	3	144

G	**Name**	$\|G\|$	$\|A\|$	**Sub**	ϵ	K_{18}	K_{27}	K_{36}	m	#
25	$[\frac{1}{2}S_3^3]3$	$2^2 3^4$		C	+	18_+	27_+	36_+		–
28	$S_3 \wr 3$	$2^3 3^4$		C	−	18_+	27_-	36_+		–
29	$[\frac{1}{2}S_3^3]S_3$	$2^3 3^4$		S	−	18_-	27_-	36_-		–
30	$\frac{1}{2}[S_3^3]S_3$	$2^3 3^4$		S	+	18_-	27_+	36_-		–
31	$S_3 \wr S_3$	$2^4 3^4$		S	−	18_-	27_-	36_-		–

The resolvents K_{18}, K_{27} and K_{36} are sometimes irreducible. In this case, the corresponding slots on Table 3.2 contain 18_+, 18_-, 27_+, 27_-, 36_+ or 36_-. In general, we indicate a field of degree > 9 by giving its degree and its parity. We indicate a field of degree ≤ 9 by giving its Galois group.

When G has just one Sylow 3-subgroup there are unique factorizations

$$K_{18} = K_{9b} \times K_{9x}, \tag{10}$$

$$K_{36} = K_{27b} \times K_{9y}, \tag{11}$$

with factors having the indicated degree and the properties

$$\begin{aligned} K_{9b}^{\text{gal}} &= K_{27b}^{\text{gal}} = K^{\text{gal}}, \\ K_{9x}^{\text{gal}} &= K_{9y}^{\text{gal}} \overset{3}{\subset} K^{\text{gal}}. \end{aligned} \tag{12}$$

Here the superscript 3 means that $[K^{\text{gal}} : K_{9y}^{\text{gal}}] = 3$. Table 3.2 indicates the structure of K_{9b}, K_{9x}, K_{27b}, and K_{9y}. In the K_{9y} column, $G_{6,18-}$ and $G_{6,36-}$ are sextic groups of order 18 and 36 respectively which are not contained in A_6.

The m heading a column in Table 3.2 stands for multiplicity. In general, we say that a k-tuplet of degree n is a complete list of non-isomorphic degree n fields $K_1, \ldots, K_k$ with K_j^{gal} all the same. One speaks of singletons, twins, triplets, and quadruplets, for $k = 1, 2, 3$, and 4 respectively. The T_i row has a k in the m column iff a nonic field K with $\mathrm{Gal}(K^{\mathrm{gal}}/F) = T_i$ belongs to a k-tuple. Here, 1 is indicated by a blank. In most cases, fields in a tuplet share the same T_i. The one exception, indicated by a $*$ in Table 3.2, is twins consisting of a T_{11} nonic field and a T_{13} nonic field. In fact, the thirty-four T_i fall into thirty-three isomorphism classes as abstract groups, the one coincidence being $T_{11} \cong T_{13}$.

Our priming convention in the resolvent columns of Table 3.2 is to distinguish different fields with the same Galois group. A T_i in the i row means a nonic field isomorphic to the original one. Note that in the instances of twinning, one can always pass from a field to its twin via K_{27}. In the five instances of triplets, it is K_{18} which lets one pass from a nonic field $K = K_a$ to a second triplet K_{9b}. Applying this degree eighteen resolvent construction to K_{9b} gives the remaining triplet K_{9c}. Finally, applying it to K_{9c} returns K_{9a}. Thus any collection of nonic triplets comes with a natural cyclic order. Finally, in the two instances of quadruplets, one can pass from a given field to the three others via K_{27}.

Table 3.2 makes clear how one can compute Galois groups of uniquely imprimitive nonic fields. The cases of one versus four Sylow 3-subgroups are distinguished by the reducible versus reducibility of K_{18} or equally well K_{36}. Within the one Sylow 3-subgroup case, $|A|$, **Sub**, ϵ, and K_{9y} suffice to distinguish groups. Within the four Sylow 3-subgroup case, $|A|$, **Sub**, and ϵ alone distinguish groups, except for T_{29} versus T_{31} for which the printed resolvent information doesn't help. In our setting of $F = \mathbf{Q}_3$, neither T_{29} or T_{31} arises, but one way to distinguish them is by the discriminant resolvent. Recall $f_{\mathrm{Disc}}(x) = f_{\mathrm{disc}}(x^2)$. In both cases, K_{Disc} factors as a degree 18 field times a degree 54 field. The degree 18 field has even parity in the case T_{29} and odd parity in the case T_{31}.

3.3 Primitive Fields

Neither subfields nor automorphisms can help with determining the Galois group of a primitive nonic field. The parity ϵ remains helpful, as shown in Table 3.3.

Table 3.3 also gives the degrees and parities of the field factors of the degree 36 resolvent K_{disc}. The information presented in Table 3.3 clearly suffices to identify Galois groups associated to nonic 3-adic fields, or to even nonic extensions of general 3-adic bases F where T_{15} might appear. For general bases, more resolvents of higher degrees would be required ([4,7]).

4 Nonic 3-adic Fields by Discriminant and Galois Group

In general, let $\mathcal{K}(p,n)$ be the set of isomorphism classes of degree n extension fields of $\mathbf{Q}_p$. The paper [14] describes how one goes about finding polynomials $f_i(x) \in \mathbf{Z}[x]$ such that $\mathbf{Q}_p[x]/f_i(x)$ runs over $\mathcal{K}(p,n)$. One key ingredient is the

Table 3.3. Nonic groups corresponding to primitive nonic fields

G	**Name**	$\|G\|$	ϵ	K_{disc}	#
9	$E(9):4$	$2^2 3^2$	$+$	18_-^2	2
14	$E(9):Q_8$	$2^3 3^2$	$+$	36_+	4
15	$E(9):8$	$2^3 3^2$	$+$	36_-	0
16	$E(9):D_8$	$2^3 3^2$	$-$	$18_-\ 18'_-$	4
19	$E(9):2D_8$	$2^4 3^2$	$-$	36_-	20
23	$E(9):2A_4$	$2^3 3^3$	$+$	36_+	$-$
26	$E(9):2S_4$	$2^4 3^3$	$-$	36_-	$-$
27	$PSL(2,8)$	$2^3 3^2 7$	$+$	36_+	$-$
32	$P\Sigma L(2,8)$	$2^3 3^3 7$	$+$	36_+	$-$
33	A_9	$2^6 3^5 5^1 7$	$+$	36_+	$-$
34	S_9	$2^7 3^5 5^1 7$	$-$	36_-	$-$

root finding algorithm of [13], mentioned already in both of the previous sections. Here, it is used to ensure that $\mathbf{Q}_p[x]/f_i(x)$ and $\mathbf{Q}_p[x]/f_j(x)$ are not isomorphic for $i \neq j$. The other key ingredient is the mass formula of [11,15], as refined in [14]. One knows that one has found enough polynomials to cover all of $\mathcal{K}(p,n)$ by the use of this formula. In this section, we present Table 4.1 summarizing the result of the calculation in the case $(p,n) = (3,9)$, and comment on several features of the table.

Given a nonic 3-adic field K, let K^u be its maximal subfield unramified over $\mathbf{Q}_3$. Let $f = [K^u : \mathbf{Q}_3]$ and $e = 9/f$. The original mass formula of [11,15] makes it natural to divide nonic fields K into three classes according to e. The unique field with $e = 1$ corresponds to the boldface entry with $c = 0$ in the row for group $T_1 = C_9$. The 41 fields with $e = 3$ correspond to italicized entries, and the remaining 753 are given in ordinary type.

The refined mass formula of [14] makes it natural to further divide nonic fields K according to their discriminant exponent c. This exponent indexes columns in Table 4.1. Mass formulae can be used to count the number of subfields of $\overline{F}$ of a given type. Recall from Section 3 that if a field K has automorphism group A, then this count will reflect the $9/|A|$ isomorphic copies of K in $\overline{F}$. Collecting these contributions to the mass formula and dividing by the degree 9 gives a count of isomorphism classes of fields, each weighted by their mass $1/|A|$. The sums of these masses for a given (e,c) are given by the refined mass formula in [14], and are presented as $M_e(c)$ towards the bottom of Table 4.1. Of course, one has $\#_e(c) \geq M_e(c)$ with $\#_e(c)$ the total number of fields for a given (e,c).

The lines corresponding to the four multiply imprimitive groups can be constructed directly by tensoring pairs of cubic fields. For example, as K_c runs over the four C_3 fields and K_s runs over the six S_3 fields, $K_c \otimes K_s$ runs over the twenty-four $T_4 = C_3 \times S_3$ fields. Similarly, the primitive groups T_9 and T_{16} are isomorphic to the sextic transitive groups $C_3^2.C_4$ and $C_3^2.D_4$ respectively. In each of these two cases, one nonic field K comes from two sextic fields K_{6a}, K_{6b} with

Table 4.1. Discriminants and Galois groups of nonic extensions of $\mathbf{Q}_3$. Fields with $e = 1, 3, 9$ are respectively indicated by bold, italic, and regular type.

G	$\|A\|$	0	9	10	12	13	15	16	18	19	20	21	22	23	24	25	26	#
2	9				*1*													1
4	3		*2*		*1*		6, *3*	3		9								24
5									1									1
8					1		2		3	3								9
1	9	**1**			*2*								9					12
3									1				1				3	5
6	3				*2*								6					8
7	3				*1*			3										4
10								6	11				8				24	49
11					2			1	8				9					20
12	3									9		27						36
13			*2*		*1*		*2*		3	3		9						20
17	3				*9*				9				18					36
18						2	4		3	12		18	9					48
20	3					6	12		9	45			27			81		180
21											27				27		54	108
22			*6*		*3*		*6*					9	9	27				60
24								6	12	9	27	9		27	27	27		144
9					1			1										2
14				1					3									4
15																		0
16				1					3									4
19			2		2	2	6	2		6								20
$\#_1$		**1**																**1**
M_1		$\mathbf{0.\overline{1}}$																
$\#_3$			*10*		*20*		*11*											*41*
M_3			$8.\overline{6}$		$8.\overline{6}$		*9*											
$\#_9$			2	2	6	10	30	22	66	96	54	72	96	54	54	108	81	753
M_9			2	2	6	6	18	18	54	54	54	54	54	54	54	54	81	
$M_{9,3}$					4	4	12	12	36	36	54							
$M_{9,4}$							4	4	12	12		36	36	54				
$M_{9,5}$									6	6		18	18		54	54	81	

$K^{\text{gal}} = K^{\text{gal}}_{6a} = K^{\text{gal}}_{6b}$. Otherwise the lines of Table 4.1 cannot be constructed from the lower degree tables of [10].

Twins, triplets, and quadruplets are visible in varying degrees on Table 4.1. In general, for a nonic 3-adic field K with discriminant 3^c, one has $c = 2s_u + 6s_v$ with the slopes $s_u \leq s_v$ discussed further in the next section. In a tuplet, s_v is always constant. For twins, s_u typically varies within a twin pair; however one can at least see that the total number of fields for T_{11} and T_{13} is the same and the total number for T_{18} is even. In a triplet, the cubic subfield is constant; if $3^{c_{\text{sub}}}$ is its discriminant then $s_u = c_{\text{sub}}/2$, so c_u and hence c is constant. This explains conceptually why all entries on rows 17, 20, 21, 22, and 24 are multiples of 3. For quadruplets, the possible c_{sub}'s are $(0, 4, 4, 4)$ in the case T_7 and $(3, 5, 5, 5)$ in the case T_{12}, explaining the structure of these rows.

The mass formulas mentioned already come from specializing mass formulas for general p-adic base fields F to the case F is an unramified extension of $\mathbf{Q}_3$ of degree 9, 3, or 1. One can also specialize these formulas to the case that F is a ramified cubic extension of $\mathbf{Q}_3$, and one gets the last three lines of Table 4.1.

The number in the $M_{9,c_{\mathrm{sub}}}$ row and the c column is the total mass of isomorphism classes of nonic fields of discriminant 3^c with a specified cubic subfield of discriminant $3^{c_{\mathrm{sub}}}$, where a pair $K_{\mathrm{sub}} \subset K$ is counted with mass $1/|A_0|$ where A_0 is the group of automorphisms of K stabilizing K_{sub}. This is simplest in the cases $c \geq 20$ where all fields are uniquely primitive and thus $|A| = |A_0|$.

5 Slopes in Nonic 3-adic Fields

The paper [10] provides background on slopes in general. Here we will keep the discussion focused on nonic 3-adic fields K and their associated Galois fields K^{gal}. The results of the calculations are summarized in Tables 5.1 and 5.2.

Suppose $\mathrm{Gal}(K^{\mathrm{gal}}/\mathbf{Q}_3)$ has order $2^a 3^b$, so that $b \in \{2,3,4\}$. Then there are b slopes to compute, which we always index in weakly increasing order, $s_1 \leq s_2 \ (\leq \cdots)$. Consider a chain of subfields

$$\mathbf{Q}_3 \overset{2^a}{\subseteq} K^{\mathrm{gal},0} \overset{3}{\subset} K^{\mathrm{gal},1} \overset{3}{\subset} K^{\mathrm{gal},2} \left(\overset{3}{\subset} \cdots \right), \tag{13}$$

with, as indicated, $K^{\mathrm{gal},j}$ having degree $2^a 3^j$ over $\mathbf{Q}_3$. There is only one choice for $K^{\mathrm{gal},0}$ as, in all 23 cases not excluded by (9), the group T_i has only one Sylow

Table 5.1. Slopes in nonic 3-adic fields with discriminant exponent $c \leq 18$. Visible slopes are in boldface and hidden slopes in ordinary type.

c	G	**Slopes**	#
0	1	**0, 0**	1
9	4	**0, 1.5**	2
9	13	**0, 1.5**, 1.5	2
9	22	**0, 1.5**, 1.5, 1.5	6
9	19	**1.125, 1.125**	2
10	14	**1.25, 1.25**	1
10	16	**1.25, 1.25**	1
12	2	**0, 2**	1
12	4	**0, 2**	1
12	8	**1.5, 1.5**	1
12	1	**0, 2**	2
12	6	**0, 2**, 2	2
12	7	**0, 2**, 2	1
12	11	0, **1.5, 1.5**	2
12	13	**0, 2**, 2	1
12	17	**0, 2**, 2, 2	9
12	22	**0, 2**, 2, 2	3
12	9	**1.5, 1.5**	1
12	19	**1.5, 1.5**	2

c	G	**Slopes**	#
13	18	**1.5**, 1.5, **1.667**	2
13	20	0, **1.5**, 1.5, **1.667**	6
13	19	**1.625, 1.625**	2
15	4	**1.5, 2**	6
15	4	**0, 2.5**	3
15	8	**1.5, 2**	2
15	13	**0**, 1.5, **2.5**	2
15	18	**1.5**, 1.5, **2**	4
15	20	0, **1.5**, 1.5, **2**	12
15	22	**0**, 1.5, 1.5, **2.5**	6
15	19	**1.875, 1.875**	6
16	4	**2, 2**	3
16	7	0, **2, 2**	3
16	10	**1.5**, 2, **2.167**	6
16	11	0, **2, 2**	1
16	24	**1.5**, 1.5, 2, **2.167**	6
16	9	**2, 2**	1
16	19	**2, 2**	2

c	G	**Slopes**	#
18	5	**1.5, 2.5**	1
18	8	**1.5, 2.5**	3
18	3	**1.5, 2.5**	1
18	10	0, **1.5, 2.5**	2
18	10	**1.5**, 2, **2.5**	9
18	11	0, **1.5, 2.5**	2
18	11	**2**, 2, **2.333**	3
18	11	**1.5**, 2, **2.5**	3
18	13	**2**, 2, **2.333**	3
18	17	0, **2**, 2, **2.333**	9
18	18	**1.5**, 2, **2.5**	3
18	20	0, **2**, 2, **2.333**	9
18	24	**1.5**, 1.5, 1.667, **2.5**	3
18	24	**1.5**, 1.5, 2, **2.5**	9
18	14	**2.25, 2.25**	3
18	16	**2.25, 2.25**	3

Table 5.2. Visible and hidden slopes in nonic 3-adic fields with discriminant exponent $c \geq 19$.

c	G	Slopes	#
19	4	$\mathbf{2}, \mathbf{2.5}$	9
19	8	$\mathbf{2}, \mathbf{2.5}$	3
19	12	$\mathbf{1.5}, 2.5, \mathbf{2.667}$	9
19	13	$1.5, \mathbf{2}, \mathbf{2.5}$	3
19	18	$1.5, \mathbf{2}, \mathbf{2.5}$	3
19	18	$\mathbf{1.5}, 2.5, \mathbf{2.667}$	9
19	20	$0, \mathbf{1.5}, 2.5, \mathbf{2.667}$	18
19	20	$\mathbf{1.5}, 2, 2.5, \mathbf{2.667}$	27
19	24	$\mathbf{1.5}, 2, 2.5, \mathbf{2.667}$	9
19	19	$\mathbf{2.375}, \mathbf{2.375}$	6
20	21	$\mathbf{1.5}, 2.5, 2.667, \mathbf{2.833}$	27
20	24	$\mathbf{1.5}, 2.5, 2.667, \mathbf{2.833}$	27
21	12	$1.5, \mathbf{2.5}, \mathbf{2.667}$	27
21	13	$\mathbf{2}, 2.5, \mathbf{2.833}$	9
21	18	$1.5, \mathbf{2.5}, \mathbf{2.667}$	9
21	18	$\mathbf{2}, 2.5, \mathbf{2.833}$	9
21	22	$1.5, \mathbf{2}, 2.5, \mathbf{2.833}$	9
21	24	$1.5, \mathbf{2}, 2.5, \mathbf{2.833}$	9

c	G	Slopes	#
22	1	$\mathbf{2}, \mathbf{3}$	9
22	3	$\mathbf{2}, \mathbf{3}$	1
22	6	$0, \mathbf{2}, \mathbf{3}$	6
22	10	$0, \mathbf{2}, \mathbf{3}$	2
22	10	$\mathbf{2}, 2, \mathbf{3}$	6
22	11	$2, \mathbf{2.5}, \mathbf{2.833}$	9
22	17	$0, \mathbf{2}, 2, \mathbf{3}$	18
22	18	$2, \mathbf{2.5}, \mathbf{2.833}$	9
22	20	$\mathbf{2}, 2, 2.333, \mathbf{3}$	27
22	22	$\mathbf{2}, 2, 2.333, \mathbf{3}$	9
23	22	$\mathbf{2}, 2.5, 2.833, \mathbf{3.167}$	27
23	24	$\mathbf{2}, 2.5, 2.833, \mathbf{3.167}$	27
24	21	$1.5, \mathbf{2.5}, 2.667, \mathbf{3.167}$	27
24	24	$1.5, \mathbf{2.5}, 2.667, \mathbf{3.167}$	27
25	20	$2, \mathbf{2.5}, 2.833, \mathbf{3.333}$	81
25	24	$2, \mathbf{2.5}, 2.833, \mathbf{3.333}$	27
26	3	$\mathbf{2.5}, \mathbf{3.5}$	3
26	10	$0, \mathbf{2.5}, \mathbf{3.5}$	6
26	10	$2, \mathbf{2.5}, \mathbf{3.5}$	18
26	21	$1.5, \mathbf{2.5}, 2.667, \mathbf{3.5}$	54

3-subgroup. There may be several choices for some of the intermediate $K^{\mathrm{gal},j}$, and then of course $K^{\mathrm{gal},b} = K^{\mathrm{gal}}$. We require that the intermediate fields be chosen such that the discriminant exponents c_j are the minimum possible for Galois subfields of degree $2^a 3^j$; the c_j are then uniquely defined. The jth slope is then given by the formula

$$s_j = \frac{c_j - c_{j-1}}{2^a 3^j - 2^a 3^{j-1}}, \tag{14}$$

an instance of Proposition 3.4 of [10]. Note that we are allowing 0 as a slope here. Otherwise the slopes are > 1 and correspond to wild ramification.

Two of the b slopes just discussed are visible in K itself in the following sense. Let 3^c be the discriminant of K. If K has a cubic subfield, let K_{sub} be a cubic subfield of minimum discriminant $3^{c_{\mathrm{sub}}}$. Then from the tower $\mathbf{Q}_3 \subset K_{\mathrm{sub}} \subset K$, one has slopes $s_u = c_{\mathrm{sub}}/2$ and $s_v = (c - c_{\mathrm{sub}})/6$, another instance of Proposition 3.4 of [10]. In this setting, $s_u \leq s_v$. If K is primitive, then the visible slopes are $s_u = s_v = c/8$. An important point is that the highest slope is always visible, i.e. $s_v = s_b$.

We call the remaining $b - 2$ slopes hidden. *A priori*, one might have expected that to calculate them, one would have to compute field discriminants of resolvents of relatively high degree, perhaps 27 say, and apply Proposition 3.4

of [10] yet again. However, as we explain next, in fact one needs to compute discriminants only of associated nonics.

First, if K has ≥ 2 cubic subfields or 0 cubic subfields, then $b = 2$ and so there are no hidden slopes to compute. Now consider the 14 possible Galois groups of a uniquely imprimitive 3-adic nonic. From Table 3.2, T_1 and T_3 have $b = 2$, while T_6, T_7, T_{10}, T_{11}, T_{12}, T_{13} and T_{18} have $b = 3$, and finally T_{17}, T_{20}, T_{21}, T_{22}, and T_{24} have $b = 4$.

In the two $b = 2$ cases, we are again done. For the seven $b = 3$ cases, we can consider the nonic resolvent K_{9x}. One has an exact sequence

$$C_3 \hookrightarrow \mathrm{Gal}(K^{\mathrm{gal}}/\mathbf{Q}_3) \twoheadrightarrow \mathrm{Gal}(K_{9x}^{\mathrm{gal}}/\mathbf{Q}_3). \tag{15}$$

Now the two nonic groups of order 27, namely T_6 and T_7, both have only one normal subgroup of order 3. So the remaining groups presently under consideration, T_{10}, T_{11}, T_{12}, T_{13} and T_{18}, also have only one normal subgroup of order 3. So it is necessarily the highest slope s_3 of K which disappears upon passage from K to K_{9x}. Only one of the remaining slopes s_1 and s_2 is visible in K, but both are visible in K_{9x}, allowing us to identify all three of s_1, s_2, and s_3. This computation would work equally well replacing K_{9x} by K_{9y}, as indeed $K_{9x}^{\mathrm{gal}} = K_{9y}^{\mathrm{gal}}$.

Finally for the five $b = 4$ cases, we can use again the nonic resolvent K_{9x} and the exact sequence (15). The slopes of K are s_1, s_2, s_3, and s_4. The 3-group T_{17} has just one normal subgroup of order 3, and hence the normal overgroups T_{20}, T_{21}, T_{22}, and T_{24} also have just one normal subgroup of order 3. So the slopes of K_{9x} must be s_1, s_2, and s_3, which are all identified by the previous paragraph. The highest slope s_4 is identified too, so all slopes of K have been identified. Again this computation would work equally well with K_{9x} replaced by K_{9y}.

6 Global Applications

The last section of [10] concerns degree 13 fields with Galois group $PSL_3(3)$ and illustrates how ramification can be analyzed completely, with the analysis at 2 requiring octic 2-adic fields and the analysis at 3 requiring nonic 3-adic fields. We are presently pursuing other applications which similarly require a number of group-theoretic preliminaries to describe globally. Here we will stay in a setting where the only group-theory we need is what we have set up in previous sections.

Proposition 6.1 A. *There are exactly thirteen isomorphism classes of solvable nonic number fields with discriminant of the form* $\pm 3^b$, *namely the fields* $K = \mathbf{Q}[x]/f(x)$ *with* $f(x) = \sum a_{9-i}x^i$ *as given in Table 6.2.*

B. *Assuming that Odlyzko's GRH lower bounds on discriminants hold, there are no non-solvable nonic number fields with discriminant of the form* $\pm 3^b$.

To establish Part A, we need to use the group theory set up in Sections 2 and 3, but not the specifically 3-adic information of Sections 4 and 5. First, one knows that there are no quartic number fields with discriminant $\pm 3^b$. This implies that there are no primitive nonic solvable number fields of discriminant $\pm 3^b$ because all seven solvable primitive groups have a quotient group of the form V_4, C_4, A_4, or S_4, according to Table 3.3. For the same reason, an imprimitive group G can only appear if its size has the form 3^b or $2^1 3^b$.

Table 6.2. The thirteen nonic solvable number fields with discriminant of the form $\pm 3^b$, sorted by increasing top slope

c	G	a_0	a_1	a_2	a_3	a_4	a_5	a_6	a_7	a_8	a_9	s_1	s_2	s_3	s_4
19	4	1	0	0	−3	0	0	−6	0	0	−1	**2**	**2.5**		
21	13	1	0	0	−3	0	0	0	0	0	1	**2**	2.5	**2.833**	
22	11	1	0	0	−3	0	0	3	0	0	8	2	**2.5**	**2.833**	
22	1	1	0	−9	0	27	0	−30	0	9	1	**2**	**3**		
23	22	1	0	0	−6	0	0	9	0	0	−3	**2**	2.5	2.833	**3.167**
23	22	1	0	0	−3	0	0	0	0	0	3	**2**	2.5	2.833	**3.167**
23	22	1	0	0	−3	0	0	−9	0	0	3	**2**	2.5	2.833	**3.167**
25	20	1	0	−9	−6	27	36	−24	−54	−9	22	2	**2.5**	2.833	**3.333**
25	20	1	0	−9	−3	27	18	−24	−27	−9	23	2	**2.5**	2.833	**3.333**
25	20	1	0	−9	−3	27	18	−15	−27	−36	−4	2	**2.5**	2.833	**3.333**
26	3	1	0	0	−9	0	0	27	0	0	−3	**2.5**	**3.5**		
26	10	1	0	0	0	0	0	0	0	0	−3	2	**2.5**	**3.5**	
26	10	1	0	0	−9	0	0	27	0	0	−24	2	**2.5**	**3.5**	

It is also known that there are exactly two cubic fields with discriminant $\pm 3^b$, namely $\mathbf{Q}[x]/(x^3 - 3x - 1)$ with Galois group C_3 and $\mathbf{Q}[x]/(x^3 - 3)$ with Galois group S_3. So we need look only at cubic extensions of these fields, using the exhaustive method described in Chapter 5 of [3]. Since neither of the two cubic fields contains cube roots of unity, the method requires us to adjoin cube roots of unity to get sextic fields K_6, and look within degree eighteen overfields of these to get the desired nonic fields. The method requires that K_{18}/K_6 be abelian, but abelianness is ensured by $\mathrm{ord}_2(|G|) \leq 1$. These computations, which of course we have only briefly sketched here, establish Part A.

Before moving on to establishing Part B, we will comment on some ways that Table 6.2 illustrates our previous sections. The T_{11} field and the T_{13} field form a twin pair. Similarly, the three T_{20} fields and the three T_{22} fields each form a triplet. Triplets have a cyclic order and if we call the top-listed field in each triplet K_a, then the next is K_b and the final one is K_c. Slopes are given in the same format as Tables 5.1 and 5.2. The fact that the small visible slope can change within twins but not triplets is illustrated.

Part B is similar to some other non-existence statements in the literature, for example the statement in [16], which says in particular that there are no $PSL_2(8)$ nonics with discriminant of the form $\pm 2^a$. To establish Part B, we use the 3-adic analysis of Sections 4 and 5, including the determination of hidden slopes. If K is a nonic 3-adic field with slopes $s_1 \leq \cdots \leq s_b$ and $[K^{\mathrm{gal},t} : K^{\mathrm{gal},u}] = t$, then the root discriminant of K^{gal} is 3^β with

$$\beta = \frac{2}{3}s_b + \cdots + \frac{2}{3^b}s_1 + \frac{1}{3^b}\frac{t-1}{t}. \tag{16}$$

This type calculation is explained further in [10]. From Table 5.1, one sees that in the cases $b = 2$, $b = 3$, and $b = 4$, the largest that β can be is respectively $53/18 = 2.9\overline{4}$, $55/18 = 3.0\overline{5}$, and $511/162 \approx 3.15432$, these bounds all being realized in the largest discriminant case $c = 26$, always with $t = 2$. The corresponding 3^β are then approximately 25.40, 28.70, and 31.99. These numbers are thus upper bounds for the root discriminant of a Galois number field with discriminant $\pm 3^b$ and Galois group $PSL(2,8)$, $P\Sigma L(2,8)$ and (A_9 or S_9) respectively. However Odlyzko's GRH bounds say that a field with root discriminant at most 25.40, 28.70, and 31.99 respectively must have degree at most 380, 1000, and 4400 [12]. These numbers are respectively less than $|PSL(2,8)| = 504$, $|P\Sigma L(2,8)| = 1512$, and $|A_9| = 9!/2$, giving Part B.

References

[1] G. Butler and J. McKay. The transitive groups of degree up to eleven. *Comm. Algebra*, 11(8):863–911, 1983.

[2] H. Cohen. *A course in computational algebraic number theory*. Springer-Verlag, Berlin, 1993.

[3] ——— *Advanced topics in computational number theory*, volume 193 of *Graduate Texts in Mathematics*. Springer-Verlag, New York, 2000.

[4] A. Colin. Relative resolvents and partition tables in Galois group computations. In *Proceedings of the 1997 International Symposium on Symbolic and Algebraic Computation (Kihei, HI)*, pages 78–84 (electronic), New York, 1997. ACM.

[5] J. H. Conway, A. Hulpke, and J. McKay. On transitive permutation groups. *LMS J. Comput. Math.*, 1:1–8 (electronic), 1998.

[6] H. Darmon and D. Ford. Computational verification of M_{11} and M_{12} as Galois groups over **Q**. *Comm. Algebra*, 17(12):2941–2943, 1989.

[7] Y. Eichenlaub. *Problèmes effectifs de théorie de Galois en degrés 8 à 11*. PhD thesis, Université Bordeaux I, 1996.

[8] A. Hulpke. Techniques for the computation of Galois groups. In *Algorithmic algebra and number theory (Heidelberg, 1997)*, pages 65–77. Springer, Berlin, 1999.

[9] J. W. Jones and D. P. Roberts. Octic 2-adic fields. In preparation.

[10] ——— A database of local fields. Submitted, 2003.

[11] M. Krasner. Remarques au sujet d'une note de J.-P. Serre: "Une 'formule de masse' pour les extensions totalement ramifiées de degré donné d'un corps local". *C. R. Acad. Sci. Paris Sér. A-B*, 288(18):A863–A865, 1979.

[12] J. Martinet. Petits discriminants des corps de nombres. In *Number theory days, 1980 (Exeter, 1980)*, volume 56 of *London Math. Soc. Lecture Note Ser.*, pages 151–193. Cambridge Univ. Press, Cambridge, 1982.

[13] P. Panayi. *Computation of Leopoldt's p-adic regulator*. PhD thesis, University of East Anglia, 1995.

[14] S. Pauli and X.-F. Roblot. On the computation of all extensions of a p-adic field of a given degree. *Math. Comp.*, 70(236):1641–1659 (electronic), 2001.
[15] J.-P. Serre. Une “formule de masse” pour les extensions totalement ramifiées de degré donné d'un corps local. *C. R. Acad. Sci. Paris Sér. A-B*, 286(22):A1031–A1036, 1978.
[16] J. Tate. The non-existence of certain Galois extensions of $\mathbf{Q}$ unramified outside 2. In *Arithmetic geometry (Tempe, AZ, 1993)*, volume 174 of *Contemp. Math.*, pages 153–156. Amer. Math. Soc., Providence, RI, 1994.

Montgomery Addition for Genus Two Curves

Tanja Lange

Institute for Information Security and Cryptology (ITSC)
Ruhr-Universität Bochum, Universitätsstraße 150, D-44780 Bochum, Germany
lange@itsc.ruhr-uni-bochum.de

Abstract. Hyperelliptic curves of low genus obtained a lot of attention in the recent past for cryptographic applications. They were shown to be competitive with elliptic curves in speed and security. In practice, one also needs to prevent from side channel analysis, a method using information leaked during the process of computing to attack the system. For elliptic curves the curve arithmetic proposed by Montgomery requires a comparably small number of field operations to perform a scalar multiplication but at the same time achieves security against non-differential side channel attacks.
This paper studies the generalization of Montgomery arithmetic for genus 2 curves. We do not give the explicit formulae here, but together with the explicit formulae for affine or projective group operations the results show how to implement it. The divisor classes can be represented using only their first polynomials, a feature that is important for actual implementations. Our method applies to arbitrary genus two curves over arbitrary fields of odd characteristic which have at least one rational Weierstraß point.

Keywords: Hyperelliptic curves, Montgomery arithmetic, fast arithmetic, cryptographic applications

1 Introduction

For integer factorization using elliptic curves, Montgomery [14] proposed the use of a special class of curves to obtain fast computation of scalar multiples. The computation of m-folds is also the main operation in public-key cryptosystems based on the discrete logarithm problem. For these applications, Montgomery arithmetic offers the additional advantage that it prevents attacks like simple power analysis which determine the secret multiplier by observing the chain of additions and doublings corresponding to the digits of the multiplier in binary expansion. There are other methods to achieve this goal but they are slower and need more storage. For many implementations, space is restricted. Therefore, these curves are attractive for implementations, and generalizations to even characteristic [12] and curves in general Weierstraß form [3] have been obtained.

The use of hyperelliptic curves in cryptography was already proposed in 1989 by Koblitz [9] but it is a rather recent result that they can compete with elliptic curves in terms of efficiency of the group law [1,11]. The security of low genus

D.A. Buell (Ed.): ANTS 2004, LNCS 3076, pp. 309–317, 2004.

hyperelliptic curves is assumed to be similar to that of elliptic curves of the same group size. Here, *small* really means genus $g \leq 3$ by [7,18], and even for $g = 3$ some care has to be taken.

In this paper we show how to extend Montgomery's ideas to genus 2 curves. We concentrate on this case at is allows to state the expressions within one line. The main problem introduced by increasing the genus appears already in this case. We show how to add two group elements given only the first part of their representation and provided that this part is also known from the difference of the two elements. We do not give the explicit formulae but provide the ingredients starting from which one can derive the recipe for implementation by following the ideas in [11]. We concentrate the efforts to show that one can reduce the number of variables for the computations, which corresponds to the fact that for Montgomery arithmetic on elliptic curves one does not use the y-coordinates. This means that the proposed arithmetic does not need more variables than usual arithmetic while preventing side-channel attacks and obtaining acceptable performance. In this paper we do not deal with special classes of curves that can offer a better performance but try to be as general as possible allowing all special choices to be included.

We first give an introduction to the arithmetic on hyperelliptic curves, including elliptic curves. Then we briefly review scalar multiplication using the Montgomery ladder and present Montgomery arithmetic for elliptic curves, giving a mathematical reason for the approach taken in [3]. The core of the paper is to show how genus two Montgomery arithmetic works. We end with an outlook on the performance of the algorithm.

2 Background on Hyperelliptic Curves

Within the scope of this paper we can only give a short introduction to the topic. More details can be found in [6,13,17].

Let $\mathbb{F}_q$ be a finite field with q elements of odd characteristic p. Assume that C is a projective curve over $\mathbb{F}_q$ without singularities. The $\mathbb{F}_q$-rational divisors of C are formal sums of points (over $\overline{\mathbb{F}_q}$) of C which are invariant under $G_{\mathbb{F}_q} := \mathrm{Aut}(\overline{\mathbb{F}_q}/\mathbb{F}_q)$. The degree of a divisor D is the sum of the multiplicities of the points occurring in it and is denoted by $\deg(D)$. A divisor is effective if all multiplicities are non negative. We define the divisor class group by the following rule: two divisors are in the same class iff their difference consists of the zeroes and poles (with multiplicity) of a function $f \in \mathbb{F}_q(C)$, i. e. they differ only by the principal divisor (f) attached to f. The $\mathbb{F}_q$-rational points of the Jacobian variety of C, $J_C(\mathbb{F}_q)$, correspond to the $\mathbb{F}_q$-rational divisor classes of degree 0 of C. An effective version of the *Theorem of Riemann-Roch* allows to turn this into an algorithmic procedure to work in $J_C(\mathbb{F}_q)$.

By Riemann-Roch, each class can be represented uniquely by a divisor of the form $D = \sum_{P_i \in C(\mathbb{F}_q)\setminus P_\infty} m_i P_i - mP_\infty$, where $m = \sum m_i \leq g$ and P_∞ is a point at infinity. These representatives are called *reduced*. If one adds two classes

$\bar{D}_1, \bar{D}_2$ one determines a function passing through the points of the representing divisors D_1, D_2 with appropriate multiplicity. The points of intersection not in the support of D_1, D_2 form a divisor D_3 such that $\bar{D}_1 + \bar{D}_2 + \bar{D}_3 = 0$, as the points on the function belong to a principal divisor. In the next step one looks for a function passing through the points of D_3 providing D_4 with $\bar{D}_3 + \bar{D}_4 = 0$, i. e. $\bar{D}_1 + \bar{D}_2 = \bar{D}_4$.

In this paper we concentrate on hyperelliptic curves, including elliptic curves. We define a hyperelliptic curve C to be a projective irreducible non singular curve of genus $g \geq 1$ with a generically étale morphism π of degree 2 to $\mathbb{P}^1$. If C has at least one $\mathbb{F}_q$-rational Weierstraß point (a fixed point under π) then the affine part of C given by

$$C_a : y^2 = f(x),\ f \in \mathbb{F}_q[x],\ \text{monic}, \deg(f) = 2g + 1.$$

The procedure can be visualized easily for genus 2 curves over the reals.

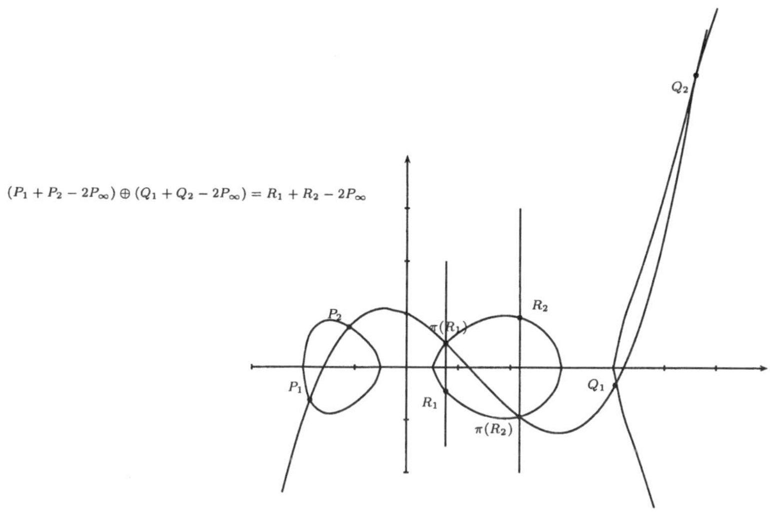

For implementations one needs to have a space-efficient representation of the divisor classes and also to be able to describe the arithmetic in terms of these representatives. One uses the Mumford representation [15][page 3.17] of the divisor classes. As the divisor class group is isomorphic to the ideal class group of the function field belonging to C one has

Theorem 1 (Mumford Representation).
Let the function field be given via the absolutely irreducible polynomial $y^2 + h(x)y - f(x)$, where $h, f \in \mathbb{F}_q[x]$, $\deg f = 2g + 1$, $\deg h \leq g$. Each nontrivial ideal class over $\mathbb{F}_q$ can be represented via a unique ideal generated by $u(x)$ and $y - v(x)$, $u, v \in \mathbb{F}_q[x]$, where

1. *u is monic,*
2. $\deg v < \deg u \le g$*,*
3. $u | v^2 + vh - f$*.*

Let $D = \sum_{i=1}^{r} P_i - rP_\infty$*, where* $P_i \ne P_\infty, P_i \ne \iota P_j$ *for* $i \ne j$ *and* $r \le g$*. Put* $P_i = (a_i, b_i)$*. Then the corresponding ideal class is represented by* $u = \prod_{i=1}^{r}(x - a_i)$ *and if* P_i *occurs* n_i *times then* $\left(\frac{d}{dx}\right)^j \left[v(x)^2 + v(x)h(x) - f(x)\right]_{|x=a_i} = 0,\ 0 \le j \le n_i - 1$.

For the arithmetic one can use Cantor's algorithm [4], which is just an algorithmic description of the steps detailed for the divisor classes.

Algorithm 1 (Composition)

INPUT: $D_1 = [u_1, v_1], D_2 = [u_2, v_2], \quad C : y^2 = f(x)$.
OUTPUT: $D = [u, v]$ semi-reduced with $D \equiv D_1 D_2$.

1. compute $d_1 = \gcd(u_1, u_2) = e_1 u_1 + e_2 u_2$;
2. compute $d = \gcd(d_1, v_1 + v_2) = c_1 d_1 + c_2(v_1 + v_2)$;
3. let $s_1 = c_1 e_1,\ s_2 = c_1 e_2,\ s_3 = c_2$;
4. $u = \frac{u_1 u_2}{d^2}$;
   $v = \frac{s_1 u_1 v_2 + s_2 u_2 v_1 + s_3(v_1 v_2 + f)}{d} \bmod u$.

Algorithm 2 (Reduction)

INPUT: $D = [u, v]$ semi-reduced.
OUTPUT: $D' = [u', v']$ reduced with $D \equiv D'$.

1. let $u' = \frac{f - v^2}{u}$, $v' = (-v) \bmod u'$;
2. if $\deg u' > g$ put $u = u', v = v'$;
   goto step 1;
3. make u' monic.

The picture corresponds to the case $\gcd(u_1, u_2) = 1$ which is the most frequent one. In practice, the other cases – except for doublings – can be neglected. To actually implement the arithmetic on hyperelliptic curves, special explicit formulae offer much better performance than Cantor's algorithm (see [1,11]).

3 Montgomery Ladder

Here we state how to compute scalar multiples using the Montgomery ladder. Details on cryptographic applications can be found in [8].

Let G be a group and $D \in G$. Assume that we want to compute nD for a positive integer $n = \sum_{i=0}^{l} n_i 2^i$. The obvious way is to do double-and-add according to the binary expansion. As mentioned above, for cryptographic applications it is desirable to have indistinguishable operations – the sequence of additions and doublings should not reveal any information about the digits of n. The easy way out is to perform a double-and-always-add algorithm. This has the disadvantage that one looses performance but an observer only sees that each doubling is followed by an addition. One needs to pay attention to modify the input for the dummy operations as otherwise one could distinguish between real and dummy additions by the power-trace. Consequently, one can just as well perform useful additions in each step that always depend on the output of the previous step. Put $E_j = \sum_{i=j}^{l} n_i 2^{i-j} D$ and $F_j = E_j + D$. We have

$$E_j = 2E_{j+1} + n_j D = E_{j+1} + F_{j+1} + n_j D - D = 2F_{j+1} + n_j D - 2D.$$

This implies

$$(E_j, F_j) = \begin{cases} (2E_{j+1}, E_{j+1} + F_{j+1}) & \text{if } n_j = 0 \\ (E_{j+1} + F_{j+1}, 2F_{j+1}) & \text{if } n_j = 1 \end{cases}$$

This technique is applicable for any group and it allows to hide the sequence of bits in n, e. g. one could always perform the doubling first and then the addition to get a uniform performance. For an integer of length l this requires l additions and l doublings. Additionally, the number of variables is doubled. For the following we note, that the difference of F_j and E_j is kept constant – in each step it is equal to D.

If affine coordinates are used in curve based cryptography the inversions in the addition and in the doubling can be performed together using Montgomery's trick. This is an obvious saving but usually inversion free systems are applied.

4 Montgomery Arithmetic on Elliptic Curves

In the previous section we obtained a method that hides the binary expansion of the multiplier. We now describe the idea behind Montgomery arithmetic, partially following [3]. We concentrate on the feature that one would like to reduce the amount of storage needed to represent a point, namely that one can get rid of the y-coordinate. We do not touch the doubling as this can obviously be done by using division polynomials (see e. g. [2,16]).

An elliptic curve is a curve of genus 1. For $p > 3$ it can be given by a Weierstraß equation $E : y^2 = x^3 + a_4 x + a_6,\ a_i \in \mathbb{F}_q$. Montgomery proposed to use curves of the form $y^2 = x^3 + a_2 x^2 + a_4 x$. Such curves can always be transformed to a Weierstraß equation but the converse is not true. Curves in Montgomery form have at least one point of order 2 leading to a non-prime group order. Therefore, it was desirable to have an analogue of Montgomery arithmetic for the general curves as well.

In Section 2 we introduced the representation of points on the Jacobian via polynomials. For elliptic curves the Jacobian is isomorphic to the group of points. If a point is given by (x_P, y_P) then the corresponding Mumford representation is $[x - x_P, y_P]$. Therefore, the arithmetic is usually given in terms of coordinates. To motivate the procedure in the genus 2 case and to obtain faster additions than [3] we stick to Cantor's algorithm for now.

Assume that we are given the first polynomials $u_1 = x - x_1, u_2 = x - x_2$ of D_1, D_2 and $u_- = x - x_-$ of $D_- = D_1 - D_2$ and that we want to compute $D_+ = D_1 + D_2, u_+ = x - x_+$. Additionally we assume that $x_1 \neq x_2$ as otherwise we would either double a class or end up with the trivial element. Therefore, $d_1 = d = 1$ in Algorithm 1 and $s_1 = e_1 = 1/(x_2 - x_1) = -e_2 = -s_2$. In Step 4 the intermediate results are $\tilde{u}_+ = \tilde{u}_- = u_1u_2$ and $\tilde{v}_+ = (u_1v_2 - u_2v_1)/(x_2 - x_1) \mod \tilde{u}_+, \tilde{v}_- = (-u_1v_2 - u_2v_1)/(x_2 - x_1) \mod \tilde{u}_-$. The modular reduction is not necessary as the degree of $\tilde{v}_\pm$ is bounded by 1. The $v_i = y_i$ are unknown, but $y_i^2 = f(x_i)$ as $(x_i, y_i) \in E$ and for the difference we know the resulting u'_-. This motivates taking the product of u'_+ with u'_- leading to

$$\begin{aligned} u'_+u'_- &= x^2 - (x_+ + x_-)x + x_+x_- = \frac{(f - \tilde{v}_-^2)(f - \tilde{v}_+^2)}{u_1^2u_2^2} \\ &= \frac{(f - (u_1^2v_2^2 + u_2^2v_1^2)/(x_2 - x_1)^2)^2 - 4u_1^2u_2^2v_1^2v_2^2/(x_2 - x_1)^4}{u_1^2u_2^2}. \end{aligned}$$

Multiplying out leads to

$$x_+ = \frac{2}{(x_1 - x_2)^2}(2a_6 + (x_1 + x_2)(a_4 + x_1x_2)) - x_-.$$

The computation of x_+ needs 4 multiplications and 1 inversion. This means a saving of 2 multiplications compared to [3]. They used the term x_+x_- instead of $x_+ + x_-$. However, in most application one avoids inversions as they are much more costly than multiplications. Allowing a further coordinate z_i per point leads to

$$\begin{aligned} x_+ &= 2(2a_6(z_1z_2)^2 + (x_1z_2 + x_2z_1)(a_4z_1z_2 + x_1x_2)) - x_-(x_1z_2 - x_2z_1)^2, \\ z_+ &= (x_1z_2 - x_2z_1)^2 \end{aligned}$$

This needs 7 multiplications and 3 multiplications by fixed constants just as the formulae in [3].

5 Montgomery Arithmetic for Genus Two Curves

In this section we try to mimic the approach taken in the previous section. We only consider the case of co-prime u_1, u_2, as the other ones are very unlikely to be needed and doubling can be done with Cantor's generali zation of the division polynomials [5]. Like for elliptic curves they only involve the u-part of the representation.

For genus two curves two main problems occur: the intermediate $\tilde{v_\pm}$ is actually reduced and one only knows $f \equiv v^2 \mod u$. Thus, v^2 is given only up to a multiple of u. Furthermore, the resulting u'_+ will not be monic, but we here know the leading coefficient.

There are obvious reasons for this behavior. Each fixed u-polynomial belongs to 4 different divisor classes depending on the signs of the y-coordinates of the points. Therefore, $f \mod u$ cannot contain all information on v. But, if the difference of two classes is known, this cancels out two possible assignments of the signs, leading to a unique result.This can be visualized in the picture above: u fixes only the x-coordinates of P_i, Q_i, hence there are 4 different groupings of the points to form the two divisors. Knowing that there is some operation leading to the x-coordinates of R_i reduces the choice to 2 possibilities – the one depicted and the one obtained as inflecting at the x-axis and both will lead to the same u-part of $P_1 + P_2 - Q_1 - Q_2 - 4P_\infty$.

To proceed we need to introduce some variables. Keep e_i from Algorithm 1 and put $u_i = x^2 + u_{i1}x + u_{i0}$, $b_i = b_{i1}x + b_{i0} \equiv (e_i v_{3-i})^2 \equiv e_i^2 f \mod u_{3-i}$ and $c_i = c_{i1}x + c_{i0} \equiv e_i v_{3-i} \mod u_{3-i}$. Then $c_i^2 = b_i + a_i u_{3-i}$ for some constant a_i with

$$\begin{aligned} c_{i1}^2 &= a_i (u \text{ is monic }) \\ 2c_{i1}c_{i0} &= b_{i1} + u_{(3-i)1}a_i \\ c_{i0}^2 &= b_{i0} + u_{(3-i)0}a_i. \end{aligned}$$

Therefore, a_i is a root of the quadratic polynomial

$$h_i(x) = (u_{(3-i)1}^2 - 4u_{(3-i)0})x^2 + (2u_{(3-i)1}b_{i1} - 4b_{i0})x + b_{i1}^2.$$

We now use that for each polynomial H one has $u_1 H \equiv u_1 \bar{H} \mod u_1 u_2$, where $H \equiv \bar{H} \mod u_2$, to simplify the expression of $u'_+ u'_-$. To compute u_+ we only need to compute the three leading terms

$$\begin{aligned} (*)\ u'_+ u'_- &= (a_1 - a_2)^2 (x^4 - (u_{+1} + u_{-1})x^3 + (u_{+1}u_{-1} + u_{+0} + u_{-0})x^2 + \ldots) \\ &= \frac{(\tilde{v}_-^2 - f)(\tilde{v}_+^2 - f)}{u_1^2 u_2^2} \\ &= \frac{(u_1^2(b_1 + a_1 u_2) + u_2^2(b_2 + a_2 u_1) - f)^2}{(u_1 u_2)^2} - 4(b_1 + a_1 u_2)(b_2 + a_2 u_1). \end{aligned}$$

Due to the construction of the b_i this last expression is alway divisible by $(u_1 u_2)^2$. To recover u_+ is thus suffices to get the a_i. In principle one could factor the above h_i and choose the right a_i from the fact that $a_i = c_{i1}^2$ is a square. But this would imply taking square-roots which is a rather costly operation in the scale we are working in here. However, we know more, namely we know that the resulting polynomial in $(*)$ of degree 4 is divisible by u_-. Formally computing $s = s_1 x + s_0 = \frac{(u_1^2(b_1 + a_1 u_2) + u_2^2(b_2 + a_2 u_1) - f)^2}{(u_1 u_2)^2} - 4(b_1 + a_1 u_2)(b_2 + a_2 u_1) \mod u_-$ leads to s_i which are quadratic in a_i. To obtain a_i one computes the resultants $r_i := \text{res}(s_1, h_{3-i})$ with respect to a_{3-i} and obtains $x - a_i = \gcd(r_i, h_i)$. In the computation of r_i one could just as well take s_0. A detailed breakdown to obtain explicit formulae will allow to choose the one which needs less operations. Experiments do not show any advantage in choosing either s_1 or s_0.

This explanation allows to derive explicit formulae like Cantor's algorithm can be made explicit. E. g. to compute $(*)$ one should not compute the numerator and then divide by $(u_1u_2)^2$ but divide by u_2 form the very beginning just like in the usual explicit formulae. However, the resulting algorithms require more operations than an addition involving the v. The main objective of this paper is to show that Montgomery arithmetic is possible for genus 2 curves. Finding curve equations that allow faster computations and determining inversion free formulae is a topic of current research.

6 Conclusion and Outlook

We have provided the mathematical background to implement Montgomery arithmetic on genus two curves. Detailing the group operations as explicit formulae is beyond the scope of this paper. We have shown that for Montgomery arithmetic the space requirements are considerably lower, namely they are just the same as for usual arithmetic where one keeps the second polynomial. To find curves where this process also allows faster arithmetic is work in progress. For an implementation it might be wise to avoid inversions. An approach similar to that for elliptic curves based on [10] will do this.

For genus two curves one can also recover the polynomial v_1 from knowing the full representation $[u_2, v_2]$ and u_1, u_-. For that we take v_{11}, v_{10} as variables and expand equation (*) including the knowledge of the coefficients of v_2. This leads to a system of equations of degree 2 in v_{11}, v_{10} which can be solved by computing resultants.

For genus g the number of unknowns a_{ij} is $2(g-1)$. Accordingly this requires more operations before u_+ can be computed.

Acknowledgments. I would like to thank Gerhard Frey and Roberto Avanzi for interesting discussions.

References

1. R. M. Avanzi. Aspects of Hyperelliptic Curves over Large Prime Fields in Software Implementations. Cryptology ePrint Archive, Report 2003/253, 2003.
2. I. F. Blake, G. Seroussi, and N. P. Smart. *Elliptic curves in cryptography.* London Mathematical Society Lecture Note Series. 265. Cambridge University Press, 1999.
3. E. Brier and M Joye. Weierstraß Elliptic Curves and Side-Channel Attacks. In *Public Key Cryptography 2002*, volume 2274 of *Lect. Notes Comput. Sci.*, pages 335–345. Springer, 2002.
4. D. G. Cantor. Computing in the Jacobian of a hyperelliptic curve. *Math. Comp.*, 48:95–101, 1987.
5. D. G. Cantor. On the analogue of the division polynomials for hyperelliptic curves. *Journal für die reine und angewandte Mathematik*, 447:91–145, 1994.
6. G. Frey and T. Lange. Mathematical Background of Public Key Cryptography. Technical Report 10, IEM Essen, 2003.
7. P. Gaudry. An algorithm for solving the discrete log problem on hyperelliptic curves. In *Advances in Cryptology – Eurocrypt'2000*, Lect. Notes Comput. Sci., pages 19–34. Springer, 2000.

8. M. Joye and Y. Sung-Ming. The Montgomery Powering Ladder. In *Cryptographic Hardware and Embedded Systems CHES 2002*, volume 2523 of *Lect. Notes Comput. Sci.*, pages 291–302. Springer, 2002.
9. N. Koblitz. Hyperelliptic cryptosystems. *J. Cryptology*, 1:139–150, 1989.
10. T. Lange. Inversion-Free Arithmetic on Genus 2 Hyperelliptic Curves. Cryptology ePrint Archive, Report 2002/147, 2002. `http://eprint.iacr.org/` or `http://www.itsc.ruhr-uni-bochum.de/tanja`.
11. T. Lange. Formulae for Arithmetic on Genus 2 Hyperelliptic Curves. `http://www.itsc.ruhr-uni-bochum.de/tanja/preprints.html`, 2003. submitted.
12. J. López and R. Dahab. Fast multiplication on elliptic curves over $GF(2^m)$ without precomputation. In *Cryptographic Hardware and Embedded Systems CHES 99*, volume 1717 of *Lect. Notes Comput. Sci.*, pages 316–327. Springer, 1999.
13. D. Lorenzini. *An Invitation to Arithmetic Geometry*, volume 9 of *Graduate studies in mathematics.* AMS, 1996.
14. P. L. Montgomery. Speeding the Pollard and elliptic curve methods of factorization. *Math. Comp.*, 48(177):243–264, 1987.
15. D. Mumford. *Tata Lectures on Theta II.* Birkhäuser, 1984.
16. J. H. Silverman. *The arithmetic of elliptic curves*, volume 106 of *Graduate texts in mathematics.* Springer, 1986.
17. H. Stichtenoth. *Algebraic Function Fields and Codes.* Springer, 1993.
18. N. Thériault. Index calculus attack for hyperelliptic curves of small genus. In *Advances in cryptology – Asiacrypt 2003*, volume 2894 of *Lect. Notes Comput. Sci.*, pages 75–92. Springer, 2003.

Numerical Evaluation at Negative Integers of the Dedekind Zeta Functions of Totally Real Cubic Number Fields

Stéphane R. Louboutin

Institut de Mathématiques de Luminy, UMR 6206
163, avenue de Luminy, Case 907
13288 Marseille Cedex 9, FRANCE
loubouti@iml.univ-mrs.fr

Abstract. Let K be a totally real number field. It is known that the values $\zeta_K(-n)$ of the Dedekind zeta function $\zeta_K(s)$ of K are rational numbers for all non-negative integers $n \geq 1$. We develop a rigorous and reasonably fast method for computing these exact values. Our method is in fact developed in the case of totally real number fields K of any degree for which $\zeta_K(s)/\zeta(s)$ is entire, which is conjecturally always the case (and holds true if K is cubic or if $K/\mathbf{Q}$ is normal).

1 Introduction

Let K be a totally real number field of degree $m > 1$. Let d_K and $\zeta_K(s)$ denote respectively its discriminant and Dedekind zeta function. Set $A_K := \sqrt{d_K/\pi^m}$. It is well known that $\Lambda_K(s) := A_K^s \Gamma^m(s/2)\zeta_K(s)$ is meromorphic, with only two poles, at $s = 1$ and $s = 0$, both simple, and satisfies the functional equation $\Lambda_K(1-s) = \Lambda_K(s)$. Hence, if $-2n \leq 0$ is even then $\zeta_K(-2n) = 0$, and if $1 - 2n \leq -1$ is odd then

$$\zeta_K(1-2n) = (-1)^{mn} d_K^{2n-1/2} \left(\frac{(2n-1)!}{2^{2n-1}\pi^{2n}}\right)^m \zeta_K(2n) \neq 0$$

(use the functional equation $\Gamma(s)\Gamma(1-s) = \pi/\sin(\pi s)$). Moreover, according the Siegel-Klingen's theorem, if $n \geq 1$ is an integer, then $\zeta_K(-n)$ is a rational number. We want to explain how one can efficiently compute these rational integers. To begin with, if K is abelian and if we let X_K denote the group (of order m) of primitive even Dirichlet characters associated with K, then $\zeta_K(s) = \prod_{\chi \in X_K} L(s,\chi)$ and well known formulae for $L(-n,\chi)$ make the computation of $\zeta_K(-n)$ straightforward. Indeed, if χ is a character modulo $f \geq 1$ then

$$L(1-2n,\chi) = -\frac{1}{2n} B_{2n,\chi} \ (n \geq 1)$$

where

$$\sum_{a=1}^{f} \frac{\chi(a)te^{at}}{e^{ft}-1} = \sum_{n=0}^{\infty} B_{n,\chi} \frac{t^n}{n!}$$

D.A. Buell (Ed.): ANTS 2004, LNCS 3076, pp. 318–326, 2004.

(e.g., see [Was, Theorem 4.2]. See also [Lou3, Formula (1) page 70] for another formula for $L(2n,\chi)$ for even Dirichlet characters χ). In particular, if χ modulo $f>1$ is even, then

$$B_{2,\chi}=\frac{1}{f}\sum_{a=1}^{f}\chi(a)a^2,$$

which in using $\zeta(-1)=-1/12$ yields

$$\zeta_K(-1)=\frac{(-1)^m}{12(2f_K)^{m-1}}\prod_{k=1}^{m-1}\Big(\sum_{a=1}^{f_K}\chi_K^k(a)a^2\Big) \qquad (1)$$

for a real cyclic field K of conductor $f_K>1$ and prime degree $m>1$ associated with a primitive modulo f_K even Dirichlet character χ_K. If K of square-free conductor $f_K>1$ is defined by some $\mathbf{Q}$-irreducible polynomial $P_K(x)$ of degree m with integral coefficients, we already explained in [Lou6, Section 2] how one can recover a character χ_K of order m generating X_K prior to using (1). In particular, we used this technique for checking the Tables published in [KK], and realized that almost all the values in these Tables are false! Indeed, we give in Table 1 below an excerpt of the computation we did for cyclic simplest cubic fields K_m associated with the $\mathbf{Q}$-irreducible cubic polynomials $P_m(x)=x^3+mx^2-(m+3)x+1$ (with $m\geq -1$ such that $D_m:=m^2+3m+9$ is square-free). We point out that whereas our values do not agree with those obtained in [KK], they do agree with the ones obtained by using the functions *zetakinit* and *zetak* of the software Pari GP , i.e. by using the functions initially written by E. Tollis according to his method developed in [Tol]

$$factor(round(-21*zetak(zetakinit(subst(x^3+mx^2-(m+3)x+1,m,\bullet)),-1))),$$

and that the authors of [KK] could have used this software to realize that there was something going wrong with their computation. In fact, after having met H. K. Kim in Tokyo last October, he told us that Theorem 3.2 in [KK] is false, which explains their problems with the results of their computation.

2 A Method for Computing $\zeta_K(1-2n)$ for Some Totally Real Number Fields K

In order to compute the exact value of the rational number $\zeta_K(1-2n)\in\mathbf{Q}$, we will compute good enough numerical approximations to it and then use some information on the size of its denominator to deduce its exact value.

Theorem 1. *(Siegel, see [Zag]). Let $n\geq 1$ and $m\geq 1$ be positive integers. Set*

$$h=2mn \text{ and } r=\begin{cases}[h/12] & \text{if } h\equiv 2 \pmod{12}\\ [h/12]+1 & \text{if } h\not\equiv 2 \pmod{12}.\end{cases}$$

There exist $b_1(h), \cdots, b_r(h) \in \mathbf{Q}$ such that for any totally real number field K of degree m we have

$$\zeta_K(1-2n) = 2^m \sum_{l=1}^{r} b_l(h) S_l^K(2n) \in \mathbf{Q},$$

where

$$S_l^K(2n) = \sum_{\substack{\nu \in \mathcal{D}_{K/\mathbf{Q}}^{-1} \\ \nu \gg 0 \text{ and } Tr_{K/\mathbf{Q}}(\nu)=l}} \sum_{\mathcal{I}|(\nu)\mathcal{D}_{K/\mathbf{Q}}} N_{K/\mathbf{Q}}(\mathcal{I})^{2n-1} \in \mathbf{Z}_{\geq 0}$$

($\mathcal{D}_{K/\mathbf{Q}}$ being the different of K and $\mathcal{I}$ ranging over the integral ideals of K dividing the integral ideal $(\nu)\mathcal{D}_{K/\mathbf{Q}}$). In particular, if K is a totally real cubic number field, then

$$-63\zeta_K(-1) = S_1^K(2) \in \mathbf{Z}_{\geq 0}.$$

From now on we assume that $\zeta_K(s)/\zeta(s)$ is entire, which holds true if K is either a cubic number field or a normal number field (and see Remark 1 at the end of the paper to see how to adapt our method in the case that $\zeta_K(s)/\zeta(s)$ is not known to be entire). Our method is based on [Lou1] and [Coh, Section 10.3]. We set

$$A_{K/\mathbf{Q}} = A_K/A_{\mathbf{Q}} = \sqrt{d_K/\pi^{m-1}} \tag{2}$$

and

$$\Lambda_{K/\mathbf{Q}}(s) := \Lambda_K(s)/\Lambda_{\mathbf{Q}}(s) = A_{K/\mathbf{Q}}^s \Gamma^{m-1}(s/2)\zeta_K(s)/\zeta(s), \tag{3}$$

which is entire and satisfies the functional equation $\Lambda_{K/\mathbf{Q}}(1-s) = \Lambda_{K/\mathbf{Q}}(s)$. Let

$$S_{K/\mathbf{Q}}(x) := \frac{1}{2\pi i}\int_{c-i\infty}^{c+i\infty} \Lambda_{K/\mathbf{Q}}(s)x^{-s}ds \quad (c>1 \text{ and } x>0) \tag{4}$$

denote the Mellin transform of $\Lambda_{K/\mathbf{Q}}(s)$. Since $\Lambda_{K/\mathbf{Q}}(s)$ is entire, it follows that $S_{K/\mathbf{Q}}(x)$ satisfies the functional equation

$$S_{K/\mathbf{Q}}(x) = \frac{1}{x}S_{K/\mathbf{Q}}(\frac{1}{x}) \tag{5}$$

(shift the vertical line of integration $\Re(s) = c > 1$ in (4) ot the left to the vertical line of integration $\Re(s) = 1-c < 0$, then use the functional equation $\Lambda_{K/\mathbf{Q}}(1-s) = \Lambda_{K/\mathbf{Q}}(s)$ to come back to the vertical line of integration $\Re(s) = c > 1$), and

$$\Lambda_{K/\mathbf{Q}}(s) = \int_0^\infty S_{K/\mathbf{Q}}(x)x^s\frac{dx}{x} = \int_1^\infty S_{K/\mathbf{Q}}(x)(x^s + x^{1-s})\frac{dx}{x} \tag{6}$$

is the inverse Mellin transform of $S_{K/\mathbf{Q}}(x)$. Finally, we set

$$H_{m-1}(x) := \frac{1}{2\pi i}\int_{c-i\infty}^{c+i\infty} \Gamma^{m-1}(s/2)x^{-s}ds \quad (c>1 \text{ and } x>0).$$

Notice that $0 < H_{m-1}(x)$ for $x > 0$ (see [Lou4, Last assertion of Theorem 20] (note however the misprint there for we should have written $\Phi_1 \star \Phi_2(x) = \int_0^\infty \Phi_1(t)\Phi_2(x/t)\frac{dt}{t}$)). Now, write

$$\zeta_K(s)/\zeta(s) = \sum_{k\geq 1} \phi_k k^{-s}.$$

Then,

$$S_{K/\mathbf{Q}}(x) = \sum_{k\geq 1} \phi_k H_{m-1}(kx/A_{K/\mathbf{Q}}),$$

by (3) and (4),

$$\Lambda_{K/\mathbf{Q}}(1-2n) = \sum_{k\geq 1} \phi_k \int_1^\infty H_{m-1}(kx/A_{K/\mathbf{Q}})(x^{1-2n} + x^{2n})\frac{dx}{x},$$

by (6), and:

Proposition 1. *For $m \geq 1$ and $n \geq 1$, set*

$$r_{m,2n} = \Gamma^{1-m}(\frac{1-2n}{2})\zeta(1-2n) = (-1)^{n(m-1)}\zeta(1-2n)\left(2^{2n-1}\sqrt{\pi}\frac{(n-1)!}{(2n-1)!}\right)^{m-1}.$$

For $A > 0$, set

$$I_{m,n}(A) := \int_1^\infty H_{m-1}(Ax)x^n\frac{dx}{x} = \frac{1}{2\pi i}\int_{c-i\infty}^{c+i\infty} \Gamma^m(s/2)A^{-s}\frac{ds}{s-n} \qquad (7)$$

(with $c > n \in \mathbf{Z}$). Then

$$\zeta_K(1-2n) = r_{m,2n}A_{K/\mathbf{Q}}^{2n-1}\sum_{k\geq 1}\phi_k\Big\{I_{m-1,1-2n}(\frac{k}{A_{K/\mathbf{Q}}}) + I_{m-1,2n}(\frac{k}{A_{K/\mathbf{Q}}})\Big\}. \quad (8)$$

In particular, if K is a totally real cubic field, then

$$\zeta_K(-1) = -\frac{A_{K/\mathbf{Q}}}{48\pi}\sum_{k\geq 1}\phi_k\Big\{I_{2,-1}(k/A_{K/\mathbf{Q}}) + I_{2,2}(k/A_{K/\mathbf{Q}})\Big\} \in \mathbf{Q}_{\leq 0}. \quad (9)$$

Let $m \geq 1$ and $n \geq 1$ be given. Let K range over a family of totally real number fields of degree m (for which $\zeta_K(s)/\zeta(s)$ is entire). Our strategy for computing the exact values of the rational numbers $\zeta_K(1-2n)$ is now clear. According to Theorem 1 there exists $d_{m,2n} \in \mathbf{Z}_{>0}$ such that $d_{m,2n}\zeta_K(1-2n)$ is a rational integer. Hence, by computing a numerical approximation $\tilde{\zeta}_K(1-2n)$ to $\zeta_K(1-2n)$ such that $|d_{m,2n}\zeta_K(1-2n) - d_{m,2n}\tilde{\zeta}_K(1-2n)| < 1/2$ we can deduce the exact value of $\zeta_K(1-2n)$. For that purpose, we use (8) where we disregard the indices $k > M$ where the integer $M \geq 1$ is chosen such that $R_{K,2n}(M) < 1/(2d_{m,2n})$ where

$$R_{K,2n}(M) := r_{m,2n}A_{K/\mathbf{Q}}^{2n-1}\left|\sum_{k>M}\phi_k\Big\{I_{m-1,1-2n}(k/A_{K/\mathbf{Q}}) + I_{m-1,2n}(k/A_{K/\mathbf{Q}})\Big\}\right|.$$

According to Proposition 2,

$$M \geq A_{K/\mathbf{Q}}((2n+o(1))\log A_{K/\mathbf{Q}})^{(m-1)/2}$$

will do (where $o(1)$ tends to zero as $d_K \longrightarrow \infty$). Then, it only remains to explain in Proposition 3 how one can compute efficiently as good as desired numerical approximations to $I_{m-1,1-2n}(k/A_{K/\mathbf{Q}})$ and $I_{m-1,2n}(k/A_{K/\mathbf{Q}})$ for $1 \leq k \leq M$.

We give in Table 2 below an excerpt of the computation of $\zeta_{K_m}(-1)$ we did for the non-normal totally real cubic fields K_m (studied in [Lou5]) associated with the $\mathbf{Q}$-irreducible polynomials $P_m(x) = x^3 - mx^2 - (m+1)x - 1$ of discriminants $D_m = (m^2+m-3)^2 - 32$. Here, we assume that (i) $4 \leq m \not\equiv 3 \pmod 7$ is such that D_m is square-free, or (ii) $4 \leq m \equiv 3 \pmod 7$ is such that $D_m/7^2$ is square-free, in which cases the set $\{1, \rho_m, \rho_m^2\}$ forms a $\mathbf{Z}$-basis of the ring of algebraic integers of $K_m = \mathbf{Q}(\rho_m)$ (see [Lou5, Theorem 3]), where ρ_m is the only positive real root of $P_m(x)$. (The computation of the ϕ_k is explained in [Lou5, Section 3], and h_m denotes the class number of K_m).

Lemma 1. *Let $I_{m,n}(A)$ be as in (7). It holds that*

$$\begin{aligned} I_{m,n}(A) &= \int\int_{t_1\cdots t_m \geq 1} e^{-(t_1+\cdots+t_m)A^{2/m}}(t_1\cdots t_m)^{n/2}\frac{dt_1\cdots dt_m}{t_1\cdots t_m} \\ &\quad \text{(which increases as } n \text{ increases)} \\ &\leq mA^{-n}\Gamma^{m-1}(n/2)\int_{A^{2/m}}^{\infty} e^{-t}t^{n/2}\frac{dt}{t}. \end{aligned}$$

Hence, setting $c_{m,n} = m((n-1)!)^m$, we have

$$I_{m,1-2n}(A) \leq I_{m,2n}(A) \leq c_{m,n}e^{-A^{2/m}}A^{-2n}\sum_{l=0}^{n-1}\frac{1}{l!}A^{2l/m},$$

which implies

$$I_{m,-1}(A) \leq I_{m,2}(A) \leq mA^{-2}e^{-A^{2/m}} \tag{10}$$

and (for $A \geq 1$)

$$I_{m,1-2n}(A) \leq I_{m,2n}(A) \leq ec_{m,n}e^{-A^{2/m}}A^{-2-2(m-1)(n-1)/m}. \tag{11}$$

Proof. For the first assertion, use (7), Fubini's theorem and (for $c > n$)

$$\frac{1}{2\pi i}\int_{c-i\infty}^{c+i\infty}\frac{x^s}{s-n}ds = \begin{cases} x^n & \text{if } x > 1 \\ 0 & \text{if } 0 < x \leq 1. \end{cases}$$

Then, notice that if $t_1 \cdots t_m \geq 1$ then at least one of the t_i is ≥ 1. For the last assertion, use

$$\frac{1}{(n-1)!}\int_x^{\infty} e^{-t}t^n\frac{dt}{t} = e^{-x}\sum_{l=0}^{n-1}\frac{1}{l!}x^l$$

(for $n \geq 1$ an integer).

Lemma 2. *See [Lou4, Lemma 26]). For $k \geq 1$ a positive integer, set*

$$d_m(k) := \prod_{d_1 \cdots d_m = k} 1.$$

It holds that $|\phi_k| \leq d_{m-1}(k)$ and

$$D_m(x) := \sum_{1 \leq n \leq x} \frac{d_m(k)}{k} \leq \Big(\sum_{1 \leq n \leq x} \frac{1}{n} \Big)^m \leq \log^m(ex) \quad (x \geq 1). \tag{12}$$

Set

$$R_{m,2n}(M, A) := \sum_{k > M} d_m(k) I_{m,2n}(k/A).$$

Then,

$$R_{K,2n}(M) \leq 2 r_{m,2n} A_{K/\mathbf{Q}}^{2n-1} R_{m-1,2n}(M, A_{K/\mathbf{Q}}),$$

by Lemma 2, and we have:

Proposition 2. *Assume that $m \geq 1$. For $n > 1$, $M \geq A \geq 1$ and $M \geq \exp(m^2/2(m-1)(n-1))$, it holds that*

$$R_{m,2n}(M, A) \leq e c_{m,n} \frac{A^2}{M} \Big(1 + \frac{2(m-1)(n-1)}{m} + \frac{2}{m}(M/A)^{2/m}\Big) \cdot e^{-(M/A)^{2/m}} (M/A)^{-2(m-1)(n-1)/m} \log^m(eM)$$

which yields

$$R_{m,2n}(M, A) \ll_{m,n,\lambda} A^{1-\lambda} (\log A)^{(3m/2)+n-mn} \quad \textit{for } M \geq A(\lambda \log A)^{m/2}.$$

For $n = 1$, $A > 0$ and $M \geq (3m/2)^{m/2}$, it holds that

$$R_{m,2}(M, A) \leq m \frac{A^2}{M} \Big(1 + \frac{2}{m}(M/A)^{2/m}\Big) e^{-(M/A)^{2/m}} \log^m(eM).$$

Proof. If $f \in \mathcal{C}^1([M, \infty), \mathbf{R})$ and $\lim_{x \to \infty} D_m(x) f(x) = 0$, then

$$\sum_{k > M} \frac{d_m(k)}{k} f(k) = -\int_M^\infty D_m(x) f'(x) dx.$$

Hence, using (11) to obtain

$$\frac{x}{A} I_{m,2n}\Big(\frac{x}{A}\Big) \leq f(x) := e c_{m,n} e^{-(x/A)^{2/m}} (x/A)^{-1-2(m-1)(n-1)/m},$$

and using (12) we obtain

$$R_{m,2n}(M, A) \leq A \sum_{k > M} \frac{d_m(k)}{k} f(k) \leq e c_{m,n} A^2 \int_M^\infty g((x/A)^{2/m}) \frac{dx}{x^2}$$

with

$$g(y) = (m/2)^m\Big(1 + \frac{2(m-1)(n-1)}{m} + \frac{2}{m}y\Big)e^{-y}y^{-(m-1)(n-1)}\log^m(By)$$

and $B = (eA)^{2/m}$. Now, if $g(y) \le g(C)$ for $y \ge C$, then for $M \ge AC^{m/2}$ we have

$$\int_M^\infty g((x/A)^{2/m})\frac{dx}{x^2} \le g((M/A)^{2/m})\int_M^\infty \frac{dx}{x^2} = \frac{g((M/A)^{2/m})}{M}$$

and

$$R_{m,2n}(M,A) \le ec_{m,n}\frac{A^2}{M}g((M/A)^{2/m}) \quad \text{for } M \ge AC^{m/2}.$$

Now, $g'(y)$ is of the same sign as

$$\frac{2}{m} - \big(1 + \frac{2(m-1)(n-1)}{m} + \frac{2}{m}y\big)\big(1 + \frac{(m-1)(n-1)}{y} - \frac{m}{y\log y}\big).$$

Hence, for $n > 1$ we have $g'(y) \le 0$ as soon as $y \ge 1$ and $(m-1)(n-1) - m/\log y \ge 0$, hence as soon as $y \ge \exp(m/((m-1)(n-1))$. For $n = 1$, $g'(y)$ which is of the same sign as $\frac{2}{m} - \big(1 + \frac{2}{m}y\big)\big(1 - \frac{m}{y\log y}\big)$ which decreases as y increases, hence is less than 0 for $y \ge 3m/2$.

Proposition 3. *For $n \notin \{-2k;\ k \in \mathbf{Z}_{\ge 0}\}$ a rational integer it holds that*

$$I_{m,n}(A) = \Gamma^m(n/2)A^{-n} + \sum_{k\ge 0}\operatorname{Res}_{s=-2k}\Big(s \mapsto \Gamma^m(s/2)\frac{A^{-s}}{s-n}\Big)$$

where this series is absolutely convergent, and for any integer $M \ge 0$ with $M \ne -(n+1)/2$ we have

$$\Big|\sum_{k>M}\operatorname{Res}_{s=-2k}\Big(s \mapsto \Gamma^m(s/2)\frac{A^{-s}}{s-n}\Big)\Big| \le \frac{c_m}{2\pi}\frac{\pi^{m/2}A^{2M+1}}{|2M+1+n|(M!/2)^m}$$

where

$$c_m = \int_{-\infty}^{\infty}\frac{dt}{(\cosh(\pi t))^{m/2}}$$

is less than or equal to $c_2 = 1$ for $m \ge 2$. Moreover, for $m = 2$ we obtain

$$I_{2,n}(A) = \Gamma^2(n/2)A^{-n} + \sum_{k\ge 0}\frac{4}{2k+n}\big(\log A + \gamma - \frac{1}{2k+n} - \sum_{l=1}^{k}\frac{1}{l}\big)\frac{A^{2k}}{(k!)^2}. \quad (13)$$

Proof. It holds that

$$\sum_{k>M}\operatorname{Res}_{s=-2k}\Big(s \mapsto \Gamma^m(s/2)\frac{A^{-s}}{s-n}\Big) = \frac{1}{2\pi i}\int_{-2M-1-i\infty}^{-2M-1+i\infty}\Gamma^m(s/2)\frac{A^{-s}}{s-n}$$

and

$$\left|\Gamma(-\frac{2M+1}{2}+it)\right|^2 = \frac{\pi}{\cosh(\pi t)}\prod_{k=0}^{M}\left|\frac{2k+1}{2}+it\right|^{-2} \leq (\frac{2}{M!})^2\frac{\pi}{\cosh(\pi t)}$$

(see [Lou2, (32)]).

Remark 1. If one does not know beforehand that $\zeta_K(s)/\zeta(s)$ is entire, then one can use the previous method starting with $\Lambda_K(s) := s(s-1)A_K^s\Gamma^m(s/2)\zeta_K(s)$ which is entire and satisfies the functional equation $\Lambda_K(1-s) = \Lambda_K(s)$.

3 Tables

Table 1. Some real cyclic cubic fields

m	D_m	$-21\zeta_{K_m}(-1)$	value in [KK]
8	97	$7\cdot 367$	$7\cdot 367$
13	$217 = 7\cdot 31$	$3^2\cdot 7\cdot 13\cdot 37$	$3\cdot 89\cdot 113$
14	$247 = 13\cdot 19$	$3\cdot 7\cdot 19\cdot 109$	$3\cdot 7\cdot 19\cdot 109$
22	$559 = 13\cdot 43$	$3^3\cdot 7\cdot 43\cdot 61$	$71\cdot 6977$
29	937	$2^2\cdot 3\cdot 7^2\cdot 3931$	$2\cdot 3^2\cdot 7\cdot 11\cdot 1667$
35	$1339 = 13\cdot 103$	$3^2\cdot 7\cdot 111091$	$2^7\cdot 3\cdot 41\cdot 443$
40	$1729 = 7\cdot 13\cdot 19$	$3^3\cdot 7\cdot 13\cdot 6073$	$3\cdot 5\cdot 241\cdot 4127$

Table 2. Some non-normal totally real cubic fields

m	D_m	h_m	$\zeta_{K_m}(-1)$
4	257	1	$-2/3$
5	$697 = 17\cdot 41$	1	$-8/3$
6	1489	1	-8
7	2777	2	-24
8	4729	1	-44
9	7537	2	-96
10	$11417 = 7^2\cdot 233$	3	-200
11	$16609 = 17\cdot 977$	2	-300
12	$23377 = 97\cdot 241$	2	$-1516/3$
13	32009	3	-896
14	$42817 = 47\cdot 911$	3	-1280
15	$56137 = 73\cdot 769$	2	-1856
16	$72329 = 151\cdot 479$	6	-3168
17	$91777 = 7^2\cdot 1873$	3	-3968
18	114889	3	-5376
19	142097	6	-8652
20	$173857 = 23\cdot 7559$	5	-10518

References

[Coh] H. Cohen. *Advanced topics in computational number theory.* Springer-Verlag, Grad. Texts Math. **193**, 2000.

[KK] H. K. Kim and J. S. Kim. Evaluation of zeta function of the simplest cubic field at odd negative integers. *Math. Comp.* **71** (2002), 1243–1262.

[Lou1] S. Louboutin. Calcul du nombre de classes des corps de nombres. *Pacific J. Math.* **171** (1995), 455–467.

[Lou2] S. Louboutin. Computation of relative class numbers of CM-fields. *Math. Comp.* **66** (1997), 1185–1194.

[Lou3] S. Louboutin. The mean value of $|L(k,\chi)|^2$ at positive rational integers $k \geq 1$. *Colloquium Math.* **90** (2001), 69–76.

[Lou4] S. Louboutin. Explicit upper bounds for residues of Dedekind zeta functions and values of L-functions at $s = 1$, and explicit lower bounds for relative class numbers of CM-fields. *Canad. J. Math.* **53** (2001), 1194–1222.

[Lou5] S. Louboutin. Class number and class group problems for some non-normal totally real cubic number fields. *Manuscripta Math.* **106** (2001), 411–427.

[Lou6] S. Louboutin. Efficient computation of class numbers of real abelian number fields. *Lectures Notes in Computer Science* **2369** (2002), 625–628.

[Tol] E. Tollis. Zeros of Dedekind zeta functions in the critical strip. *Math. Comp.* **66** (1997), 1295–1321.

[Tsu] S. Tsuyumine On values of L-functions of totally real algebraic number fields at integers. *Acta Arith.* **76** (1996), 359–392.

[Was] L. C. Washington. *Introduction to Cyclotomic Fields.* Springer-Verlag, Grad. Texts Math. **83**, Second Edition, 1997.

[Zag] D. Zagier On the values at negative integers of the zeta function of a real quadratic field. *Enseignement Math.* **22** (1976), 55–95.

Salem Numbers of Trace −2 and Traces of Totally Positive Algebraic Integers

James McKee[1] and Chris Smyth[2]

[1] Department of Mathematics, Royal Holloway, University of London, Egham Hill, Egham, Surrey TW20 0EX, UK
James.McKee@rhul.ac.uk

[2] School of Mathematics, University of Edinburgh, James Clerk Maxwell Building, King's Buildings, Mayfield Road, Edinburgh EH9 3JZ, Scotland, UK
C.Smyth@ed.ac.uk

Abstract. Until recently, no Salem numbers were known of trace below -1. In this paper we provide several examples of trace -2, including an explicit infinite family. We establish that the minimal degree for a Salem number of trace -2 is 20, and exhibit all Salem numbers of degree 20 and trace -2. Indeed there are just two examples.
We also settle the closely-related question of the minimal degree d of a totally positive algebraic integer such that its trace is $\leq 2d-2$. This minimal degree is 10, and there are exactly three conjugate sets of degree 10 and trace 18. Their minimal polynomials enable us to prove that all except five conjugate sets of totally positive algebraic integers have absolute trace greater than $16/9$.
We end with a speculative section where we prove that, if a single polynomial with certain properties exists, then the trace problem for totally positive algebraic integers can be solved.

1 Introduction

A *Salem number* is a real algebraic integer greater than 1 whose other conjugates all lie in the closed disc $|z| \leq 1$, with at least one on the circle $|z| = 1$. Salem numbers have a long history, making their appearance in surprisingly diverse areas of mathematics. These include: polynomials having small Mahler measure, questions concerning uniform distribution, harmonic analysis, dynamical systems, growth series of Coxeter groups, pretzel knots, and special values of L-functions. As detailed below, the problem of finding Salem numbers that have negative trace is closely connected to the old problem ([9] appeared in 1918) of finding totally positive algebraic integers of unusually small trace (in the sense that the trace divided by the degree, the *absolute trace*, is small). Until recently, no examples were known of Salem numbers having trace below -1. Here we settle the question: what is the smallest possible degree for a Salem number of trace -2?

D.A. Buell (Ed.): ANTS 2004, LNCS 3076, pp. 327–337, 2004.

The problem is related to that of finding totally positive algebraic integers of degree d and trace $2d - 2$. For suppose that

$$f(x) = x^d - (2d - 2)x^{d-1} + \cdots$$

is the minimal polynomial of a totally positive algebraic integer. Then we apply the transformation $x = z + 1/z + 2$, and clear denominators, to produce a reciprocal polynomial

$$F(z) = z^{2d} + 2z^{2d-1} + \cdots + 2z + 1$$

which is the minimal polynomial of an algebraic integer of degree $2d$ and trace -2. The reverse transformation is a little more complicated: in $F(z)/z^d$, replace each $z^j + 1/z^j$ by $T_j(x - 2)$, where T_j is the j-th Chebyshev polynomial, defined by $T_j(z + 1/z) = z^j + 1/z^j$. Any roots of $f(x)$ in the interval $0 < x < 4$ are mapped to pairs of roots of $F(z)$ on the unit circle. Any roots of $f(x)$ in the interval $x > 4$ are mapped to pairs of reciprocal real positive roots of $F(z)$. We see that the problem of finding all Salem numbers of degree $2d$ and trace -2 is equivalent to that of finding all totally positive algebraic integers θ of degree d and trace $2d - 2$ such that both (i) $\theta > 4$; and (ii) all other conjugates of θ are in the interval $0 < x < 4$.

The similar problem for trace -1 was settled some time ago: the smallest degree for a Salem number of trace -1 is 8, and there is just one such Salem number, having minimal polynomial

$$z^8 + z^7 - z^6 - 4z^5 - 5z^4 - 4z^3 - z^2 + z + 1 \ .$$

In [14] it is shown that there are infinitely many Salem numbers of trace -1, with examples of degree $2d$ for every $d \geq 4$. At that time, no examples of trace below -1 were known. We now know (see [5]) that there are infinitely many Salem numbers of every trace. In this paper we give a simpler proof that there are infinitely many Salem numbers of trace -2, using techniques from [14].

Some examples of Salem numbers of trace -2 are given in the next section, including one of degree only 26. These examples were obtained using a graphical construction described in [6] (generalising that in [4]), and using an interlacing construction described in [5] (which is greatly generalised in [7]). Bounds obtained in [11] show that to achieve trace -2 the degree must be at least 18. Further computations, announced in [14], showed that to achieve trace -2 the degree must be at least 20. This is confirmed by the computations of Sect. 3. However, we need no longer rely on these computations, as a direct proof of this is given in Sect. 4. The gap between 20 and 26 seemed tantalisingly narrow, and an improved search algorithm, detailed below, was set to work on degree 20. Luckily for us, we did not need to go up to degree 22! There are two examples at degree 20, and their minimal polynomials are given in Table 1.

The search in fact found all totally positive algebraic integers of degree 10 and trace 18. There are just three conjugate sets of these, and their minimal polynomials are displayed in Table 2. The first two of these polynomials yield

Table 1. Minimal polynomials of the Salem numbers of degree 20 and trace −2

$$z^{20} + 2z^{19} + z^{18} - 3z^{17} - 9z^{16} - 15z^{15} - 18z^{14} - 18z^{13} - 16z^{12} - 14z^{11}$$
$$-13z^{10} - 14z^9 - 16z^8 - 18z^7 - 18z^6 - 15z^5 - 9z^4 - 3z^3 + z^2 + 2z + 1$$

$$z^{20} + 2z^{19} - 8z^{17} - 22z^{16} - 40z^{15} - 58z^{14} - 74z^{13} - 87z^{12} - 96z^{11}$$
$$-99z^{10} - 96z^9 - 87z^8 - 74z^7 - 58z^6 - 40z^5 - 22z^4 - 8z^3 + 2z + 1$$

Table 2. Minimal polynomials of the totally positive algebraic integers of degree 10 and trace 18

$$f_1(x) = x^{10} - 18x^9 + 135x^8 - 549x^7 + 1320x^6 - 1920x^5$$
$$+1662x^4 - 813x^3 + 206x^2 - 24x + 1$$

$$f_2(x) = x^{10} - 18x^9 + 134x^8 - 538x^7 + 1273x^6 - 1822x^5$$
$$+1560x^4 - 766x^3 + 200x^2 - 24x + 1$$

$$f_3(x) = x^{10} - 18x^9 + 134x^8 - 537x^7 + 1265x^6 - 1798x^5$$
$$+1526x^4 - 743x^3 + 194x^2 - 24x + 1$$

Salem numbers via the transformation $x = z + 1/z + 2$; the third does not, since it has two roots greater than 4.

In [11] a lower bound is given for the absolute trace of totally positive algebraic integers. Using all three of the polynomials in Table 2, we are able to improve this bound (Sect. 4). The paper then concludes with a speculative section on the trace problem for totally positive algebraic integers.

2 Examples of Salem Numbers of Trace −2

2.1 Examples from Graphs

In [4], Salem numbers were constructed using star-like trees. It is known which star-like trees have exactly one eigenvalue $\lambda > 2$, and for any such tree we define $\tau > 1$ by $\sqrt{\tau} + 1/\sqrt{\tau} = \lambda$. Then τ is a Salem number, unless λ is a rational integer. It was shown in particular that for any integer $r \geq 2$, and any integers $a_1, \ldots, a_r$, all at least 2 (and excluding certain exceptional choices), the only solutions to the equation

$$\sum_{i=1}^{r} \frac{z^{a_i - 1} - 1}{z^{a_i} - 1} = 1 + \frac{1}{z} \tag{1}$$

are a certain Salem number (or perhaps a reciprocal Pisot number), its conjugates, and possibly some roots of unity. The corresponding star-like tree has a_1, …, a_r vertices on its r arms.

Applying the method in [3], if one takes $r = 10$, $a_1 = 390$, $a_2 = 462$, $a_3 = 1190$, $a_4 = 1938$, $a_5 = 1995$, $a_6 = 2090$, $a_7 = 2805$, $a_8 = 4641$, $a_9 = 4862$, and $a_{10} = 5005$, one produces a Salem number of degree 23838 and trace −2. It might seem a daunting task to test a polynomial $f(z)$ of degree 23838 for irreducibility. Luckily we need only check that no roots of unity are roots of $f(z)$, and it is sufficient to test that

$$\gcd(f(z), f(-z)) = \gcd(f(z), f(z^2)) = \gcd(f(z), f(-z^2)) = 1 \ ,$$

since if ω is a root of unity, then ω is conjugate to one of $-\omega$, ω^2, $-\omega^2$.

The degree can be reduced greatly by exploiting other graphs. For example, by adding forks to the ends of some of the branches of a star-like tree, the Salem formula (1) can be generalised (again with some exclusions) to

$$\sum_{i=1}^{r} \frac{z^{a_i-1}-1}{z^{a_i}-1} + \sum_{j=1}^{s} \frac{z^{b_j-1}+1}{z^{b_j}+1} = 1 + \frac{1}{z} \ . \tag{2}$$

Taking $r = s = 3$, $a_1 = 66$, $a_2 = 130$, $a_3 = 238$, $b_1 = 255$, $b_2 = 273$, and $b_3 = 385$ in (2), one obtains a Salem number of trace -2 and degree 1278.

For more on producing Salem numbers from graphs, see [6]. The current record low degree for a Salem number τ of trace -2 obtained from graphs is degree 460, with τ being a root of

$$\frac{z^{69}-1}{z^{70}-1} + \frac{(z^{13}+1)(z^{182}-1)}{(z-1)(z^{195}+1)} + \frac{(z^{11}+1)(z^{220}-1)}{(z-1)(z^{231}+1)} = 1 + \frac{1}{z} \ .$$

2.2 Examples via Interlacing

Let $Q(z)$ and $P(z)$ be relatively prime polynomials with integer coefficients, and with all their roots on the unit circle. Suppose further that $P(z)$ is monic, $Q(z)$ has positive leading coefficient, and that the roots of P and Q interlace on the unit circle. This last condition means that as you progress clockwise around the unit circle, you encounter a zero of P and a zero of Q alternately. Finally we suppose that either $P(1) = 0$, or $Q(1) = 0$ and $2P(1) - Q'(1) < 0$. Then part of Proposition 4 of [5] states that $(z^2-1)P(z) - zQ(z)$ is the minimal polynomial of a Salem number (or perhaps a reciprocal Pisot number), possibly multiplied by a cyclotomic polynomial (i.e., a polynomial all of whose roots are roots of unity).

In the next section, we use this interlacing construction to produce an infinite family of Salem numbers of trace -2. The smallest degree of any member of this family is 38, with the Salem number being a root of

$$\frac{z^5-1}{(z^2-1)(z^3-1)} + \frac{z^{12}-1}{(z^5-1)(z^7-1)} + \frac{z^{24}-1}{(z^{11}-1)(z^{13}-1)} = z - \frac{1}{z} \ .$$

The current record via interlacing is of degree only 26. Define polynomials

$$\begin{aligned}P(z) = {} & z^{24}+4z^{23}+9z^{22}+15z^{21}+21z^{20}+26z^{19}+29z^{18}+29z^{17}+26z^{16} \\ & +21z^{15}+15z^{14}+8z^{13}-8z^{11}-15z^{10}-21z^{9}-26z^{8}-29z^{7}-29z^{6} \\ & -26z^{5}-21z^{4}-15z^{3}-9z^{2}-4z-1 \ ,\end{aligned}$$

$$\begin{aligned}Q(z) = {} & 2z^{24}+7z^{23}+14z^{22}+21z^{21}+27z^{20}+31z^{19}+33z^{18}+33z^{17}+32z^{16} \\ & +31z^{15}+31z^{14}+31z^{13}+31z^{12}+31z^{11}+31z^{10}+31z^{9}+32z^{8}+33z^{7} \\ & +33z^{6}+31z^{5}+27z^{4}+21z^{3}+14z^{2}+7z+2 \ .\end{aligned}$$

Then $P(z)$ and $Q(z)$ satisfy all the required conditions, and the polynomial $(z^2-1)P(z)-zQ(z)$ (which has trace -2) is in fact irreducible, and is the minimal polynomial of a Salem number of degree 26. For an explanation of the construction of this remarkable pair of polynomials, see [7].

2.3 Infinitely Many Examples

For any integer p that is coprime to $2\cdot 3\cdot 5\cdot 7\cdot 11$, clearing denominators in the equation

$$\frac{z^5-1}{(z^2-1)(z^3-1)}+\frac{z^{12}-1}{(z^5-1)(z^7-1)}+\frac{z^{p+11}-1}{(z^{11}-1)(z^p-1)}=z-\frac{1}{z} \ . \tag{3}$$

gives a polynomial of trace -2. From Proposition 4 of [5], this is the minimal polynomial of a Salem number, possibly multiplied by a cyclotomic polynomial. We now show that in fact this polynomial is irreducible for all $p>11$ coprime to $2\cdot 3\cdot 5\cdot 7\cdot 11$, giving infinitely many examples of Salem numbers of trace -2. All that is required is to show that no root of unity satisfies (3).

Putting $y=z^p$, (3) reads $h(z,y)=0$, where

$$h(z,y)=\frac{z^5-1}{(z^2-1)(z^3-1)}+\frac{z^{12}-1}{(z^5-1)(z^7-1)}+\frac{yz^{11}-1}{(z^{11}-1)(y-1)}-z+\frac{1}{z} \ .$$

We again apply the trick that if a root of unity ω satisfies (3), then so does one of $-\omega$, ω^2, $-\omega^2$. (See [1] for more applications of this idea.) Our restriction on p implies in particular that p is odd, and hence $(-\omega)^p=-\omega^p$.

Eliminating z between $h(z,y)=0$ and $h(-z,-y)=0$ (by computing the resultant of the numerators of $h(z,y)$ and $h(-z,-y)$, treating z as the variable) yields

$$(y^2+1)^2 f(y)=0 \ ,$$

where $f(y)$ has no cyclotomic factors. Eliminating y instead yields

$$(z^4-z^2+1)g(z)=0 \ ,$$

where $g(z)$ has no cyclotomic factors. If both $z=\omega$ and $z=-\omega$ were to satisfy (3), then we would need $y=\omega^p$ to be a primitive fourth root of unity, with ω a primitive twelfth root of unity. It would follow that p is divisible by 3.

Similarly, eliminating first y and then z between $h(z,y)=0$ and $h(z^2,y^2)=0$ yields that for both $z=\omega$ and $z=\omega^2$ to satisfy (3) requires that ω is a primitive second, third, fifth, seventh, or eleventh root of unity, and that $\omega^p=1$. It would follow that p is divisible by at least one of 2, 3, 5, 7, or 11.

Finally, considering $h(z,y)=0$ and $h(-z^2,-y^2)=0$ simultaneously shows that for both $z=\omega$ and $z=-\omega^2$ to satisfy (3) requires that p is divisible by 3.

We see that no roots of unity can satisfy (3) provided that p is prime to $2\cdot 3\cdot 5\cdot 7\cdot 11$. There are infinitely many such p, giving infinitely many Salem numbers of trace -2.

3 Totally Positive Algebraic Integers of Given Degree and Trace

3.1 The Old Search Algorithm

For ease of exposition, we consider the search for totally positive algebraic integers of degree 10 and trace 18. The algorithm clearly generalises to arbitrary degree and trace.

We seek all positive integers $a_2, \ldots, a_{10}$ such that

$$f(x)=x^{10}-18x^9+a_2x^8-a_3x^7+a_4x^6-a_5x^5+a_6x^4-a_7x^3+a_8x^2-a_9x+a_{10}$$

has 10 distinct positive real roots. It is extremely difficult for a totally positive algebraic integer of degree 10 to have trace as small as 18, so having found all suitable a_i, nearly all of the corresponding polynomials $f(x)$ will be reducible.

For $1\le i\le 10$, we let $f_i(x)$ be the $(10-i)$th derivative of $f(x)$. If $f(x)$ has 10 distinct positive roots, then for each i, $f_i(x)$ will have i distinct positive roots. We find all possibilities for $f(x)$ by building up from below: we list all values of a_2 such that $f_2(x)$ has 2 distinct positive roots; for each suitable a_2, we list all values of a_3 such that $f_3(x)$ has 3 distinct positive roots; and so on. Many of our candidates for the higher derivatives of $f(x)$ do not survive this lifting process: we frequently find that for some $f_i(x)$ having i positive real roots there is no choice of a_{i+1} that makes $f_{i+1}(x)$ have $i+1$ positive roots.

Given a candidate for $f_i(x)$, having i distinct positive roots, the technique of Robinson [8] (used also in [11]) to find all suitable values of a_{i+1} was to observe that the *real* values of a_{i+1} such that $f_{i+1}(x)$ has $i+1$ distinct positive roots form an interval (possibly empty), with endpoints determined by considering the values of $f_{i+1}(x)$ at its local maxima and minima. Although much more efficient than a naive brute-force search, this method requires the computation of the roots of a huge number of polynomials, using floating-point arithmetic.

3.2 The New Search Algorithm

Our new algorithm still builds up $f(x)$ from its derivatives, as in the previous section. But the endpoints of the interval for a_i are determined in a different manner, removing the need for floating-point arithmetic, and hugely speeding the search.

With notation as in the previous section, we suppose that we have a candidate for $f_i(x)$ having i distinct positive roots. We wish to identify the (possibly empty) range of values for a_{i+1} such that $f_{i+1}(x)$ has $i+1$ distinct positive roots. For ease of notation, we put $a = a_{i+1}$. We observe that $D(a)$, the discriminant of $f_{i+1}(x)$ (which is a polynomial in a, given that all higher coefficients have been selected), vanishes at the (real) endpoints of the desired interval for a. In fact *all* its roots are real: they are the values of $(-1)^i \int_0^\beta f_i$, where β is a root of f_i.

Indeed the required interval is marked by the middle two roots of $D(a)$ (with the interpretation that if D has odd degree, then we take the middle zero and the one to the left of it). For when $a = a_{i+1}$ is large and negative, f_{i+1} has either one or two real roots (depending on the parity of i), and as a is increased the number of real roots of f_{i+1} jumps by two as we pass each root of $D(a)$. When a is large and positive, the number of real roots is either one or none. The only possible interval in which the number of real roots can be as large as $i+1$ is that bounded by the middle two roots. For these roots of f_{i+1} all to be positive we require also that $a > 0$.

It might appear that the problem of root-finding has simply been transferred to a different polynomial, but note that we only need to find the roots of $D(a)$ to the nearest integer. To this end, we take some crude initial approximations to the middle two roots, then refine these using Sturm sequences to pin the roots down to the nearest integer. (The initial approximation that we used was simply to try the endpoints of the previous interval.) Then a further Sturm sequence computation for a *single* value of a_{i+1} in the interval will reveal whether or not f_{i+1} has the full $i+1$ real roots for *all* a_{i+1} in the interval. We also require $a_{i+1} > 0$, to ensure that all these roots are positive.

A further improvement is to use non-trivial lower bounds for the a_i, based on known lower bounds for the traces of totally positive algebraic integers given in [11]—see also Sect. 4. This prunes out many hopeless $f_i(x)$ with i small.

The full search took 150 hours on a 1.2GHz PC, using PARI/GP, and produced three irreducible polynomials of degree 10, trace 18, with 10 distinct positive roots, as listed in Table 2. Some 4065 reducible polynomials of degree 10 and trace 18 were found. Studying the irreducible factors of this output provides the necessary information to find all Salem numbers of trace -1 and degree ≤ 18, and also confirms that to achieve degree d and trace $\leq 2d-2$ requires $d \geq 10$ (this also follows from the result of the next section).

4 Improving the Lower Bound for the Absolute Trace of Totally Positive Algebraic Integers

In [12] it was shown that all except five conjugate sets of totally positive algebraic integers α have absolute (also called mean) trace $\mathrm{tr}(\alpha)/\deg(\alpha) > 1.7719$ ($\mathrm{tr}(\alpha)$ and $\deg(\alpha)$ being the trace and degree respectively). We can now use the three newly discovered polynomials of degree 10 to improve this bound to

1.778378 > 16/9. The proof employs the same method as [12]: semi-infinite linear programming is used to produce the following inequality, valid for all $x > 0$

$$\begin{gathered} x - .5455833645\log|x| - .4958676072\log|x-1| - .05892353929\log|x-2| \\ -.1846627119\log|x^2-3x+1| - .002613011520\log|x^2-4x+1| \\ -.008163503307\log|x^2-4x+2| - .09063100904\log|x^3-5x^2+6x-1| \\ -.01899914258\log|x^3-6x^2+9x-1| - .008696349375\log|x^3-6x^2+9x-3| \\ -.05794447530\log|x^4-7x^3+13x^2-7x+1| \\ -.03510719518\log|x^4-7x^3+14x^2-8x+1| \\ -.008492128216\log|x^5-9x^4+28x^3-35x^2+15x-1| \\ -.01082775244\log|x^5-9x^4+27x^3-31x^2+12x-1| \\ -.0008908117930\log|x^6-11x^5+43x^4-72x^3+51x^2-14x+1| \\ -.005949580568\log|x^7-13x^6+63x^5-143x^4+158x^3-80x^2+16x-1| \\ -.008478368652\log|f_1(x)| - .007206449910\log|f_2(x)| \\ -.01019001634\log|f_3(x)| > 1.7783786\ , \end{gathered} \tag{4}$$

where $f_1(x)$, $f_2(x)$, $f_3(x)$ are the three degree 10 polynomials displayed in Table 2. To prove the existence of the lower bound $\mathrm{tr}(\alpha)/\deg(\alpha) > 1.7783786$ for a totally positive nonexceptional α, we substitute for x each conjugate α_j of α, and average. Then if the minimal polynomial of α does not appear in the inequality, we get that $\mathrm{tr}(\alpha)/\deg(\alpha) > 1.7783786 + \sum c_k \log|R_k|$, where the c_k are positive, and the R_k are nonzero integer resultants. Hence $\mathrm{tr}(\alpha)/\deg(\alpha) > 1.7783786$, as claimed.

The exceptional α are those of absolute trace less than 1.7783786 whose minimal polynomial *does* appear in the above inequality, namely α having minimal polynomial $x-1$, x^2-3x+1, x^3-5x^2+6x-1, $x^4-7x^3+13x^2-7x+1$ or $x^4-7x^3+14x^2-8x+1$.

Note that, if $d = \deg(\alpha)$ and $\mathrm{tr}(\alpha) \le 2d-2$ then, as this inequality excludes the five exceptional polynomials, we must have $16d/9 < \mathrm{tr}(\alpha) \le 2d-2$, so that $d \ge 10$. This confirms again the computation at the end of the previous section, and checks too that there are no totally positive algebraic integers of degree 9 and trace 16.

5 A Polynomial That Would Solve the Trace Problem

5.1 Background

The trace problem for totally positive algebraic integers (called the "Schur-Siegel-Smyth trace problem" by Peter Borwein in his very nice recent book [2]), is the following.

Problem 1. Fix $\rho < 2$. Then show that all but finitely many totally positive algebraic integers β have $\mathrm{tr}(\beta)/\deg(\beta) > \rho$.

Thus here β is a zero of an irreducible monic polynomial of degree $\deg(\beta)$ with integer coefficients, whose roots are all positive, and whose sum is $\mathrm{tr}(\beta)$.

In 1918 I. Schur [9] solved the problem for $\rho < \sqrt{e} = 1.6487\cdots$. In 1943 C.L. Siegel [10] solved it for $\rho < 1.737$. In [11] (see also [12]) the problem was solved for $\rho < 1.7719$, while in the previous section we solve it for $\rho < 1.7783786$. In [13] it was shown that there was no inequality of the type (4) having a lower bound ρ for any ρ larger than $2-10^{-41}$. Shortly afterwards J.-P. Serre (personal communication, see "Note added in proof" in [13]), showed that there was no such inequality for any ρ larger than $1.8983021\cdots$. Here we present possible further evidence against this problem being solvable for all $\rho < 2$. We prove that the existence of a single polynomial f with properties given below would imply that the problem cannot be solved for ρ sufficiently close to 2. The result is, however, highly speculative, as such a polynomial may not exist!

5.2 The Polynomial

Suppose that f is a monic polynomial of degree at least 2 with integer coefficients and all positive distinct roots such that

- $|f(0)| \geq 2$;
- between every pair of distinct roots of f there is an x with $|f(x)| \geq 2$.

Then we claim that the set of all totally positive algebraic integers contains infinitely many β whose absolute trace is no greater than $\operatorname{tr}(f)/\deg(f)$.

We now prove the claim. Let $p > 2$ be prime, ω_p a primitive p-th root of unity, and $\alpha = \omega_p + 1/\omega_p$, with conjugates α_i, and let Q be the minimal polynomial of α. Then

$$F(x) = \prod_i (f(x) - \alpha_i) = Q(f(x))$$

is a polynomial of trace $\operatorname{tr}(f)\deg(Q)$, degree $\deg(f)\deg(Q)$ and, and so of absolute trace $:= \operatorname{tr}(F)/\deg(F) = \operatorname{tr}(f)/\deg(f)$. Note that, as the α_i are in $(-2,2)$, it is clear from the graph of f that all the roots of F are real, positive and distinct. Let β be any one of them. Then $f(\beta) = \alpha_i$ for some i, so that the field $\mathbb{Q}(\beta)$ contains α_i, and hence $\deg(\beta) \geq \frac{1}{2}(p-1)$. Let $m(\beta)$ be the absolute trace of β. Then, taking the β_j to be the representatives of the conjugate sets of the roots of F, we have

$$\operatorname{tr}(F) = \sum_j \operatorname{tr}(\beta_j) = \sum_j m(\beta_j)\deg(\beta_j)$$

and so

$$\operatorname{tr}(F)/\deg(F) = \operatorname{tr}(f)/\deg(f) = \sum_j m(\beta_j) w_j$$

where the weights $w_j := \deg(\beta_j)/\deg(F)$ sum to 1. Hence at least one of the $m(\beta_j)$, $m(\beta^{(p)})$ say, is at most $\operatorname{tr}(f)/\deg(f)$.

Now let $p \to \infty$ through a sequence of primes. Then $\deg(\beta^{(p)}) \to \infty$, so that there must be infinitely many different $\beta^{(p)}$. Thus the $\beta^{(p)}$ give the required algebraic integers. This proves the claim.

Now we see that if there is an f as above with also $\mathrm{tr}(f)/\deg(f) < 2$, then the trace problem has no solution when $\rho > \mathrm{tr}(f)/\deg(f)$.

We do not know whether such a polynomial f exists. Any such f clearly must have absolute trace $\mathrm{tr}(f)/\deg(f) > 1.7783786$. Indeed, its absolute trace must be larger than any lower bound of any inequality (which might be found in the future) of the type (4). For a monic integral polynomial with all positive roots and absolute trace less than 2, the crucial quantity is

$$M_f := \min\left(|f(0)|, |f(\gamma_1)|, |f(\gamma_2)|, \dots, |f(\gamma_{n-1})|\right) ,$$

where n is the degree of f, and $\gamma_1, \gamma_2, \dots, \gamma_{n-1}$ are the roots of f'. We need $M_f \geq 2$.

There are many variants on this idea. For example, if f is a monic polynomial of degree at least 3 with integer coefficients such that between every pair of distinct roots of f there is an x with $|f(x)/x| \geq 2$, then again one can show that the set of all totally positive algebraic integers contains infinitely many β whose absolute trace is no greater than $\mathrm{tr}(f)/\deg(f)$. The argument is entirely similar, but using the polynomial $F(x) = \prod(f(x) - \alpha_i x)$.

Acknowledgment. The authors are duly grateful for the helpful suggestions and comments of an anonymous referee.

References

1. Beukers, F., Smyth, C. J.: Cyclotomic points on curves. Number theory for the millennium, I (Urbana, IL, 2000), 67–85, A K Peters, Natick, MA, 2002.
2. Borwein, Peter: Computational excursions in analysis and number theory. CMS Books in Mathematics, 10. Springer-Verlag, New York, 2002.
3. McKee, James: Families of Pisot numbers with negative trace. Acta Arith. **93** (2000), no. 4, 373–385.
4. McKee, J. F., Rowlinson, P., Smyth, C. J.: Salem numbers and Pisot numbers from stars. Number theory in progress, Vol. 1 (Zakopane-Kościelisko, 1997), 309–319, de Gruyter, Berlin, 1999.
5. McKee, J. F., Smyth, C.J.: There are Salem numbers of every trace. Bull. London Math. Soc., to appear.
6. McKee, J. F., Smyth, C.J.: Salem numbers, Pisot numbers, Mahler measure and graphs, in preparation.
7. McKee, J.F., Smyth, C.J.: Salem numbers and Pisot numbers via interlacing, in preparation.
8. Robinson, Raphael M.: Algebraic equations with span less than 4. Math. Comp. **18** (1964), 547–559.
9. Schur, Issai: Über die Verteilung der Wurzeln bei gewissen algebraischen Gleichungen mit ganzzahligen Koeffizienten. Math. Z. 1(1918), 377-402.
10. Siegel, Carl Ludwig: The trace of totally positive and real algebraic integers. Ann. of Math. (2) **46** (1945), 302–312.
11. Smyth, Christopher: Totally positive algebraic integers of small trace. Ann. Inst. Fourier (Grenoble) **34** (1984), no. 3, 1–28.

12. Smyth, C. J.: The mean values of totally real algebraic integers. Math. Comp. **42** (1984), 663–681.
13. Smyth, C. J.: An inequality for polynomials. Number theory (Ottawa, ON, 1996), 315–321, CRM Proc. Lecture Notes, 19, Amer. Math. Soc., Providence, RI, 1999.
14. Smyth, C. J.: Salem numbers of negative trace. Math. Comp. **69** (2000), no. 230, 827–838.

Low-Dimensional Lattice Basis Reduction Revisited

Phong Q. Nguyen[1] and Damien Stehlé[2]

[1] CNRS/École normale supérieure, Département d'informatique, 45 rue d'Ulm, 75230 Paris Cedex 05, France.
`pnguyen@di.ens.fr`
`http://www.di.ens.fr/\char126\relaxpnguyen/`

[2] LORIA/INRIA Lorraine, 615 rue du J. botanique, 54602 Villers-lès-Nancy, France.
`stehle@loria.fr`
`http://www.loria.fr/\char126\relaxstehle/`

Abstract. Most of the interesting algorithmic problems in the geometry of numbers are NP-hard as the lattice dimension increases. This article deals with the low-dimensional case. We study a greedy lattice basis reduction algorithm for the Euclidean norm, which is arguably the most natural lattice basis reduction algorithm, because it is a straightforward generalization of the well-known two-dimensional Gaussian algorithm. Our results are two-fold. From a mathematical point of view, we show that up to dimension four, the output of the greedy algorithm is optimal: the output basis reaches all the successive minima of the lattice. However, as soon as the lattice dimension is strictly higher than four, the output basis may not even reach the first minimum. More importantly, from a computational point of view, we show that up to dimension four, the bit-complexity of the greedy algorithm is quadratic without fast integer arithmetic: this allows to compute various lattice problems (*e.g.* computing a Minkowski-reduced basis and a closest vector) in quadratic time, without fast integer arithmetic, up to dimension four, while all other algorithms known for such problems have a bit-complexity which is at least cubic. This was already proved by Semaev up to dimension three using rather technical means, but it was previously unknown whether or not the algorithm was still polynomial in dimension four. Our analysis, based on geometric properties of low-dimensional lattices and in particular Voronoï cells, arguably simplifies Semaev's analysis in dimensions two and three, unifies the cases of dimensions two, three and four, but breaks down in dimension five.

1 Introduction

A *lattice* is a discrete subgroup of $\mathbb{R}^n$. Any lattice L has a *lattice basis*, i.e. a set $\{\boldsymbol{b}_1, \ldots, \boldsymbol{b}_d\}$ of linearly independent vectors such that the lattice is the set of all integer linear combinations of the $\boldsymbol{b}_i$'s: $L[\boldsymbol{b}_1, \ldots, \boldsymbol{b}_d] = \left\{\sum_{i=1}^{d} x_i \boldsymbol{b}_i : x_i \in \mathbb{Z}\right\}$. A lattice basis is usually not unique, but all the bases have the same number of elements, called the *dimension* or *rank* of the lattice. In dimension higher than

D.A. Buell (Ed.): ANTS 2004, LNCS 3076, pp. 338–357, 2004.

one, there are infinitely many bases, but some are more interesting than others: they are called *reduced*. Roughly speaking, a reduced basis is a basis made of reasonably short vectors which are almost orthogonal. Finding good reduced bases has proved invaluable in many fields of computer science and mathematics, particularly in cryptology (see for instance the survey [18]); and the computational complexity of lattice problems has attracted considerable attention in the past few years (see for instance the book [16]), following Ajtai's discovery [1] of a connection between the worst-case and average-case hardness of certain lattice problems.

There exist many different notions of reduction, such as those of Hermite [8], Minkowski [17], Korkine-Zolotarev (KZ) [9,11], Venkov [19], Lenstra-Lenstra-Lovász [13], *etc.* Among these, the most intuitive one is perhaps Minkowski's reduction; and up to dimension four, it is arguably optimal compared to all other known reductions, because it reaches all the so-called successive minima. However, finding a Minkowski-reduced basis or a KZ-reduced basis is NP-hard under randomized reductions as the dimension increases, because such bases contain a shortest lattice vector, and the shortest vector problem is NP-hard under randomized reductions [2,15]. In order to better understand lattice reduction, it is tempting to study the low-dimensional case. Improvements in low-dimensional lattice reduction may lead to significant running-time improvements in high-dimensional lattice reduction, as the best lattice reduction algorithms known in theory and in practice for high-dimensional lattices, namely Schnorr's blockwise KZ-reduction [20] and its heuristic variants [21,22], are based on a repeated use of low-dimensional KZ-reduction.

The classical Gaussian algorithm [5] computes in quadratic time (without fast integer arithmetic [23]) a Minkowski-reduced basis of any two-dimensional lattice. This algorithm was extended to dimension three by Vallée [27] in 1986 and Semaev [24] in 2001: Semaev's algorithm is quadratic without fast integer arithmetic, whereas Vallée's algorithm has cubic complexity. More generally, Helfrich [7] showed in 1986 by means of the LLL algorithm [13] how to compute in cubic time a Minkowski-reduced basis of any lattice of fixed (arbitrary) dimension, but the hidden complexity constant grows very fast with the dimension.

In this paper, we generalize the Gaussian algorithm to arbitrary dimension. Although the obtained greedy algorithm is arguably the simplest lattice basis reduction algorithm known, its analysis becomes remarkably more and more complex as the dimension increases. Semaev [24] was the first to prove that the algorithm was still polynomial-time in dimension three, but the polynomial-time complexity remained open for higher dimension. We show that up to dimension four, the greedy algorithm computes a Minkowski-reduced basis in quadratic time without fast arithmetic. This implies that a shortest vector and a KZ-reduced basis can be computed in quadratic time up to dimension four. Independently of the running time improvement, we hope our analysis will help to design new lattice reduction algorithms. The main novelty of our approach compared to previous work is that we use geometrical properties of low-dimensional lattices. In dimension two, the method is very close to the argument given by

Semaev in [24], which is itself very different from previous analyses of the Gaussian algorithm [12,28,10]. In dimension three, Semaev's analysis [24] is based on a rather exhaustive analysis of all the possible behaviors of the algorithm, which involves quite a few computations and makes it difficult to extend to higher dimension. We replace the main technical arguments by geometrical considerations on two-dimensional lattices. This makes it possible to extend the analysis to dimension four, by carefully studying geometrical properties of three-dimensional lattices, although a few additional difficulties appear. However, it is still unknown whether or not the greedy algorithm remains polynomial-time beyond dimension four. Besides, we show that the output basis may not even reach the shortest vector beyond dimension four.

The paper is organized as follows. In Section 2, we recall useful facts about lattices. In Section 3, we recall Gauss' algorithm and describe its natural greedy generalization. In Section 4, we analyze Gauss' algorithm and extend the analysis to dimensions three and four, using geometrical results. We explain why our analysis breaks down in dimension five. In Section 5, we prove geometrical results on low-dimensional lattices which are useful to prove the so-called gap lemmata, an essential ingredient of the complexity analysis of Section 4.

Important remark: Due to lack of space, this extended abstract contains few proofs, and we only show that the algorithm is polynomial-time. The proof of the quadratic complexity will appear in the full version of the paper.
Notations: $\|.\|$ and $\langle .,. \rangle$ denote respectively the Euclidean norm and inner product of $\mathbb{R}^n$; variables in bold are vectors; whenever the notation $[\boldsymbol{b}_1, \ldots, \boldsymbol{b}_d]$ is used, we have $\|\boldsymbol{b}_1\| \leq \ldots \leq \|\boldsymbol{b}_n\|$ and in that case we say that the $\boldsymbol{b}_i$'s are *ordered*. Besides, the complexity model we use is the RAM model, and the computational cost is measured in elementary operations on bits. In any complexity statement, we assume that the underlying lattice L is integral ($L \subseteq \mathbb{Z}^n$). If $x \in \mathbb{R}$, then $\lfloor x \rceil$ denotes a nearest integer to x.

2 Preliminaries

We assume the reader is familiar with geometry of numbers (see [4,6,14,25]).
Gram-Schmidt orthogonalization. Let $\boldsymbol{b}_1, \ldots, \boldsymbol{b}_d$ be vectors. The Gram matrix $G(\boldsymbol{b}_1, \ldots, \boldsymbol{b}_d)$ of $\boldsymbol{b}_1, \ldots, \boldsymbol{b}_d$ is the $d \times d$ matrix $(\langle \boldsymbol{b}_i, \boldsymbol{b}_j \rangle)_{1 \leq i,j \leq d}$ formed by all the inner products. $\boldsymbol{b}_1, \ldots, \boldsymbol{b}_d$ are linearly independent if and only if the determinant of $G(\boldsymbol{b}_1, \ldots, \boldsymbol{b}_d)$ is not zero. The volume $\mathrm{vol}(L)$ of a lattice L is the square root of the determinant of the Gram matrix of any basis of L. The orthogonality-defect of a basis $(\boldsymbol{b}_1, \ldots, \boldsymbol{b}_d)$ of L is defined as $(\prod_{i=1}^{d} \|\boldsymbol{b}_i\|)/\mathrm{vol}(L)$: it is always greater than 1, with equality if and only if the basis is orthogonal. Let $(\boldsymbol{b}_1, \ldots, \boldsymbol{b}_d)$ be linearly independent vectors. The *Gram-Schmidt orthogonalization* $(\boldsymbol{b}_1^*, \ldots, \boldsymbol{b}_d^*)$ is defined as follows: $\boldsymbol{b}_i^*$ is the component of $\boldsymbol{b}_i$ orthogonal to the subspace spanned by $\boldsymbol{b}_1, \ldots, \boldsymbol{b}_{i-1}$.
Successive minima and Minkowski reduction. Let L be a d-dimensional lattice in $\mathbb{R}^n$. For $1 \leq i \leq d$, the *i-th minimum* $\lambda_i(L)$ is the radius of the smallest

closed ball centered at the origin containing at least i linearly independent lattice vectors. The most famous lattice problem is the *shortest vector problem* (SVP): given a basis of a lattice L, find a lattice vector of norm $\lambda_1(L)$. There always exist linearly independent lattice vectors $\boldsymbol{v}_i$'s such that $\|\boldsymbol{v}_i\| = \lambda_i(L)$ for all i. Surprisingly, as soon as $d \geq 4$, such vectors do not necessarily form a lattice basis, and when $d \geq 5$, there may not even exist a lattice basis reaching all the minima. A basis $[\boldsymbol{b}_1, \ldots, \boldsymbol{b}_d]$ of L is *Minkowski-reduced* if for all $1 \leq i \leq d$, $\boldsymbol{b}_i$ has minimal norm among all lattice vectors $\boldsymbol{b}_i$ such that $[\boldsymbol{b}_1, \ldots, \boldsymbol{b}_i]$ can be extended to a basis of L. Surprisingly, up to dimension six, one can easily decide if a given basis is Minkowski-reduced or not, by checking a small number of explicit norm inequalities, known as Minkowski's conditions. A basis reaching all the minima must be Minkowski-reduced, but a Minkowski-reduced basis may not reach all the minima, except the first four ones (see [29]): if $[\boldsymbol{b}_1, \ldots, \boldsymbol{b}_d]$ is a Minkowski-reduced basis of L, then for all $1 \leq i \leq \min(d, 4)$, $\|\boldsymbol{b}_i\| = \lambda_i(L)$. Thus, a Minkowski-reduced basis is optimal in a natural sense up to dimension four. A classical result (see [29]) states that the orthogonality-defect of a Minkowski-reduced basis can be upper-bounded by a constant which only depends on the lattice dimension.

Voronoï cell and Voronoï vectors. The *Voronoï cell* [30] of $L = L[\boldsymbol{b}_1, \ldots, \boldsymbol{b}_d]$, denoted by $\mathrm{Vor}(\boldsymbol{b}_1, \ldots, \boldsymbol{b}_d)$, is the set of vectors $\boldsymbol{x}$ in the linear span of L which are closer to 0 than to any other lattice vector: for all $\boldsymbol{v} \in L$, $\|\boldsymbol{x} - \boldsymbol{v}\| \geq \|\boldsymbol{x}\|$, that is $\|\boldsymbol{v}\|^2 \geq 2|\langle \boldsymbol{v}, \boldsymbol{x} \rangle|$. The Voronoï cell is a finite polytope which tiles the linear span of L by translations by lattice vectors. We extend the notation $\mathrm{Vor}(\boldsymbol{b}_1, \ldots, \boldsymbol{b}_d)$ to the case where the first vectors may be zero (the remaining vectors being linearly independent): $\mathrm{Vor}(\boldsymbol{b}_1, \ldots, \boldsymbol{b}_d)$ denotes the Voronoï cell of the lattice spanned by the non-zero $\boldsymbol{b}_i$'s. A lattice vector $\boldsymbol{v} \in L$ is called a *Voronoï vector* if $\boldsymbol{v}/2$ belongs to the Voronoï cell (in which case $\boldsymbol{v}/2$ will be on the boundary of the Voronoï cell). $\boldsymbol{v} \in L$ is a *strict Voronoï vector* if $\boldsymbol{v}/2$ is contained in the interior of a $(d-1)$-dimensional face of the Voronoï cell. A classical result states that Voronoï vectors correspond to the minima of the cosets of $L/2L$. We say that $(x_1, \ldots, x_d) \in \mathbb{Z}^d$ is a *possible Voronoï coord* if there exists a Minkowski-reduced basis $[\boldsymbol{b}_1, \ldots, \boldsymbol{b}_d]$ such that $x_1\boldsymbol{b}_1 + \ldots + x_d\boldsymbol{b}_d$ is a Voronoï vector. In some parts of the article, we will deal with Voronoï coordinates with respect to other types of reduced bases: the kind of reduction considered will be clear from the context. The *covering radius* $\rho(L)$ of a lattice L is half of the diameter of the Voronoï cell. The *closest vector problem* (CVP) is a non-homogeneous version of SVP: given a basis of a lattice and an arbitrary vector $\boldsymbol{x}$ of $\mathbb{R}^n$, find a lattice vector $\boldsymbol{v}$ minimizing the distance $\|\boldsymbol{v} - \boldsymbol{x}\|$; in other words, if $\boldsymbol{y}$ denotes the orthogonal projection of $\boldsymbol{x}$ over the linear span of L, find $\boldsymbol{v} \in L$ such that $\boldsymbol{v} - \boldsymbol{y}$ belongs to the Voronoï cell of L.

3 A Greedy Generalization of Gauss' Algorithm

In dimension two, there is a simple and efficient lattice basis reduction algorithm due to Gauss. We view Gauss' algorithm as a greedy algorithm based on

Fig. 1. Gauss' algorithm.

Input: A basis $[\boldsymbol{u}, \boldsymbol{v}]$ with its Gram matrix.
Output: A reduced basis of $L[\boldsymbol{u}, \boldsymbol{v}]$, together with its Gram matrix.
1. Repeat
2. $\quad \boldsymbol{r} := \boldsymbol{v} - x\boldsymbol{u}$ where $x := \left\lfloor \frac{\langle \boldsymbol{u}, \boldsymbol{v}\rangle}{\|\boldsymbol{u}\|^2} \right\rceil$,
3. $\quad \boldsymbol{v} := \boldsymbol{u}$,
4. $\quad \boldsymbol{u} := \boldsymbol{r}$,
5. $\quad$ Update the Gram matrix of $(\boldsymbol{u}, \boldsymbol{v})$,
6. Until $\|\boldsymbol{u}\| \geq \|\boldsymbol{v}\|$.
7. Return $[\boldsymbol{v}, \boldsymbol{u}]$ and its Gram matrix.

the one-dimensional CVP, which suggests a natural generalization to arbitrary dimension that we call the greedy reduction algorithm. We study properties of the bases output by the greedy algorithm by defining a new type of reduction and comparing it to Minkowski reduction.

3.1 Gauss' Algorithm

Gauss' algorithm – described in Figure 1 – can be seen as a two-dimensional generalization of the centered Euclidean algorithm [28]. At Step 2 of each loop iteration, $\boldsymbol{u}$ is shorter than $\boldsymbol{v}$, and one would like to shorten $\boldsymbol{v}$ rather than $\boldsymbol{u}$, while preserving the fact that $[\boldsymbol{u}, \boldsymbol{v}]$ is a basis of L. This can be achieved by subtracting to $\boldsymbol{v}$ a multiple $x\boldsymbol{u}$ of $\boldsymbol{u}$, because such a transformation is unimodular. The optimal choice is when $x\boldsymbol{u}$ is the closest vector to $\boldsymbol{v}$, in the one-dimensional lattice spanned by $\boldsymbol{u}$, which gives rise to $x := \left\lfloor \frac{\langle \boldsymbol{u}, \boldsymbol{v}\rangle}{\|\boldsymbol{u}\|^2} \right\rceil$. The values $\langle \boldsymbol{u}, \boldsymbol{v}\rangle$ and $\|\boldsymbol{u}\|^2$ are extracted from $G(\boldsymbol{u}, \boldsymbol{v})$, which is updated at Step 5 of each loop iteration. The complexity of Gauss' algorithm is given by the following classical result:

Theorem 1. *Given as input a basis $[\boldsymbol{u}, \boldsymbol{v}]$ of a lattice L, Gauss' algorithm outputs a Minkowski-reduced basis of L in time $O(\log \|\boldsymbol{v}\| \cdot [1 + \log \|\boldsymbol{v}\| - \log \lambda_1(L)])$.*

Note that this result is not trivial to prove. It is not even clear *a priori* why Gauss' algorithm outputs a Minkowski-reduced basis.

3.2 The Greedy Reduction Algorithm

Gauss' algorithm suggests the general greedy algorithm described in Figure 2, which uses reduction and closest vectors in dimension $d - 1$, to reduce bases in dimension d. We make a few remarks on the description of the algorithm. If the Gram matrix is not given, we may compute it in time $O(\log^2 \|\boldsymbol{b}_d\|)$ for fixed d. The algorithm updates the Gram matrix each time the basis changes.

Name: $\mathsf{Greedy}(\boldsymbol{b}_1, \ldots, \boldsymbol{b}_d)$.
Input: A basis $[\boldsymbol{b}_1, \ldots, \boldsymbol{b}_d]$ together with its Gram matrix.
Output: An ordered greedy-reduced basis of $L[\boldsymbol{b}_1, \ldots, \boldsymbol{b}_d]$ with its Gram matrix.
1. If $d = 1$, return $\boldsymbol{b}_1$.
2. Repeat
3. Order $(\boldsymbol{b}_1, \ldots, \boldsymbol{b}_d)$ by increasing lengths and update the Gram matrix,
4. $[\boldsymbol{b}_1, \ldots, \boldsymbol{b}_{d-1}] := \mathsf{Greedy}(\boldsymbol{b}_1, \ldots, \boldsymbol{b}_{d-1})$,
5. Compute a vector $\boldsymbol{c}$ closest to $\boldsymbol{b}_d$, in $L[\boldsymbol{b}_1, \ldots, \boldsymbol{b}_{d-1}]$,
6. $\boldsymbol{b}_d := \boldsymbol{b}_d - \boldsymbol{c}$ and update the Gram matrix,
7. Until $\|\boldsymbol{b}_d\| \geq \|\boldsymbol{b}_{d-1}\|$.
8. Return $[\boldsymbol{b}_1, \ldots, \boldsymbol{b}_d]$ and its Gram matrix.

Fig. 2. The greedy lattice basis reduction algorithm in dimension d.

Step 3 is easy: if this is the first iteration of the loop, the basis is already ordered; otherwise, $[\boldsymbol{b}_1, \ldots, \boldsymbol{b}_{d-1}]$ is already ordered, and only $\boldsymbol{b}_d$ has to be inserted among $\boldsymbol{b}_1, \ldots, \boldsymbol{b}_{d-1}$. At Step 4, the greedy algorithm calls itself recursively in dimension $d-1$: $G(\boldsymbol{b}_1, \ldots, \boldsymbol{b}_{d-1})$ does not need to be computed before calling the algorithm, since $G(\boldsymbol{b}_1, \ldots, \boldsymbol{b}_d)$ is already known. At this point, we do not explain how Step 5 (the computation of closest vectors) is performed: this issue is postponed to subsection 3.4. Note that for $d = 2$, the greedy algorithm is exactly Gauss' algorithm. From a geometrical point of view, the goal of Steps 5 and 6 is to make sure that the orthogonal projection of $\boldsymbol{b}_d$ over the lattice spanned by $[\boldsymbol{b}_1, \ldots, \boldsymbol{b}_{d-1}]$ lies in the Voronoï cell of that lattice.

An easy proof by induction on d shows that the algorithm terminates. Indeed, the new vector $\boldsymbol{b}_d$ of Step 6 is strictly shorter than $\boldsymbol{b}_{d-1}$ if the loop does not end at Step 7. Thus the product of the norms of the $\boldsymbol{b}_i$'s decreases strictly at each iteration of the loop which is not the last one. But for all B, the number of lattice vectors of norm less than B is finite, which completes the proof.

Although the description of the greedy algorithm is fairly simple, analyzing its bit complexity seems very difficult. Even the two-dimensional case of the Gaussian algorithm is not trivial.

3.3 Greedy Reduction

In this subsection, we study properties of the bases output by the greedy algorithm. As previously mentioned, it is not clear why Gauss' algorithm outputs a Minkowski-reduced basis. But it is obvious that the output basis $[\boldsymbol{u}, \boldsymbol{v}]$ satisfies: $\|\boldsymbol{u}\| \leq \|\boldsymbol{v}\| \leq \|\boldsymbol{v} - x\boldsymbol{u}\|$ for all $x \in \mathbb{Z}$. This suggests the following definition:

Definition 2. *An ordered basis* $[\boldsymbol{b}_1, \ldots, \boldsymbol{b}_d]$ *is* greedy-reduced *if for all* $2 \leq i \leq d$ *and for all* $x_1, \ldots, x_{i-1} \in \mathbb{Z}$: $\|\boldsymbol{b}_i\| \leq \|\boldsymbol{b}_i + x_1\boldsymbol{b}_1 + \ldots + x_{i-1}\boldsymbol{b}_{i-1}\|$.

In other words, we have the following recursive definition: a one-dimensional basis is always greedy-reduced, and an ordered basis $[\boldsymbol{b}_1, \ldots, \boldsymbol{b}_d]$ is greedy-reduced if and only if $[\boldsymbol{b}_1, \ldots, \boldsymbol{b}_{d-1}]$ is greedy-reduced and the projection of $\boldsymbol{b}_d$ over the linear span of $\boldsymbol{b}_1, \ldots, \boldsymbol{b}_{d-1}$ lies in the Voronoï cell $\mathrm{Vor}(\boldsymbol{b}_1, \ldots, \boldsymbol{b}_{d-1})$. The greedy algorithm outputs a greedy-reduced basis, and if the input basis is greedy-reduced, the output basis will be equal to the input basis.

The fact that Gauss' algorithm outputs Minkowski-reduced bases is a particular case of the following result, which compares greedy-reduction with Minkowski reduction:

Lemma 3. *The following statements hold:*
1. Any Minkowski-reduced basis is greedy-reduced.
2. A basis of $d \leq 4$ vectors is Minkowski-reduced if and only if it is greedy-reduced.
3. If $d \geq 5$, there exists a basis of d vectors which is greedy-reduced, but not Minkowski-reduced.

As a consequence, the greedy algorithm outputs a Minkowski-reduced basis up to dimension four, thus reaching all the successive minima of the lattice; but beyond dimension four, the greedy algorithm outputs a greedy-reduced basis which may not be Minkowski-reduced. The following lemma shows that greedy-reduced bases may considerably differ from Minkowski-reduced bases beyond dimension four:

Lemma 4. *Let $d \geq 5$. For all $\varepsilon > 0$, there exists a lattice L and a greedy-reduced basis $[\boldsymbol{b}_1, \ldots, \boldsymbol{b}_d]$ of L such that: $\lambda_1(L)/\|\boldsymbol{b}_1\| \leq \varepsilon$ and $\mathrm{vol}(L)/\prod_{i=1}^{d} \|\boldsymbol{b}_i\| \leq \varepsilon$.*

Such properties do not hold for Minkowski-reduced bases. The first phenomenon shows that greedy-reduced bases may be arbitrarily far from the first minimum, while the second one shows that a greedy-reduced basis may be far from being orthogonal.

3.4 Computing Closest Vectors from Minkowski-Reduced Bases

We now explain how Step 5 of the greedy algorithm can be implemented efficiently up to $d = 5$. Step 5 is trivial only when $d \leq 2$. Otherwise, note that after Step 4, the $(d-1)$-dimensional basis $[\boldsymbol{b}_1, \ldots, \boldsymbol{b}_{d-1}]$ is greedy-reduced, and therefore Minkowski-reduced as long as $d \leq 5$. And we know the Gram matrix of $[\boldsymbol{b}_1, \ldots, \boldsymbol{b}_{d-1}, \boldsymbol{b}_d]$.

Theorem 5. *Let $d \geq 1$ be an integer. There exists an algorithm which, given as input a Minkowski-reduced basis $[\boldsymbol{b}_1, \ldots, \boldsymbol{b}_{d-1}]$, a target vector $\mathbf{t}$ longer than all the $\mathbf{b}_i$'s, and the Gram matrix of $[\boldsymbol{b}_1, \ldots, \boldsymbol{b}_{d-1}, \boldsymbol{t}]$, outputs a closest lattice vector $\mathbf{c}$ to $\mathbf{t}$ (in the lattice spanned by the $[\boldsymbol{b}_1, \ldots, \boldsymbol{b}_{d-1}]$), and the Gram matrix of $(\boldsymbol{b}_1, \ldots, \boldsymbol{b}_{d-1}, \boldsymbol{t}-\boldsymbol{c})$, in time $O(\log ||\boldsymbol{t}|| \cdot [1 + \log ||\boldsymbol{t}|| - \log ||\boldsymbol{b}_\alpha||])$, where $1 \leq \alpha \leq d$ is any integer such that $[\boldsymbol{b}_1, \ldots, \boldsymbol{b}_{\alpha-1}, \boldsymbol{t}]$ is Minkowski-reduced.*

Intuitively, the algorithm works as follows: an approximation of the coordinates (with respect to the $\boldsymbol{b}_i$'s) of the closest vector is computed using linear algebra, and the approximation is then corrected by a suitable exhaustive search.

Let $\boldsymbol{h}$ be the orthogonal projection of $\boldsymbol{t}$ over the linear span of $\boldsymbol{b}_1, \ldots, \boldsymbol{b}_{d-1}$. There exist $y_1, \ldots, y_{d-1} \in \mathbb{R}$ such that $\boldsymbol{h} = \sum_{i=1}^{d-1} y_i \boldsymbol{b}_i$. If $c = \sum_{i=1}^{d-1} x_i \boldsymbol{b}_i$ is a closest vector to $\boldsymbol{t}$, then $\boldsymbol{h} - \boldsymbol{c}$ belongs to $\mathrm{Vor}(\boldsymbol{b}_1, \ldots, \boldsymbol{b}_{d-1})$. However, for any $C > 0$, the coordinates (with respect to any basis of orthogonality-defect $\leq C$) of any point inside the Voronoï cell can be bounded independently from the lattice (see [26]). It follows that if we know an approximation of the y_i's with sufficient precision, then $\boldsymbol{c}$ can be derived from a $O(1)$ exhaustive search, since the coordinates $y_i - x_i$ of $\boldsymbol{h} - \boldsymbol{c}$ are bounded, and so is the orthogonality-defect of a Minkowski-reduced basis.

To approximate the y_i's, we use linear algebra. Let $G = G(\boldsymbol{b}_1, \ldots, \boldsymbol{b}_{d-1})$ and $H = \left(\frac{\langle \boldsymbol{b}_i, \boldsymbol{b}_j \rangle}{\|\boldsymbol{b}_i\|^2}\right)_{1 \leq i,j \leq d-1}$. We have:

$$G \cdot \begin{bmatrix} y_1 \\ \vdots \\ y_n \end{bmatrix} = - \begin{bmatrix} \langle \boldsymbol{b}_1, \boldsymbol{t} \rangle \\ \vdots \\ \langle \boldsymbol{b}_n, \boldsymbol{t} \rangle \end{bmatrix} \text{ therefore } \begin{bmatrix} y_1 \\ \vdots \\ y_n \end{bmatrix} = -H^{-1} \cdot \begin{bmatrix} \frac{\langle \boldsymbol{b}_1, \boldsymbol{t} \rangle}{\|\boldsymbol{b}_1\|^2} \\ \vdots \\ \frac{\langle \boldsymbol{b}_n, \boldsymbol{t} \rangle}{\|\boldsymbol{b}_n\|^2} \end{bmatrix}.$$

We use the latter formula to compute the y_i's with a one-bit accuracy, in the expected time. Let $r = \max_i \lceil \log \frac{\langle \boldsymbol{b}_i, \boldsymbol{t} \rangle}{\|\boldsymbol{b}_i\|^2} \rceil$. Notice that $r = O(1 + \log \|\boldsymbol{t}\| - \log \|\boldsymbol{b}_{\alpha-1}\|)$ by bounding $\langle \boldsymbol{b}_i, \boldsymbol{t} \rangle$ depending on whether $i \geq \alpha$. Notice also that the entries of H are all ≤ 1 in absolute value (because $[\boldsymbol{b}_1, \ldots, \boldsymbol{b}_{d-1}]$ is Minkowski-reduced), and $\det(H)$ is lower bounded by some universal constant (because the orthogonality-defect of $[\boldsymbol{b}_1, \ldots, \boldsymbol{b}_{d-1}]$ is bounded). It follows that one can compute the entries of H^{-1} with a $\Omega(r)$-bit accuracy, in $O(r^2)$ binary operations. One eventually derives the y_i's with a one-bit accuracy,

4 Complexity Analysis of the Greedy Algorithm

4.1 A Geometric Analysis of Gauss' Algorithm

We provide yet another proof of the classical result that Gauss' algorithm has quadratic complexity. Compared to other proofs, our method closely resembles the recent one of Semaev [24], itself relatively different from [12,28,10,3]. The analysis is not optimal (as opposed to [28]), but its basic strategy can be extended up to dimension four. Consider the value of x at Step 2:

- If $x = 0$, this must be the last iteration of the loop.
- If $|x| = 1$, there are two cases:
  - If $\|\boldsymbol{v} - x\boldsymbol{u}\| \geq \|\boldsymbol{u}\|$, then this is the last loop iteration.
  - Otherwise, the inequality can be rewritten as $\|\boldsymbol{u} - x\boldsymbol{v}\| < \|\boldsymbol{u}\|$, which means that $\boldsymbol{u}$ can be shortened with the help of $\boldsymbol{v}$, which can only happen if this is the first loop iteration, because of the greedy strategy.
- Otherwise, $|x| \geq 2$, which implies that $x\boldsymbol{u}$ is not a Voronoï vector of the lattice spanned by $\boldsymbol{u}$. Intuitively, this means that $x\boldsymbol{u}$ is far away from $\mathrm{Vor}(\boldsymbol{u})$, so that $\boldsymbol{v} - x\boldsymbol{u}$ is considerably shorter than $\boldsymbol{v}$. More precisely, one can show that $\|\boldsymbol{v}\|^2 \geq \|\boldsymbol{v} - x\boldsymbol{u}\|^2 + 2\|\boldsymbol{u}\|^2$, which is therefore $> 3\|\boldsymbol{v} - x\boldsymbol{u}\|^2$ if this is not the last loop iteration.

This shows that the product of the basis vectors norms decreases by a factor at least $\sqrt{3}$ every loop iteration except possibly the first and last ones. Thus, the number τ of loop iterations is bounded by: $\tau \leq 2 + \log_{\sqrt{3}} \|\boldsymbol{v}\| - \log_{\sqrt{3}} \lambda_1(L)$.

It remains to estimate the cost of each Step 2, which is the cost of computing x. Because $\frac{|\langle \boldsymbol{u}, \boldsymbol{v}\rangle|}{\|\boldsymbol{u}\|^2} \leq \frac{\|\boldsymbol{v}\|}{\|\boldsymbol{u}\|}$, one can see that the bit complexity of Step 2 is $O(\log \|\boldsymbol{v}\| \cdot [1 + \log \|\boldsymbol{v}\| - \log \|\boldsymbol{u}\|])$. If we denote by $\boldsymbol{u}_i$ and $\boldsymbol{v}_i$ the values of $\boldsymbol{u}$ and $\boldsymbol{v}$ at the i-th iteration, then $\boldsymbol{v}_{i+1} = \boldsymbol{u}_i$ and we obtain that the bit complexity of Gauss' algorithm is bounded by:

$$\begin{aligned} &O\left(\textstyle\sum_{i=1}^{\tau} \log \|\boldsymbol{v}_i\| \cdot [1 + \log \|\boldsymbol{v}_i\| - \log \|\boldsymbol{u}_i\|]\right) \\ &\qquad = O\left(\log \|\boldsymbol{v}\| \cdot \textstyle\sum_{i=1}^{\tau} [1 + \log \|\boldsymbol{v}_i\| - \log \|\boldsymbol{v}_{i+1}\|]\right) \\ &\qquad = O\left(\log \|\boldsymbol{v}\| \cdot [\tau + \log \|\boldsymbol{v}\| - \log \lambda_1(L)]\right). \end{aligned}$$

This completes the proof of Theorem 1.

4.2 A Geometric Analysis up to Dimension Four

The main result of the paper is the following:

Theorem 6. *Let $1 \leq d \leq 4$. Given as input an ordered basis $[\boldsymbol{b}_1, \ldots, \boldsymbol{b}_d]$, the greedy algorithm of Figure 2 based on the algorithm of Theorem 5 outputs a Minkowski-reduced basis of $L[\boldsymbol{b}_1, \ldots, \boldsymbol{b}_d]$, using a number of bit operations bounded by $O(\log \|\boldsymbol{b}_d\| \cdot [1 + \log \|\boldsymbol{b}_d\| - \log \lambda_1(L)])$.*

However, due to lack of space, we only prove the following weaker result in this extended abstract:

Theorem 7. *Let $1 \leq d \leq 4$. Given as input an ordered basis $[\boldsymbol{b}_1, \ldots, \boldsymbol{b}_d]$, the greedy algorithm of Figure 2 based on the algorithm of Theorem 5 outputs a Minkowski-reduced basis of $L[\boldsymbol{b}_1, \ldots, \boldsymbol{b}_d]$, using a number of bit operations bounded by a polynomial in $\log \|\boldsymbol{b}_d\|$.*

The result asserts the polynomial-time complexity of the greedy algorithm, which is by far the hardest part of Theorem 6. Both theorems are proved iteratively: the case $d = 4$ is based on the case $d = 3$, which is itself based on the case $d = 2$. The analysis of Gauss' algorithm (Section 4.1) was based on the fact that if $|x| \geq 2$, $x\boldsymbol{u}$ is far away from the Voronoï cell of the lattice spanned by $\boldsymbol{u}$. The proof of Theorem 7 relies on a similar phenomenon in dimensions two and three. However, the situation is considerably more complex, as the following basic remarks hint:

- For $d = 2$, we considered the value of x, but if $d \geq 3$, there will be several coefficients instead of a single x, and it is not clear which coefficient will be useful in the analysis.
- For $d = 2$, Step 4 cannot change the basis, as there are only two bases in dimension one. If $d \geq 3$, Step 4 may completely change the vectors, and it could be hard to keep track of what is going on.

In order to prove Theorem 7, we introduce a few notations. Consider the i-th loop iteration. Let $[\boldsymbol{a}_1^i, \ldots, \boldsymbol{a}_d^i]$ denote the basis $[\boldsymbol{b}_1, \ldots, \boldsymbol{b}_d]$ at the beginning of the i-th loop iteration. The basis $[\boldsymbol{a}_1^i, \ldots, \boldsymbol{a}_d^i]$ becomes $[\boldsymbol{b}_1^i, \ldots, \boldsymbol{b}_{d-1}^i, \boldsymbol{a}_d^i]$ with $\|\boldsymbol{b}_1^i\| \leq \ldots \leq \|\boldsymbol{b}_{d-1}^i\|$ after Step 4, and $(\boldsymbol{b}_1^i, \ldots, \boldsymbol{b}_d^i)$ after Step 6, where $\boldsymbol{b}_d^i = \boldsymbol{a}_d^i - \boldsymbol{c}^i$ and $\boldsymbol{c}^i$ is the closest vector $\boldsymbol{c}$ found at Step 5. Let p_i be the number of integers $1 \leq j \leq d$ such that $\|\boldsymbol{b}_j^i\| \leq \|\boldsymbol{b}_d^i\|$. Let π_i be the rank of $\boldsymbol{b}_d^i$ once $(\boldsymbol{b}_1^i, \ldots, \boldsymbol{b}_d^i)$ is sorted by length: for example, $\pi_i = 1$ if $\|\boldsymbol{b}_d^i\| < \|\boldsymbol{b}_1^i\|$. Clearly, $1 \leq \pi_i \leq p_i \leq d$, if $p_i = d$ then the loop terminates, and otherwise $\|\boldsymbol{a}_{\pi_i}^{i+1}\| = \|\boldsymbol{a}_{p_i}^{i+1}\|$. Note that π_i may not be equal to p_i because there may be several choices when sorting the vectors by length in case of equalities.

Now consider the $(i+1)$-th loop iteration for some $i \geq 1$. Recall that by definition of π_i, we have $\boldsymbol{a}_{\pi_i}^{i+1} = \boldsymbol{b}_d^i = \boldsymbol{a}_d^i - \boldsymbol{c}^i$, while $[\boldsymbol{a}_j^{i+1}]_{j \neq \pi_i} = [\boldsymbol{b}_j^i]_{1 \leq j \leq d-1}$. The closest vector $\boldsymbol{c}^{i+1}$ belongs to $L[\boldsymbol{b}_1^{i+1}, \ldots, \boldsymbol{b}_{d-1}^{i+1}] = L[\boldsymbol{a}_1^{i+1}, \ldots, \boldsymbol{a}_{d-1}^{i+1}]$: there exist integers $x_1^{i+1}, \ldots, x_{d-1}^{i+1}$ such that $\boldsymbol{c}^{i+1} = \sum_{j=1}^{d-1} x_j^{i+1} \boldsymbol{a}_j^{i+1}$.

Suppose we know that Theorem 7 is correct in dimension $d-1$ with $2 \leq d \leq 4$; we are to prove that it is still valid in dimension d. Because of this induction hypothesis and of Theorem 5, the number of bit operations performed in the i-th loop iteration is bounded by a polynomial in $\log \|\boldsymbol{a}_d^i\|$. Since $\|\boldsymbol{a}_d^i\| \leq \|\boldsymbol{a}_d^1\|$ for any i, it is sufficient to prove that the number of loop iterations is bounded by a polynomial in $\log \|\boldsymbol{a}_d^1\|$. Indeed, we show that there exist a universal constant $C_d > 1$ such that for any execution of the d-dimensional greedy algorithm, in any d consecutive iterations of the loop, the product of the lengths of the current vectors decreases by some factor higher than C_d:

$$\frac{\|\boldsymbol{a}_1^i\| \ldots \|\boldsymbol{a}_d^i\|}{\|\boldsymbol{a}_1^{i+d}\| \ldots \|\boldsymbol{a}_d^{i+d}\|} \geq C_d. \tag{1}$$

This automatically ensures that the number of loop iterations is at most proportional to $\log \|\boldsymbol{a}_d^1\|$, and that the total number of bit operations is bounded by a polynomial in $\log \|\boldsymbol{a}_d^1\|$.

We deal with the first difficulty mentioned: which coefficient will be useful? The trick is to consider the value of $x_{\pi_i}^{i+1}$, that is, the coefficient of $\boldsymbol{a}_{\pi_i}^{i+1} = \boldsymbol{a}_d^i - \boldsymbol{c}^i$ in $\boldsymbol{c}^{i+1}$, and to use the greedy properties of the algorithm.

Lemma 8. *Among d consecutive iterations of the loop of the greedy algorithm of Figure 2, there is at least one iteration of index $i+1$ such that $p_{i+1} \leq p_i$. Moreover, for such a loop iteration, we have $|x_{\pi_i}^{i+1}| \geq 2$.*

Proof. The first statement is obvious. Consider one such loop iteration $i+1$. Suppose we have a small $|x_{\pi_i}^{i+1}|$, that is $|x_{\pi_i}^{i+1}| = 0$ or $|x_{\pi_i}^{i+1}| = 1$.

- If $x_{\pi_i}^{i+1} = 0$, $\boldsymbol{c}^{i+1} \in L[\boldsymbol{a}_j^{i+1}]_{j \neq \pi_i, j \leq d-1} = L[\boldsymbol{b}_1^i, \ldots, \boldsymbol{b}_{d-2}^i]$. We claim that the $(i+1)$-th iteration must be the last one. Because the i-th loop iteration was not terminal, we have $\boldsymbol{a}_d^{i+1} = \boldsymbol{b}_{d-1}^i$. Moreover, $[\boldsymbol{b}_1^i, \ldots, \boldsymbol{b}_{d-1}^i]$ is greedy-reduced because of Step 4 of the i-th loop iteration. These two facts imply that $\boldsymbol{c}^{i+1}$ must be zero, and the $(i+1)$-th loop iteration is the last one.

– If $|x_{\pi_i}^{i+1}| = 1$, we claim that $p_{i+1} > p_i$. We have $\boldsymbol{c}^{i+1} = \sum_{j=1}^{d-1} x_j^{i+1} \boldsymbol{a}_j^{i+1}$ where $\boldsymbol{a}_{\pi_i}^{i+1} = \boldsymbol{a}_d^i - \boldsymbol{c}^i$ and $[\boldsymbol{a}_j^{i+1}]_{j \neq \pi_i} = [\boldsymbol{b}_j^i]_{1 \leq j \leq d-1}$. Thus, $\boldsymbol{c}^{i+1}$ can be written as $\boldsymbol{c}^{i+1} = \pm(\boldsymbol{a}_d^i - \boldsymbol{c}^i) + \boldsymbol{e}$ where $\boldsymbol{e} \in L[\boldsymbol{b}_1^i, \ldots, \boldsymbol{b}_{d-2}^i]$. Therefore $\boldsymbol{a}_d^{i+1} - \boldsymbol{c}^{i+1} = \boldsymbol{b}_{d-1}^i \pm (\boldsymbol{a}_d^i - \boldsymbol{c}^i) + \boldsymbol{e}$. In other words, $\|\boldsymbol{a}_d^{i+1} - \boldsymbol{c}^{i+1}\| = \|\boldsymbol{a}_d^i - \boldsymbol{f}\|$ for some $\boldsymbol{f} \in L[\boldsymbol{b}_1^i, \ldots, \boldsymbol{b}_{d-1}^i]$. It follows that $p_{i+1} \geq 1 + p_i$, which achieves the claim.

□

We will see that in dimension three, any such loop iteration $i+1$ implies that at least one of the basis vectors significantly decreases in the $(i+1)$-th loop iteration, or had significantly decreased in the i-th loop iteration. This is only "almost" true in dimension four: fortunately, we will be able to isolate the bad cases, and to show that when a bad case occurs, the number of remaining loop iterations can be bounded by some universal constant.

We now deal with the second difficulty. Recall that $\boldsymbol{c}^{i+1} = \sum_{j=1}^{d-1} x_j^{i+1} \boldsymbol{a}_j^{i+1}$ but the basis $[\boldsymbol{a}_1^{i+1}, \ldots, \boldsymbol{a}_{d-1}^{i+1}]$ is not necessarily greedy-reduced. We distinguish two cases:

– 1) The basis $[\boldsymbol{a}_1^{i+1}, \ldots, \boldsymbol{a}_{d-1}^{i+1}]$ is somehow far from being greedy-reduced. Then $\boldsymbol{b}_d^i$ was significantly shorter than $\boldsymbol{a}_d^i$. Note that this length decrease concerns the i-th loop iteration and not the $(i+1)$-th.
– 2) Otherwise, the basis $[\boldsymbol{a}_1^{i+1}, \ldots, \boldsymbol{a}_{d-1}^{i+1}]$ is almost greedy-reduced. The fact that $|x_{\pi_i}^{i+1}| \geq 2$ roughly implies that $\boldsymbol{c}^{i+1}$ is somewhat far away from the Voronoï cell $\mathrm{Vor}(\boldsymbol{a}_1^{i+1}, \ldots, \boldsymbol{a}_{d-1}^{i+1})$: this phenomenon will be precisely captured by the so-called Gap Lemma. When this is the case, the new vector $\boldsymbol{b}_d^{i+1}$ will be significantly shorter than $\boldsymbol{a}_d^{i+1}$.

To capture the property that a set of vectors is almost greedy-reduced, we introduce the so-called *ε-reduction* where $\varepsilon \geq 0$, which is defined as follows:

Definition 9. *A single vector $\boldsymbol{b}_1$ is always ε-reduced; for $d \geq 2$, a d-tuple $(\boldsymbol{b}_1, \ldots, \boldsymbol{b}_d)$ is ε-reduced if $(\boldsymbol{b}_1, \ldots, \boldsymbol{b}_{d-1})$ is ε-reduced, $\|\boldsymbol{b}_{d-1}\| \leq \|\boldsymbol{b}_d\|$, and the orthogonal projection of $\boldsymbol{b}_d$ over the linear span of $(\boldsymbol{b}_1, \ldots, \boldsymbol{b}_{d-1})$ belongs to $(1+\varepsilon)\,Vor(\boldsymbol{b}_1, \ldots, \boldsymbol{b}_{d-1})$.*

With this definition, a greedy-reduced basis is ε-reduced for any $\varepsilon \geq 0$. In the definition of ε-reduction, we did not assume that the $\boldsymbol{b}_i$'s were nonzero nor linearly independent. This is because the Gap Lemma is essentially based on compactness properties: the set of ε-reduced d-tuples needs to be closed (from a topological point of view), while a limit of bases may not be a basis.

We can now give the precise statements of the two cases described above. Lemma 10 corresponds to case 1), and Lemma 11 to case 2).

Lemma 10. *Let $2 \leq d \leq 4$. There exists a constant $\varepsilon_1 > 0$ such that for any $\varepsilon \leq \varepsilon_1$ there exists $C_\varepsilon > 1$ such that the following statement holds. Consider the $(i+1)$-th loop iteration of an execution of the d-dimensional greedy algorithm. If $[\boldsymbol{a}_1^{i+1}, \ldots, \boldsymbol{a}_{d-1}^{i+1}]$ is not ε-reduced, then $\|\boldsymbol{a}_d^i\| \geq C_\varepsilon \|\boldsymbol{b}_d^i\|$.*

Lemma 11. *Let $2 \leq d \leq 4$. There exist two constants $\varepsilon_2 > 0$ and $D > 0$ such that the following statement holds. Consider the $(i+1)$-th loop iteration of an execution of the d-dimensional greedy algorithm. Suppose that $[\boldsymbol{a}_1^{i+1}, \ldots, \boldsymbol{a}_{d-1}^{i+1}]$ is ε_2-reduced, and that $||\boldsymbol{a}_k^{i+1}|| \geq (1-\varepsilon_2)||\boldsymbol{a}_d^{i+1}||$ for some $1 \leq k \leq d-1$. Then, if $|x_k| \geq 2$ and if we are not in the 211-case, we have:*

$$||\boldsymbol{b}_d^{i+1}||^2 + D||\boldsymbol{b}_k^{i+1}||^2 \leq ||\boldsymbol{a}_d^{i+1}||^2,$$

where the 211-case is: $d = 4$, $|x_k| = 2$ and the other $|x_j|$'s are all equal to 1.

This last lemma is a direct consequence of Pythagore and the Gap Lemma (which is crucial to our analysis, and to which the next section is devoted):

Theorem 12 (Gap Lemma). *Let $2 \leq d \leq 4$. There exist two constants $\varepsilon_2 > 0$ and $D > 0$ such that the following statement holds. Let $[\boldsymbol{a}_1, \ldots, \boldsymbol{a}_{d-1}]$ be ε-reduced vectors, $\boldsymbol{u}$ be a vector of $Vor(\boldsymbol{a}_1, \ldots, \boldsymbol{a}_{d-1})$ and $x_1, \ldots, x_{d-1}$ be integers. If $||\boldsymbol{a}_k|| \geq (1-\varepsilon)||\boldsymbol{a}_{d-1}||$ for some $k \leq d-2$, then:*

$$||\boldsymbol{u}||^2 + D||\boldsymbol{b}_k||^2 \leq ||\boldsymbol{u} + \sum_{j=1}^{d-1} x_j \boldsymbol{b}_j||^2,$$

where $|x_k| \geq 2$, and if $d = 4$ the two other $|x_j|$'s are not all equal to 1.

This completes the overall description of the proof of Theorem 7. Indeed, choose three constants $\varepsilon, D > 0$ and $C > 1$ such that we can apply Lemmata 10 and 11. We prove that Equation (1) holds for $C_d = \min(C, \sqrt{1+D}, \frac{1}{1-\varepsilon}) > 1$. Consider a loop iteration $i+1$ such that $p_{i+1} \leq p_i$. Recall that among any d consecutive iterations of the loop, there is at least one such iteration. For such an iteration, we have $|x_{\pi_i}^{i+1}| \geq 2$. We distinguish four cases:

- $[\boldsymbol{a}_1^{i+1}, \ldots, \boldsymbol{a}_{d-1}^{i+1}]$ is not ε-reduced: then Lemma 10 gives the result through the i-th loop iteration.
- $||\boldsymbol{a}_{\pi_i}^{i+1}|| < (1-\varepsilon)||\boldsymbol{a}_d^{i+1}||$: because $p_{i+1} \leq p_i$, we have the inequalities $||\boldsymbol{b}_d^{i+1}|| < ||\boldsymbol{a}_{p_i}^{i+1}|| = ||\boldsymbol{a}_{\pi_i}^{i+1}|| < (1-\varepsilon)||\boldsymbol{a}_d^{i+1}||$.
- We are in the 211-case, i.e. $d = 4$ with $|x_{\pi_i}| = 2$ and the other $|x_j|$'s are all equal to 1, we refer to the detail analysis of subsection 4.3.
- Otherwise, we apply Lemma 11, which gives the expected result through the $(i+1)$-th loop iteration.

We described our strategy to prove that the greedy algorithm is polynomial-time up to dimension four. One can further prove that the bit complexity is in fact quadratic, by carefully assessing the costs of each loop iteration and combining them, but the proof is much more technical than in the two-dimensional case.

4.3 Concluding in Dimension Four

In the previous subsections, we showed that the greedy algorithm is polynomial-time in dimensions two and three, but we noticed that a new difficulty arose in dimension four: the Gap Lemma is useless in the so-called 211-case. This is because there are three-dimensional Minkowski-reduced bases $[\boldsymbol{b}_1, \boldsymbol{b}_2, \boldsymbol{b}_3]$ for which $2\boldsymbol{b}_i + s_1\boldsymbol{b}_j + s_2\boldsymbol{b}_k$ – with $\{i,j,k\} = \{1,2,3\}$ and $|s_1| = |s_2| = 1$ – is a Voronoï vector. Indeed consider the lattice spanned by the columns $\boldsymbol{b}_1, \boldsymbol{b}_2, \boldsymbol{b}_3$ of the following matrix:

$$M = \begin{bmatrix} 1 & 1 & -1 \\ 1 & -1 & 0 \\ 0 & 0 & 1 \end{bmatrix}.$$

This basis is Minkowski-reduced and $\|\boldsymbol{b}_1 + \boldsymbol{b}_2 + 2\boldsymbol{b}_3\| = \|\boldsymbol{b}_1 + \boldsymbol{b}_2\| \leq \|(2k_1 + 1)\boldsymbol{b}_1 + (2k_2 + 1)\boldsymbol{b}_1 + 2k_3\boldsymbol{b}_3\|$ for any $k_1, k_2, k_3 \in \mathbb{Z}$. Therefore, a vector in the Voronoï cell centered in $\boldsymbol{b}_1 + \boldsymbol{b}_2 + 2\boldsymbol{b}_3$ can avoid being significantly shortened when translated inside the Voronoï cell centered in 0.

The Gap Lemma cannot tackle this problem. However, we note that $(1,1,2)$ is rarely a Voronoï coordinate (with respect to a Minkowski-reduced basis), and when it is, it cannot be a strict Voronoï coord: we can prove that if $(1,1,2)$ is a Voronoï coord, then $\|\boldsymbol{b}_1+\boldsymbol{b}_2\| = \|\boldsymbol{b}_1+\boldsymbol{b}_2+2\boldsymbol{b}_3\|$, which tells us that $\boldsymbol{b}_1+\boldsymbol{b}_2+2\boldsymbol{b}_3$ is not the only vector in its coset of $L/2L$ reaching the minimum. It turns out that the lattice spanned by the columns of M is essentially the only one for which $(1,1,2)$ – modulo any change of sign and permutation of coordinates – can be a Voronoï coord. More precisely, if $(1,1,2)$ – modulo any change of sign and permutation of coordinates – is a Voronoï coord for a lattice basis, then the basis matrix can be written as rUM where r is any non-zero real number and U is any orthogonal matrix. Since a basis can be arbitrarily close to one of these without being one of them, we need to consider a small compact set of normalized bases around the annoying ones. More precisely, this subset is:

$$\{[\boldsymbol{b}_1, \boldsymbol{b}_2, \boldsymbol{b}_3] \text{ } \varepsilon\text{-reduced} \ / \ \exists \sigma \in \mathcal{S}_3, \| \frac{1}{\|\boldsymbol{b}_3\|^2} |G(\boldsymbol{b}_{\sigma(1)}, \boldsymbol{b}_{\sigma(2)}, \boldsymbol{b}_{\sigma(3)})| - |M^tM| \, \|_\infty \leq \varepsilon\},$$

for some sufficiently small $\varepsilon > 0$, where $\|M\|_\infty$ is the maximum of the absolute values of the matrix M and $|M|$ is the matrix of the absolute values.

Now, consider we are in the 211-case at some loop iteration $i+1$. We distinguish three cases:

- $[\boldsymbol{a}_1^{i+1}, \boldsymbol{a}_2^{i+1}, \boldsymbol{a}_3^{i+1}]$ is outside the compact. In this case, a variant of the Gap Lemma (Lemma 29) proved in Section 5 is valid, and can be used to show that $\boldsymbol{b}_4^{i+1}$ is significantly shorter than $\boldsymbol{a}_4^{i+1}$.
- $[\boldsymbol{a}_1^{i+1}, \boldsymbol{a}_2^{i+1}, \boldsymbol{a}_3^{i+1}]$ is inside the compact, but $\boldsymbol{a}_4^{i+1}$ is far from the Voronoï cell $\mathrm{Vor}(\boldsymbol{a}_1^{i+1}, \boldsymbol{a}_2^{i+1}, \boldsymbol{a}_3^{i+1})$. In this case, $\boldsymbol{b}_4^{i+1}$ is significantly shorter than $\boldsymbol{a}_4^{i+1}$.
- Otherwise the overall geometry of $[\boldsymbol{a}_1^{i+1}, \boldsymbol{a}_2^{i+1}, \boldsymbol{a}_3^{i+1}, \boldsymbol{a}_4^{i+1}]$ is very precisely known, and we can show that there remain at most $O(1)$ loop iterations.

More precisely, by using Lemma 29, we show that:

Lemma 13. *There exist three constants $C, \varepsilon > 0$, and $c \in \mathbb{Z}$ such that the following holds. Consider an execution of the four-dimensional greedy algorithm, and a loop iteration $i+1$ for which $p_{i+1} \leq p_i$, $[\boldsymbol{a}_1^{i+1}, \boldsymbol{a}_2^{i+1}, \boldsymbol{a}_3^{i+1}]$ is ε-reduced, $\|\boldsymbol{a}_{p_i}^{i+1}\| \geq (1-\varepsilon)\|\boldsymbol{a}_4^{i+1}\|$, and $(|x_{\sigma(1)}|, |x_{\sigma(2)}|, |x_{\sigma(3)}|) = (1,1,2)$ for some permutation σ of $\{1,2,3\}$. Then either $\|\boldsymbol{a}_4^{i+1}\| \geq (1+C)\|\boldsymbol{b}_4^{i+1}\|$ or:*

$$\left\|\frac{1}{\|\boldsymbol{a}_4^{i+1}\|^2}|G(\boldsymbol{a}_{\sigma(1)}^{i+1}, \boldsymbol{a}_{\sigma(2)}^{i+1}, \boldsymbol{a}_{\sigma(3)}^{i+1}, \boldsymbol{a}_4^{i+1})| - A\right\|_\infty \leq \varepsilon, \text{ with} : A = \begin{bmatrix} 1 & 0 & \frac{1}{2} & 0 \\ 0 & 1 & \frac{1}{2} & 0 \\ \frac{1}{2} & \frac{1}{2} & 1 & \frac{1}{2} \\ 0 & 0 & \frac{1}{2} & 1 \end{bmatrix}.$$

In order to prove this result, we restrict more and more the possible geometry of $[\boldsymbol{a}_1^i, \boldsymbol{a}_2^i, \boldsymbol{a}_3^i, \boldsymbol{a}_4^i]$. Note that this critical geometry corresponds to the root lattice D_4. We treat this last case by applying the following lemma, which roughly says that if the Gram matrix of a basis is sufficiently close to some invertible matrix, then the number of short vectors generated by the basis remains bounded.

Lemma 14. *Let A be an invertible $d \times d$ matrix, and $B > 0$. Then there exists $\varepsilon, N > 0$ such that for any basis $(\boldsymbol{b}_1, \ldots, \boldsymbol{b}_d)$, if $\|G(\boldsymbol{b}_1, \ldots, \boldsymbol{b}_d) - A\|_\infty \leq \varepsilon$, then:*

$$|\{(x_1, \ldots, x_d) \ / \ \|x_1\boldsymbol{b}_1 + \ldots + x_d\boldsymbol{b}_d\| \leq B\}| \ \leq \ N.$$

4.4 Failure in Dimension Five

In this subsection, we explain why the analysis of the greedy algorithm breaks down in dimension five. First of all, the basis returned by the algorithm is not necessarily Minkowski reduced, since greedy and Minkowski reductions differ in dimension five. Consider the lattice spanned by the columns of the following matrix:

$$\begin{bmatrix} 2 & 0 & 0 & 0 & 1 \\ 0 & 2 & 0 & 0 & 1 \\ 0 & 0 & 2 & 0 & 1 \\ 0 & 0 & 0 & 2 & 1 \\ 0 & 0 & 0 & 0 & \varepsilon \end{bmatrix},$$

where $0 < \varepsilon < 1$. This basis is clearly greedy reduced, but the vector $(0,0,0,0,\varepsilon)^t$ belongs to the lattice. Moreover, for a small ε, this shows that a greedy reduced basis can be arbitrarily far from the first minimum, and can have an arbitrarily large orthogonality defect: $\|\boldsymbol{b}_5^*\|$ is very small towards $\|\boldsymbol{b}_5\|$. For ε close to 1, this basis shows that the length decrease factor through one loop iteration of the five-dimensional greedy algorithm can be arbitrarily close to 1.

Nevertheless, the greedy algorithm is well-defined in dimension five (the four-dimensional greedy algorithm can be used for Step 4, and since it returns a Minkowski reduced basis, the CVP algorithm of Theorem 5 can be used in Step 5). Despite the fact that the algorithm does not return a Minkowski reduced basis, one may wonder if the analysis remains valid, and if the number of loop iterations of the 5-dimensional greedy algorithm is linear in $\log\|\boldsymbol{b}_5\|$.

The analysis in dimensions two, three and four essentially relies on the fact that if one of the x_j's found at Step 5 has absolute value higher than 2, then $||\boldsymbol{b}_d^i||$ is significantly shorter than $||\boldsymbol{a}_d^i||$. This fact is derived from the so-called Gap Lemma. In dimension four, this was only partly true, but the exception (the 211-case) happened in very few cases and could be dealt by considering the very specific shape of the lattices for which it could go wrong. Things worsen in dimension five. Indeed, for Minkowski-reduced bases, $(1, 1, 1, 2)$ and $(1, 1, 2, 2)$ – modulo any change of sign and permutation of coordinates – are possible Voronoï coords. Here is an example of a lattice where $(1, 1, 2, 2)$ is a Voronoï coord:

$$\begin{bmatrix} 1 & -1 & 0 & 0 \\ 1 & 1 & -1 & 0 \\ 0 & 0 & 1 & -1 \\ 0 & 0 & 0 & 1 \end{bmatrix}.$$

The lattice basis given by the columns is Minkowski-reduced, but:

$$||\boldsymbol{b}_1+\boldsymbol{b}_2+2\boldsymbol{b}_3+2\boldsymbol{b}_4|| = 2 = ||\boldsymbol{b}_1+\boldsymbol{b}_2|| \leq ||(2k_1+1)\boldsymbol{b}_1+(2k_2+1)\boldsymbol{b}_2+2k_3\boldsymbol{b}_3+2k_4\boldsymbol{b}_4||,$$

for any $k_1, k_2, k_3, k_4 \in \mathbb{Z}$. Note that $(1, 1, 2, 2)$ cannot be a strict Voronoï coord: if $\boldsymbol{b}_1 + \boldsymbol{b}_2 + 2\boldsymbol{b}_3 + 2\boldsymbol{b}_4$ reaches the length minimum of its coset of $L/2L$, then so does $\boldsymbol{b}_1 + \boldsymbol{b}_2$. Thus it might be possible to work around the difficulty coming from $(1, 1, 2, 2)$ like in the previous subsection. However, the case $(1, 1, 1, 2)$ would still remain, and this possible Voronoï coordinate can be strict.

5 The Geometry of Low-Dimensional Lattices

In this section, we give some results about Voronoï cells in dimensions two and three, which are crucial to our complexity analysis of the greedy algorithm described in Section 3. More precisely, the analysis is based on the Gap Lemma (subsection 5.3), which is derived from the study of Voronoï cells in the case of ε-reduced vectors (subsection 5.2), itself derived from the study of Voronoï cells for Minkowski-reduced bases (subsection 5.1).

5.1 Voronoï Cells in the Case of Minkowski-Reduced Bases

We give simple bounds on the diameter of the Voronoï cell and on the Gram-Schmidt orthogonalization of a Minkowski-reduced basis:

Lemma 15. *Let $d \geq 1$. Let $[\boldsymbol{b}_1, \ldots, \boldsymbol{b}_d]$ be a basis of a lattice L. Then $\rho(L) \leq \frac{\sqrt{d}}{2}||\boldsymbol{b}_d||$. As a consequence, if $d \leq 4$ and if $[\boldsymbol{b}_1, \ldots, \boldsymbol{b}_d]$ is a Minkowski-reduced basis, then $||\boldsymbol{b}_d^*|| \geq \frac{\sqrt{5-d}}{2}||\boldsymbol{b}_d||$.*

The following lemma provides the possible Voronoï vectors of a two-dimensional lattice given by a Minkowski-reduced basis. Such a basis confines the coordinates of Voronoï vectors:

Lemma 16. *In dimension two, the possible Voronoï coords are* $(1,0)$ *and* $(1,1)$, *modulo any change of signs and permutation of coordinates, i.e. any nonzero* $(\varepsilon_1,\varepsilon_2)$ *where* $|\varepsilon_1|,|\varepsilon_2| \leq 1$.

The proof relies on a detailed study of the expression $||(2x_1+\varepsilon_1)\cdot \boldsymbol{b}_1+(2x_2+\varepsilon_2)\cdot \boldsymbol{b}_2||^2 - ||\varepsilon_1\boldsymbol{b}_1 + \varepsilon_2\boldsymbol{b}_2||^2$, where $[\boldsymbol{b}_1,\boldsymbol{b}_2]$ is Minkowski-reduced, $\varepsilon_1,\varepsilon_2 \in \{0,1\}$ and $x_1,x_2 \in \mathbb{Z}$. Indeed, since Voronoï coords of a lattice L are given by the minima of the non-zero cosets of $L/2L$, it suffices to show that if $x_1x_2 \neq 0$, then this expression is strictly positive. We do this by a rather technical study.

We generalize this analysis to the three-dimensional case. The underlying ideas of the proof are the same, but because of the increasing number of variables, the analysis becomes more tedious.

Lemma 17. *In dimension three, the possible Voronoï coordinates are* $(1,0,0)$, $(1,1,0)$, $(1,1,1)$ *and* $(2,1,1)$, *modulo any change of signs and permutation of coordinates.*

The possible Voronoï coord $(2,1,1)$ creates difficulties when analyzing the greedy algorithm in dimension four, because it contains a two, which cannot be handled with the greedy argument used for the ones. We tackle this problem as follows: we show that when $(2,1,1)$ happens to be a Voronoï coord, the lattice has a very specific shape, for which the behavior of the algorithm is well-understood.

Lemma 18. *Suppose* $[\boldsymbol{b}_1,\boldsymbol{b}_2,\boldsymbol{b}_3]$ *is a Minkowski-reduced basis.*
1. If any of $(s_1,s_2,2)$ *is a Voronoï coord with* $s_i = \pm 1$ *for* $i \in \{1,2\}$, *then* $||\boldsymbol{b}_1|| = ||\boldsymbol{b}_2|| = ||\boldsymbol{b}_3||$, $\langle \boldsymbol{b}_1,\boldsymbol{b}_2\rangle = 0$ *and* $\langle \boldsymbol{b}_i,\boldsymbol{b}_3\rangle = -s_i||\boldsymbol{b}_1||^2/2$ *for* $i=1,2$.
2. If any of $(s_1,2,s_3)$ *is a Voronoï coord with* $s_i = \pm 1$ *for* $i \in \{1,3\}$, *then* $||\boldsymbol{b}_1|| = ||\boldsymbol{b}_2||$. *Moreover, if* $||\boldsymbol{b}_1|| = ||\boldsymbol{b}_2|| = ||\boldsymbol{b}_3||$, *then* $\langle \boldsymbol{b}_1,\boldsymbol{b}_3\rangle = 0$ *and* $\langle \boldsymbol{b}_i,\boldsymbol{b}_2\rangle = -s_i||\boldsymbol{b}_1||^2/2$ *for* $i=1,3$.
3. If any of $(2,s_2,s_3)$ *is a Voronoï coord with* $s_i = \pm 1$ *for* $i \in \{2,3\}$ *and* $||\boldsymbol{b}_1|| = ||\boldsymbol{b}_2|| = ||\boldsymbol{b}_3||$, *then* $\langle \boldsymbol{b}_2,\boldsymbol{b}_3\rangle = 0$ *and* $\langle \boldsymbol{b}_i,\boldsymbol{b}_1\rangle = -s_i||\boldsymbol{b}_1||^2/2$ *for* $i=2,3$.

5.2 Voronoï Cells in the Case of ε-Reduced Vectors

We extend the results of the previous subsection to the case of ε-reduced vectors. The idea is that if we compact the set of Minkowski-reduced bases and slightly enlarge it, the possible Voronoï coords remain the same. Unfortunately, by doing so, some of the vectors we consider may be zero, and this creates an infinity of possible Voronoï coords: for example, if $\boldsymbol{b}_1 = 0$, any pair $(x_1,0)$ is a Voronoï coord of $[\boldsymbol{b}_1,\boldsymbol{b}_2]$. To tackle this problem, we restrict to $\boldsymbol{b}_i$ with "similar" lengths. More precisely, we use the so-called Topological Lemma: if we can guarantee that the possible Voronoï coords of the enlargement of the initial compact set of bases are bounded, then for a sufficiently small enlargement, the possible Voronoï coords remain the same. We first give rather simple results on ε-reduced vectors and their Gram-Schmidt orthogonalization, then we introduce the Topological Lemma (Lemma 21), from which we finally derive the relaxed versions of Lemmata 16, 17 and 18.

Lemma 19. *There exists a constant $c > 0$ such that for any sufficiently small $\varepsilon > 0$, if $[\boldsymbol{b}_1, \boldsymbol{b}_2, \boldsymbol{b}_3]$ are ε-reduced, then the following inequalities hold:*

$$|\langle \boldsymbol{b}_i, \boldsymbol{b}_j \rangle| \leq \frac{1+c\varepsilon}{2} \|\boldsymbol{b}_i\|^2 \qquad \text{for any } i < j,$$
$$|\langle \boldsymbol{b}_3, \varepsilon_1 \boldsymbol{b}_1 + \varepsilon_2 \boldsymbol{b}_2 \rangle| \leq \frac{1+c\varepsilon}{2} \|\varepsilon_1 \boldsymbol{b}_1 + \varepsilon_2 \boldsymbol{b}_2\|^2 \quad \text{for any } \varepsilon_1, \varepsilon_2 \in \{-1, 1\}.$$

This result implies that if $[\boldsymbol{b}_1, \ldots, \boldsymbol{b}_d]$ are ε-reduced, the only case for which the $\boldsymbol{b}_i$'s can be linearly dependent is when some of them are zero, but this case cannot be avoided since we need compacting the set of Minkowski-reduced bases. The following lemma generalizes Lemma 15. It shows that even with ε-reduced vectors, if the dimension is below four, then the Gram-Schmidt orthogonalization process cannot arbitrarily decrease the lengths of the initial vectors.

Lemma 20. *There exists $C > 0$ such that for any $1 \leq d \leq 4$ and any sufficiently small $\varepsilon > 0$, if $[\boldsymbol{b}_1, \ldots, \boldsymbol{b}_d]$ are ε-reduced vectors, then we have $\|\boldsymbol{b}_d^*\| \geq C\|\boldsymbol{b}_d\|$.*

The Topological Lemma is the key argument when extending the results on possible Voronoï coords from Minkowski-reduced bases to ε-reduced vectors. When applying it, X_0 will correspond to the x_i's, K_0 to the $\boldsymbol{b}_i$'s, X to the possible Voronoï coordinates, and f to the continuous function of real variables $f : (y_i)_i, (\boldsymbol{b}_i)_i \longrightarrow \|y_1 \boldsymbol{b}_1 + \ldots + y_d \boldsymbol{b}_d\|$.

Lemma 21 (Topological Lemma). *Let $n, m \geq 1$. Let X_0 and K_0 be compact sets of $\mathbb{R}^n$ and $\mathbb{R}^m$. Let f be a continuous function from $K_0 \times X_0$ to $\mathbb{R}$. For any $a \in K_0$ we define $M_a = \{x \in X_0 \cap \mathbb{Z}^n \ / \ f(a, x) = min_{x' \in X_0 \cap \mathbb{Z}^n}(f(a, x'))\}$. Let $K \subset K_0$ be a compact and $X = \cup_{a \in K} M_a \subset X_0 \cap \mathbb{Z}^n$. With these notations, there exists $\varepsilon > 0$ such that if $b \in K_0$ satisfies $dist(b, K) \leq \varepsilon$, we have $M_b \subset X$.*

In order to apply the Topological Lemma, we need to map the relaxed bases into a compact set. For any $\varepsilon \geq 0$ and any $\alpha \in [0, 1]$, we define:

$$K_2(\varepsilon, \alpha) = \{(\boldsymbol{b}_1, \boldsymbol{b}_2) / \boldsymbol{b}_1, \boldsymbol{b}_2 \ \varepsilon\text{-reduced}, \ \alpha \leq \|\boldsymbol{b}_1\| \leq \|\boldsymbol{b}_2\| = 1\}$$
$$K_3(\varepsilon, \alpha) = \{(\boldsymbol{b}_1, \boldsymbol{b}_2, \boldsymbol{b}_3) / \boldsymbol{b}_1, \boldsymbol{b}_2, \boldsymbol{b}_3 \ \varepsilon\text{-reduced}, \ \alpha \leq \|\boldsymbol{b}_1\| \leq \|\boldsymbol{b}_2\| \leq \|\boldsymbol{b}_3\| = 1\}.$$

Lemma 22. *If $\varepsilon \geq 0$ and $\alpha \in [0, 1]$, $K_2(\varepsilon, \alpha)$ and $K_3(\varepsilon, \alpha)$ are compact sets.*

The following lemma is the relaxed version of Lemma 16. It can also be viewed as a reciprocal to Lemma 19.

Lemma 23. *For any $\alpha \in \,]0, 1]$ and any sufficiently small $\varepsilon > 0$, the possible Voronoï coords of $[\boldsymbol{b}_1, \boldsymbol{b}_2] \in K_2(\varepsilon, \alpha)$ are the same as for Minkowski-reduced bases, i.e. $(1, 0)$ and $(1, 1)$, modulo any change of signs and permutation of coordinates.*

We now relax Lemma 17 in the same manner.

Lemma 24. *For any $\alpha \in]0,1]$ and any sufficiently small $\varepsilon > 0$, the possible Voronoï coords of $[\boldsymbol{b}_1, \boldsymbol{b}_2, \boldsymbol{b}_3] \in K_3(\varepsilon, \alpha)$ are the same as for Minkowski-reduced bases.*

The following result generalizes Lemma 18 about the possible Voronoï coord $(1,1,2)$. As opposed to the two previous results, there is no need to use the Topological Lemma in this case, because only a finite number of (x_1, x_2, x_3)'s is considered.

Lemma 25. *There exists a constant $c' > 0$ such that for any sufficiently small $\varepsilon > 0$, if $[\boldsymbol{b}_1, \boldsymbol{b}_2, \boldsymbol{b}_3]$ are ε-reduced and $\|\boldsymbol{b}_3\| = 1$, then:*
1. If any of $(s_1, s_2, 2)$ is a Voronoï coord with $s_i = \pm 1$ for $i \in \{1, 2\}$, then: $\|\boldsymbol{b}_1\| \geq 1 - c'\varepsilon$, $|\langle \boldsymbol{b}_1, \boldsymbol{b}_2 \rangle| \leq c'\varepsilon$ and $|\langle \boldsymbol{b}_i, \boldsymbol{b}_3 \rangle + s_i \frac{\|\boldsymbol{b}_1\|^2}{2}| \leq c'\varepsilon$ for $i = 1, 2$.
2. If any of $(s_1, 2, s_3)$ is a Voronoï coord with $s_i = \pm 1$ for $i \in \{1, 3\}$, then $(1 - c'\varepsilon)\|\boldsymbol{b}_2\| \leq \|\boldsymbol{b}_1\| \leq \|\boldsymbol{b}_2\|$. Moreover, if $\|\boldsymbol{b}_1\| \geq 1 - \varepsilon$, then: $|\langle \boldsymbol{b}_1, \boldsymbol{b}_3 \rangle| \leq c'\varepsilon$ and $|\langle \boldsymbol{b}_i, \boldsymbol{b}_2 \rangle + s_i \frac{\|\boldsymbol{b}_1\|^2}{2}| \leq c'\varepsilon$ for $i = 1, 3$.
3. If any of $(2, s_2, s_3)$ is a Voronoï coord with $s_i = \pm 1$ for $i \in \{2, 3\}$ and if $\|\boldsymbol{b}_1\| \geq 1 - \varepsilon$, then: $|\langle \boldsymbol{b}_2, \boldsymbol{b}_3 \rangle| \leq c'\varepsilon$ and $|\langle \boldsymbol{b}_i, \boldsymbol{b}_1 \rangle + s_i \frac{\|\boldsymbol{b}_1\|^2}{2}| \leq c'\varepsilon$ for $i = 2, 3$.

5.3 The Gap Lemma

The goal of this subsection is to prove that even with relaxed bases, if one adds a lattice vector with not too small coordinates to a vector of the Voronoï cell, this vector becomes significantly longer. This result will be used the other way round: if the x_i's found at Step 5 of the greedy algorithm are not too small, then $\boldsymbol{b}_d$ is significantly shorter than $\boldsymbol{a}_d$. We first need to generalize the compact sets K_2 and K_3. For any $\varepsilon \geq 0$ and any $\alpha \in [0,1]$, we define:

$$K_2'(\varepsilon, \alpha) = \{(\boldsymbol{b}_1, \boldsymbol{b}_2, \boldsymbol{u}) / (\boldsymbol{b}_1, \boldsymbol{b}_2) \in K_2(\varepsilon, \alpha),\ \boldsymbol{u} \in \mathrm{Vor}(\boldsymbol{b}_1, \boldsymbol{b}_2)\}$$
$$K_3'(\varepsilon, \alpha) = \{(\boldsymbol{b}_1, \boldsymbol{b}_2, \boldsymbol{b}_3, \boldsymbol{u}) / (\boldsymbol{b}_1, \boldsymbol{b}_2, \boldsymbol{b}_3) \in K_3(\varepsilon, \alpha),\ \boldsymbol{u} \in \mathrm{Vor}(\boldsymbol{b}_1, \boldsymbol{b}_2)\}.$$

Lemma 26. *If $\varepsilon > 0$ and $\alpha \in [0,1]$, $K_2'(\varepsilon, \alpha)$ and $K_3'(\varepsilon, \alpha)$ are compact sets.*

The next result is the two-dimensional version of the Gap Lemma.

Lemma 27. *There exist two constants $\varepsilon, C > 0$ such that for any ε-reduced vectors $[\boldsymbol{b}_1, \boldsymbol{b}_2]$ and any $\boldsymbol{u} \in Vor(\boldsymbol{b}_1, \boldsymbol{b}_2)$, if at least one of the following conditions holds, then: $\|\boldsymbol{u} + x_1\boldsymbol{b}_1 + x_2\boldsymbol{b}_2\|^2 \geq \|\boldsymbol{u}\|^2 + C\|\boldsymbol{b}_2\|^2$.*

- (1) $|x_2| \geq 2$,
- (2) $|x_1| \geq 2$ *and* $\|\boldsymbol{b}_1\|^2 \geq \|\boldsymbol{b}_2\|^2/2$.

We now give the three-dimensional Gap Lemma, on which relies the analysis of the four-dimensional greedy algorithm.

Lemma 28. *There exist two constants $\varepsilon, C > 0$ such that for any ε-reduced vectors $[\boldsymbol{b}_1, \boldsymbol{b}_2, \boldsymbol{b}_3]$ and any $\boldsymbol{u} \in Vor(\boldsymbol{b}_1, \boldsymbol{b}_2, \boldsymbol{b}_3)$, if at least one of the following conditions holds, then:*

$$\|\boldsymbol{u} + x_1\boldsymbol{b}_1 + x_2\boldsymbol{b}_2 + x_3\boldsymbol{b}_3\|^2 \geq \|\boldsymbol{u}\|^2 + C\|\boldsymbol{b}_3\|^2.$$

– (1) $|x_3| \geq 3$, *or* $|x_3| = 2$ *and* $(|x_1|, |x_2|) \neq (1, 1)$;
– (2) $||b_2|| \geq ||b_3||/2$ *and:* $|x_2| \geq 3$, *or* $|x_2| = 2$ *with* $(|x_1|, |x_3|) \neq (1, 1)$;
– (3) $||b_1|| \geq ||b_3||/2$ *and:* $|x_1| \geq 3$, *or* $|x_1| = 2$ *with* $(|x_2|, |x_3|) \neq (1, 1)$;

Like in the previous subsections, we now consider the case of the possible Voronoï coords $(\pm 1, \pm 1, \pm 2)$ modulo any permutation of coordinates.

Lemma 29. *There exist two constants* $\varepsilon, C > 0$ *such that for any* ε*-reduced vectors* $[b_1, b_2, b_3]$ *and any* $u \in Vor(b_1, b_2, b_3)$*, if at least one of the following conditions holds, then:*

$$||u + x_1 b_1 + x_2 b_2 + x_3 b_3||^2 \geq ||u||^2 + C||b_3||^2.$$

– *1-* $(x_1, x_2, x_3) = (s_1, s_2, 2)$, $|s_i| = 1$ *for* $i \in \{1, 2\}$ *and at least one of the following conditions holds:*
$||b_1|| \leq (1-\varepsilon)||b_3||$, *or* $||b_2|| \leq (1-\varepsilon)||b_3||$, *or* $|\langle b_1, b_2\rangle| \geq \varepsilon||b_3||^2$, *or* $|\langle b_1, b_3\rangle + s_1\frac{||b_1||^2}{2}| \geq \varepsilon||b_3||^2$, *or* $|\langle b_2, b_3\rangle + s_2\frac{||b_1||^2}{2}| \geq \varepsilon||b_3||^2$.
– *2a-* $(x_1, x_2, x_3) = (s_1, 2, s_3)$, $|s_i| = 1$ *for* $i \in \{1, 3\}$ *and* $||b_1|| \leq (1-\varepsilon)||b_3||$.
– *2b-* $(x_1, x_2, x_3) = (s_1, 2, s_3)$, $|s_i| = 1$ *for* $i \in \{1, 3\}$, $||b_1|| \geq (1-\varepsilon)||b_3||$ *and at least one of the following conditions holds:* $|\langle b_1, b_3\rangle| \geq \varepsilon||b_3||^2$, *or* $|\langle b_1, b_2\rangle + s_1\frac{||b_1||^2}{2}| \geq \varepsilon||b_3||^2$, *or* $|\langle b_3, b_2\rangle + s_3\frac{||b_1||^2}{2}| \geq \varepsilon||b_3||^2$.
– *3-*$(x_1, x_2, x_3) = (2, s_2, s_3)$, $|s_i| = 1$ *for* $i \in \{2, 3\}$, $||b_1|| \geq (1-\varepsilon)||b_3||$ *and at least one of the following conditions holds:* $|\langle b_2, b_3\rangle| \geq \varepsilon||b_3||^2$, *or* $|\langle b_2, b_1\rangle + s_2\frac{||b_1||^2}{2}| \geq \varepsilon||b_3||^2$, *or* $|\langle b_3, b_1\rangle + s_3\frac{||b_1||^2}{2}| \geq \varepsilon||b_3||^2$.

Acknowledgements. We thank Ali Akhavi, Florian Heß, Igor Semaev and Jacques Stern for helpful discussions and comments.

References

1. M. Ajtai. Generating hard instances of lattice problems (extended abstract). In *Proc. of the 28th Symposium on the Theory of Computing*, pages 99–108. ACM Press, 1996.
2. M. Ajtai. The shortest vector problem in L_2 is NP-hard for randomized reductions (extended abstract). In *Proc. of the 30th Symposium on the Theory of Computing*, pages 10–19. ACM Press, 1998.
3. A. Akhavi and C. Moreira dos Santos. Another view of the Gaussian algorithm. In *Proc. of LATIN '04*, Lecture Notes in Computer Science. Springer-Verlag, 2004.
4. J.W.S. Cassels. *An Introduction to the Geometry of Numbers.* Springer-Verlag, Berlin, 1959.
5. C.F. Gauss. *Disquisitiones Arithmeticæ.* Leipzig, 1801.
6. M. Gruber and C. G. Lekkerkerker. *Geometry of Numbers.* North-Holland, Amsterdam, 1987.
7. B. Helfrich. Algorithms to construct Minkowski reduced and Hermite reduced lattice bases. *Th. Computer Science*, 41:125–139, 1985.
8. C. Hermite. Extraits de lettres de M. Hermite à M. Jacobi sur différents objets de la théorie des nombres, deuxième lettre. *J. Reine Angew. Math.*, 40:279–290, 1850.

9. C. Hermite. *Œuvres*. Gauthier-Villars, Paris, 1905.
10. M. Kaib and C. P. Schnorr. The generalized Gauss reduction algorithm. *J. of Algorithms*, 21(3):565–578, 1996.
11. A. Korkine and G. Zolotareff. Sur les formes quadratiques. *Math. Ann.*, 6:336–389, 1873.
12. J. C. Lagarias. Worst-case complexity bounds for algorithms in the theory of integral quadratic forms. *J. of Algorithms*, 1:142–186, 1980.
13. A. K. Lenstra, H. W. Lenstra, Jr., and L. Lovász. Factoring polynomials with rational coefficients. *Math. Ann.*, 261:513–534, 1982.
14. J. Martinet. *Perfect Lattices in Euclidean Spaces*. Springer-Verlag, Heidelberg, 2002.
15. D. Micciancio. The shortest vector problem is NP-hard to approximate to within some constant. In *Proc. of the 39th Symposium on the Foundations of Computer Science*, pages 92–98. IEEE, 1998.
16. D. Micciancio and S. Goldwasser. *Complexity of lattice problems: A cryptographic perspective*. Kluwer Academic Publishers, Boston, 2002.
17. H. Minkowski. *Geometrie der Zahlen*. Teubner-Verlag, Leipzig, 1896.
18. P. Q. Nguyen and J. Stern. The two faces of lattices in cryptology. In *Proc. of CALC '01*, volume 2146 of *Lecture Notes in Computer Science*, pages 146–180. Springer-Verlag, 2001.
19. S. S. Ryskov. On Hermite, Minkowski and Venkov reduction of positive quadratic forms in n variables. *Soviet Math. Doklady*, 13:1676–1679, 1972.
20. C. P. Schnorr. A hierarchy of polynomial lattice basis reduction algorithms. *Th. Computer Science*, 53:201–224, 1987.
21. C. P. Schnorr and M. Euchner. Lattice basis reduction: improved practical algorithms and solving subset sum problems. *Math. Programming*, 66:181–199, 1994.
22. C. P. Schnorr and H. H. Hörner. Attacking the Chor-Rivest cryptosystem by improved lattice reduction. In *Proc. of Eurocrypt '95*, volume 921 of *Lecture Notes in Computer Science*, pages 1–12. Springer-Verlag, 1995.
23. A. Schönhage and V. Strassen. Schnelle Multiplikation grosser Zahlen. *Computing*, 7:281–292, 1971.
24. I. Semaev. A 3-dimensional lattice reduction algorithm. In *Proc. of CALC '01*, volume 2146 of *Lecture Notes in Computer Science*, pages 181–193. Springer-Verlag, 2001.
25. C. L. Siegel. *Lectures on the Geometry of Numbers*. Springer-Verlag, Berlin, Heidelberg, 1989.
26. M. I. Stogrin. *Regular Dirichlet-Voronoï partitions for the second triclinic group*. American Mathematical Society, 1975. English translation of the Proceedings of the Steklov Institute of Mathematics, Number 123 (1973).
27. B. Vallée. *Une Approche Géométrique de la Réduction de Réseaux en Petite Dimension*. PhD thesis, Université de Caen, 1986.
28. B. Vallée. Gauss' algorithm revisited. *J. of Algorithms*, 12(4):556–572, 1991.
29. B. L. van der Waerden. Die Reduktionstheorie der positiven quadratischen Formen. *Acta Mathematica*, 96:265–309, 1956.
30. G. Voronoï. Nouvelles applications des paramètres continus à la théorie des formes quadratiques. *J. Reine Angew. Math.*, 134:198–287, 1908.

Computing Order Statistics in the Farey Sequence

Corina E. Pătraşcu[1] and Mihai Pătraşcu[2]

[1] Harvard University, Department of Mathematics
patrascu@fas.harvard.edu

[2] MIT, Computer Science and Artificial Intelligence Laboratory
mip@mit.edu

Abstract. We study the problem of computing the k-th term of the Farey sequence of order n, for given n and k. Several methods for generating the entire Farey sequence are known. However, these algorithms require at least quadratic time, since the Farey sequence has $\Theta(n^2)$ elements. For the problem of finding the k-th element, we obtain an algorithm that runs in time $O(n \lg n)$ and uses space $O(\sqrt{n})$. The same bounds hold for the problem of determining the rank in the Farey sequence of a given fraction. A more complicated solution can reduce the space to $O(n^{1/3}(\lg \lg n)^{2/3})$, and, for the problem of determining the rank of a fraction, reduce the time to $O(n)$. We also argue that an algorithm with running time $O(\text{poly}(\lg n))$ is unlikely to exist, since that would give a polynomial-time algorithm for integer factorization.

1 Introduction

For any positive integer n, the Farey sequence of order n is the set of all irreducible fractions $\frac{p}{q}$, with $0 < p < q \le n$, arranged in increasing order. An alternative definition could include $\frac{0}{1}$ and $\frac{1}{1}$ as special fractions. For example, the Farey sequence for $n = 5$ is: $\frac{1}{5}, \frac{1}{4}, \frac{1}{3}, \frac{2}{5}, \frac{1}{2}, \frac{3}{5}, \frac{2}{3}, \frac{3}{4}, \frac{4}{5}$. The number of elements of the Farey sequence is asymptotically $\frac{3}{\pi^2}n^2 + O(n \lg n)$ [5].

The Farey sequence is a well-known concept in number theory, whose exploration has lead to a number of interesting results. However, from an algorithmic point of view, very little is known. In particular, the only problem that appears to be investigated is that of generating the entire sequence for a given n. Several interesting solutions exist for this problem, though none of them presents any algorithmic challenge:

- Sort all unreduced fractions $\frac{p}{q}$, and remove duplicates. The running time is $O(n^2 \lg n)$, which is almost optimal, but the space is $O(n^2)$. (We assume that we are only interested in generating the fractions, not storing them; otherwise, quadratic space is clearly the best possible.)
- The space in the above algorithm can be reduced to $O(n)$, without changing the running time [7]. This uses a priority queue to merge n sequences, where the i-th such sequence is $\frac{1}{i}, \frac{2}{i}, \dots, \frac{i-1}{i}$.

D.A. Buell (Ed.): ANTS 2004, LNCS 3076, pp. 358–366, 2004.

- To obtain the sequence of order $n+1$ from the sequence of order n, consider all consecutive fractions $\frac{a}{b}$ and $\frac{c}{d}$ from the sequence of order n, and insert the mediant fraction $\frac{a+c}{b+d}$ between them, if the denominator is $n+1$ [5]. This surprising construction is based on the initial observation made by Farey in 1816 [4]. The resulting algorithm is the worst so far: the running time is $O(n^3)$, and the space is $O(n^2)$.
- Combining several properties satisfied by Farey sequence, one can get a trivial iterative algorithm, which generates the next Farey fraction, based on the previous two ([5], problem 4-61). If $\frac{p}{q}$ and $\frac{p'}{q'}$ are the last two fractions, the next one is given by:

$$p'' = \left\lfloor \frac{q+n}{q'} \right\rfloor p' - p, \quad q'' = \left\lfloor \frac{q+n}{q'} \right\rfloor q' - q$$

  This is an ideal algorithm: it uses $O(n^2)$ time, and $O(1)$ space.
- The Stern-Brocot tree is obtained by starting with $\frac{0}{1}$ and $\frac{1}{1}$, and repeatedly inserting the mediant between any two fractions that are consecutive in the in-order traversal of the tree [5]. Farey fractions form a subtree of the Stern-Brocot tree, often called the Farey tree. One can generate the Farey fractions in order, by recursively exploring the tree. The algorithm requires quadratic time, and $O(n)$ memory (corresponding to the maximum depth of the Farey fractions of order n).

The model of computation assumed by these algorithms, as well as the remaining algorithms from this paper, is the standard word RAM (Random Access Machine). Such a machine can access words of $O(\lg n)$ bits, and can perform usual arithmetic operations on such words in unit time. Space is also counted in words.

In this work, we consider the most natural question in addition to that of generating the entire Farey sequence. For given n and k, our problem is to generate just the k-th element of the Farey sequence of order n (often called the k-th order statistic [2]). Our motivation is not based on any practical application of this problem (we are aware of none), but rather on the algorithmic challenges it presents. It seems impossible to obtain good, or even just subquadratic time bounds for these problems by modifying the algorithms listed above (the obvious choice, and the one to which we devoted the most attention was the solution involving the Stern-Brocot tree). Instead, our algorithms will be based on a set of rather different ideas.

Our solution involves algorithms for another natural problem: given a fraction, determine its rank in the Farey sequence. The bounds we obtain are usually identical for both problems. In section 2, we describe a reduction between these problems, and design an initial algorithm that runs in time $O(n \lg^2 n)$ and uses space $O(n)$. In section 3, we improve the time complexity of this algorithm to $O(n \lg n)$. Finally, section 4 improves the space complexity to $O(\sqrt{n})$, while preserving the running time.

We have implemented the final algorithm, as well as the methods described above for generating the entire sequence, and found ours to be very efficient, both in terms of time, and space. Experimental results or source code can be obtained from the authors.

This result leaves two natural questions unanswered. The first one is how little memory suffices for an algorithm with roughly linear, say $O(n \cdot \text{poly}(\lg n))$, running time. In particular, reducing the memory to $\text{poly}(\lg n)$ would be interesting. In section 5, we present a more complex algorithm, based on a result of Deléglise and Rivat [3], which uses space $O(n^{1/3}(\lg\lg n)^{2/3})$. The algorithm can determine the rank of a fraction in $O(n)$ time; for our original problem (finding a fraction of a given rank) the $O(n \lg n)$ time bound is not improved.

The second question is concerned with time, rather than space. Note that the input to the problem consists of just two words, or $O(\lg n)$ bits, namely n and k. Therefore, there is nothing that prohibits the existence of an algorithm with running time sublinear in n. For somewhat related problems, such as computing the number of primes less than a certain value, sublinear time algorithms are known [1]. It seems reasonable to hope that the running time of the algorithm from section 5 can be improved to $O(n^{1-\varepsilon})$, for some constant $\varepsilon > 0$. We describe two subproblems which would be sufficient to obtain such a result, but which we cannot solve. In section 6 we argue instead that a much faster algorithm, with running time $O(\text{poly}(\lg n))$, is unlikely to exist. More precisely, we show that such a polynomial algorithm for our problem would immediately imply a polynomial-time algorithm for integer factorization.

2 An Initial Algorithm

We now describe a first attempt to solve the problem, which will give an algorithm running in $O(n \lg^2 n)$ time and $O(n)$ space. This algorithm forms the basis of our improved algorithms from the following sections. Our solution uses as a subroutine an algorithm for determining the rank of a given fraction (not necessarily in reduced form) in the Farey sequence. The subroutine developed in this section will run in $O(n \lg n)$ time and $O(n)$ space.

We begin by a reduction from the order statistic problem to the fraction rank problem. Assume we want to compute the k-th order statistic. We first determine a number j such that the answer lies in the interval $\left[\frac{j}{n}, \frac{j+1}{n}\right)$. This can be done by binary search, as follows. Assume we have a guess for the value of j. Determine the rank of $\frac{j}{n}$ in the Farey sequence. If this rank is at most k we know the correct value of j is not smaller than the current guess. Otherwise, we should continue searching below j. This stage of the algorithm uses $O(\lg n)$ calls to the fraction rank subroutine.

To solve the problem, we must now determine the fraction with rank equal to $k - rank(\frac{j}{n})$ among all irreducible fractions from the interval $\left[\frac{j}{n}, \frac{j+1}{n}\right)$. Notice that there is at most one fraction in this interval for any denominator less than n. This follows from the fact that the length of the interval is $\frac{1}{n}$, and consecutive fractions with denominators $q < n$ are separated by $\frac{1}{q} > \frac{1}{n}$. In addition, for a given q, this fraction can be found in constant time, since the numerator must be $\left\lfloor \frac{(j+1)q-1}{n} \right\rfloor$. There might not be any fraction in the range for a certain denominator, so we also need to check that the fraction we obtain in this way lies in the feasible interval.

Given these properties, it is not hard to find efficient algorithms for this subproblem. A particularly easy one would consist of generating all (up to n) fractions from the feasible interval, sorting them, and eliminating duplicates. Alternatively, one can make a list of just the irreducible fractions, and then use a linear time order statistic algorithm [2] on this list. However, we describe a more interesting solution. This solution has the advantage that it runs in $O(n)$ time, and uses just $O(1)$ memory, which will prove important in the following section. First, generate the fractions in the range by considering all possible denominators. As fractions are generated, keep just the minimum fraction found that is strictly greater than $\frac{j}{n}$. At the end, reduce both $\frac{j}{n}$ and the minimum fraction greater than it. We now have two consecutive fractions from the Farey sequence. As mentioned in the introduction, there is a simple constant-time algorithm that can generate the next fraction in the Farey sequence based on the previous two. This means that we can iterate through the fractions in the range in increasing order, remembering just the previous two fractions. All we have to do is count up to the desired rank, and return the corresponding fraction.

We are now left with giving an algorithm for the fraction rank problem. We will actually solve a slightly more general problem: for any real number x, determine the number of irreducible fractions $\frac{p}{q} \leq x$, with $q \leq n$. Let A_q be the number of such irreducible fractions with denominator equal to q. Any fraction with denominator q is either irreducible, in which case it should be counted in A_q, or has a unique reduced representation. The denominator of the reduced representation is a divisor of q. This transformation is, in fact, reversible: given any irreducible fraction $\frac{c}{d}$, where d is a divisor of q, we can multiply both the numerator and the denominator by q/d to get a fraction with denominator equal to q. So we have a bijection between the set of all fractions in $(0, x]$ with denominator q and the set of reduced fractions with denominator d, for all divisors d of q. This gives a recursive formula for A_q, which leads to the solution of our problem:

$$rank(x) = \sum_{q=1}^{n} A_q, \qquad A_q = \lfloor x \cdot q \rfloor - \sum_{d<q,\ d|q} A_d \tag{1}$$

The way to translate this formula into an efficient algorithm is obviously not the most direct one, since there is no fast way to iterate over all divisors of a number. Instead, begin by initializing an array $A[1..n]$ by $A[q] = \lfloor x \cdot q \rfloor$. Then consider all numbers q in increasing order from 1 to n. For each q, consider all multiples mq, and subtract $A[q]$ from $A[mq]$. At step number q, we always have $A[q] = A_q$, and this algorithm is a simple reformulation of the recursive formula from above. The rank can be computed by summing the final values of all $A[q]$. The running time is $O\left(n + \sum_{q=1}^{n} \frac{n}{q}\right) = O\left(n \sum_{q=1}^{n} \frac{1}{q}\right) = O(n \lg n)$. So we have an $O(n \lg n)$ algorithm for the rank problem, and thus an $O(n \lg^2 n)$ algorithm for the order statistic problem. Both algorithms use $O(n)$ space.

3 Improving the Running Time

We now improve the time complexity of the algorithm from the previous section. Since the running time of our order statistic algorithm is dominated by the running time of the fraction rank algorithm, we concentrate on the latter. As noted before, the order statistic algorithm makes $O(\lg n)$ calls to the fraction rank subroutine. The key idea suggested by this fact is that one should try to introduce some amount of preprocessing, which gives a one-time cost, in order to improve the running time of every call to the fraction rank subroutine.

To find a way to trade preprocessing for running time of the rank subroutine, let us re-examine relation 1. After recursive expansions of all A_q, the resulting $rank(x)$ will be a linear combination of terms of the form $\lfloor x \cdot q \rfloor$, for all $q \leq n$. Except for these terms, the recursive formula is independent of x. Therefore, we can precalculate the coefficient of every $\lfloor x \cdot q \rfloor$. Now the rank routine becomes trivial: for all q, calculate $\lfloor x \cdot q \rfloor$, and add this value to the rank, weighted by the appropriate coefficient. The numbers appearing at intermediate steps in the computation, and even the coefficients themselves, may become large. However, the end result (the rank) is obviously bounded by n^2, so if all computations are performed modulo a number greater than that, the result will be correct (this is important because we can only manipulate in constant time numbers that have $O(\lg n)$ bits). This issue is transparent to the implementation, since normally all computations are carried out modulo a power of two, such as 2^{32} for the usual 32-bit machines.

The algorithm for precalculating the coefficients of $\lfloor x \cdot q \rfloor$ is symmetric to our old algorithm for calculating the rank. It is based on the following recursion defining the coefficients:

$$C_q = 1 - \sum_{t>q,\ q|t} C_t, \quad \text{for all } q \leq n \tag{2}$$

The correctness of the formula follows from "reverse induction", since the calculation of C_q uses only coefficients C_t with $t > q$. Indeed, the term $\lfloor x \cdot q \rfloor$ appears initially in A_q. Then, A_q is subtracted from all its multiples t. At that point A_t contains $\lfloor x \cdot t \rfloor$ with coefficient 1, and $\lfloor x \cdot q \rfloor$ with coefficient -1. All subsequent operations involving A_t contribute to the total coefficient of $\lfloor x \cdot q \rfloor$ exactly by minus the coefficient of $\lfloor x \cdot t \rfloor$: since A_t is the only one that contains $\lfloor x \cdot t \rfloor$ initially, all operations involving A_t are described by the final coefficient of $\lfloor x \cdot t \rfloor$.

The algorithm follows immediately from the formula, and calculates the coefficients from C_n down to C_1. The running time is $O\left(\sum_{q=1}^{n} \frac{n}{q}\right) = O(n \lg n)$. This cost is paid once, and every call to the rank subroutine can be answered in $O(n)$ time, so the total running time of the order statistic algorithm is also $O(n \lg n)$.

4 Improving the Space Complexity

The key observation for improving the space complexity is that coefficients C_q and $C_{q'}$ are identical whenever $\lfloor n/q \rfloor = \lfloor n/q' \rfloor$. This is not hard to see, and also offers a good intuition for the new algorithm. Consider some value q. The term $\lfloor x \cdot q \rfloor$ is added initially to A_q. Then A_q, which includes $\lfloor x \cdot q \rfloor$, is subtracted from $m_1 q$, for all possible m_1. The values $A_{m_1 q}$, which now contain $\lfloor x \cdot q \rfloor$ with a coefficient of -1, are then subtracted from $A_{m_2 m_1 q}$, and so on. The branches of this recursion are only trimmed off when $q \prod m_i > n$, which is equivalent to $\prod m_i \geq \lfloor n/q \rfloor$. So the coefficient C_q only depends on $\lfloor n/q \rfloor$.

Based on this fact, we observe that there are only $\sqrt{n}$ distinct values for the coefficients among all C_q with $q > \sqrt{n}$. To avoid such repetitions, we break the coefficients into two groups. The coefficients $C_1, \ldots, C_{\lfloor \sqrt{n} \rfloor}$ are stored as before. Instead of storing C_q for $q > \sqrt{n}$, we store an array D_r, such that $C_q = D_{\lfloor n/q \rfloor}$ for any $q > \sqrt{n}$. Notice that both arrays have $O(\sqrt{n})$ elements. The fraction rank algorithm remains trivial. For all $q \leq \sqrt{n}$, we calculate $\lfloor x \cdot q \rfloor$ and add it to the rank, weighted by C_q. For the remaining q's, we instead use the weight $D_{\lfloor n/q \rfloor}$.

It remains to show how to precompute the sequences D_r and C_q using just $O(\sqrt{n})$ space. The computation of D_r is based on the following recursive formula:

$$D_r = 1 - \sum_{t=2}^{r} D_{\lfloor r/t \rfloor} \tag{3}$$

The formula can be obtained by careful relabeling of formula 2. Take a q, such that $r = \lfloor n/q \rfloor$. Now consider consider our previous recursive formula for C_q (slightly rewritten here):

$$C_q = 1 - \sum_{t=2}^{\lfloor n/q \rfloor} C_{tq}$$

By definition, we have $D_r = C_q$ and $\lfloor n/q \rfloor = r$. Also, $C_{tq} = D_{\lfloor n/tq \rfloor}$. The index on the right-hand side can be rewritten as $\left\lfloor \frac{n}{tq} \right\rfloor = \left\lfloor \frac{\lfloor n/q \rfloor}{t} \right\rfloor = \lfloor r/t \rfloor$, finalizing the transformation into formula 3.

Once we have the values D_r, computing the array C_q can be done as before, using relation 2. The only difference is that whenever the algorithm needs C_t for $t > \sqrt{n}$, is should instead use $D_{\lfloor n/t \rfloor}$, since we are only computing the values C_t for $t \leq \sqrt{n}$. The time required by the computation of the sequence D_r is quadratic in the size of the table, which is $O(n)$. Computing C_q takes $O(n \lg n)$ time, as before. Finally, rank queries still require linear time, so the overall running time is unchanged, and the space is reduced to $O(\sqrt{n})$.

5 A Better Way to Calculate Coefficients

This section describes a better way to calculate the coefficients D_r from the previous section, resulting in improved, but more complicated, algorithms. It can be seen that D_r is precisely equal to $M(r)$, where M is the summatory function of the Möbius function, $M(r) = \sum_{t=1}^{r} \mu(t)$. The Möbius function is defined by:

- $\mu(1) = 1$;
- $\mu(t) = 0$ if t has a squared prime factor;
- $\mu(t) = (-1)^k$ if $t = p_1 \cdot \ldots \cdot p_k$, where all p_i's are distinct and prime.

The identity between our coefficients and $M(r)$ is immediate, since $M(r)$ satisfies the recursive formula 3 defining our coefficients ([5], relation 4.61) – this is one of the fundamental properties of the Möbius function. Given this identity, we can use an algorithm by Deléglise and Rivat [3], which calculates $M(r)$ using $O(r^{2/3}(\lg\lg r)^{1/3})$ time and $O(r^{1/3}(\lg\lg r)^{2/3})$ space.

Remember that our algorithm needs the coefficients $C_q, q \in \{1, \ldots, n\}$ and that $C_q = D_{\lfloor n/q \rfloor}$. For all $q < n^{1/6}$, we will use the algorithm of [3] to calculate $M(\lfloor n/q \rfloor)$. This first stage will require $n^{1/6} \cdot n^{2/3+o(1)} = n^{5/6+o(1)}$ time. Calculating C_q for all $q \geq n^{1/6}$ can be done by calculating D_r for all $r \leq n^{5/6}$. Since $D_r = \sum_{t=1}^{r} \mu(t)$, calculating the D_r's in the increasing order of r reduces to calculating $\mu(t)$ for all $t \leq n^{5/6}$. In turn, calculating $\mu(t)$ is trivial if we have the prime factorization of t. There exist several factorization algorithms which can factor t in $t^{o(1)}$ time [6], so we obtain a running time of $n^{5/6+o(1)}$ for calculating all coefficients. The dominant factor in space usage is the space used by a call to the subroutine for calculating $M(r)$, which is $O(n^{1/3}(\lg\lg n)^{2/3})$.

The dominant factor in the running time is no longer calculating the coefficients. As each C_q becomes available, we need to multiply it by $\lfloor x \cdot q \rfloor$ and add the result to an accumulator. Thus, the running time for computing the rank of a fraction is $O(n)$. The time needed to compute the fraction of a given rank is still $O(n \lg n)$, because we run a binary search on top of the rank computation, as explained in section 2.

Since the algorithm of this section can compute all coefficients in sublinear time, there is hope of obtaining a sublinear running time overall. We now describe a possible approach; however, since we cannot solve the two necessary subproblems, this remains speculation. First, to compute ranks in sublinear time, we could observe that many consecutive terms of $\lfloor x \cdot q \rfloor$ have the same coefficient C_q when $q > \sqrt{n}$. Thus, we should look for an algorithm to evaluate $\sum_{q=a}^{b} \lfloor x \cdot q \rfloor$ in time $O((b-a)^{1-\varepsilon})$, for some constant $\varepsilon > 0$. Even this, however, would not imply a sublinear time algorithm for finding the fraction of a given rank. This is because the binary search from section 2 can only narrow down the range of possible fractions, and does not actually produce an output fraction in the form p/q. As described, the range is reduced to $\left[\frac{x}{n}, \frac{x+1}{n}\right)$; we do not know how to find a fraction of a given rank from this range without enumerating all $O(n)$ fractions. Alternatively, we could reduce the range to an interval of size $O(1/n^2)$, and then we would need to find the unique fraction left in this range.

6 Relation to Factorization

We now show that a polynomial-time algorithm (i.e. one running in $O(\text{poly}(\lg n))$ time) for our problem is unlikely to exist, since that would immediately give a polynomial algorithm for integer factorization. It is a well-known conjecture that such an algorithm does not exist.

Given a polynomial time algorithm for finding an order statistic, we can construct a polynomial time algorithm for finding the rank of a given fraction (this is the reverse of the reduction used for our actual algorithms). Observe that we can test whether a guessed rank is too high or too low, using only one oracle query to the order statistic algorithm: simply determine the fraction with that rank, and compare it with the input fraction. Therefore, we can use galloping binary search to solve the problem. We begin by trying powers of two until we either obtain a fraction greater than the input fraction, or we exceed the number of fractions in the Farey sequence (as reported by the order statistic algorithm). Then, we do a binary search for the correct rank in the interval between the last powers of two tried. This algorithm makes $O(\lg n)$ calls to the order statistic algorithm, and no other expensive computations, so it runs in polynomial time.

Our algorithm for factorization is based on yet another problem: given a number n, and $k \leq n$ that is relatively prime to n, report the number of integers in $[2, k]$ that are relatively prime to n. Given a polynomial time algorithm for this problem, we can use binary search to find a factor of n. Assume k is relatively prime to n (otherwise, we immediately get a factor), and we know the number of integers in $[2, k]$ that are relatively prime to n. If this number is $k-1$, we know that the smallest factor of n is greater than k; otherwise, there is at least one factor below k.

It remains to describe the relation between the Farey sequence and this problem. Observe that all numbers $i \in [2, k]$ that are relatively prime to n give fractions $\frac{i}{n}$ in the Farey sequence of order n. To count these numbers, we begin by finding the rank of $\frac{k}{n}$ in the Farey sequence of order n. Then, we determine the largest fraction, strictly smaller than $\frac{k}{n}$. This can be done by one call to the order statistic algorithm, since we already know the rank of $\frac{k}{n}$. Since $\frac{k}{n}$ is irreducible (by assumption), and it is the mediant of neighboring fractions from the Farey sequence, it follows that the preceding fraction has a denominator strictly smaller than n. We now find the rank of this fraction, in the Farey sequence of order $n-1$. Observe that the difference between the rank of $\frac{k}{n}$ and this rank is equal to the number of irreducible fractions $\frac{i}{n} \leq \frac{k}{n}$, which is what we wanted to count.

The reduction from above uses the order statistic oracle in one place, namely to find the largest fraction smaller than $\frac{k}{n}$. One may wonder whether the reduction holds if we assume just a polynomial time algorithm for the problem of ranking a fraction, proving also the hardness of this problem. The following smarter reduction answers this question in the affirmative. Consider the fractions $\frac{k+1}{n}$ and $\frac{k-1}{n}$. Since their difference is $\frac{2}{n}$, there exists exactly one fraction in this range with a denominator of $n-1$. Find this fraction (with $O(1)$ arithmetic operations), and reduce it, which takes $O(\lg n)$ time. Now find the ranks of this fraction in the Farey sequences of order n and $n-1$. The difference in ranks is exactly the number of irreducible fractions $\frac{i}{n} < \frac{k}{n}$, possibly plus one due to $\frac{k}{n}$. This nonuniformity is easily fixed by testing whether our fraction with a denominator of $n-1$ is smaller or larger than $\frac{k}{n}$, so we can again count the number of integers from $[2, k]$ that are relatively prime to n.

Acknowledgements. We thank an anonymous referee for valuable comments leading to the results from section 5. This problem, with time and space limitations corresponding to our initial algorithm from section 2, was proposed by the authors as a competition problem for the Balkan Olympiad in Informatics 2003. We thank all people who expressed positive comments on that occasion. In addition, we thank Erik Demaine and Noam Elkies for discussions and encouragement.

References

1. E. Bach and J. Shallit, *Algorithmic Number Theory, Volume I: Efficient Algorithms*, MIT Press, 1996.
2. T. H. Cormen, C. E. Leiserson, R. L. Rivest and C. Stein, *Introduction to Algorithms*, 2nd edition, MIT Press and McGraw-Hill, 2001.
3. M. Deléglise and J. Rivat, *Computing the Summation of the Möbius Function*, Experimental Mathematics 5, 291-295, 1996.
4. J. Farey, *On a Curious Property of Vulgar Fractions*, London, Edinburgh and Dublin Phil. Mag. 47, 385, 1816.
5. R. L. Graham, D. E. Knuth and O. Patashnik, *Concrete Mathematics: A Foundation for Computer Science*, 2nd edition, Addison-Wesley, 1994.
6. D. E. Knuth, *The Art of Computer Programming, volume 2: Seminumerical Algorithms*, 2nd edition, Addison-Wesley, 1981.
7. V. Mitrana, *Provocarea algoritmilor* (in Romanian), Ed. Agni, Bucharest, Romania, 1995.

The Discrete Logarithm in Logarithmic ℓ-Class Groups and Its Applications in K-theory

Sebastian Pauli and Florence Soriano-Gafiuk

Institut für Mathematik, Technische Universität Berlin and
Département de Mathématiques, Université de Metz

Abstract. We present an algorithm for the computation of the discrete logarithm in logarithmic ℓ-Class Groups. This is applied to the calculation of the ℓ-rank of the wild kernel $WK_2(F)$ of a number field F. In certain cases it can also be used to determine generators of the ℓ-part of $WK_2(F)$.

1 Introduction

A new invariant of number fields, called group of logarithmic classes, was introduced by J.-F. Jaulent in 1994 [J3]. The arithmetic of logarithmic classes is interesting because of its applicability to K-Theory. Indeed for a given prime number ℓ, the ℓ-rank of the logarithmic ℓ-class group of a number field F containing the 2ℓ-th roots of unity equals the ℓ-rank of the wild kernel.

We present a method for the computation of the ℓ-rank of the wild kernel for the case where ℓ is an odd prime number and F does not contain the ℓ-th roots of unity.

For the case where ℓ^m is the exponent of the logarithmic ℓ-class group $\widetilde{\mathcal{C}\ell}_{F,\ell}$ of F and F contains the ℓ^{m+1}-th roots of unity we give a complete logarithmic description of the ℓ-part of the wild kernel.

First we recall the most important definitions from the theory of logarithmic ℓ-class groups and the algorithm for their computation. This is followed by an algorithm for the computation of discrete logarithms in these groups (section 2). In section 3 we give a short introduction to the wild kernel and derive the algorithms for the computation of its ℓ-rank in a general setting. Section 4 contains the complete description of the ℓ-part of the wild kernel through the logarithmic ℓ-class group. This is followed by some examples.

In the following, ℓ denotes a fixed prime number and $\mathbb{Z}_\ell$ the completion of $\mathbb{Z}$ with respect to the non-archimedian exponential valuation v_ℓ. F denotes a number field.

2 The Logarithmic ℓ-Class Group

For a detailed presentation of the logarithmic theory see [J3]. A first algorithm for the computation of the group of logarithmic classes of a number field F was

D.A. Buell (Ed.): ANTS 2004, LNCS 3076, pp. 367–378, 2004.

developed by F. Diaz y Diaz and F. Soriano in 1999 [DS]. We use the algorithm from [DJ+] as it removes the restriction to Galois extensions of $\mathbb{Q}$. This algorithm uses the ideal theoretic description of the logarithmic ℓ-class groups. Before we discuss it we need some definitions.

Let p be a prime number and let $\mathfrak{p}$ be a prime ideal of F over p. For $a \in \mathbb{Q}_p^\times \cong p^{\mathbb{Z}} \times \mathbb{F}_p^\times \times (1+2p\mathbb{Z}_p)$ denote by $\langle a \rangle$ the projection of a to $(1+2p\mathbb{Z}_p)$. Let $F_\mathfrak{p}$ be the completion of F with respect to $\mathfrak{p}$. For $\alpha \in F^*$ we define

$$g_\mathfrak{p}(\alpha) := \frac{\operatorname{Log}_p \langle \mathrm{N}_{F_\mathfrak{p}/\mathbb{Q}_p}(\alpha)\rangle}{[F_\mathfrak{p} : \mathbb{Q}_p] \cdot \deg_p p} \quad \text{where} \quad \deg_\ell p = \begin{cases} \operatorname{Log}_\ell p & \text{for } p \neq \ell; \\ \ell & \text{for } p = \ell \neq 2; \\ 4 & \text{for } p = \ell = 2. \end{cases}$$

The logarithmic ramification index $\widetilde{e}_\mathfrak{p}$ can be described as follows. The p-part of the logarithmic ramification index $\widetilde{e}_\mathfrak{p}$ is $[g_\mathfrak{p}(F_\mathfrak{p}^*) : \mathbb{Z}_p]$. For all primes q with $q \neq p$ the q part of $\widetilde{e}_\mathfrak{p}$ is the q part of the ramification index $e_\mathfrak{p}$ of $\mathfrak{p}$. The logarithmic inertia degree $\widetilde{f}_\mathfrak{p}$ is defined by the relation $\widetilde{e}_\mathfrak{p}\widetilde{f}_\mathfrak{p} = e_\mathfrak{p} f_\mathfrak{p} = \deg(F/\mathbb{Q})$, where $f_\mathfrak{p}$ is the classic inertia degree. We use it for the definition of the logarithmic degree of a place $\mathfrak{p}$:

$$\widetilde{\deg}_\ell \mathfrak{p} := \widetilde{f}_\mathfrak{p} \deg_\ell p.$$

Furthermore we set

$$\widetilde{v}_\mathfrak{p}(x) := -\frac{\operatorname{Log}_\ell(\mathrm{N}_{F_\mathfrak{p}/\mathbb{Q}_p}(x))}{\widetilde{\deg}_\ell \mathfrak{p}} \text{ for } x \in \mathcal{R}_F = \mathbb{Z}_\ell \otimes_\mathbb{Z} F^*.$$

We define the group of ℓ-ideals

$$\mathcal{I}d_{F,\ell} := \left\{ \mathfrak{a} = \textstyle\prod_{\mathfrak{p}\nmid\ell} \mathfrak{p}^{\alpha_\mathfrak{p}} \mid \alpha_\mathfrak{p} = 0 \text{ for almost all } \mathfrak{p} \right\},$$

and for $\mathfrak{a} = \prod_{\mathfrak{p}\nmid\ell} \mathfrak{p}^{\alpha_\mathfrak{p}} \in \mathcal{I}d_{F,\ell}$ set $\deg_\ell \mathfrak{a} := \sum_{\mathfrak{p}\nmid\ell} \alpha_\mathfrak{p} f_\mathfrak{p} \deg_\ell p$. We denote by

$$\widetilde{\mathcal{I}d}_{F,\ell} := \left\{ \mathfrak{a} \in \mathcal{I}d_{F,\ell} \mid \deg_\ell \mathfrak{a} = 0 \right\}$$

the subgroup of ℓ-ideals of degree 0 and by

$$\widetilde{\mathcal{P}r}_{F,\ell} := \left\{ \textstyle\prod_{\mathfrak{p}\nmid\ell} \mathfrak{p}^{v_\mathfrak{p}(a)} \mid a \in \mathcal{R}_F \text{ and } \widetilde{v}_\mathfrak{p}(a) = 0 \ \forall \mathfrak{p} \mid \ell \right\}$$

the subgroup of principal ℓ-ideals having logarithmic valuations 0 at all ℓ-adic places. The group of logarithmic ℓ-classes is isomorphic to the quotient of the latter two:

$$\widetilde{\mathcal{C}\ell}_{F,\ell} \cong \widetilde{\mathcal{I}d}_{F,\ell} / \widetilde{\mathcal{P}r}_{F,\ell}.$$

The generalized Gross conjecture (for the field F and the prime ℓ) asserts that the logarithmic class group $\widetilde{\mathcal{C}\ell}_{F,\ell}$ is finite (cf. [J3]). This conjecture, which is a consequence of the p-adic Schanuel conjecture, was only proved in the abelian case and a few others (cf. [FG,J4]). Nevertheless, since $\widetilde{C\ell}_{F,\ell}$ is a $\mathbb{Z}_\ell$-module of finite type (by ℓ-adic class field theory), the Gross' conjecture just claims the existence of an integer m such that ℓ^m kills the logarithmic class group. In [DJ+] we present a method for the computation of an upper bound for m. That algorithm does not terminate in general if Gross' conjecture is false. This upper bound can be used as the ℓ-adic precision in the computation of the logarithmic class group.

2.1 Generators and Relations of $\widetilde{C\ell}_{F,\ell}$ [DJ+]

Let $\mathfrak{a}_1, \ldots, \mathfrak{a}_t$ be a basis of the ideal classgroup $C\ell_F$ of F with $\gcd(\mathfrak{a}_i, \ell) = 1$ for all $1 \le i \le t$. Denote by $\mathfrak{p}_1, \ldots, \mathfrak{p}_s$ the ℓ-adic places of F. Let $\alpha_1, \ldots, \alpha_s$ be elements of $\mathcal{R}_F = \mathbb{Z}_\ell \otimes F^*$ with $\widetilde{v}_{\mathfrak{p}_i}(\alpha_j) = \delta_{i,j}$ $(i, j = 1, \ldots, s)$ and $\gcd((\alpha_i), \ell) = 1$ for all $1 \le i \le s$. Set $\mathfrak{a}_{t+i} := (\alpha_i)$ for $1 \le i \le s$. For an ideal $\mathfrak{a}$ of F denote by $\bar{\mathfrak{a}}$ the projection of $\mathfrak{a}$ from $\bigoplus_{\mathfrak{p}} \mathfrak{p}^{\mathbb{Z}_\ell}$ to $\bigoplus_{\mathfrak{p}\nmid(\ell)} \mathfrak{p}^{\mathbb{Z}_\ell}$. We distinguish two cases:

I. If $\deg_\ell(\mathfrak{a}_i) = 0$ for all $1 \le i \le t+s$ then set $\mathfrak{b}_i := \mathfrak{a}_i$. The group $\widetilde{C\ell}_{F,\ell}$ is generated by $\bar{\mathfrak{b}}_1, \ldots, \bar{\mathfrak{b}}_{t+s}$.

II. Otherwise let $1 \le j \le t+s$ such that $v_\ell(\deg_\ell(\mathfrak{a}_j)) = \min_{1 \le i \le t+s} v_\ell(\deg_\ell(\mathfrak{a}_i))$. If we have $\mathfrak{a} = \bar{\mathfrak{a}} \equiv \bar{\mathfrak{a}}_1^{a_1} \cdot \cdots \cdot \bar{\mathfrak{a}}_{t+s}^{a_{t+s}} \bmod \widetilde{\mathcal{P}r}$ for an ideal $\mathfrak{a} \in \widetilde{\mathcal{I}d}$ then $0 = \deg(\bar{\mathfrak{a}}) = \sum_{i=1}^{t+s} a_i \deg_\ell(\bar{\mathfrak{a}}_i)$, thus $-a_j = \sum_{i \ne j}^{t+s} a_i \deg_\ell(\bar{\mathfrak{a}}_i)/\deg_\ell(\bar{\mathfrak{a}}_j)$. Set $\mathfrak{b}_i := \mathfrak{a}_i/\mathfrak{a}_j^{d_i}$ with $d_i \equiv \frac{\deg_\ell(\mathfrak{a}_i)}{\deg_\ell(\mathfrak{a}_j)} \bmod \ell^m$ where $\ell^m > \exp(\widetilde{C\ell}_{F,\ell})$. The group $\widetilde{C\ell}_{F,\ell}$ is generated by $\bar{\mathfrak{b}}_1, \ldots, \bar{\mathfrak{b}}_{j-1}, \bar{\mathfrak{b}}_{j+1}, \ldots, \bar{\mathfrak{b}}_{t+s}$.

Obviously the ideals $\bar{\mathfrak{a}}_1, \ldots, \bar{\mathfrak{a}}_t$ are representatives of generators of the group $C\ell' := C\ell_F/\langle \mathfrak{p}_1, \ldots, \mathfrak{p}_s \rangle$. Let $(a_{i,j})_{i,j}$ be the corresponding relation matrix. The relations between the generators $\bar{\mathfrak{a}}_1, \ldots, \bar{\mathfrak{a}}_t$ of $C\ell'$ are of the form $\prod_{i=1}^t \bar{\mathfrak{a}}_i^{a_i} = \overline{(\alpha)}$ with $\alpha \in \mathcal{R}_F$. There exist integers $c_1, \ldots, c_s$ such that $\overline{(\alpha)} \equiv \prod_{i=1}^s \overline{(\alpha_i)}^{c_i} \bmod \widetilde{\mathcal{P}r}$. This yields the relation $\prod_{i=1}^t \bar{\mathfrak{a}}_i^{a_i} \equiv \prod_{i=1}^s \overline{(\alpha_i)}^{c_i} \bmod \widetilde{\mathcal{P}r}$. We can derive all relations involving the generators $\bar{\mathfrak{a}}_i + \widetilde{\mathcal{P}r}$ from their relations as generators of the group $C\ell'$ in this way.

The other relations between the generators of $\widetilde{C\ell}$ are obtained as follows: A relation between the generators $\bar{\alpha}_i$ is of the form $\prod_{i=1}^s \overline{(\alpha_i)}^{v_i} \equiv (1) \bmod \widetilde{\mathcal{P}r}$ or equivalently $\prod_{i=1}^s (\alpha_i)^{v_i} \cdot \prod_{i=1}^s \mathfrak{p}_i^{w_i} = (\alpha)$ for some $\alpha \in \mathcal{R}_F$. The last equality is fulfilled if and only if $\prod_{i=1}^s \mathfrak{p}_i^{w_i}$ is principal, i.e., if $\prod_{i=1}^s \mathfrak{p}_i^{w_i}$ is an (ℓ)-unit. Let $\gamma_1, \ldots, \gamma_r$ be a basis of the (ℓ)-units of $\mathcal{R}_F$. set $v_{i,j} := \widetilde{v}_{\mathfrak{p}_j}(\gamma_i)$ $(1 \le i \le r, 2 \le j \le s)$. We obtain the relation matrix

$$M := \begin{pmatrix} b_{1,1} & \cdots & b_{1,t} & -c_{1,2} & \cdots & -c_{1,s} \\ \vdots & \ddots & \vdots & \vdots & \ddots & \vdots \\ b_{t,1} & \cdots & b_{t,t} & -c_{t,2} & \cdots & -c_{t,s} \\ 0 & \cdots & 0 & v_{1,2} & \cdots & v_{1,s} \\ \vdots & \ddots & \vdots & \vdots & \ddots & \vdots \\ 0 & \cdots & 0 & v_{r,2} & \cdots & v_{r,s} \end{pmatrix}.$$

For the two cases we obtain:

I. $((\overline{\mathfrak{b}}_1, \ldots, \overline{\mathfrak{b}}_{t+s}), M)$ are generators and relations of $\widetilde{C\ell}$.
II. Let j be chosen as above. Denote by N the matrix obtained by removing the j-th column from M. Then $((\overline{\mathfrak{b}}_1, \ldots, \overline{\mathfrak{b}}_{j-1}, \overline{\mathfrak{b}}_{j+1}, \ldots, \overline{\mathfrak{b}}_{t+s}), N)$ are generators and relations of $\widetilde{C\ell}$.

This gives the following algorithm:

Algorithm 1 (Logarithmic Classgroup)
Input: a number field F and a prime number ℓ
Output: generators g and and a relation matrix H for $\widetilde{C\ell}_{F,\ell}$

- Determine a bound ℓ^m for the exponent of $\widetilde{C\ell}_{F,\ell}$ and use it as the precision for the rest of the algorithm.
- Compute generators $\mathfrak{a}_1, \ldots, \mathfrak{a}_t$ of $C\ell' = C\ell_F / \langle \mathfrak{p}_1, \ldots, \mathfrak{p}_s \rangle$, where $\mathfrak{p}_1, \ldots, \mathfrak{p}_s$ are the ideals of F over ℓ.
- Determine $\mathfrak{a}_{t+1} = (\alpha_1), \ldots, \mathfrak{a}_{t+s} = (\alpha_s)$ with $\widetilde{v}_{\mathfrak{p}_i}(\alpha_j) = \delta_{i,j}$.
- Compute generators $g := (\overline{\mathfrak{b}}_1, \ldots, \overline{\mathfrak{b}}_{t+s})^T$ with $\deg(\mathfrak{b}_i) = 0$ from $\mathfrak{a}_1, \ldots, \mathfrak{a}_{t+s}$ $(i = 1, \ldots, t+s)$.
- Compute a relation matrix M between the generators g.
- In case **II.** remove the j-th column from M and the j-th generator from g.
- Compute the ℓ-adic Hermite normal form H of M.
- Return (g, H).

The Smith normal form of H and the respective transformations of the generators yield a basis representation of $\widetilde{C\ell}_{F,\ell}$.

2.2 The Discrete Logarithm in $\widetilde{C\ell}_{F,\ell}$

Let $\mathfrak{a} \in \widetilde{\mathcal{I}d}$. Let $g = (\overline{\mathfrak{b}}_1, \ldots, \overline{\mathfrak{b}}_r)^T$ be a vector of generators of $\widetilde{C\ell}$. The discrete logarithm algorithm returns a vector $c = (c_1, \ldots, c_r)$ such that

$$c^T g = \overline{\mathfrak{b}}_1^{c_1} \ldots \overline{\mathfrak{b}}_r^{c_r} \equiv \mathfrak{a} \bmod \widetilde{\mathcal{P}r}.$$

We use the notation from above and proceed as follows:

Let $\mathfrak{a} \in \widetilde{\mathcal{I}d}$. There exist $\gamma \in \mathcal{R}_F$ and $a_1, \ldots, a_t \in \mathbb{Z}_\ell$ such that $\mathfrak{a} = \prod_{i=1}^t \mathfrak{a}_i^{a_i} \cdot (\gamma)$. Set $g_i := \widetilde{v}_{\mathfrak{p}_i}(\gamma)$ for $1 \leq i \leq s$. Now

$$\mathfrak{a} = \prod_{i=1}^s \mathfrak{a}_i^{a_i} \cdot \left((\gamma) \cdot \prod_{j=1}^s (\alpha_i)^{-g_i}\right) \cdot \left(\prod_{j=1}^s (\alpha_i)^{g_i}\right).$$

By the definition of $\widetilde{\mathcal{I}d}$ we have

$$\mathfrak{a} = \overline{\mathfrak{a}} = \prod_{i=1}^t \overline{\mathfrak{a}}_i^{a_i} \cdot \overline{\left((\gamma) \cdot \prod_{j=1}^s (\alpha_j)^{-g_j}\right)} \cdot \left(\prod_{j=1}^s \overline{(\alpha_j)^{g_j}}\right)$$

As $\widetilde{v}_{\mathfrak{p}_i}\left((\gamma) \prod_{j=1}^s (\alpha_j)^{-g_j}\right) = 0$ for $i = 1, \ldots, s$ we obtain

$$\mathfrak{a} \equiv \prod_{i=1}^t \mathfrak{a}_i^{a_i} \cdot \left(\prod_{j=1}^s \overline{(\alpha_j)}^{g_j}\right) \bmod \widetilde{\mathcal{P}r}.$$

For the two cases we obtain:

I. $(a_1, \ldots, a_t, g_1, \ldots, g_s)$ is a representation of $\mathfrak{a}$ in $\widetilde{C\ell}_{F,\ell}$.
II. Let $(c_1, \ldots, c_{t+s}) = (a_1, \ldots, a_t, g_1, \ldots, g_s)$ then $(c_1, \ldots, c_{j-1}, c_{j+1}, \ldots, c_{t+s})$ is a representation of $\mathfrak{a}$ in $\widetilde{C\ell}_{F,\ell}$.

3 The Wild Kernel

In the following we develop an algorithm for the computation of the wild kernel of a number field F. The study of the K_2-group of F, which can be defined as

$$K_2(F) = F^* \otimes_{\mathbb{Z}} F^* \Big/ \big\langle x \otimes_{\mathbb{Z}} (1-x) \mid x \in F \setminus \{0,1\} \big\rangle,$$

is difficult [T1,T2]. In order to understand the structure of $K_2(F)$ one constructs a morphism from $K_2(F)$ to a known abelian group whose kernel is finite. Let $\mathfrak{p}$ be a non-complex place of F and let $F_\mathfrak{p}$ be the completion of F at $\mathfrak{p}$. Denote by $\mu_\mathfrak{p}$ the torsion subgroup of $F_\mathfrak{p}^*$. We define

$$h_\mathfrak{p} : F_\mathfrak{p}^* \times F_\mathfrak{p}^* \to \mu_\mathfrak{p},\ (x, y) \mapsto \sqrt[m_\mathfrak{p}]{x}^{\,\omega_\mathfrak{p}(y)-1}$$

where $m_\mathfrak{p} = |\mu_\mathfrak{p}|$ and where $\omega_\mathfrak{p}$ is the Artin map. It follows from the multiplicativity of the norm residue symbol and Kummer theory [Gr, pp. 195-197] that the map $h_\mathfrak{p}$ is a $\mathbb{Z}$-linear map which is trivial for elements of the form $(x, 1-x)$ where $x \in F \setminus \{0,1\}$, *i.e.*, $h_\mathfrak{p}$ is a symbol. $h_\mathfrak{p}$ gives us a map from $K_2(F)$ to $\mu_\mathfrak{p}$, which we also denote by $h_\mathfrak{p}$. The wild kernel of K_2 is

$$WK_2(F) = \big\{\mathcal{X} \in K_2(F) \mid h_\mathfrak{p}(\mathcal{X}) = 1 \text{ for all non-complex places } \mathfrak{p} \text{ of } F\big\}.$$

Garlands theorem [Ga] states that $WK_2(F)$ is finite. There exist idelic [Ko] and cohomologic methods for studying the wild kernel. We chose to use logarithmic methods as they allow for the use of an algorithmic approach. Another approach, which could be exploited algorithmically, is F. Keunes description of the ℓ-part of $WK_2(F)$ in terms of the ideal class group of ℓ-power cyclotomic extensions.

The following theorem by Jaulent [J2] establishes the relationship between the wild kernel and the logarithmic ℓ-class groups $\widetilde{C\ell}_{F,\ell}$ for the case where F contains a primitive $2\ell^q$-th root of unity $\zeta_{2\ell^q}$.

Theorem 2. *Assume that $\zeta_{2\ell^q} \in F^*$. Let $q \in \mathbb{N}$, $q \geq 1$. For every divisor $\mathfrak{a} = \sum_\mathfrak{p} a_\mathfrak{p} \mathfrak{p}$ of degree 0 there exists $\mathcal{X} \in K_2(F)$ such that $h_\mathfrak{p}(\mathcal{X}) = \zeta_{\ell^q}^{a_\mathfrak{p}}$. If $\zeta_{2\ell^q} \in F^*$ then the map*

$$\phi : \mu_{\ell^q} \otimes_{\mathbb{Z}} \widetilde{C\ell}_{F,\ell} \to WK_2(F)/WK_2(F)^{\ell^q}$$

defined by

$$\zeta_{\ell^q} \otimes \mathfrak{a} \mapsto \mathcal{X}^{\ell^q}$$

is an isomorphism.

Moore's reciprocity law [Mo] assures the existance of such an $\mathcal{X}$.

Corollary 3. *If F contains the 2ℓ-th roots of unity, then*

$$\operatorname{rank}_\ell WK_2(F) = \operatorname{rank}_\ell \widetilde{\mathcal{C}\ell}_{F,\ell}.$$

The algorithm in [DJ+] computes the structure of $\widetilde{\mathcal{C}\ell}_F$, and therefore the ℓ-rank of $\widetilde{\mathcal{C}\ell}_F$. Thus by the theorem above, the ℓ-rank of the wild kernel is known if F contains the 2ℓ-th roots of unity.

3.1 F Does Not Contain the 2ℓ-th Roots of Unity

If $\ell = 2$ and $i \notin F$ the group of positive divisor classes can be used for the description of the 2-rank of the §wild kernel [JS2]. We deal with the remaining case and therefore assume in the following that ℓ is odd.

Let ζ_ℓ be a primitive ℓ-th root of unity. Let F' be the Galois extension $F(\zeta_\ell)$. Let $d = |\mathrm{Gal}(F'/F)|$. We have $d \mid (\ell - 1)$ and therefore $\gcd(\ell, d) = 1$. In other words $d \in \mathbb{Z}_\ell^*$.

There is an idempotent $e_2 \in \mathbb{Z}_\ell[\mathrm{Gal}(F'/F)]$ with $e_2 = \frac{1}{d}\sum_{\sigma\in\mathrm{Gal}(F'/F)} k_\sigma \sigma$ where $k_\sigma \in \mathbb{Z}_\ell$ such that $\zeta^\sigma = \zeta^{k_\sigma}$ for all $\sigma \in \mathrm{Gal}(F'/F)$. We construct such an element e_2 in the next section.

Proposition 4 ([JS1]). *If ℓ is odd and F does not contain the 2ℓ-th roots of unity then*

$$\operatorname{rank}_\ell WK_2(F) = \operatorname{rank}_\ell \widetilde{\mathcal{C}\ell}^{e_2}_{F(\zeta_\ell),\ell}.$$

For a better understanding we give a more detailed proof than in [JS1].

Proof. Let $F' := F(\zeta_\ell)$. Set $\Delta := \mathrm{Gal}(F'/F)$. Because F' contains the 2ℓ-th roots of unity the isomorphism

$$\mu_\ell \otimes_{\mathbb{Z}} \widetilde{\mathcal{C}\ell}_{F'} \cong WK_2(F')/WK_2(F')^\ell$$

holds (Theorem 2). As Δ acts on $K_2(F')$ such that $\{x,y\}^\sigma = \{x^\sigma, y^\sigma\}$ for all $\sigma \in \Delta$ and all $(x,y) \in \left(F'^*\right)^2$ it follows that

$$\left(\mu_\ell \otimes_{\mathbb{Z}} \widetilde{\mathcal{C}\ell}_{F'}\right)^{e_1} \cong \left(WK_2(F')/WK_2(F')^\ell\right)^{e_1}$$

for $e_1 = \frac{1}{d}\sum_{\sigma\in\Delta}\sigma$. As ℓ does not divide d the idempotent e_1 induces a surjective morphism $\frac{1}{d}\,\mathrm{Tr}$ where Tr is called transfer from the ℓ-part of $K_2(F')$ to the ℓ-part of $K_2(F)$. Therefore $WK_2(F)/WK_2(F)^\ell$ is the image of $WK_2(F')/WK_2(F')^\ell$ under the restriction of the transfer map Tr. Hence

$$\left(\mu_\ell \otimes_{\mathbb{Z}} \widetilde{\mathcal{C}\ell}_{F'}\right)^{e_1} \cong WK_2(F)/WK_2(F)^\ell.$$

For $\mathfrak{a} \in \mathcal{D}\ell_{F'}$ we have

$$(\zeta \otimes \mathfrak{a})^{d\cdot e_1} = \prod_{\sigma\in\Delta} (\zeta \otimes \mathfrak{a})^\sigma = \prod_{\sigma\in\Delta} \zeta^\sigma \otimes \mathfrak{a}^\sigma = \prod_{\sigma\in\Delta} \zeta^{k_\sigma} \otimes \mathfrak{a}^\sigma = \prod_{\sigma\in\Delta} \zeta \otimes \mathfrak{a}^{k_\sigma\sigma}$$

and

$$(\zeta \otimes a)^{e_1} = \left(\zeta \otimes \prod_{\sigma \in \Delta} \mathfrak{a}^{k_\sigma \sigma}\right)^{d^{-1}} = \zeta \otimes a^{e_2}.$$

Therefore

$$\left(\mu_\ell \otimes_{\mathbb{Z}} \widetilde{C\ell}_{F'}\right)^{e_1} = \mu_\ell \otimes \widetilde{C\ell}_{F'}^{e_2}.$$

Remark 5. It is possible to generalize this method to the computation of the ℓ-part of the higher analogs $WK_{2i-2}(F)$ $(1 \geq 2)$ of $WK_2(F)$ which were introduced by G. Banaszak [Ba]. We will treat them in a future article.

Example 6 ([JS1]). If $\ell = 3$ and $F = \mathbb{Q}(\sqrt{d})$ with $d \in \mathbb{Z}$ squarefree then $F' = F(\sqrt{-3})$ with cyclic Galois group $\mathrm{Gal}(F'/F) = \langle \tau \rangle$ and $\zeta_3 = \frac{-1+\sqrt{-3}}{2} \in F'$. Because $\zeta_3^\tau = \zeta_3^{-1}$ we set $e_2 = 1/2(1-\tau)$. We have

$$\mathrm{rank}_3\, WK_2(F) = \mathrm{rank}_3\, \widetilde{C\ell}_{F'}^{e_2}$$

Because $e_2 = 1/2(1-\tau) = 1/2(1+\sigma)$ where $\langle \sigma \rangle = \mathrm{Gal}(F'/F_*)$ with $F_* = \mathbb{Q}(\sqrt{-3d})$ we obtain

$$\mathrm{rank}_3\, WK_2(F) = \mathrm{rank}_3\, \widetilde{C\ell}_{F'}^{1+\sigma} = \mathrm{rank}_3\, \widetilde{C\ell}_{F_*}$$

and

$$\mathrm{rank}_3\, WK_2(\mathbb{Q}(\sqrt{d})) = \mathrm{rank}_3\, \widetilde{C\ell}_{\mathbb{Q}(\sqrt{-3d})}.$$

This is particularly interesting as we do not need any computations in the extension $F(\zeta_3)$.

3.2 Computing e_2

Let $d := |\mathrm{Gal}(F'/F)|$ and let σ be a generator of $\mathrm{Gal}(F'/F)$. We are looking for an element $e \in \mathbb{Z}_\ell[\mathrm{Gal}(F'/F)]$ with $e = e^2$. The element e is of the form $e = \frac{1}{d}\sum_{i=0}^{d-1} k_i \sigma^i$ with $k_i \in \mathbb{Z}_\ell$ $(0 \leq i < d)$. Hence the condition $e = e^2$ becomes

$$\left(\sum_{u=0}^{d-1} k_u \sigma^u\right)\left(\sum_{v=0}^{d-1} k_v \sigma^v\right) = d \sum_{i=0}^{d-1} k_i \sigma^i.$$

Let ℓ^m be the exponent of $\widetilde{C\ell}_{F,\ell}$. It is obvious that it suffices to compute e up to a precision of m ℓ-adic digits. Set

$$S_i := \left\{(u,v) \in \mathbb{Z}^2 \mid u,v \in \{0,\dots,d-1\}, u+v \equiv i \bmod d\right\}.$$

For $0 \le i \le d-1$ we solve the congruences

$$\sum_{(u,v)\in S_i} k_u \cdot k_v \equiv dk_i \bmod \ell^m.$$

We write k_i as $\sum_{j=0}^{m-1} x_{i,j}\ell^j$ with unknown $x_{i,j} \in \{0,\ldots,\ell-1\}$ ($0 \le i < d$, $0 \le j < m$). Thus our congruences become

$$\sum_{(u,v)\in S_i} \left(\sum_{j=0}^{m-1} x_{u,j}\ell^j\right)\left(\sum_{j=0}^{m-1} x_{v,j}\ell^j\right) \equiv d\left(\sum_{j=0}^{m-1} x_{i,j}\ell^j\right) \bmod \ell^m. \quad (1)$$

We start by solving them modulo ℓ:

$$\sum_{(u,v)\in S_i} x_{u,0}x_{v,0} \equiv dx_{i,0}.$$

Let $\alpha \in \mathbb{F}_\ell$ be a generator of the cyclic group $\mathbb{F}_\ell^*$. Set $\delta = \frac{\ell-1}{d}$ then α^δ has order d in $\mathbb{F}_\ell^*$. Let a be a representative of α^δ in $\mathbb{Z}_\ell$. The elements $a_{0,0} = 1$, $a_{1,0} = a$, $a_{2,0} = a^2, \ldots, a_{d-1,0} = a^{d-1}$ are solutions for $x_{0,0}, \ldots, x_{d-1,0}$.

Assume that we have found $a_{i,j} \in \{1\ldots,\ell-1\}$ ($0 \le i < d$, $0 \le j < w < m$) such that

$$A_{i,w} := -\sum_{(u,v)\in S_i} \left(\sum_{j=0}^{w-1} a_{u,j}\ell^j\right)\left(\sum_{j=0}^{w-1} a_{v,j}\ell^j\right) + d\left(\sum_{j=0}^{w-1} a_{i,j}\ell^j\right) \equiv 0 \bmod \ell^w.$$

With (1) we obtain

$$\sum_{(u,v)\in S_i} x_{u,w}\ell^w a_{v,0} + x_{v,w}\ell^w a_{u,0} \equiv dx_{i,w}\ell^w + A_{i,w} \bmod \ell^{w+1}.$$

and as $A_{i,w} \equiv 0 \bmod \ell^w$ this becomes

$$\sum_{(u,v)\in S_i} x_{u,w}a_{v,0} + x_{v,w}\, a_{u,0} - dx_{i,w} \equiv \frac{A_{i,w}}{\ell^w} \bmod \ell \quad (2)$$

for $i = 1,\ldots,d-1$ which is a system of d linear equations in d variables over $\mathbb{F}_\ell$.

Therefore we obtain a solution to (1) by first computing $a_{0,0},\ldots,a_{d-1,0}$ as described above and then solving systems of linear equations (2) inductively for $w = 1,\ldots,m-1$ to obtain values $a_{0,w},\ldots,a_{d-1,w}$ for $x_{0,w},\ldots,x_{d-1,w}$.

3.3 Computing the ℓ-Rank of the Wild Kernel

By proposition 4 the ℓ-rank of the wild kernel of F equals the ℓ-rank of $\widetilde{C\ell}^{e_2}_{F(\zeta_\ell),\ell}$. Let $\bar{\mathfrak{b}}_1,\ldots,\bar{\mathfrak{b}}_r$ be a basis of $\widetilde{C\ell}_{F(\zeta_\ell),\ell}$ and let ℓ^{b_i} be the order of $\bar{\mathfrak{b}}_i$ in $\widetilde{C\ell}_{F(\zeta_\ell),\ell}$ ($1 \le i \le r$), i.e.,

$$\widetilde{C\ell}_{F(\zeta_\ell),\ell} = \bigoplus_{i=1}^{r} \mathbb{Z}/\ell^{b_i}\mathbb{Z}[\bar{\mathfrak{b}}_i].$$

The elements $\overline{\mathfrak{b}}_1^{e_2}, \ldots, \overline{\mathfrak{b}}_r^{e_2}$ are generators of $\widetilde{C\ell}_{F(\zeta_\ell),\ell}^{e_2}$. For $1 \leq i \leq r$ the discrete logarithm in $\widetilde{C\ell}_{F(\zeta_\ell),\ell}$ gives representations $(n_{i,1}, \ldots, n_{i,r})$ of the $\overline{\mathfrak{b}}_i^{e_2}$ with

$$\overline{\mathfrak{b}}_i^{e_2} \equiv \overline{\mathfrak{b}}_1^{n_{i,1}} \cdots \overline{\mathfrak{b}}_r^{n_{i,r}} \bmod \widetilde{\mathcal{P}r}.$$

Let $A \in \mathbb{Z}_\ell^{r \times 2r}$ such that

$$\begin{pmatrix} \ell^{b_1} & & 0 & n_{1,1} & \cdots & n_{r,1} \\ & \ddots & & \vdots & \ddots & \vdots \\ 0 & & \ell^{b_r} & n_{1,r} & \cdots & n_{r,r} \end{pmatrix} A = 0.$$

We write $A = \begin{pmatrix} A_1 \\ A_2 \end{pmatrix}$ where $A_1, A_2 \in \mathbb{Z}_\ell^{r \times r}$. A_2 is a relation matrix of the subgroup $\widetilde{C\ell}_{F(\zeta_\ell),\ell}^{e_2}$ generated by $\overline{\mathfrak{b}}_1, \ldots, \overline{\mathfrak{b}}_r$ which are represented by $(n_{i,1}, \ldots, n_{i,r})$ $(1 \leq i \leq r)$. Denote by $\left(h_{i,j}\right)_{i,j}$ the ℓ-adic Hermite normal form of A_2. Then

$$\operatorname{rank}_\ell WK_2(F) = \operatorname{rank}_\ell \widetilde{C\ell}_{F(\zeta_\ell),\ell}^{e_2} = \#\{h_{i,i} \mid 1 \leq i \leq r,\ h_{i,i} \neq 1\}.$$

4 A Complete Description of the ℓ-part of the Wild Kernel

Let F be a number field. Let ℓ be a prime number. Let $\ell^m := \exp \widetilde{C\ell}_{F,\ell}$. If F contains a primitive ℓ^{m+1}-th root of unity $\zeta_{\ell^{m+1}}$ then we obtain a complete description of the ℓ-part of the wild kernel of F as follows:

Assume that $\widetilde{C\ell}_{F,\ell}$ is not trivial, then

$$\widetilde{C\ell}_{F,\ell} = \bigoplus_{i=1}^{r} \mathbb{Z}/\ell^{n_i}\mathbb{Z}[\mathfrak{a}_i].$$

Therefore there exist a family $(\alpha_i)_i \subset \mathcal{R}_F = \mathbb{Z}_\ell \otimes_{\mathbb{Z}} F^*$ such that $\ell^{n_i}\mathfrak{a}_i = (\alpha_i)$ for $1 \leq i \leq r$. We denote by $\{\cdot,\cdot\}$ the canonical map

$$\{\cdot,\cdot\} : F^* \times F^* \to K_2(F);$$

it is called Steinberg's symbol. The ℓ-part of the wild kernel is [So]

$$\bigoplus_{i=1}^{r} \mathbb{Z}/\ell^{n_i}\mathbb{Z}\{\zeta_{\ell^{n_i}}, \alpha_i\}.$$

Let $\alpha \in \mathcal{R}_F$. We denote by $\overline{\alpha}$ the approximation of α to a precision of m ℓ-adic digits. As Steinberg's symbol is $\mathbb{Z}_\ell$-bilinear we have $\{\zeta_{\ell^{n_i}}, \alpha\} = \{\zeta_{\ell^{n_i}}, \overline{\alpha}\}$ for all $\alpha \in \mathcal{R}_F$. Therefore the ℓ-part of the wild kernel is

$$\bigoplus_{i=1}^{r} \mathbb{Z}/\ell^{n_i}\mathbb{Z}\{\zeta_{\ell^{n_i}}, \overline{\alpha}_i\}.$$

5 Examples

All algorithms presented here have been implemented in the computer algebra system Magma [C+]. The groups are given as lists of the orders of their cyclic factors. By i we denote a root of x^2+1, by ζ_m we denote a primitive m-th root of unity.

K. Belabas and H. Gangl have developed an algorithm for the computation of the tame kernel $K_2\mathcal{O}_F$ [BG]. The following table contains the structure of $K_2\mathcal{O}_F$ as computed by Belabas and Gangl and the ℓ-rank of the wild kernel $WK_2(F)$ calculated with our methods. The starred entry is a conjectural result.

F	$K_2\mathcal{O}_F$	ℓ	$\widetilde{C\ell}_{F(\zeta_\ell),\ell}$	$\widetilde{C\ell}^{e_2}_{F(\zeta_\ell),\ell}$	$\mathrm{rank}_\ell(WK_2(F))$
$\mathbb{Q}(\sqrt{-331})$	[3]	3	[3,3]	[3]	1
$\mathbb{Q}(\sqrt{-367})$	[3]	3	[3,9]	[3]	1
$\mathbb{Q}(\sqrt{-472})$	[5]	5	[5,5]	[5]	1
$\mathbb{Q}(\sqrt{-571})$	[5]	5	[5,5]	[5]	1
$\mathbb{Q}(\sqrt{-696})$	[42]	3	[3]	[1]	0
		7	[7,7]	[7]	1
$\mathbb{Q}(\sqrt{-759})$	$[2,18]^*$	3	[3,3]	[3]	1

The next table contains more fields together with the main data needed for the computation of the ℓ-rank of WK_2. χ_α denotes the minimal polynomial of α over $\mathbb{Q}$.

F	ℓ	$\widetilde{C\ell}_{F(\zeta_\ell),\ell}$	$\widetilde{C\ell}^{e_2}_{F(\zeta_\ell),\ell}$	$\mathrm{rank}_\ell(WK_2(F))$
$\mathbb{Q}(\sqrt{-7307})$	5	[5,25]	[1]	0
$\mathbb{Q}(\sqrt{-356467})$	3	[3,3,27]	[3]	1
$\mathbb{Q}(\alpha)$, $\chi_\alpha = x^3+x^2-9x-365$	3	[9]	[9]	1
$\mathbb{Q}(\alpha)$, $\chi_\alpha = x^3+x^2-133x-1937$	3	[3,3]	[3]	1
$\mathbb{Q}(\alpha)$, $\chi_\alpha = x^3+x^2-65x+1875$	3	[3,3,3]	[3,3]	2
$\mathbb{Q}(\alpha)$, $\chi_\alpha = x^3+x^2-65x+1875$	3	[3,3,3]	[3,3]	2
$\mathbb{Q}(\alpha)$, $\chi_\alpha = x^4+9x^2+125$	3	[3,3]	[3]	1

Our last table gives examples of the 2-part of the wild kernel together with the generators of the cyclic factors. We made extensive use of the discrete logarithm in $\widetilde{C\ell}_{F,2}$ in order to find small generators for it.

F	$\widetilde{\mathcal{C}\ell}_{F,2}$	2-part of $WK_2(F)$
$\mathbb{Q}(i,\sqrt{85})$	[2,2]	$\mathbb{Z}/2\mathbb{Z}\{-1,i-2\}\oplus\mathbb{Z}/2\mathbb{Z}\left\{-1,\frac{\sqrt{85}+11}{2}\right\}$
$\mathbb{Q}(i,\sqrt{357})$	[2,2,2]	$\mathbb{Z}/2\mathbb{Z}\{-1,3\}\oplus\mathbb{Z}/2\mathbb{Z}\left\{-1,\frac{i\sqrt{357}+21i+2}{2}\right\}\oplus$ $\mathbb{Z}/2\mathbb{Z}\left\{-1,\frac{(i+4)\sqrt{357}+19i+76}{2}\right\}$
$\mathbb{Q}(i,\sqrt{1173}$	[2,2,2]	$\mathbb{Z}/2\mathbb{Z}\{-1,3\}\oplus\mathbb{Z}/2\mathbb{Z}\left\{-1,\frac{(4i+16)\sqrt{1173}+137i+548}{2}\right\}\oplus$ $\mathbb{Z}/2\mathbb{Z}\left\{-1,\frac{(-927i-3300)\sqrt{1173}-31749i-13022}{2}\right\}$
$\mathbb{Q}(\zeta_8,\sqrt{561})$	[4,4,4]	$\mathbb{Z}/4\mathbb{Z}\left\{i,(2\zeta_8^3+3\zeta_8^2+2\zeta_8)\sqrt{561}-80\zeta_8^3+80\zeta_8+114\right\}\oplus$ $\mathbb{Z}/4\mathbb{Z}\left\{i,\frac{(15\zeta_8^3+12\zeta_8^2+38\zeta_8+12)\sqrt{561}-93\zeta_8^3+12\zeta_8^2-330\zeta_8-372}{2}\right\}\oplus$ $\mathbb{Z}/4\mathbb{Z}\left\{i,(-\zeta_8^3+\zeta_8^2-\zeta_8)\sqrt{561}+13\zeta_8^3-28\zeta_8^2+15\zeta_8+2\right\}$

References

[Ba] Banaszak, Grzegorz, *Generalization of the Moore exact sequence and the wild kernel for higher K-groups*, Compos. Math. 86, No.3, 281-305 (1993).

[BT] H. Bass and J. Tate, *The Milnor ring of a global field*, in H. Bass (ed.), Algebraic K-theory, Lecture Notes in Math. **342**, Springer Verlag Berlin (1973), 349–446.

[BG] K. Belabas and H. Gangl, *Generators and Relations for $K_2\mathcal{O}_F$*, preprint.

[C+] J.J. Canon et al., The computer algebra system Magma, University of Sydney (2003), `http://magma.maths.usyd.edu.au/magma/`.

[DJ+] F. Diaz y Diaz, J.-F. Jaulent, S. Pauli, M.E. Pohst, and F. Soriano, *A new Algorithm for the Computation of logarithmic ℓ-class groups of number fields*, submitted to Experimental Mathematics.

[DS] F. Diaz y Diaz and F. Soriano, *Approche algorithmique du groupe des classes logarithmiques,* J. Number Theory **76** (1999), 1–15.

[FG] L.J. Federer and B.H. Gross, (with an appendix by W. Sinnot) *Regulators and Iwasawa modules,* Invent. Math. **62** (1981), 443–457.

[Ga] H. Garland, *A finiteness theorem for K_2 of a number field*, Ann. of Math. **94** (1971), 534–548.

[Gr] G. Gras, *Class Field Theory*, Springer Monographs in Mathematics (2003).

[J1] J.-F. Jaulent, *Introduction au K_2 des corps de nombres*, Publ. Math. Fac. Sci. Besançon (Théorie des nombres), Anèes 1984/85–1985/86.

[J2] J.-F. Jaulent, *Sur le noyau sauvage des corps de nombres*, Acta Arithmetica **67** (1994), 335–348.

[J3] J.-F. Jaulent, *Classes logarithmiques des corps de nombres*, J. Théor. Nombres Bordeaux **6** (1994), 301–325.

[J4] J.-F. Jaulent, *Classes logarithmiques des corps de nombres totalement réels*, Acta Arithmetica **103** (2002), 1–7.

[JS1] J.-F. Jaulent and F. Soriano-Gafiuk, *Sur le noyau sauvage des corps de nombres et le groupe des classes logarithmiques*, Math. Z. **238** (2001), 335–354.

[JS2] J.-F. Jaulent and F. Soriano-Gafiuk, *2-Groupe des classes positives et noyau sauvage de la K_2-Théorie*, to appear in J. Number Theory.

[Ke] F. Keune, *On the structure of the K_2 of the ring of integers in a number field*, K-Theory 2, No.5, 625-645 (1989). MSC 2000:

[Ko] M. Kolster, *An idelic Approach to the Wild Kernel*, Invent. Math. 103 (1991), 9-24.

[Mi] J. Milnor, *Introduction to Algbraic K-Theory*, Ann. of Math. Studies **72**, Princeton Univ. Press, Princeton (1971).

[Mo] C. Moore *Group extensions of p-adic and adelic linear groups*, Publ. Math. I.H.E.S. **35** (1969), 5–74.

[So] F. Soriano-Gafiuk, *Sur le noyau hilbertien d'un corps de nombres*, C. R. Acad. Sci. Paris, t. 330, Série I (2000), 863-866.

[T1] J. Tate, *Symbols in arithmetic*, Actes Congrès Int. de Math. 1970, Tome 1, 201–211.

[T2] J. Tate, *Relations between K_2 and Galois cohomology*, Invent. Math. **36** (1976), 257–274.

Point Counting on Genus 3 Non Hyperelliptic Curves

Christophe Ritzenthaler*

IEM Essen, Germany,
ritzenth@math.jussieu.fr
http://www.math.jussieu.fr/~ritzenth

Abstract. We propose an algorithm to compute the Frobenius polynomial of an ordinary non hyperelliptic curve of genus 3 over $\mathbb{F}_{2^N}$. The method is a generalization of Mestre's AGM-algorithm for hyperelliptic curves and leads to a quasi quadratic time algorithm for point counting.

The current methods for point counting on curves over finite fields of small characteristic rely essentially on a p-adic approach. They split up in three classes: those based on cohomology (Kedlaya), those based on deformation theory (Lauder) and those based on the canonical lift, proposed initially by Satoh. The AGM-algorithm developed by Mestre [Mes02] for elliptic curves belongs to this last category. It is an elegant and natural variant in characteristic 2 of Satoh's one using, in analogy with the complex field, the machinery of theta functions. Mestre generalized it later to the hyperelliptic case and Lercier-Lubicz's implementations of these algorithms are the current records for point counting in characteristic 2 [LL03].
In this paper, we propose a generalization of the AGM-algorithm to the genus 3 non hyperelliptic case, as described in the author's PhD thesis [Rit03]. Fast algorithms for point counting on non hyperelliptic curves exist in the literature. However, as far as we know, the present one is the first which is fast enough for problems of cryptographic size in characteristic 2. The aim of the paper is to give all the details for an implementation. With this end in view, we shall give an algorithm based on Weber's work for the computation of the initial theta constants. This computation, which is fairly easy for hyperelliptic curves, requires some work for this case. The other issue is to find a good lift in order that the computations take place in the field of definition of the lift. It is the main achievement of this paper that it shows how find good lifts. Once this is achieved, we are able to use the optimized iteration process of [LL03]. We shall illustrate the method with an example over $\mathbb{F}_{2^{100}}$.

* C. Ritzenthaler acknowledges the financial support provided through the European Community's Human Potential Programme under contract HPRN-CT-2000-00114, GTEM

D.A. Buell (Ed.): ANTS 2004, LNCS 3076, pp. 379–394, 2004.

1 The Principles of the AGM-algorithms

A fundamental tool in p-adic methods is the canonical lift of an abelian variety. However, the canonical lift exists only for ordinary abelian varieties which form a dense open set of the moduli space.

Definition 1. *Let k be a field of characteristic $p > 0$. We say that an abelian variety A/k of dimension g is* ordinary *if $|A[p](\overline{k})| = p^g$. We say that a curve C/k is ordinary if $\mathrm{Jac}(C)$ is ordinary.*

Let $k = \mathbb{F}_q$ with $q = 2^N$ and let $\mathbb{Q}_q$ be the unramified extension of $\mathbb{Q}_2$ of degree N, $\mathbb{Z}_q$ its ring of integers, π an uniformizer and σ the Frobenius substitution. We consider A/k an ordinary principally polarized abelian variety. The key construction is the canonical lift [Mes72, Chap. V].

Theorem 1. *There exists a unique (up to isomorphism) principally polarized abelian scheme $A^{\uparrow}$ over Spec($\mathbb{Z}_q$) (called the* canonical lift *of A) characterized by the following properties :*

1. *its special fiber is isomorphic to A.*
2. $\mathrm{End}_{\mathbb{Q}_q}(A^{\uparrow}) \simeq \mathrm{End}_k(A)$ *or equivalently there exists an isogeny* $\mathrm{Fr}^{\uparrow} : A^{\uparrow} \to {}^{\sigma}(A^{\uparrow})$ *lifting the little Frobenius* $\mathrm{Fr} : A \to A^{(2)}$.

The idea of AGM-algorithms is to produce sequences of 2-adic elements of which subsequences converge to some 'invariants' associated to $A^{\uparrow}$. The formulas involved in the recurrence are a generalization of the Arithmetic Geometric Mean $((a+b)/2, \sqrt{ab})$ (see Formula (2)). They correspond to 2-isogenies $A_i \to A_{i+1} = A_i/A_i[2]^{\mathrm{loc}}$ where A_0 is a lift of A and $A_i[2]^{\mathrm{loc}}$ are the 2-torsion points in the kernel of the reduction. The convergence of the subsequences is then a consequence of a result of Carls [Car02].
When we are near enough (in 2-adic metric) to $A^{\uparrow}$, the second step is to determine the Frobenius polynomial. One can prove that the determinant of the action of the lift of Frobenius on differentials is linked to a quotient of elements in the sequence (see Step 4 in Section 4). The proof relies on the transformation formula for theta constants [Mum83]. On the other hand, this determinant is related up to a sign to the algebraic integer obtained as the product of the roots of the Frobenius polynomial which are 2-adic units. The last step is then to recover the minimal polynomial of this product (with LLL) and then the Frobenius polynomial.

Let now A be the Jacobian of an ordinary curve C/k. One of the advantages of the AGM method is that the recurrence and the formula to compute the Frobenius are 'universal'. In particular they are the same for hyperelliptic and non hyperelliptic curves of the same genus. Thus the specificities of the computation relies only on the initialization. The initial terms are p-adic algebraic analogues of (quotients of powers of) theta constants and no general formula seems to be known (at least in the non hyperelliptic case). In Section

2, we sum up the results of Riemann and Weber which lead to these formulas for the non hyperelliptic genus 3 case.

For a actual implementation we need to keep all these computations in $\mathbb{Q}_q$. This leads us to find a good model for C and for its lift. In particular, for genus 3 non hyperelliptic case, we ask for the rationality of 28 specific lines (the bitangents of the lift of C). This is carried out in Section 3.

2 Computation of the Initial Theta Constants

In this section, we review some results of Riemann [Rie98] and Weber [Web76] about bitangents and the computation of theta constants. At the end of this section, we add a new result (Prop. 3) that is useful to our algorithm.

Let K be an algebraically closed field of characteristic $p \geq 0$ and C/K a genus three non hyperelliptic curve. We suppose that C is embedded as a plane quartic and we denote x_1, x_2, x_3 (or sometimes x, y, z) the coordinates in the projective plane. As C is canonically embedded, the canonical divisors K_C on C can be described as the intersection of the lines with C. We will center on these divisors, especially on the following ones :

Definition 2. *A line l is called a* bitangent *of C if the intersection divisor $(l \cdot C)$ is of the form $2P + 2Q$ for some not necessarily distinct points P, Q of C.*

Remark 1. If there exists a line l with $(l \cdot C) = 4P$, we call P a *hyperflex*. It is known that a generic quartic does not have any hyperflex.

For the rest of this section, we suppose that the characteristic of K is different from 2 (the characteristic 2 case is postponed to Section 3.1).

We denote $\Sigma = \{L \in \mathrm{Pic}^2(C), L^2 = \mathcal{K}\}$ the set of *theta characteristic bundles* (where $\mathcal{K}$ is the canonical bundle). This set splits into two disjoint subsets Σ_i, $i = 0, 1$, defined by $\Sigma_i = \{L \in \Sigma, h^0(L) := \dim H^0(L) = i\}$.

Proposition 1. *There is a canonical bijection between the set of bitangents and Σ_1.*

Proof. Let l a bitangent. We write $(l \cdot C) = 2(P+Q)$. We define $L = L(P+Q)$. It is obvious that $L \in \Sigma$ and, since C is non hyperelliptic, Riemann-Roch theorem implies that $h^0(L) = 1$. Therefore $L \in \Sigma_1$.
Conversely, if $L \in \Sigma_1$, there exists a regular section $s \in H^0(L)$ different from 0. Let $D = (s)$. Since $L = L(D) \in \mathrm{Pic}^2(C)$, we have $D \sim P + Q$ and since $L \in \Sigma$, $2D \sim K_C$. We conclude that D is the divisor of a bitangent and these two processes are inverse one of each other.

The set Σ is a principal homogeneous space for $\mathrm{Jac}(C)[2]$ and if we choose a symplectic basis (e_i, f_i) of $\mathrm{Jac}(C)[2]$ (for the Weil paring) we can represent each 2-torsion point as $\begin{pmatrix} \varepsilon \\ \varepsilon' \end{pmatrix} = \sum \varepsilon'_i e_i + \sum \varepsilon_i f_i$.
This choice also fixes a $L_0 \in \Sigma$ and so we have an explicit bijection between $\mathrm{Jac}(C)[2]$ and Σ. We denote by L_ϵ the associated theta characteristic bundle with $[\epsilon] = \begin{bmatrix} \varepsilon \\ \varepsilon' \end{bmatrix}$ a 2×3 matrix called a *characteristic* (we refer to [GH01] for a more general setting).
We can also split the set of characteristics into two disjoint subsets. We say that a characteristic $\begin{bmatrix} \varepsilon \\ \varepsilon' \end{bmatrix}$ is even (resp. odd) if the scalar product $\varepsilon \cdot \varepsilon' = 0$ (resp. 1).

Proposition 2 ([GH01]). *The bijection between Σ and the set of characteristics restricts to a bijection between Σ_1 and the set of odd characteristics.*

In particular, counting the number of odd characteristics, we recover the classical result on quartics :

Corollary 1. *A smooth plane quartic has exactly 28 bitangents.*

In order to understand the combinatorics of the bitangents, we introduce the following notion :

Definition 3. *Let $S = ([\epsilon_i])_{i=1\ldots7}$ a subset of characteristics. The subset is called a* principal set *if*

- *every odd characteristic can be written as $[\epsilon_i]$ or $[\epsilon_i] + [\epsilon_j]$, $i \neq j$ and*
- *every even characteristic can be written as $[0]$ or $[\epsilon_i] + [\epsilon_j] + [\epsilon_k]$ with i, j, k distinct.*

We may choose S as

$$[\epsilon_1] = \begin{bmatrix} 0\,0\,1 \\ 1\,0\,1 \end{bmatrix}, [\epsilon_2] = \begin{bmatrix} 0\,1\,1 \\ 1\,1\,0 \end{bmatrix}, [\epsilon_3] = \begin{bmatrix} 0\,1\,0 \\ 1\,1\,1 \end{bmatrix}, [\epsilon_4] = \begin{bmatrix} 1\,1\,1 \\ 0\,0\,1 \end{bmatrix},$$
$$[\epsilon_5] = \begin{bmatrix} 1\,0\,0 \\ 1\,0\,0 \end{bmatrix}, [\epsilon_6] = \begin{bmatrix} 1\,0\,1 \\ 0\,1\,1 \end{bmatrix}, [\epsilon_7] = \begin{bmatrix} 1\,1\,0 \\ 0\,1\,0 \end{bmatrix}$$

We denote β_i the bitangent associated to $[\epsilon_i]$ and β_{ij} the bitangent associated to $[\epsilon_i] + [\epsilon_j]$ by Propositions 1 and 2.

Definition 4. *The set $(\beta_i)_{i=1\ldots7}$ is called an* Aronhold system.

Remark 2. Geometrically, on Aronhold system is characterized by the fact that the points of contact of any three bitangents in the set are never on a conic. There are 288 Aronhold systems for a given quartic (see [Dol03]).

After a linear transformation, we may suppose

$$\begin{cases} \beta_1 : x_1 = 0 & \beta_5 = a_1x_1 + a_2x_2 + a_3x_3 \\ \beta_2 : x_2 = 0 & \beta_6 = a_1'x_1 + a_2'x_2 + a_3'x_3 \\ \beta_3 : x_3 = 0 & \beta_7 = a_1''x_1 + a_2''x_2 + a_3''x_3 \\ \beta_4 : x_1 + x_2 + x_3 = 0 & \end{cases}$$

It was already known to Riemann how to construct a quartic for which the $(\beta_i)_{i=1\ldots7}$ are one of its Aronhold system. Recently, Caporaso and Sernesi in [CS00] and Lehavi in [Leh02] proved that this quartic is uniquely defined by the set (β_i). So, Riemann's construction below shows how to recover the curve from an Aronhold system and Propositions 3 and 4 show how to find an Aronhold system from C. Let us first recall Riemann's construction under this point of view [Rie98].

Theorem 2 (Riemann). *The curve C is isomorphic to the curve (which we call a* Riemann model*)*

$$\sqrt{x_1u_1} + \sqrt{x_2u_2} + \sqrt{x_3u_3} = 0$$

where u_1, u_2, u_3 are given by

$$\begin{cases} u_1 + u_2 + u_3 + x_1 + x_2 + x_3 = 0 \\ \frac{u_1}{a_1} + \frac{u_2}{a_2} + \frac{u_3}{a_3} + ka_1x_1 + ka_2x_2 + ka_3x_3 = 0 \\ \frac{u_1}{a_1'} + \frac{u_2}{a_2'} + \frac{u_3}{a_3'} + k'a_1'x_1 + k'a_2'x_2 + k'a_3'x_3 = 0 \\ \frac{u_1}{a_1''} + \frac{u_2}{a_2''} + \frac{u_3}{a_3''} + k''a_1''x_1 + k''a_2''x_2 + k''a_3''x_3 = 0 \end{cases}$$

with k, k', k'' solutions of

$$\begin{pmatrix} \frac{1}{a_1} & \frac{1}{a_1'} & \frac{1}{a_1''} \\ \frac{1}{a_2} & \frac{1}{a_2'} & \frac{1}{a_2''} \\ \frac{1}{a_3} & \frac{1}{a_3'} & \frac{1}{a_3''} \end{pmatrix} \begin{pmatrix} \lambda \\ \lambda' \\ \lambda'' \end{pmatrix} = \begin{pmatrix} -1 \\ -1 \\ -1 \end{pmatrix}, \begin{pmatrix} \lambda a_1 & \lambda' a_1' & \lambda'' a_1'' \\ \lambda a_2 & \lambda' a_2' & \lambda'' a_2'' \\ \lambda a_3 & \lambda' a_3' & \lambda'' a_3'' \end{pmatrix} \begin{pmatrix} k \\ k' \\ k'' \end{pmatrix} = \begin{pmatrix} 1 \\ -1 \\ -1 \end{pmatrix}.$$

Then we can express all the bitangents :

$$\beta_1 : x_1 = 0 \;\; \beta_2 : x_2 = 0 \;\; \beta_3 : x_3 = 0$$
$$\beta_{23} : u_1 = 0 \; \beta_{13} : u_2 = 0 \; \beta_{12} : u_3 = 0$$

$$\beta_4 : x_1 + x_2 + x_3 = 0 \qquad \beta_5 : a_1x_1 + a_2x_2 + a_3x_3 = 0$$
$$\beta_6 : a_1'x_1 + a_2'x_2 + a_3'x_3 = 0 \; \beta_7 : a_1''x_1 + a_2''x_2 + a_3''x_3 = 0$$

$$\beta_{14} : u_1 + x_2 + x_3 = 0 \qquad \beta_{15} : \frac{u_1}{a_1} + ka_2x_2 + ka_3x_3 = 0$$
$$\beta_{16} : \frac{u_1}{a_1'} + k'a_2'x_2 + k'a_3'x_3 = 0 \; \beta_{17} : \frac{u_1}{a_1''} + k''a_2''x_2 + k''a_3''x_3 = 0$$

$$\beta_{24} : x_1 + u_2 + x_3 = 0 \qquad \beta_{25} : ka_1x_1 + \frac{u_2}{a_2} + ka_3x_3 = 0$$
$$\beta_{26} : k'a_1'x_1 + \frac{u_2}{a_2'} + k'a_3'x_3 = 0 \; \beta_{27} : k''a_1''x_1 + \frac{u_2}{a_2''} + k''a_3''x_3 = 0$$

$$\beta_{34} : x_1 + x_2 + u_3 = 0 \qquad \beta_{35} : ka_1x_1 + ka_2x_2 + \tfrac{u_3}{a_3} = 0$$
$$\beta_{36} : k'a_1'x_1 + k'a_2'x_2 + \tfrac{u_3}{a_3'} = 0 \; \beta_{37} : k''a_1''x_1 + k''a_2''x_2 + \tfrac{u_3}{a_3''} = 0$$

$$\beta_{45} : \frac{u_1}{a_1(1-ka_2a_3)} + \frac{u_2}{a_2(1-ka_3a_1)} + \frac{u_3}{a_3(1-ka_1a_2)} = 0$$
$$\beta_{46} : \frac{u_1}{a_1'(1-k'a_2'a_3')} + \frac{u_2}{a_2'(1-k'a_3'a_1')} + \frac{u_3}{a_3'(1-k'a_1'a_2')} = 0$$
$$\beta_{47} : \frac{u_1}{a_1''(1-ka_2''a_3'')} + \frac{u_2}{a_2''(1-k''a_3''a_1'')} + \frac{u_3}{a_3''(1-k''a_1''a_2'')} = 0$$

$$\beta_{56} : \frac{u_1}{1-k''a_2''a_3''} + \frac{u_2}{1-k''a_3''a_1''} + \frac{u_3}{1-k''a_1''a_2''} = 0$$
$$\beta_{57} : \frac{u_1}{1-k'a_2'a_3'} + \frac{u_2}{1-k'a_3'a_1'} + \frac{u_3}{1-k'a_1'a_2'} = 0$$
$$\beta_{67} : \frac{u_1}{1-ka_2a_3} + \frac{u_2}{1-ka_3a_1} + \frac{u_3}{1-ka_1a_2} = 0$$

Following Riemann, Weber [Web76] performed computations over $\mathbb{C}$ to evaluate algebraically the theta constants and express them in terms of the bitangents. Recall that $\vartheta[\epsilon](0,\Omega)$ is the theta constant defined (here with $g = 3$) by

$$\vartheta\begin{bmatrix}\varepsilon\\ \varepsilon'\end{bmatrix}(0,\Omega) = \sum_{n\in\mathbb{Z}^g} \exp(i\pi(n+\varepsilon/2)\,\Omega\,{}^t(n+\varepsilon/2) + 2i\pi(n+\varepsilon/2)\,{}^t(\varepsilon'/2))$$

where Ω is a Riemann matrix associated to $(\mathrm{Jac}(C),\Theta)$ and to the Aronhold system.

Theorem 3 (Weber). *Let $[\chi] = [\epsilon_i] + [\epsilon_j] + [\epsilon_k]$ an even characteristic. Then*

$$\left(\frac{\vartheta[\chi](0,\Omega)}{\vartheta[0](0,\Omega)}\right)^4 = \frac{|\beta_i,\beta_j,\beta_{ij}||\beta_{ik},\beta_{jk},\beta_{ij}||\beta_j,\beta_{jk},\beta_k||\beta_i,\beta_{ik},\beta_k|}{|\beta_j,\beta_{jk},\beta_{ij}||\beta_i,\beta_{ik},\beta_{ij}||\beta_i,\beta_j,\beta_k||\beta_{ik},\beta_{jk},\beta_k|}$$

where $|\beta_{l_1},\beta_{l_2},\beta_{l_3}| = \det(\beta_{l_1},\beta_{l_2},\beta_{l_3})$.

To complete the understanding of quartics, we have to show how to find an Aronhold system starting from a Riemann model. We give two solutions. The first (Prop. 3) is more useful for theoretical purposes and will be used in Section 3.2. The second (Prop. 4) is more suited for explicit computations.

Proposition 3. *Let $C : \sqrt{x_1u_1} + \sqrt{x_2u_2} + \sqrt{x_3u_3} = 0$ be a smooth quartic. An Aronhold system for the curve is given by seven lines $(x_i)_{i=1\dots7}$, the four last being computed by the following algorithm :*

1. *Let $D_1(\lambda)$ be the determinant of the Hessian of the family of conics*

$$Q_1(\lambda) = \lambda^2(x_2u_3) + \lambda(x_1u_1 - x_2u_2 - x_3u_3) + (x_3u_2).$$

2. *We compute the resultant $R_1(x_1,x_2,x_3) = Res(D_1,Q_1,\lambda)/\lambda$.*
3. *In the same way, we compute $R_2(x_1,x_2,x_3)$, relatively to*

$$Q_2(\lambda) = \lambda^2(x_1u_3) + \lambda(x_2u_2 - x_1u_1 - x_3u_3) + (x_3u_1).$$

4. *Then $R = gcd(R_1,R_2) = \prod_{i=4}^7 x_i$.*

Proof. We only sketch the proof. For details see [Rit03, Sec. 3.5]. If l is a bitangent with intersection divisor $2D$, we denote $\sqrt{l}$ a section of $L(D) \in \Sigma_1$. After a choice of a symplectic basis for $\mathrm{Jac}(C)[2]$, we may assign to each $\sqrt{l}$ a characteristic which we denote $[\sqrt{l}]$. Then, we can find a symplectic basis such that the following decompositions holds (and we say for example that (x_2) and (u_3) are paired relatively to the first following decomposition) :

$$\begin{aligned}[\sqrt{x_1}] &= [\sqrt{x_2}] + [\sqrt{u_3}] = [\sqrt{x_3}] + [\sqrt{u_2}] = [\sqrt{x_4}] + [\sqrt{u_4}] \\ &= [\sqrt{x_5}] + [\sqrt{u_5}] = [\sqrt{x_6}] + [\sqrt{u_6}] = [\sqrt{x_7}] + [\sqrt{u_7}] \\ [\sqrt{x_2}] &= [\sqrt{x_1}] + [\sqrt{u_3}] = [\sqrt{x_3}] + [\sqrt{u_1}] = [\sqrt{x_4}] + [\sqrt{u_4'}] \\ &= [\sqrt{x_5}] + [\sqrt{u_5'}] = [\sqrt{x_6}] + [\sqrt{u_6'}] = [\sqrt{x_7}] + [\sqrt{u_7'}]\end{aligned}$$

We generalize the notion of theta characteristic bundle by considering bundles of degree $2n$ such that $L^2 = \mathcal{K}^n$. For each n, this set is again a principal homogeneous space of $\mathrm{Jac}(C)[2]$ and we can associate to each bundle a unique characteristic $[\epsilon]$. We denote $L_\epsilon^{(n)}$ the corresponding bundle. Riemann-Roch theorem implies that

$$h^0(L_\epsilon^{(n)}) = \begin{cases} 0 & \text{if } n = 1 \text{ and } [\epsilon] \text{ even} \\ 1 & \text{if } n = 1 \text{ and } [\epsilon] \text{ odd} \\ 3 & \text{if } [\epsilon] = [0] \text{ and } n = 2 \\ 2n-2 & \text{otherwise.} \end{cases}$$

In particular, for $n = 2$ and $[\epsilon] \neq [0]$, the dimension of $H^0(L_\epsilon^{(2)})$ is 2. A basis for $L_{\sqrt{x_1}}^{(2)}$ is given by $(\sqrt{x_2u_3}, \sqrt{x_3u_2})$. Hence, every $s \in L_{\sqrt{x_1}}^{(2)}$ may be written as $s = \lambda\sqrt{x_2u_3} + \sqrt{x_3u_2}$. Therefore

$$s^2 = \lambda^2(x_2u_3) + \lambda(x_1u_1 - x_2u_2 - x_3u_3) + (x_3u_2).$$

Thus, the sections s^2 are parametrized by a family of conics which are tangent to C and which degenerate for 6 values of λ (including $\lambda = \infty$) for which s^2 splits into two linear factors which are pair of bitangents with characteristics given by the decomposition of $[\sqrt{x_1}]$. Indeed, it is obvious that the family splits only when the conic is singular i.e. its Hessian matrix has a zero determinant. We can proceed in the same way with $s \in L_{\sqrt{x_2}}^{(2)}$ and we obtain

$$s^2 = \lambda^2(x_1u_3) + \lambda(x_3u_3 - x_2u_2 - x_1u_1) + (x_3u_1).$$

Looking at the decompositions of $[\sqrt{x_1}]$ and $[\sqrt{x_2}]$, the bitangents common to both decompositions are x_3, u_3 (for resp. $\lambda = 0$, $\lambda = \infty$) and $(x_i)_{i=4\ldots7}$. So the algorithm gives the product of the four last.

The second construction is due to Aronhold. We refer to [Sal79] for details. If $u = ax_1 + bx_2 + cx_3$, $v = a'x_1 + b'x_2 + c'x_3$ and $w = a''x_1 + b''x_2 + c''x_3$ are three non concurrent lines, then we form the three conics

$$x_2v - x_3w = 0, \; x_3w - x_1u = 0, \; x_1u - x_2v = 0,$$

which intersect in four points. We consider these intersection points as lines $x_4, \dots, x_7$ in the dual plane. Then $(x_i)_{i=1\dots7}$ form an Aronhold system for the quartic $\sqrt{x_1u_1} + \sqrt{x_2u_2} + \sqrt{x_3u_3} = 0$ with

$$\begin{pmatrix} u_1 \\ u_2 \\ u_3 \end{pmatrix} = \begin{pmatrix} a & b & c \\ a' & b' & c' \\ a'' & b'' & c'' \end{pmatrix}^{\text{adj}} \begin{pmatrix} x_1 \\ x_2 \\ x_3 \end{pmatrix}$$

where M^{adj} is the adjoint matrix.

Proposition 4. *Let $C : \sqrt{x_1u_1} + \sqrt{x_2u_2} + \sqrt{x_3u_3} = 0$ be a smooth quartic. We denote $U_1 = (u_2 \cap u_3)^*$, $U_2 = (u_3 \cap u_1)^*$ and $U_3 = (u_1 \cap u_2)^*$ viewed as lines in the dual plane. Let $(x_i)_{i=4\dots7}$ be the lines viewed in the dual plane as the intersection points of the conics $x_1U_1 - x_2U_2$, $x_2U_2 - x_3U_3$, $x_3U_3 - x_1U_1$. Then the set $(x_i)_{i=1\dots7}$ is an Aronhold system for C.*

3 Model over $K = \overline{\mathbb{F}}_2$ and Lift

3.1 Ordinary Quartics in Characteristic 2

Proposition 5. *A smooth quartic $\tilde{C}/K$ is ordinary if and only if it has 7 bitangents.*

Proof. By [Ser58], we may split $\mathrm{H}^1(\tilde{C}, \mathcal{O}_{\tilde{C}})$ into two parts $V^{\text{nil}} \oplus V^{\text{uni}}$ where the Frobenius F is an isomorphism on V^{uni} and nilpotent on V^{nil}. By Serre duality, the dimension of V^{nil} is equal to the dimension of the space of regular exact differentials (i.e. differentials of the form df, $f \in K(\tilde{C})$). The dimension of V^{uni} is equal to the order of the group of regular logarithmic differentials (i.e. differentials of the form df/f). Moreover this dimension is also the 2-rank $\gamma_{\tilde{C}}$ of $\tilde{C}$.
If $f \in K(\tilde{C})$ then $(df) = 2D_0$ for some divisor D_0 which does not depend on f. We call $L(D_0)$ the canonical theta characteristic bundle. By [SV87, Prop. 3.1] we know that the dimension $h^0(L(D_0))$ equals the dimension of regular exact differentials. So $L(D_0)$ belongs to Σ_1 if and only if $\gamma_{\tilde{C}} \leq 2$.
According to [SV87, Prop. 3.3], there is a bijection between regular logarithmic differentials and non zero non canonical theta characteristic bundles which sends $\omega = (df/f)$ to $L(\omega/2)$. In particular, every non zero non canonical theta characteristic bundle is in Σ_1.
Combining these two results, we see that the number of bitangents is $7, 4, 2$ and 1 when the $\gamma_{\tilde{C}}$ is respectively $3, 2, 1$ and 0. In particular, $\tilde{C}$ is ordinary if and only if $\tilde{C}$ has 7 bitangents.

Thanks to this proposition, it is easy to show [Rit03, Prop. 3.38] :

Proposition 6. *An ordinary quartic is isomorphic over $\overline{\mathbb{F}}_2$ to the model $(*)$*

$$(ax^2 + by^2 + cz^2 + dxy + exz + fyz)^2 = xyz(x + y + z)$$

with

$$abc(a+b+d)(a+c+e)(b+c+f)(a+b+c+d+e+f+1) \neq 0.$$

Conversely, every curve which satisfies this condition is an ordinary non hyperelliptic curve of genus 3.

Remark 3. One may find in [NR03] a complete classification of smooth quartics over a finite field of characteristic 2 using the work of [Wal95].
Note that, if $\tilde{C}$ has all its bitangents defined over k (i.e. all the 2-torsion points of the Jacobian are rational) then $\tilde{C}$ is isomorphic over k to a model $(*)$.

3.2 Good Lift of the Model (*)

Let $k = \mathbb{F}_q$ with $q = 2^N$ and let $\mathbb{Q}_q$ be the unramified extension of degree N of $\mathbb{Q}_2$ with ring of integers $\mathbb{Z}_q$, π a uniformizer and v the valuation such that $v(\pi) = 1$.

In this section, we want to relate the model $(*)$ in characteristic 2 to the Riemann model which is defined in characteristic 0. The reason is that we want to use the result on the bitangents of the Riemann model. Moreover we want that the 28 bitangents are defined over $\mathbb{Q}_q$. Of course, this is not true if we choose the lift arbitrarily. In order to do that, we would like to imitate the elliptic case where $y^2 + yx = f(x)$ is lifted to $Y^2 = (2y+x)^2 = x^2 + 4f(x)$. But it is not clear how to do that on the model $(*)$ and on the Riemann model. The trick is to study a cover of the Riemann model on which it is easy to add 'Artin-Schreier terms' yx.

Let $C : \sqrt{x_1u_1} + \sqrt{x_2u_2} + \sqrt{x_3u_3} = 0$, over a field of characteristic 0. Consider the curve D

$$\begin{cases} Y_1^2 = x_1u_1 \\ Y_2^2 = x_2u_2 \\ Y_3^2 = x_3u_3 \\ Y_1 + Y_2 + Y_3 = 0 \end{cases}$$

The curve D is a genus 5 unramified cover of C. The cover is given by the map $\pi : (Y_1 : Y_2 : Y_3 : x_1 : x_2 : x_3) \mapsto (x_1 : x_2 : x_3)$.

If $\tilde{C}/k : (ax^2 + by^2 + cz^2 + dxy + exz + fyz)^2 - xyz(x+y+z) = 0$ is given, we want to add Artin Schreier terms Y_il_i (with l_i linear in x, y, z) to the left members of the equations above in order that the new curve $\tilde{D}$ is a 2-cover of the curve $\tilde{C}$. This is just the result of some formal computations :

Proposition 7. *$\tilde{C}$ is equal to the quotient of*

$$\begin{cases} Y_1^2 + xY_1 = cx(by + (c+f)z) \\ Y_2^2 + yY_2 = cy(ax + dy + (c+e)z) \\ Y_3^2 + (x+y)Y_3 = c(x+y)(ax + (d+b)y + (1+c+e+f)z) \\ Y_1 + Y_2 + Y_3 = cz \end{cases}$$

by the map $(Y_1 : Y_2 : Y_3 : x : y : z) \mapsto (x : y : z)$

We lift identically the expressions over $\mathbb{Q}_q$ except for the $x+y$ term which is lifted to $-2cz-x-y$. Then we complete the squares on the left and since $x+y+(-2cz-x-y) = -2cz = -2(Y_1+Y_2+Y_3)$, we obtain the genus 5 curve

$$\begin{cases} (2Y_1+l_1)^2 = l_1 \cdot (l_1+4v_1) \\ (2Y_2+l_2)^2 = l_2 \cdot (l_2+4v_2) \\ (2Y_3+l_3)^2 = l_3 \cdot (l_3+4v_3) \\ (2Y_1+l_1)+(2Y_2+l_2) \\ \quad +(2Y_3+l_3) = 0 \end{cases} \text{with} \begin{cases} l_1 = x,\ l_2 = y,\ l_3 = -2cz-x-y \\ v_1 = bcy + c(c+f)z \\ v_2 = acx + dcy + c(c+e)z \\ v_3 = acx + (d+b)cy \\ \quad +c(1+c+e+f)z) \end{cases}$$

The quotient curve is a model for $\tilde{C}$ over $\mathbb{Q}_q$ and by a change of coordinates $x = x_1$, $y = x_2$ and $z = -(x_1+x_2+x_3)/(2c)$, we obtain the curve

$$C/\mathbb{Q}_q : \sqrt{x_1\underbrace{(4v_1+l_1)}_{=u_1}} + \sqrt{x_2\underbrace{(4v_2+l_2)}_{=u_2}} + \sqrt{x_3\underbrace{(4v_3+l_3)}_{=u_3}} = 0 \qquad (1)$$

Theorem 4. *The model (1) has all its bitangents defined over* $\mathbb{Q}_q$.

Proof. It is enough to show that there exists an Aronhold system which can be defined over $\mathbb{Q}_q$. We show this on the model preceding the change of coordinates by using the construction of Proposition 3. We have

$$C : \sqrt{l_1\underbrace{(4v_1+l_1)}_{=u_1}} + \sqrt{l_2\underbrace{(4v_2+l_2)}_{=u_2}} + \sqrt{l_3\underbrace{(4v_3+l_3)}_{=u_3}} = 0$$

Call $(\beta_i)_{i=1\dots7}$ the bitangents of the Aronhold system. We may suppose that $\beta_1 = l_1$, $\beta_2 = l_2$ and $\beta_3 = l_3$. Consider now the family of conics

$$Q_3(\lambda) = l_1u_2\lambda^2 + (l_3u_3 - u_2l_2 - u_1l_1)\lambda + l_2u_1.$$

If we denote $D_3(\lambda)$ the determinant of the Hessian of the family Q_3, we denote $P(\lambda) = D_3(\lambda)/(8\lambda)$. We want to study the roots λ_i of this polynomial. To do that, we let $\mu = \lambda+1$ and $P(\lambda) \equiv c^2\mu^4 \pmod{\pi}$. By building up the Newton polygon of $P(\mu)$, we see that all roots are congruent to 0 modulo π, so $\lambda_i \equiv 1 \pmod{\pi}$. Let $\lambda = -1+\pi\mu$ for some multiple μ. Plugging this into P, we obtain $P(\lambda)/16 = c^2\mu^2(\mu+c)^2 \pmod{\pi}$. Hence two roots are of the form $-1-\pi c \pmod{\pi^2}$ and the two others are of the form $-1 \pmod{\pi^2}$. These two expressions show that

$$Q_3(\lambda_i)/4 = \begin{cases} c^2(y+z)(x+z) & \pmod{\pi} \text{ when } \lambda_i = -1-\pi c \pmod{\pi^2} \\ c^2z(x+y+z) & \pmod{\pi} \text{ when } \lambda_i = -1 \pmod{\pi^2} \end{cases}$$

We do the same computation with the family $Q_1(\lambda) = u_3l_2\lambda^2 + (u_1l_1 - u_2l_2 - u_3l_3)\lambda + l_3u_2$. We obtain the decomposition by pairs

$$Q_1(\lambda_i)/4 = \begin{cases} c^2(x+y+z)(y+z) & \pmod{\pi} \text{ when } \lambda_i = -1-\pi c \pmod{\pi^2} \\ c^2z(x+z) & \pmod{\pi} \text{ when } \lambda_i = -1 \pmod{\pi^2} \end{cases}$$

and at last with the family $Q_2(\lambda) = u_3 l_2 \lambda^2 + (u_1 l_1 - u_2 l_2 - u_3 l_3)\lambda + l_3 u_2$:

$$Q_2(\lambda_i)/4 = \begin{cases} c^2(x+y+z)(x+z) & \pmod{\pi} \text{ when } \lambda_i = -1 - \pi c \pmod{\pi^2} \\ c^2 z(y+z) & \pmod{\pi} \text{ when } \lambda_i = -1 \pmod{\pi^2} \end{cases}$$

The bitangents $\beta_4, \beta_5, \beta_6, \beta_7$ must be in the three families *but not* in the same pair, as it appears on the decompositions in the proof of Proposition 3. Suppose β_4 reduces to $x+y+z$. Is it possible that β_5 also reduces to $x+y+z$? The family Q_3 shows then that neither β_6 nor β_7 reduces to z. The family Q_2 leads to $y+z$ is the reduction of β_6 or β_7. But in this case, Q_1 shows that β_4 and (β_6 or β_7) or β_5 and (β_6 or β_7) are in the same pair : excluded.
Therefore, the bitangents reduce to four distinct lines. The algorithm of Proposition 3 constructs a homogeneous polynomial of degree 4 with coefficients in $\mathbb{Q}_q$ having the $(\beta_i)_{i=4\ldots7}$ as linear factors. Now we know that, modulo π, the polynomial splits into distinct factors. By a version of Hensel's lemma with several variables, we can conclude that the (β_i) are defined over $\mathbb{Q}_q$.

Remark 4. It is easy to find the kernel of the reduction : it is generated by the divisor $((l)-(l'))/2'$ for two bitangents l, l' which reduce to the same bitangent. In particular, if we choose the following reduction for the Aronhold system

β_1	β_2	β_3	β_4	β_5	β_6	β_7
x	y	$x+y$	$x+y+z$	$x+z$	z	$y+z$

and the principal set of the example, then the kernel is generated by

$$[\epsilon_1]+[\epsilon_2]+[\epsilon_3] = \begin{pmatrix} 0\,0\,0 \\ 1\,0\,0 \end{pmatrix};\ [\epsilon_1]+[\epsilon_5]+[\epsilon_6] = \begin{pmatrix} 0\,0\,0 \\ 0\,1\,0 \end{pmatrix};\ [\epsilon_3]+[\epsilon_5]+[\epsilon_7] = \begin{pmatrix} 0\,0\,0 \\ 0\,0\,1 \end{pmatrix}.$$

By [Dol03, th. 6.19], to specify an ordered Aronhold system is equivalent to give a symplectic basis of $\mathrm{Jac}(C)[2]$ for the Weil pairing. Thus, it determines the 'local part' of the AGM sequence (see Section 4 Step 2).

4 The Algorithm

Let $\tilde{C}/k$ be a genus g ordinary curve. We assume that its Jacobian is absolutely simple and has all its 2^g points of order 2 defined over k. We give here the general AGM-algorithm with specific details concerning the non hyperelliptic genus 3 case. Several different proofs and implementations can be found for elliptic curves ([Gau02], [Koh03], [Rit03, Sec. 2.1], [Ver03]). For hyperelliptic curves we refer to [LL03]. For the general case we refer to [Rit03, Part. 3].

1. We lift the curve $\tilde{C}$ to $\mathbb{Q}_q$. For non hyperelliptic curves of genus 3 we consider $\tilde{C}/k$ given by $(*)$ and we lift it as in Section 3.2.

2. We compute (in terms of the coefficients of C) 2^g constants which are square roots of the algebraic expressions of the corresponding complex quotients (for a certain Riemann matrix Ω):

$$\left(\frac{\vartheta\begin{bmatrix}\varepsilon\\ \varepsilon'\end{bmatrix}(0,\Omega)}{\vartheta\begin{bmatrix}0\\ 0\end{bmatrix}(0,\Omega)}\right)^4_{\varepsilon,\varepsilon'\in(\mathbb{Z}/2\mathbb{Z})^g}.$$

Jac(C) is a principally polarized abelian variety. As $\tilde{C}$ is ordinary, one can prove that Jac(C)$[2]^{\mathrm{loc}}$ is a maximal isotropic $\mathbb{F}_2$-vector space for the Weil pairing and therefore it is always possible to find a symplectic basis of Jac(C)[2] such that the 2^g quotients we need to build the sequence are those with $\varepsilon = 0$ (for the genus 3 non hyperelliptic case, we have showed how to find this basis in Remark 4). We denote these quotients by $(\vartheta_e^{(0)})_{e\in(\mathbb{Z}/2\mathbb{Z})^g}$. The computation of $(\vartheta_e^{(0)})_{e\in(\mathbb{Z}/2\mathbb{Z})^3}$ for genus 3 non hyperelliptic curves is given in Section 2. Note that it uses 'ternary invariants' (bitangents) whereas in the hyperelliptic case the computation is based on Thomae formula which involves only 'binary invariants' (Weierstrass points) [Fay73].
3. We build a sequence by using formulas which are a generalization of the Arithmetic Geometric Mean. In complex setting, they are

$$\vartheta\begin{bmatrix}0\\ \varepsilon'\end{bmatrix}(0,2\Omega)^2 = \frac{1}{2^g}\sum_{d\in(\mathbb{Z}/2\mathbb{Z})^g}\vartheta\begin{bmatrix}0\\ \varepsilon'+d\end{bmatrix}(0,\Omega)\cdot\vartheta\begin{bmatrix}0\\ d\end{bmatrix}(0,\Omega).$$

In 2-adic context, these formulas are replaced by

$$\vartheta_e^{(i+1)} = \frac{1}{2^g}\sum_{f\in(\mathbb{Z}/2\mathbb{Z})^g}\vartheta_e^{(i)}\sqrt{\frac{\vartheta_{e+f}^{(i)}}{\vartheta_e^{(i)}}} \qquad (2)$$

The square root of an element $x \in 1 + 8\mathbb{Z}_q$ is chosen to be congruent to 1 modulo 4.
4. Let π_i be the roots of the Frobenius polynomial that are 2-adic units (there are exactly g such roots since $\tilde{C}$ is ordinary). The fundamental proposition of the AGM theory is that the quotient $\vartheta_e^{(n+N)}/\vartheta_e^{(n)}$ converges linearly to $\alpha = \pm\pi_1\ldots\pi_g$.
5. We compute the minimal polynomial of $\beta = \alpha + 2^{gN}/\alpha$. For genus 1 (resp. 2), this polynomial is essentially given by β [Mes02] thus it is enough to know this number with a precision of $3N/2$ (resp. $2N$) bits. This is no longer true for the genus 3 case and Mestre proposed to use LLL to recover this polynomial. We need now to know β with precision $10N$ bits. Then, it is easy to find the characteristic polynomial of Frobenius up to a sign problem which we solve with fast addition in the Jacobian. This last step is illustrated in the following example.

Remark 5. The complexity of the algorithm is the same as in the hyperelliptic case and it depends essentially on the lift phase. Following [LL03], the complexity in time is $O(N^{2\mu}\log N)$ where $O(N^{\mu})$ is the number of bit operations for the multiplication of two N bit length integers. With Karatsuba $\mu = \log_2(3)$, with FFT multiplication $\mu = 1 + \epsilon$. The complexity in space is $O(N^2)$.

5 Example

Let $\tilde{C}$ over $k = \mathbb{F}_q$, $q = 2^N$ with $N = 100$, be defined by

$$(\omega x^2 + (\omega^3+1)y^2 + \omega^2 z^2 + \omega^4 xy + (\omega^3+\omega^2)xzc + \omega^6 yz)^2 - xyz(x+y+z) = 0$$

where ω is a root of $(X^{101}-1)/(X-1)$. The choice of this generator is due to the use of a Gaussian normal basis in [LL03]. Note that cryptographic sizes are about $N = 60$. The following computations are realized on a 731 MHz DEC alpha computer using C libraries and MAGMA 2.10.

We compute the v_1, v_2, v_3 from the proof of Proposition 7 over an unramified extension of degree 100 of $\mathbb{Q}_2$:

$$\begin{cases} v_1 = & (w^5+w^2)y + (w^4+w^8)z \\ v_2 = & w^3x + w^6y + (w^5+\pi w^4)z \\ v_3 = & w^3x + (w^6+w^5+w^2)y + (w^8+w^5+\pi w^4+w^2)z \end{cases}$$

Then we compute the Aronhold system by Proposition 4 and all the bitangents by Theorem 2. Note that Theorem 2 requires to make a change of variables to send the line x_4 of Proposition 4 to $x_1 + x_2 + x_3$. We obtain the initial theta constants $(\vartheta_e^{(0)})_{e\in(\mathbb{Z}/2\mathbb{Z})^3}$ by Theorem 3 (**time** : less than 1 minute)

$$\begin{cases} 1 + O(\pi^5) \\ 1 + (w^{17}+w^{14}+w^{13}+w^9)\pi^4 + O(\pi^5) \\ 1 + (w^{19}+w^{18}+w^{17}+w^{14}+w^{12}+w^{11}+w^{10}+w^5+w^4+w)\pi^4 + O(\pi^5) \\ 1 + (w^{19}+w^{18}+w^{13}+w^{12}+w^{11}+w^{10}+w^9+w^5+w^4+w)\pi^4 + O(\pi^5) \\ 1 + (w^{14}+w^{13}+w^{10}+w^9+w^7+w^6+w^4+w^2)\pi^4 + O(\pi^5) \\ 1 + (w^{17}+w^{10}+w^7+w^6+w^4+w^2)\pi^4 + O(\pi^5) \\ 1 + (w^{19}+w^{18}+w^{17}+w^{13}+w^{12}+w^{11}+w^9+w^7+w^6+w^5+w^2+w)\pi^4 + O(\pi^5) \\ 1 + (w^{19}+w^{18}+w^{14}+w^{12}+w^{11}+w^7+w^6+w^5+w^2+w)\pi^4 + O(\pi^5) \end{cases}$$

We use Lercier-Lubicz's doubling algorithm [LL03] which is shared by the hyperelliptic and non hyperelliptic cases. It is a optimized computation of the formula (2). We find (**time** : 1 minute)

$$\alpha = \frac{\vartheta_1^{(1000)}}{\vartheta_1^{(1100)}} + O(2^{1000}) = \begin{array}{l} 1 + 2^5 + 2^9 + 2^{12} + 2^{13} + 2^{18} + 2^{20} + 2^{22} + 2^{24} \\ +2^{25} + 2^{27} + 2^{29} + 2^{31} + 2^{32} + 2^{34} + 2^{38} + 2^{40} \\ +2^{45} + 2^{47} + 2^{49} + 2^{51} + 2^{53} + 2^{55} + 2^{60} + 2^{61} \\ +2^{62} + 2^{63} + 2^{65} + 2^{67} + 2^{68} + 2^{69} + 2^{71} + 2^{72} \\ +2^{73} + \ldots + 2^{991} + 2^{993} + 2^{995} + 2^{996} + \\ 2^{997} + 2^{999} + O(2^{1000}) \end{array}$$

We compute the minimal polynomial of $\beta = \alpha + 2^{3N}/\alpha$ with an adapted version of LLL (see [LL03] or [Rit03, sec. 4.2] for details, **time** : 2 seconds). It is generically a polynomial of degree 4.

$$\begin{aligned}P_{\text{sym}} = {} & X^4 - 267153656730945219545193916024670843780592 97 \cdot X^3 \\ & -198242842884307905775923471070871362337924115854035368470774 1 \cdot 2^{100} \cdot X^2 + \\ & 248550842369414498721759586691744404695893277906335586587728273970726713469 \cdot 2^{200} \cdot X + \\ & 1750580292573481691132980376309834722405905550149560762854072836850377018127363799426601 61 \cdot 2^{300}\end{aligned}$$

The formulas linking P_{sym} and $\chi_{\tilde{C}}$ are the following : if $\chi_{\tilde{C}} = X^6 - aX^5 + bX^4 - cX^3 + bqX^2 - aq^2X + q^3$ then

$$\begin{aligned}P_{\text{sym}} = {} & X^4 - c_1X^3 + q(b_1^2 - 2a_1c_1 - 2q(a_1^2 - 2b_1))X^2 \\ & -q^2c_1(a_1^2 - 2b_1 - 8q)X + q^3(c_1^2 + qa_1(a_1^3 - 4a_1b_1 + 8c_1))\end{aligned}$$

with $a_1 = a, b_1 = b - 3q, c_1 = c - 2qa$.
Therefore, we recover the Frobenius polynomial up to a sign (i.e. $\chi_{\tilde{C}}(\pm X)$) because α determines $\prod \pi_i$ only up to a sign. This step requires only one fourth root in $\mathbb{Z}$.

$$\begin{aligned}\chi_{\tilde{C}} = {} & X^6 + 377276036264709 \cdot X^5 + 34553510611690458388942279374 03 \cdot X^4 + \\ & 92979302197227669130776666646461687227769187 1 \cdot X^3 \\ & +34553510611690458388942279374 03 \cdot 2^{100} \cdot X^2 \\ & +377276036264709 \cdot 2^{200} \cdot X + 2^{300}.\end{aligned}$$

The determination of the sign can be done by answering the question : $\chi_{\tilde{C}}(1) \cdot D \overset{?}{\sim} 0$ where D is a generic degree 0 divisor. With the algorithms developed in [FOR03], we can prove that the present $\chi_{\tilde{C}}$ has the correct sign in 4 seconds.

Remark 6. With the same ideas, we managed to compute the characteristic polynomial of a quartic over $\mathbb{F}_{2^{5002}}$ in two weeks (the result may be found on `http://www.math.jussieu.fr/~ritzenth`).
We may choose special forms for the model $(*)$ in order that the sign problem is more obvious. For example, the curves

$$C : (ax + by + cz)^4 - xyz(x + y + z) = 0$$

with a, b, c non zero and distinct. On this model, we have hyperflexes P_1, P_2, P_3, P_∞, respectively intersections of C with $x = 0, y = 0, x + y + z = 0$ and $z = 0$. The divisors $D_1 = P_1 - P_\infty$ and $D_2 = P_2 - P_\infty$ are linearly independent (but $P_1 - P_\infty + P_2 - P_\infty + P_3 - P_\infty \sim ((ax + by + cz)/z)$). Now $\text{Jac}(C)[4](\overline{k}) \simeq (\mathbb{Z}/4\mathbb{Z})^3$ and the action of ϕ on $< D_1, D_2 >$ is trivial since they are rational divisors. So at least two roots of χ_C are congruent to 1 modulo 4. If we change the sign at most one would be. So to decide the sign, we have only to check the 2-adic valuations of the roots of $\chi_C(X + 1)$. Note that since these models have a hyperflex, they are $C_{3,4}$ curves for which the algorithms of [FOR03] for addition in the Jacobian are even better.

Conclusion

We have described a generalization of the AGM-algorithm of Mestre for non hyperelliptic curves of genus 3. For higher genus, there are several difficulties : first, it seems to be quite difficult to find in general (i.e. for a non hyperelliptic curve) an algebraic expression of the initial theta constants. Moreover the complexity of the algorithm grows exponentially with the genus and owing to the use of LLL, the constant is quite bad (even for $g = 3$). A first improvement would be to find not only the product of eigenvalues of Frobenius but the whole matrix of the action. A possible approach is maybe to find suitable bases for the invariant differentials of Mumford equations (see [Mum83]).

References

[Car02] R. Carls : Approximation of canonical lifts, available on http://www.math.leidenuniv.nl/ carls/, (2002).

[CS00] L. Caporaso & E. Sernesi : Recovering plane curves from their bitangents, `http://arxiv.org/abs/math.AG/0008239`, (2000).

[Dol03] I. Dolgachev : *topics in classical algebraic geometry.*, Part I, available on `http://www.math.lsa.umich.edu/~idolga/lecturenotes.html`, (2003).

[Fay73] J.D. Fay : *Theta Functions on Riemann Surfaces*, Lecture Notes in Math. **352**, Springer-Verlag, (1973).

[FOR03] S. Flon, R. Oyono & C. Ritzenthaler : Arithmetic on Jacobians of genus 3 quartic curves, work in progress, (2003).

[Gau02] P. Gaudry : A comparison and a combination of SST and AGM algorithms for counting points of elliptic curves in characteristic 2, *Advances in Cryptology*, Asiacrypt'2002, Springer-Verlag, LNCS 2501, 311-327, (2002).

[GH01] B.H. Gross & J. Harris : On some geometric constructions related to theta characteristics, available on `http://abel.math.harvard.edu/~gross/preprints/` (2001).

[Koh03] D. Kohel : The AGM-$X_0(N)$ Heegner point lifting algorithm and elliptic curve point counting, to appear in the Proceedings of Asiacrypt 2003.

[Leh02] D. Lehavi : *Bitangents and two level structure for curves of genus* 3, PhD thesis available on `http://www.ma.huji.ac.il/~dlehavi/` (2002).

[LL03] R. Lercier & D. Lubicz : A quasi quadratic time algorithm for hyperelliptic curve point counting, available on `http://www.medicis.polytechnique.fr/~lercier/preprints/riemann.pdf`, (2003).

[Mes72] W. Messing : *The crystals associated to Barsotti-Tate groups : with Applications to Abelian Schemes*, Lect. Notes in Math., **264**, Berin-Heidelberg-New-York, Springer (1972).

[Mes02] J.-F. Mestre : Algorithmes pour compter des points en petite caractéristique en genre 1 et 2, available on `www.maths.univ-rennes1.fr/crypto/2001-02/mestre.ps` (2002).

[Mum83] D. Mumford : *Tata Lectures on Theta*, Vol **1,2**, Birkhäuser, (1983).

[NR03] E. Nart & C. Ritzenthaler : *Non hyperelliptic curves of genus three over finite fields of characteristic two*, available on `http://front.math.ucdavis/math.NT/0312366`, submitted (2003).

[Rie98] B. Riemann : Sur la théorie des fonctions abéliennes, *Oeuvres de Riemann*, second edition, p. 487, (1898).

[Rit03] C. Ritzenthaler : *Problèmes arithmétiques relatifs à certaines familles de courbes sur les corps finis*, PhD thesis, Université Paris 7 - Denis Diderot, June 2003 available on `http://www.math.jussieu.fr/~ritzenth`.

[Sal79] G. Salmon : *A treatise on the higher plane curves*, third edition, Chelsea, (1879).

[Ser58] J.-P. Serre : Sur la topologie des variétés algébriques en caractéristique p, Symp. Int. Top. Alg., Mexico City, (1958), 24-53 (or *Oeuvres* **1**, 501-530).

[SV87] K.-O. Stöhr & J.-F. Voloch : A formula for the Cartier operator on plane algebraic curves, J. für die Reine und Ang. Math., **377**, (1987), 49-64.

[Ver03] F. Vercauteren : *Computing zeta functions of curves over finite fields*, PhD thesis, Katholieke Universiteit Leuven, November 2003.

[Wal95] C.T.C Wall : Quartic curves in characteristic 2, Math. Proc. Cambridge Phil. Soc. **117**, (1995), 393-414.

[Web76] H. Weber : *Theorie der abelschen Functionen vom Geschlecht 3*, (1876).

Algorithmic Aspects of Cubic Function Fields

R. Scheidler*

Department of Mathematics & Statistics
University of Calgary
2500 University Drive N.W.
Calgary, Alberta T2N 1N4, Canada
rscheidl@math.ucalgary.ca

Abstract. This paper presents an investigative account of arbitrary cubic function fields. We present an elementary classification of the signature of a cubic extension of a rational function field of finite characteristic at least five; the signature can be determined solely from the coefficients of the defining curve. We go on to study such extensions from an algorithmic perspective, presenting efficient arithmetic of reduced ideals in the maximal order as well as algorithms for computing the fundamental unit(s) and the regulator of the extension.

1 Introduction

The arithmetic of algebraic curves over finite fields is a subject of considerable interest, due to its mathematical importance as well as its applications to cryptography. Since general-purpose methods tend to be computationally inefficient, the discussion of fast algorithms has so far predominantly focused on elliptic and hyperelliptic curves. In addition, the arithmetic of purely cubic curves $y^3 = D(x)$ has been investigated in considerable detail [12,10,11,2]; Picard curves represent a special case thereof. Several other particular classes of of curves have also been studied from an algorithmic point of view, such as superelliptic curves (curves of the form $y^n = D(x)$) [5] and C_{ab} curves [1].

In this paper, we investigate arbitrary cubic extensions of a rational function field of finite characteristic. We give a simple technique for finding the signature (and thus the unit rank) of a cubic extension when the characteristic is at least five; the signature can be determined solely from the coefficients of the defining curve. We also investigate efficient arithmetic of reduced fractional ideals in the maximal order of the field and show how to use this arithmetic to find the fundamental unit(s) and the regulator of the extension. Our method is based on a procedure that was originally developed by Voronoi for cubic number fields [14] and was recently adapted to purely cubic function fields [12,6].

* Research supported by NSERC of Canada.

D.A. Buell (Ed.): ANTS 2004, LNCS 3076, pp. 395–410, 2004.

2 Cubic Function Fields and Curves

Let $k = \mathbb{F}_q$ be a finite field of characteristic not equal to 3, and denote by $\overline{k}$ its algebraic closure. Consider an absolutely irreducible nonsingular affine plane curve $\mathcal{C}_0$ defined by an equation $H(x,y) = 0$ where $H \in k[x][Y]$ is a bivariate polynomial of degree 3 in Y which is irreducible over $\overline{k}(x)$; write $H(x,Y) = SY^3 + UY^2 + VY + W$ with $S,U,V,W \in k[x]$, $SW \neq 0$. Then the function field K of $\mathcal{C}_0$ over k is a cubic extension of the rational function field $k(x)$ with minimal polynomial $H(x,Y)$; that is, $K = k(x,y)$.

It is easy to verify that under the transformation $(x,y) \rightarrow (x, S^{-1}(y-U/3))$, $\mathcal{C}_0$ is birationally equivalent to the curve $\mathcal{C}_1 : y^3 - Ay + B = 0$ where

$$A = \frac{U^2}{3} - SV, \quad B = S^2W - \frac{SUV}{3} + \frac{2U^3}{27}.$$

Furthermore, the singular points on $\mathcal{C}_1$ are exactly the points $(a, U(a)/3) \in \overline{k}^2$ where $S(a) = 0$. If Q^2 divides A and Q^3 divides B for some $Q \in k[x]$, then all points of the form $(a,0)$ with $Q(a) = 0$ are singular, and $\mathcal{C}_1$ is birationally equivalent to the curve $y^3 - (A/Q^2)y + (B/Q^3) = 0$. For brevity, we call a (possibly singular) curve $\mathcal{C}$ a *standard model* for $K/k(x)$ if $\mathcal{C}$ is of the form $y^3 - Ay + B = 0$ with $A, B \in k[x]$, $B \neq 0$, and for no $Q \in k[x]$ does Q^2 divide A and Q^3 divide B. We also say that such a curve, and its function field, are in *standard form.* Clearly, every absolutely irreducible nonsingular affine plane curve over k of degree 3 in y is birationally equivalent to a standard model.

A standard model is *purely cubic* if $A = 0$. Note that if $q \equiv 1 \pmod 3$ — this can always be accomplished by adjoining a primitive cube root of unity to k if necessary — then by Kummer theory, a cubic extension $K/k(x)$ has a purely cubic model if and only if it is a Galois extension (see Lemma 2.1 of [6]). If $q \equiv -1 \pmod 3$, it is not clear which cubic extensions over the field $\mathbb{F}_q(x)$ have purely cubic representations.

For a curve $y^3 - Ay + B = 0$ in standard form with function field K, the polynomial $f = f(Y) = Y^3 - AY + B \in k[x][Y]$ is the minimal polynomial of $K/k(x)$. It has three distinct roots $y_0 = y, y_1 = y', y_2 = y''$ in an algebraic extension of $k(x)$ of degree at most 6. For any $\alpha = a + by + cy^2 \in K$ with $a, b, c \in k(x)$, denote by $\alpha' = a + by' + c(y')^2$ and $\alpha'' = a + by'' + c(y'')^2$ the *conjugates* of α. The *norm* of α is $N(\alpha) = \alpha\alpha'\alpha'' \in k(x)$ and the *trace* of α is $Tr(\alpha) = \alpha + \alpha' + \alpha''$; both are rational functions (i.e. in $k(x)$). The *discriminant* of f is the nonzero polynomial $D = (y-y')^2(y'-y'')^2(y''-y)^2 = 4A^3 - 27B^2 \in k[x]$. We recall that if k has odd characteristic, then $K/k(x)$ is a Galois extension if and only if D is a square; in particular, a purely cubic extension is Galois if and only if $q \equiv 1 \pmod 3$.

We have the following simple characterization of singular points:

Lemma 2.1. *Let $\mathcal{C} : y^3 - Ay + B = 0$ be a standard model of a cubic extension $K/k(x)$ where k has characteristic at least 5. Set $D = 4A^3 - 27B^2$, and let $a \in \overline{k}$. Then (a,b) is a singular point of $\mathcal{C}$ for some $b \in \overline{k}$ if and only if $D(a) =$*

$\frac{\partial D}{\partial x}(a) = 0$ *and either* $B(a) \neq 0$ *or* $B(a) = \frac{\partial B}{\partial x}(a) = 0$. *In the latter case, we have* $b = A(a) = \frac{\partial^2 D}{\partial x^2}(a) = 0$.

Proof. $(a, b) \in \overline{k}^2$ is a singular point of $\mathcal{C}$ if and only if

$$b^3 - A(a)b + B(a) = 0, \tag{2.1}$$

$$3b^2 - A(a) = 0, \tag{2.2}$$

$$\frac{\partial A}{\partial x}(a)b - \frac{\partial B}{\partial x}(a) = 0. \tag{2.3}$$

Suppose (2.1) – (2.3) hold, then $A(a) = 3b^2$ and $B(a) = 2b^3$, $D(a) = 0$, and $\frac{\partial D}{\partial x}(a) = 54B(a)\left(\frac{\partial A}{\partial x}(a)b - \frac{\partial B}{\partial x}(a)\right) = 0$. Furthermore, if $B(a) = 0$, then $b = 0$, so $A(a) = 0$. In this case, (2.3) yields $\frac{\partial B}{\partial x}(a) = 0$, and hence $\frac{\partial^2 D}{\partial x^2}(a) = 0$.

Conversely, suppose that $D(a) = \frac{\partial D}{\partial x}(a) = 0$ and either $B(a) \neq 0$ or $B(a) = \frac{\partial B}{\partial x}(a) = 0$. Then $4A(a)^3 = 27B(a)^2$, so there exists $b \in \overline{k}$ with $A(a) = 3b^2$ and $B(a) = 2b^3$. It follows that (2.1) and (2.2) hold. Now $\frac{\partial D}{\partial x}(a) = 0$ implies $B(a)\left(\frac{\partial A}{\partial x}(a)b - \frac{\partial B}{\partial x}(a)\right) = 0$, so if $B(a) \neq 0$, then (2.3) holds, and if $B(a) = \frac{\partial B}{\partial x}(a) = 0$, then $b = 0$ and (2.3) holds as well; furthermore, in the latter case, $A(a) = 0$ and $\frac{\partial^2 D}{\partial x^2}(a) = 0$. □

For $G, P \in k[x]$, let $v_P(G)$ denote the maximal power of P dividing G. By Lemma 2.1, the curve $\mathcal{C}$ is nonsingular if and only if $v_P(D) \geq 2$ implies $v_P(B) = 1$ for every irreducible divisor $P \in k[x]$ of D. This implies the following:

Corollary 2.2. *Let* $\mathcal{C} : y^3 - Ay + B = 0$ *be a standard model of a cubic extension* $K/k(x)$ *where* k *has characteristic at least 5. Set* $D = 4A^3 - 27B^2$. *Then* $\mathcal{C}$ *is nonsingular if and only* $\gcd(D, B)$ *is squarefree.*

If Δ is the discriminant of $K/k(x)$ (unique up to nonzero constant square factors), then there exist $I \in k[x]$ (the *index* or *conductor* of y) such that $D = I^2\Delta$. The curve $\mathcal{C}$ is nonsingular if and only if $I \in k^* = k \setminus \{0\}$, i.e. if and only if D and Δ agree up to a square factor in k. Using a result due to Llorente and Nart (see Theorem 2 of [8]) that is is readily extendable from cubic number fields to their function field analogue, one can easily compute Δ and I from D:

Lemma 2.3. *Let* $\mathcal{C} : y^3 - Ay + B = 0$ *be a standard model of a cubic extension* $K/k(x)$ *where* k *has characteristic different from 3. If* Δ *is the discriminant of* $K/k(x)$ *and* $P \in k[x]$ *is any irreducible divisor of* $D = 4A^3 - 27B^2$, *then*

- $v_P(\Delta) = 2$ *if and only if* $v_P(A) \geq v_P(B) \geq 1$;
- $v_P(\Delta) = 1$ *if and only if* $v_P(D)$ *is odd;*
- $v_P(\Delta) = 0$ *otherwise, i.e. if and only if* $v_P(D)$ *is even and* $v_P(A) = v_P(B) = 0$.

The characterization of the "otherwise" case stems from the condition $v_Q(A) \geq 2$ forcing $v_Q(B) \leq 2$ for all $Q \in k[x]$. The same condition implies in the case where $v_P(\Delta) = 2$ for any $P \mid D$ that $1 \leq v_P(B) \leq 2$. Note also that if $v_P(D)$ is odd, then either $v_P(A) = v_P(B) = 0$ or $1 = v_P(A) < v_P(B)$.

3 Integral Bases

Let $f(x, y) = 0$ with $f = f(Y) = Y^3 - AY + B \in k[x, Y]$ be the standard model of an affine plane curve defining a cubic function field $K = k(x, y)$. As before, let $D = 4A^3 - 27B^2 = I^2\Delta$ where $I \in k[x]$ is the index of y and Δ is the discriminant of $K/k(x)$. The integral closure of $k[x]$ in K is the *ring of regular functions* or *maximal order* of $K/k(x)$ and is denoted by $\mathcal{O}$. $\mathcal{O}$ is a $k[x]$-module of rank 3, and any $k[x]$-basis of $\mathcal{O}$ is called an *integral basis of* $K/k(x)$.

Every nonzero ideal $\mathfrak{a}$ in $\mathcal{O}$ is a $k[x]$-submodule of $\mathcal{O}$ of rank 3; write $\mathfrak{a} = [\lambda, \phi, \psi]$ where $\{\lambda, \phi, \psi\}$ is any $k[x]$-basis of $\mathfrak{a}$. The *norm* $N(\mathfrak{a})$ is the (finite) group index $[\mathcal{O} : \mathfrak{a}]$; it is a nonzero constant multiple of the determinant of the 3 by 3 transformation matrix with polynomial entries that maps any integral basis to any $k[x]$-basis of $\mathfrak{a}$. The *discriminant* of $\mathfrak{a}$ is $\Delta(\mathfrak{a}) = N(\mathfrak{a})^2\Delta$; it is unique up to nonzero constant factors. We have $\Delta(\mathcal{O}) = \Delta$, and since $D([1, y, y^2]) = D = I^2\Delta$, the norm of the ideal $k[x, y] = [1, y, y^2]$ is a constant multiple of I.

Our goal is to find an integral basis of $K/k(x)$ that is suitable for computation. Voronoi (see [4, pp. 108-112]) first proposed how to do this for cubic number fields.

Lemma 3.1. *For any integral basis of* K *of the form* $\{1, \phi, \psi\}$ *where* $\phi = y + S$ *and* $\psi = (y^2 + Ty + U)/I$ *with* $S, T, U \in k[x]$, *we have* $3T^2 - A \equiv 0 \pmod{I}$ *and* $T^3 - AT + B \equiv 0 \pmod{I^2}$.

Proof. Since $\phi\psi, \psi^2 \in \mathcal{O}$, there must exist $r, s, t, u, v, w \in k[x]$ such that $\phi\psi = r\psi + s\phi + t$ and $\psi^2 = u\psi + v\phi + w$. An easy but tedious calculation reveals that $s = (T^2 - U - A)/I$, $u = (T^2 + 2U + A)/I$, and $v = (AT - T^3 - B)/I^2$. So $2s + u = (3T^2 - A)/I \in k[x]$ and $-v = (T^3 - AT + B)/I^2 \in k[x]$. □

It is clear that a basis of the form described in Lemma 3.1 — and hence a polynomial T with $3T^2 - A \equiv 0 \pmod{I}$ and $T^3 - AT + B \equiv 0 \pmod{I^2}$ — always exists.

Corollary 3.2. *Let* $T \in k[x]$ *with* $3T^2 - A \equiv 0 \pmod{I}$ *and* $T^3 - AT + B \equiv 0 \pmod{I^2}$. *Then the set* $\{1, \rho, \omega\}$ *with*

$$\rho = y - T, \quad \omega = \frac{1}{I}(y^2 + Ty + T^2 - A)$$

is an integral basis of $K/k(x)$ *with* $\rho\omega \in k[x]$.

Proof. Let $3T^2 - A = EI$ and $T^3 - AT + B = FI^2$ with $E, F \in k[x]$. We have $\rho^3 + 3T\rho^2 + EI\rho + FI^2 = 0$ and $\omega^3 - E\omega^2 + 3FIT - F^2I = 0$, so ρ and ω are integral over $k[x]$ and hence lie in $\mathcal{O}$. Now $[1, \rho, I\omega] = [1, y, y^2]$, so $I^2\Delta([1, \rho, \omega]) = \Delta([1, \rho, I\omega]) = \Delta([1, y, y^2]) = D = I^2\Delta$ and hence $\Delta([1, \rho, \omega]) = \Delta = \Delta(O)$. It follows that $[1, \rho, \omega] = \mathcal{O}$. Finally, $\rho\omega = -FI \in k[x]$. □

Note that we can always choose T so that $\deg(T) < \deg(I)$, in which case the above basis is polynomially bounded in the size of the coefficients A, B of the standard model. We call a basis of the type given in Corollary 3.2 such that $\deg(T) < \deg(I)$ a *canonical* basis of $K/k(x)$. The following identities are easily verified and show that canonical bases are indeed very suitable for computation. Here, $3T^2 - A = EI$ and $T^3 - AT + B = FI^2$ with $E, F \in k[x]$.

$$\begin{array}{lll} \rho^2 = I\omega - 3T\rho - EI, & Tr(\rho) = -3T, & N(\rho) = -FI^2, \\ \omega^2 = E\omega - F\rho - 3FT, & Tr(\omega) = E, & N(\omega) = F^2 I, \\ \rho\omega = -FI. & & \end{array}$$

Note that when our curve is nonsingular, i.e. $I \in k^*$, then we may take $I = 1$ and $T = 0$, in which case $E = -A$, $F = B$, $\rho = y$, and $\omega = y^2 - A$. If $K/k(x)$ is purely cubic, then $A = 0$, so we may once again take $T = 0$. In this case $E = 0$, so I is the square part and F the squarefree part of B. Here, $\rho = y$ and $\omega = y^2/I$.

4 Signatures

In order to determine the behavior at infinity of a cubic function field extension (which in turn will reveal the signature), we first require some notation and a simple lemma. For any finite field $\mathbb{F}_q$, we denote by $\mathbb{F}_q\langle x^{-1/e}\rangle$ the field of Laurent series in $x^{-1/e}$ ($e \in \mathbb{N}$); note that $k\langle x^{-1}\rangle$ is the completion with respect to the infinite place of $\mathbb{F}_q(x)$. If $\alpha = \sum_{i=-\infty}^{m} a_i x^{i/e}$ is any nonzero element in $\mathbb{F}_q\langle x^{-1/e}\rangle$ with $m \in \mathbb{Z}$, $a_i \in \mathbb{F}_q$ for $i \leq m$, and $a_m \neq 0$, then $\operatorname{sgn}(\alpha) = a_m$ is the *sign*, $\deg(\alpha) = m$ the *degree* (in x^{-e}), and $|\alpha| = q^m = q^{\deg(\alpha)}$ the *absolute value* of α.

The following simple lemma will prove useful.

Lemma 4.1. *Let q be any prime power, p a prime not dividing q, and α a nonzero element in $\mathbb{F}_q\langle x^{-1}\rangle$. If $\deg(\alpha)$ is divisible by p, then α has a p-th root in $\mathbb{F}_q(\operatorname{sgn}(\alpha)^{1/p})\langle x^{-1}\rangle$, otherwise α has a p-th root in $\mathbb{F}_q\langle x^{-1/p}\rangle$, but in no subfield of Laurent series of $\mathbb{F}_q\langle x^{-1/p}\rangle$.*

Proof. Let $\beta = \sum_{i=-\infty}^{n} b_i x^i \in \overline{\mathbb{F}}_q\langle x^{-1}\rangle$. Then $\beta^p = \sum_{i=-\infty}^{pn} c_i x^i$ where $c_{pn} = b_n^p$ and for $i \in \mathbb{N}$, $c_{pn-i} = pb_n^{p-1}b_{n-i} + f_i$ where f_i is a homogeneous polynomial of degree p in $b_{n-i+1}, b_{n-i+2}, \ldots, b_n$ with coefficients in $\mathbb{F}_q$. In particular, if $\beta^p \in \mathbb{F}_q\langle x^{-1}\rangle$, i.e. $c_i \in \mathbb{F}_q$ for $i \leq pn$, then inductively, $b_i \in \mathbb{F}_q(b_n)$ for $i = n, n-1, n-2, \ldots$

Now let $\alpha \in \mathbb{F}_q\langle x^{-1}\rangle$ and write $\deg(\alpha) = pn + r$ with $0 \leq r \leq p-1$. Set $\gamma = x^{-r}\alpha$, so $\gamma \in \mathbb{F}_q\langle x^{-1}\rangle$ with $\deg(\gamma) = pn$. Write $\gamma = \sum_{i=-\infty}^{pn} c_i x^i$. Let b_n be any p-th root of c_{pn} and recursively define $b_{n-i} = (c_{pn-i} - f_i)/pb_n^{p-1} \in \mathbb{F}_q(b_n)$ for $i \in \mathbb{N}$, where f_i is the polynomial in $b_{n-i+1}, b_{n-i+2}, \ldots b_n$ described above. If we set $\beta = \sum_{i=-\infty}^{n} b_i x^i$, then $\beta \in \mathbb{F}_q(b_n)\langle x^{-1}\rangle$ and $\beta^p = \gamma$. Therefore $\alpha = (x^{r/p}\beta)^p$. If $r = 0$, then α has a p-th root in $\mathbb{F}_q(\operatorname{sgn}(\alpha)^{1/p})\langle x^{-1}\rangle$, otherwise the smallest field of Laurent series containing a p-th root of α is $L = \mathbb{F}_q(b_n)\langle x^{-1}\rangle(x^{r/p})$. Since r is coprime to p, we have $L = \mathbb{F}_q\langle x^{-1}\rangle(x^{1/p})$. Clearly $L \subseteq \mathbb{F}_q\langle x^{-1/p}\rangle$, and since both fields are extensions of degree p of $\mathbb{F}_q\langle x^{-1}\rangle$, they must be equal. $\square$

If F is any finite algebraic extension of $\mathbb{F}_q(x)$ of degree n, then the place at infinity of $\mathbb{F}_q(x)$ splits in F as

$$(\infty) = \mathfrak{p}_1^{e_1}\mathfrak{p}_2^{e_2}\cdots\mathfrak{p}_s^{e_s}, \tag{4.1}$$

where $s \in \mathbb{N}$, and for $1 \leq i \leq s$, $\mathfrak{p}_i$ is a place of F of residue degree $f_i \in \mathbb{N}$ and ramification index $e_i \in \mathbb{N}$ with $\sum_{i=1}^{s} e_i f_i = n$. Then the completion of F with respect to the place $\mathfrak{p}_i$ is $F_{\mathfrak{p}_i} = \mathbb{F}_{q^{f_i}}\langle x^{-e_i}\rangle$. If we sort the pairs (e_i, f_i), $1 \leq i \leq s$, in lexicographical order, then the $2s$-tuple $(e_1, f_1, e_2, f_2, \ldots, e_s, f_s)$ is the *signature* of $F/\mathbb{F}_q(x)$.

We are now ready to determine the signature of a cubic extension. Note that transforming such an extension into standard form as described in Section 1 does not affect the signature.

Theorem 4.2. *Let $\mathcal{C} : f(x,y) = 0$ with $f(Y) = Y^3 - AY + B \in k[x][Y]$ be a standard model of a cubic extension $K/k(x)$ where $k = \mathbb{F}_q$ is a finite field of characteristic at least 5. Set $D = 4A^3 - 27B^2$. Then $K/k(x)$ has signature*

- $(1,1,1,1,1,1)$ *if*
  - $|A|^3 > |B|^2$, $\deg(A)$ *even, and* $\operatorname{sgn}(A)$ *is a square in* k, *or*
  - $|A|^3 < |B|^2$, $\deg(B) \equiv 0 \pmod 3$, $\operatorname{sgn}(B)$ *is a cube in* k, *and* $q \equiv 1 \pmod 3$, *or*
  - $|A|^3 = |B|^2$, $4\operatorname{sgn}(A)^3 \neq 27\operatorname{sgn}(B)^2$, *and the equation* $t^3 - \operatorname{sgn}(A)t + \operatorname{sgn}(B) = 0$ *has three roots in* k, *or*
  - $|A|^3 = |B|^2$, $4\operatorname{sgn}(A)^3 = 27\operatorname{sgn}(B)^2$, $\deg(D)$ *is even, and* $\operatorname{sgn}(D)$ *is a square in* k;
- $(1,1,1,2)$ *if*
  - $|A|^3 > |B|^2$, $\deg(A)$ *even, and* $\operatorname{sgn}(A)$ *is not a square in* k, *or*
  - $|A|^3 < |B|^2$, $\deg(B) \equiv 0 \pmod 3$, $\operatorname{sgn}(B)$ *is a cube in* k, *and* $q \equiv -1 \pmod 3$, *or*
  - $|A|^3 = |B|^2$, $4\operatorname{sgn}(A)^3 \neq 27\operatorname{sgn}(B)^2$, *and the equation* $t^3 - \operatorname{sgn}(A)t + \operatorname{sgn}(B) = 0$ *has one root in* k, *or*
  - $|A|^3 = |B|^2$, $4\operatorname{sgn}(A)^3 = 27\operatorname{sgn}(B)^2$, $\deg(D)$ *is even, and* $\operatorname{sgn}(D)$ *is not a square in* k;
- $(1,3)$ *if*
  - $|A|^3 < |B|^2$, $\deg(B) \equiv 0 \pmod 3$, *and* $\operatorname{sgn}(B)$ *is not a cube in* k, *or*
  - $|A|^3 = |B|^2$, $4\operatorname{sgn}(A)^3 \neq 27\operatorname{sgn}(B)^2$, *and the equation* $t^3 - \operatorname{sgn}(A)t + \operatorname{sgn}(B) = 0$ *has no roots in* k;
- $(1,1,2,1)$ *if*
  - $|A|^3 > |B|^2$ *and* $\deg(A)$ *is odd, or*
  - $|A|^3 = |B|^2$ *and* $\deg(D)$ *is odd (so* $4\operatorname{sgn}(A)^3 = 27\operatorname{sgn}(B)^2$*);*
- $(3,1)$ *if* $|A|^3 < |B|^2$ *and* $\deg(B) \not\equiv 0 \pmod 3$.

Proof. Let $l = \max\{\lceil\deg(A)/2\rceil, \lceil\deg(B)/3\rceil\}$ and consider the polynomial $f_\infty(Y) = x^{-3l}f(x, Yx^l) = Y^3 - A(x)x^{-2l}Y + B(x)x^{-3l} \in k[x^{-1}, Y]$. If

$$f_\infty(Y) = P_1(Y)^{e_1}P_2(Y)^{e_2}\cdots P_s(Y)^{e_s}$$

is the factorization of $f_\infty(Y)$ into powers of distinct monic irreducible polynomials in $k\langle x^{-1}\rangle[Y]$, where $s \in \mathbb{N}$, $e_1, e_2, \ldots, e_s \in \mathbb{Z}^{\geq 0}$, and P_i has degree f_i in Y for $1 \leq i \leq s$, then with the proper ordering, $(e_1, f_1, e_2, f_2, \ldots, e_s, f_s)$ is the signature of $K/k(x)$.

Clearly, $\theta = \theta(x)$ is a root of $f_\infty(Y) = 0$ if and only if θx^l is a root of $f(Y) = 0$. Hence, in order to determine the signature of $K/k(x)$, it suffices to find for each zero α of f minimal positive integers e and f such that $\alpha \in \mathbb{F}_{q^f}\langle x^{-e}\rangle$.

If k has characteristic at least 5, then the the zeros $\{y, y', y''\} = \{y_0, y_1, y_2\}$ of f are given by Cardano's formulae

$$y_i = \frac{1}{3}(u^i\delta_+ + u^{-i}\delta_-) \quad (i = 0, 1, 2), \tag{4.2}$$

where u is a primitive cube root of unity and $\delta_+ = y_0 + u^2 y_1 + u y_2$, $\delta_- = y_0 + u y_1 + u^2 y_2$. Here

$$\delta_+ = \sqrt[3]{-\frac{3}{2}\left(9B + \sqrt{-3D}\right)}, \quad \delta_- = \sqrt[3]{-\frac{3}{2}\left(9B - \sqrt{-3D}\right)}, \tag{4.3}$$

where the cube roots are taken so that $\delta_+\delta_- = 3A$ (note that this leaves three choices for the cube root of δ_+, but different choices for this cube root only lead to a different ordering of the roots y_0, y_1, y_2).

For brevity, set $m = \deg(A)$, $n = \deg(B)$, $a = \operatorname{sgn}(a)$, and $b = \operatorname{sgn}(B)$.

Case $|A|^3 < |B|^2$: By Lemma 4.1, $\sqrt{-3D} \in k\langle x^{-1}\rangle$, implying $\delta_+^3, \delta_-^3 \in k\langle x^{-1}\rangle$ and hence again by Lemma 4.1, $\delta_+, \delta_- \in k\langle x^{-1/3}\rangle$. From (4.3), $|\delta_+| = q^{n/3} > |\delta_-|$, $\operatorname{sgn}(\delta_+) = -3b^{1/3}$ for some cube root $b^{1/3}$ of b, and for $i = 0, 1, 2$: $|y_i| = q^{n/3}$ and $\operatorname{sgn}(y_i) = -u^i b^{1/3}$ from (4.2).

If $n \not\equiv 0 \pmod 3$, then $y_i \in k\langle x^{-1/3}\rangle \backslash k\langle x^{-1}\rangle$ for $i = 0, 1, 2$, so $K/k(x)$ has signature $(3, 1)$, whereas if $n \equiv 0 \pmod 3$, then by Lemma 4.1, $y_i \in k(b^{1/3}, u)\langle x^{-1}\rangle$ for $i = 0, 1, 2$, so any ramification index e_i in the signature must be 1. In this case, if b is not a cube in k, then $q \equiv 1 \pmod 3$ (as otherwise, every element in k is a cube), so $u \in k$, $[k(b^{1/3}, u) : k] = 3$, and the signature is $(1, 3)$. On the other hand, if b is a cube in k, then $y_0 \in k\langle x^{-1}\rangle$ and $y_1, y_2 \in k(u)\langle x^{-1}\rangle$. Hence, if $q \equiv 1 \pmod 3$, or equivalently, $u \in k$, then the signature is $(1, 1, 1, 1, 1, 1)$, whereas if $q \equiv -1 \pmod 3$, then $[k(u) : k] = 2$, and since $\operatorname{sgn}(y_1), \operatorname{sgn}(y_2) \in k(u) \setminus k$, the signature must be $(1, 1, 1, 2)$.

Case $|A|^3 > |B|^2$: Here, $\delta_+^3, \delta_-^3 \in k\langle x^{1/2}\rangle$, $|\delta_+|^3, |\delta_-|^3 = q^{3m/2}$ and $\operatorname{sgn}(\delta_+^3) = -\operatorname{sgn}(\delta_-^3) = (-3a)^{3/2}$. By Lemma 4.1, $\delta_+, \delta_- \in k\langle x^{-1/2}\rangle$, so $y_i \in k\langle x^{-1/2}\rangle$ for $i = 0, 1, 2$. Choose the cube root of δ_+ so that $\operatorname{sgn}(\delta_+) = -\operatorname{sgn}(\delta_-) = (-3a)^{1/2}$. Then $|y_1| = |y_2| = q^{m/2} > |y_0|$.

If m is odd, then by Lemma 4.1, $y_1, y_2 \in k\langle x^{-1/2}\rangle \setminus k\langle x^{-1}\rangle$, so at least one of the ramification indices in the signature is 2, forcing signature $(1, 1, 2, 1)$. Suppose now that m is even, then $\delta_+, \delta_- \in k((-3a)^{1/2})\langle x^{-1}\rangle$. Write $\delta_+ = \beta + \gamma s$ with $\beta, \gamma \in k\langle x^{-1}\rangle$ and $s^2 = -3a$. Since $\delta_+^3 = \overline{\delta_-^3} = (\overline{\delta_-})^3$ where the map $\bar{\ } : k(s) \to k(s)$ takes s to $-s$, we have $\delta_- = u^j\overline{\delta_+} = u^j(\beta - \gamma s)$ for some $j \in \{0, 1, 2\}$. Since $3A = \delta_+\delta_- = u^j(\beta^2 - 3a\gamma^2)$, we have $u^j \in k$. If $j = 0$, then it is simple to deduce $y_0 \in k\langle x^{-1}\rangle$ and $y_1, y_2 \in k(a^{1/2})\langle x^{-1}\rangle$. If $j \neq 0$, then we must have $q \equiv 1 \pmod 3$, so $u \in k$, and it is once again easy to see that exactly one among y_0, y_1, y_2 is in $k\langle x^{-1}\rangle$, while the other two are in $k(a^{1/2})\langle x^{-1}\rangle$. In any case, this yields signature $(1,1,1,1,1,1)$ if a is a square in k and $(1,1,1,2)$ otherwise.

Case $|A|^3 = |B|^2$: Then $|A|^3 = |B|^2 = q^{6j}$ for some $j \in \mathbb{N}$.

Assume first that $\deg(D)$ is even. Then δ_+^3 and δ_-^3 are Laurent series in x^{-1} of degree $3j$ with coefficients in $\overline{k}$. By Lemma 4.1, $\delta_+, \delta_- \in \overline{k}\langle x^{-1}\rangle$, so the same holds for y_0, y_1, y_2. Furthermore, at least two among these roots have degree j, and since $|y_0 y_1 y_2| = |B| = q^{3j}$, they all have degree j.

Suppose first that $4a^3 \neq 27b^2$. Let $y \in \{y_0, y_1, y_2\}$ and set $s = \operatorname{sgn}(y)$. Then $s^3 - as + b = 0$. We note that $3s^2 \neq a$, as otherwise $b = as - s^3 = 2s^3$, implying $4a^3 = 27b^2$. For $i \in \mathbb{N}$, let s_{j-i} be the coefficient of x^{j-i} in y. By considering the coefficient of x^{3j-i} in the equation $y^3 - Ay + B = 0$, we see that $s_{j-i} = (3s^2 - a)^{-1} g$ where g is a linear combination of products involving the coefficients of A and B as well as $s_{j-i+1}, s_{j-i+2}, \ldots, s_{j-1}, s$ $(i \in \mathbb{N})$. A simple induction argument thus shows that $s_{j-i} \in k(s)$ for all $i \in \mathbb{N}$. Hence, $y_i \in k(\operatorname{sgn}(y_i))\langle x^{-1}\rangle$ where $\operatorname{sgn}(y_i)$ is a root of the equation $t^3 - at + b = 0$ for $i = 0, 1, 2$. This equation has 0, 1, or 3 distinct roots, yielding respective signatures $(1,3)$, $(1,1,1,2)$, and $(1,1,1,1,1,1)$.

Now suppose that $4a^3 = 27b^2$. Then $a = 3e^2, b = 2e^3$ where $e = 3b/2a \in k^*$, and $\operatorname{sgn}(\delta_+^3) = \operatorname{sgn}(\delta_-^3) = -27e^3$. Let s be a square root of $-3\operatorname{sgn}(D)$ in some suitable extension of k. Then by Lemma 4.1, $\sqrt{-3D} \in k(s)\langle x^{-1}\rangle$, so $\delta_+^3, \delta_-^3 \in k(s)\langle x^{-1}\rangle$. Again by Lemma 4.1, $\delta_+, \delta_- \in k(s)\langle x^{-1}\rangle$. Write $\delta_+ = \beta + \gamma s$ with $b, \gamma \in k\langle x^{-1}\rangle$. Then we reason completely analogous to the case $3m > 2n$, m even, that $K/k(x)$ has signature $(1,1,1,1,1,1)$ if $\operatorname{sgn}(D)$ is a square in k and $(1,1,1,2)$ otherwise.

Assume now that $\deg(D)$ is odd. Then $\sqrt{-3D} \in k\langle x^{-1/2}\rangle \setminus k\langle x^{-1}\rangle$, so $\delta_+^3, \delta_-^3 \in k\langle x^{-1/2}\rangle \setminus k\langle x^{-1}\rangle$, and hence $\delta_+, \delta_- \notin \overline{k}\langle x^{-1}\rangle$. If follows that at least one of the roots does not lie in $\overline{k}\langle x^{-1}\rangle$, so the signature is $(1,1,2,1)$ or $(3,1)$. But for signature $(3,1)$, we have $y_i \in k\langle x^{-1/3}\rangle$ for $i = 0, 1, 2$, so $\delta_+ \in k(y_0, y_1, y_2, u) = k(u)\langle x^{-1/3}\rangle$, and hence $\delta_+^3 \in k(u)\langle x^{-1/3}\rangle \cap k\langle x^{-1/2}\rangle = k\langle x^{-1}\rangle$, which is a contradiction. So $K/k(x)$ must have signature $(1,1,2,1)$ in this case. □

We point out that Lee [7] provided an elegant proof of the above theorem that uses the Hilbert class field of $k(x, \sqrt{\Delta})$, but it is restricted to square-free Δ. One can also apply the transformation $x \to x^{-1}$ and investigate the polynomial $x^{3l} F(Y, x^{-1}) \pmod x$, but this will be inconclusive in certain cases (when $(0,0)$ is a singular point of the resulting curve, i.e. $\deg(A)$ is odd and $\deg(B) \equiv 1 \pmod 3$).

Using the signature description for purely cubic function fields given in Theorem 2.1 of [12] and the well-known characterization of hyperelliptic function fields (see for example Proposition 14.6 on p. 248 of [9]), we can reformulate and summarize Theorem 4.2 as follows.

Corollary 4.3. *Let $C : f(x,y) = 0$ with $f(Y) = Y^3 - AY + B \in k[x][Y]$ be a standard model of a cubic extension $K/k(x)$ where $k = \mathbb{F}_q$ is a finite field of characteristic at least 5. Set $D = 4A^3 - 27B^2$. Then the following holds:*

- *If $|D| \neq |B|^2$ — this is exactly the case if either $|A|^3 > |B|^2$ or $|A|^3 = |B|^2$ and $4\,\mathrm{sgn}(A)^3 = 27\,\mathrm{sgn}(B)^2$ — then the signature of $K/k(x)$ is $(1,1,S)$ where S is the signature of the hyperelliptic extension $k(x)(\sqrt{D})/k(x)$.*
- *If $|D| = |B|^2$, then there are two cases:*
    - *If $|A|^3 < |B|^2$, then the signature of $K/k(x)$ is equal to the signature of the purely cubic extension $k(x)(\sqrt[3]{D})/k(x)$.*
    - *If $|A|^3 = |B|^2$ and $4\,\mathrm{sgn}(A)^3 \neq 27\,\mathrm{sgn}(B)^2$, then $K/k(x)$ is unramified (i.e. all the e_i in the signature of $K/k(x)$ are equal to 1), and the f_i in the signature are the degrees (with respect to the indeterminate t) of the irreducible factors of the equation $t^3 - \mathrm{sgn}(A)\,t + \mathrm{sgn}(B) = 0$ over k.*

5 Unit Group and Regulator

Let $K = k(x,y)$ be a cubic function field of characteristic different from 3 in standard form with minimal polynomial $f(Y) = Y^3 - AY + B \in k[x][Y]$. As before, denote by $y = y_0, y' = y_1, y'' = y_2$ the roots of $f(Y)$ (given by (4.2) if k has odd characteristic). For any $\theta = a + by + cy^2 \in K$, write $\theta^{(i)} = a + by_i + cy_i^2$ for the i-th conjugate of θ $(0 \leq i \leq 2)$. The *unit group* of $K/k(x)$ is the group of units $\mathcal{O}^*$ of the maximal order $\mathcal{O}$ of K. By Dirichlet's Unit Theorem, $\mathcal{O}^*$ is an infinite Abelian group whose torsion part is k^* and whose torsion-free part has rank $s-1$ where s is the number of places at infinity in $K/k(x)$. The quantity $r = s - 1$ is called the *unit rank* of $K/k(x)$. The following table outlines the possible unit rank scenarios for cubic function fields.

Signature	*Unit Rank*
$(1,3)$ or $(3,1)$	0
$(1,1,1,2)$ or $(1,1,2,1)$	1
$(1,1,1,1,1,1)$	2

A set of generators $\{\epsilon_1, \epsilon_2, \ldots, \epsilon_r\}$ of $\mathcal{O}^*/k^*$ is a system of *fundamental units.* Let $\{\mathfrak{p}_1, \mathfrak{p}_2, \ldots \mathfrak{p}_s\}$ be the set of divisors in K lying above the place at infinity in $k(x)$ as described in (4.1). For $1 \leq i \leq s$, let f_i denote the residue degree of $\mathfrak{p}_i$ and ν_i the additive valuation associated with $\mathfrak{p}_i$. Consider the $r \times s$ integer matrix

$$M = \begin{pmatrix} -f_1\nu_1(\epsilon_1) & -f_2\nu_2(\epsilon_1) & \ldots & -f_s\nu_s(\epsilon_1) \\ -f_1\nu_1(\epsilon_2) & -f_2\nu_2(\epsilon_2) & \ldots & -f_s\nu_s(\epsilon_2) \\ \vdots & \vdots & & \vdots \\ -f_1\nu_1(\epsilon_r) & -f_2\nu_2(\epsilon_r) & \ldots & -f_s\nu_s(\epsilon_r) \end{pmatrix}.$$

Rosen [9, p. 245] defines the *regulator* $R_S^{(q)}$ to be the absolute value of the determinant of any of the $r \times r$ minors obtained by deleting the j-th column from M $(1 \leq j \leq s)$; it is easy to show that this definition is independent of the minor and the set of fundamental units chosen. While this definition is consistent with the definition of the regulator for an algebraic number field, Schmidt [13] presents a slightly different definition.

Denote by $\mathcal{S}$ the group generated by $\mathfrak{p}_1, \mathfrak{p}_2, \ldots, \mathfrak{p}_s$, and let $\mathcal{S}_0 = \{\mathfrak{d} \in \mathcal{S} \mid f_{\mathfrak{d}} = 0\}$ where $f_{\mathfrak{d}}$ is the residue degree of $\mathfrak{d}$. If $\mathcal{P}$ denotes the group of principal divisors, then $\mathcal{S}_0 \cap \mathcal{P}$ is generated by the divisors (ϵ_i) of the fundamental units ϵ_i for $1 \le i \le s$. According to Schmidt, the *regulator* of $K/k(x)$ is the group index $R = [\mathcal{S}_0 : \mathcal{P} \cap \mathcal{S}_0]$. By Lemma 4.13 of [9], the two regulators are related via the identity

$$R_S^{(q)} = \frac{f_1 f_2 \cdots f_s}{\gcd(f_1, f_2, \ldots, f_s)} R. \tag{5.1}$$

Furthermore, if h is the class number, i.e. the order of the Jacobian, of K/k, and h' the ideal class number of $K/k(x)$, then by Theorem 25 of [13],

$$h = \frac{R}{\gcd(f_1, f_2, \ldots, f_s)} h' = \frac{R_S^{(q)}}{f_1 f_2 \cdots f_s} h'. \tag{5.2}$$

Specifically, for cubic function fields:

Theorem 5.1. *Let $C : y^3 - Ay + B = 0$ be a standard model of a cubic extension $K/k(x)$. Let h be the order of the Jacobian of K/k, and let h', $R_S^{(q)}$, and R denote the ideal class number, the regulator à la Rosen, and the regulator à la Schmidt, of $K/k(x)$, respectively. Let $\{\epsilon_1, \epsilon_2, \ldots, \epsilon_r\}$ be a system of fundamental units of $K/k(x)$. If $K/k(x)$ has signature*

- *$(1,1,1,1,1,1)$, then $R = R_S^{(q)} = \left| \det \begin{pmatrix} \deg(\epsilon_1^{(i)}) & \deg(\epsilon_2^{(i)}) \\ \deg(\epsilon_1^{(j)}) & \deg(\epsilon_2^{(j)}) \end{pmatrix} \right|$ where $i, j \in \{0, 1, 2\}$ with $i \neq j$, and $h = Rh'$;*
- *$(1,1,1,2)$, then $R = R_S^{(q)}/2 = |\deg(\epsilon_1)|/2$, and $h = Rh'$;*
- *$(1,3)$ then $R = R_S^{(q)}/3 = 1$, and $h = h'/3$;*
- *$(1,1,2,1)$, then $R = R_S^{(q)} = |\deg(\epsilon_1)|$, and $h = Rh'$;*
- *$(3,1)$, then $R = R_S^{(q)} = 1$, and $h = h'$.*

Proof. The relationships between $R_S^{(q)}$ and R as well as h and h' follow from (5.1) and (5.2), respectively. For the rest, we can reason as in the proof of Theorem 2.1 of [12]: in the cases where there is only one (inert or totally ramified) place at infinity in K, $\mathcal{S}_0$ is trivial and hence $R = 1$. For signature (1, 1, 1, 1, 1, 1), the formula for R follows from the fact that the map that permutes the three roots of $f(Y)$ also permutes the three places at infinity. Finally, if there are two places at infinity in K, of respective degrees $f_1 = 1$ and $f_2 = 1$ or 2, then $R = |\nu_1(\epsilon_1)|/f_2$. Since $f_1 = 1$, the completion of K with respect to ν_1 is equal to $k\langle x^{-1}\rangle$, so $|\nu_1(\epsilon_1)| = |\deg(\epsilon_1)|$. □

If $K/k(x)$ has signature $(3,1)$, i.e. the place at infinity in $k(x)$ is totally ramified in K, then the Jacobian of K/k is in fact isomorphic to the ideal class group of $K/k(x)$.

6 Fundamental Units

In order to compute the regulator and/or fundamental unit(s) of a cubic function field of nonzero unit rank, we require the notion of ideal reduction. Once again, we let $K(x,y)$ be a cubic function field of characteristic at least 5 in standard form with minimal polynomial $f(Y) = Y^3 - AY + B \in k[x,Y]$ and unit rank $r > 0$. Possible signatures for $K/k(x)$ are $(1,1,1,1,1,1)$ if $r = 2$ and $(1,1,1,2)$ or $(1,1,2,1)$ if $r = 1$. Let the roots of f be $y_0 = y, y_1 = y', y_2 = y''$. We henceforth write the embedding(s) of K into $k\langle x^{-1}\rangle$ multiplicatively. If $r = 2$, then there are three such embeddings given by three valuations $|\cdot|_i$ $(0 \le i \le 2)$. We write $|\cdot|_0 = |\cdot|$ and number the valuations so that $|y|_i = |y_i| = q^{\deg(y_i)}$, so $|\theta|_i = |\theta^{(i)}|$ for all $\theta \in K$ and $0 \le i \le 2$. If $r = 1$, then there is just one embedding of K into $k\langle x^{-1}\rangle$ which we write as $|\cdot|$. To unify the notation for both unit rank scenarios, we set $|\theta|_0 = |\theta|$, $|\theta|_1 = |\theta|_2 = |\theta'\theta''|^{1/2}$ for all $\theta \in K$ in the unit rank 1 case.

The regulator and fundamental unit(s) of $K/k(x)$ can be computed exactly as described in [12] for the unit rank 1 case and [6] for the case of unit rank 2, so we only give the minimal necessary background here and recall the algorithm for completeness. We focus our discussion on *fractional* ideals of $\mathcal{O}$, i.e. subsets $\mathfrak{f}$ of K such that $d\mathfrak{f}$ is an ideal in $\mathcal{O}$ for some nonzero $d \in k[x]$. In our context, fractional ideals are always nonzero — so they are $k[x]$-submodules of K of rank 3 — and contain 1. A fractional ideal $\mathfrak{f}$ is *reduced* if for any $\theta \in \mathfrak{f}$, the inequalities $|\theta|_i \le 1$ for $i = 0,1,2$ imply $\theta \in k$. Note that $\mathcal{O}$ is reduced.

Let $\{1, \rho, \omega\}$ be a canonical basis of $K/k(x)$. For $\alpha = a + b\rho + c\omega \in K$ with $a, b, c \in k(t)$, we let[1]

$$\begin{aligned}
\zeta_\alpha &= \alpha' + \alpha'' &&= Tr(\alpha) - \alpha &&= (2a - 3bT + cE) - b\rho - c\omega, \\
\xi_\alpha &= \alpha - \frac{1}{3}Tr(\alpha) &&= \frac{1}{3}(2\alpha - \zeta_\alpha) &&= (bT - \frac{1}{3}Ec) + b\rho + c\omega, \\
\eta_\alpha &= \alpha' - \alpha'' && &&= (y' - y'')\left(b - \frac{c}{I}\rho\right),
\end{aligned} \tag{6.1}$$

where E, T, and I (the index of y) are as in Corollary 3.2. Note that ζ_α, ξ_α, $\eta_\alpha/(y' - y'') \in K$.

Let $\{1, \mu, \nu\}$ be a $k[x]$-basis of some non-zero reduced fractional ideal $\mathfrak{f}$ of $\mathcal{O}$. Then it is easy to verify that

$$\det\begin{pmatrix} \xi_\mu & \eta_\mu \\ \xi_\nu & \eta_\nu \end{pmatrix}^2 = (\xi_\mu\eta_\nu - \xi_\nu\eta_\mu)^2 = s\,\Delta(\mathfrak{f}) \tag{6.2}$$

for some $s \in k^*$. For $i \in \{0,1,2\}$, the basis $\{1, \mu, \nu\}$ is said to be *i-reduced* if

$$|\xi_\mu|_i > |\xi_\nu|_i, \quad |\eta_\mu|_i < 1 \le |\eta_\nu|_i, \quad |\zeta_\mu|_i < 1, |\zeta_\nu|_i < 1. \tag{6.3}$$

[1] The definitions of $\zeta_\alpha, \xi_\alpha, \eta_\alpha$ can be modified in such a way that the reduction algorithm given below also works in fields of even characteristic, other than $\mathbb{F}_2$. Since we excluded the characteristic 2 case up to now, we omit the details here.

Such a basis always exists and is unique up to nonzero constant factors (Theorem 4.4 in [6]). Since $|\eta_\mu|_i, |\zeta_\mu|_i < 1$, we have $|\mu'|_i, |\mu''|_i < 1$, so $|\mu| > 1$ since $\mathfrak{f}$ is reduced and $\mu \notin k$. It follows that $|\xi_\mu|_i = |\mu|_i > 1$ and hence from (6.2) and (6.3), $|\Delta(\mathfrak{f})| > 1$ and $|\nu|_i < |\mu|_i \leq |\Delta(\mathfrak{f})|$. Furthermore, $|\nu'|_i = |\nu''|_i \geq 1$.

The following theorem gives the connection between reduced bases and fundamental units. For purely cubic extensions, the relevant discussion of the unit rank 1 case can be found on p. 1255 of [12]; see also Theorem 3.7 in [6] for unit rank 2. The result was proved for arbitrary number fields of unit rank 1 and 2 in [3], and the proofs in that source carry over completely to the function field setting.

Theorem 6.1.

1. *Suppose* $r = 1$ *and set* $\mathfrak{f}_0 = \mathcal{O}, \mathfrak{f}_{n+1} = (\mu_n)^{-1}\mathfrak{f}_n$ *for* $n \geq 1$ *where* $\{1, \mu_n, \nu_n\}$ *is a 0-reduced basis of the reduced fractional ideal* $\mathfrak{f}_n$. *Let* $l \in \mathbb{N}$ *be the minimal index such that* $\mathfrak{f}_l = \mathfrak{f}_0$. *Then*
$$\epsilon = \prod_{i=0}^{l-1} \mu_i$$
is a fundamental unit of $K/k(x)$.
2. *Suppose* $r = 2$ *and set* $\mathfrak{f}_0 = \mathcal{O}, \mathfrak{f}_{n+1} = (\alpha_n^{-1})\mathfrak{f}_n$ *for* $n \geq 0$ *where*
$$\alpha_n = \begin{cases} \mu_n & \text{if } |\nu_n|_1 > 1, \\ \nu_n - \operatorname{sgn}(\nu_n') & \text{if } |\nu_n|_1 = 1, \end{cases}$$
and $\{1, \mu_n, \nu_n\}$ *is a 0-reduced basis of the reduced fractional ideal* $\mathfrak{f}_n$. *Let* $p \in \mathbb{Z}^{\geq 0}$ *and* $l \in \mathbb{N}$ *be minimal such that* $\mathfrak{f}_{p+l} = \mathfrak{f}_p$ *and set*
$$\epsilon_1 = \prod_{i=p}^{p+l-1} \alpha_i .$$
Now set $\mathfrak{g}_0 = \mathfrak{f}_p, \mathfrak{g}_{n+1} = (\beta_n^{-1})\mathfrak{g}_n$ *for* $n \geq 1$ *where*
$$\beta_n = \begin{cases} \sigma_n & \text{if } |\tau_n|_0 > 1, \\ \tau_n - \operatorname{sgn}(\tau_n) & \text{if } |\tau_n|_0 = 1. \end{cases}$$
and $\{1, \sigma_n, \tau_n\}$ *is a 2-reduced basis of the reduced fractional ideal* $\mathfrak{g}_n$. *Let* $m, h \in \mathbb{Z}^{\geq 0}$ *be minimal such that* $\mathfrak{g}_m = \mathfrak{f}_{p+h}$ *and set*
$$\epsilon_2 = \prod_{j=0}^{m-1} \beta_j \left(\prod_{i=p}^{p+h-1} \alpha_i \right)^{-1} .$$
Then $\{\epsilon_1, \epsilon_2\}$ *is a pair of fundamental units of* $K/k(x)$.

To find the regulator instead of the fundamental unit(s), we can avoid evaluating the (computationally expensive) products given above. Instead, we simply sum over the degrees of the μ_i (for $r = 1$), respectively, the $\alpha_i, \alpha_i', \beta_j$, and β_j' (for $r = 2$). Note that ideal equality as required in the above theorem can be tested by comparing appropriately normalized 0-reduced bases.

Theorem 6.1 implies that in order to determine the fundamental unit(s) of $K/k(x)$, we require a way to compute for $i \in \{0, 1, 2\}$ an i-reduced basis of a reduced fractional ideal $\mathfrak{f}$, where $\mathfrak{f}$ is given in terms of a $k[x]$-basis of the form $\{1, \tilde{\mu}, \tilde{\nu}\}$ with

$$\begin{aligned} &\{\tilde{\mu}, \tilde{\nu}\} = \{\rho, \omega\} \text{ or} \\ &\{\tilde{\mu}, \tilde{\nu}\} = \{\mu^{-1}, \nu\mu^{-1}\} \text{ or} \\ &\{\tilde{\mu}, \tilde{\nu}\} = \{\theta^{-1}, \mu\theta^{-1}\} \text{ where } \theta = \nu - \operatorname{sgn}(\nu^{(i+1)}), \end{aligned} \tag{6.4}$$

where the last case only occurs for unit rank 2 and $i = 0$ or $i = 2$ (in the latter case, $i+1$ is taken to be 0). Then the desired reduced bases can be computed using the following algorithm (see also Algorithm 7.1 of [12] with the simplification of Algorithm 6.3 in [10] for $r = 1$, and Algorithm 4.6 of [6] for $r = 2$):

Algorithm 6.2.

Input: $(i, \tilde{\mu}, \tilde{\nu})$ *where* $i \in \{0, 1, 2\}$ *and* $\tilde{\mu}, \tilde{\nu}$ *are given by (6.4).*
Output: (μ, ν) *where* $\{1, \mu, \nu\}$ *is an* i*-reduced basis of* $\mathfrak{f}$.
Algorithm:

1. *Set* $\mu = \tilde{\mu}$, $\nu = \tilde{\nu}$.
2. *If* $|\xi_\mu|_i < |\xi_\nu|_i$ *or if* $|\xi_\mu|_i = |\xi_\nu|_i$ *and* $|\eta_\mu|_i < |\eta_\nu|_i$, *replace* $\begin{pmatrix}\mu\\ \nu\end{pmatrix}$ *by* $\begin{pmatrix}0 & 1\\ -1 & 0\end{pmatrix}\begin{pmatrix}\mu\\ \nu\end{pmatrix}$.
3. *If* $|\eta_\mu|_i \geq |\eta_\nu|_i$ *then*
   3.1. $(r = 2$ *only.) While* $|\xi_\nu\eta_\nu|_i > |\Delta(\mathfrak{f})|^{1/2}$, *replace* $\begin{pmatrix}\mu\\ \nu\end{pmatrix}$ *by* $\begin{pmatrix}0 & 1\\ -1 & \lfloor \xi_{\mu^{(i)}}/\xi_{\nu^{(i)}} \rfloor\end{pmatrix}\begin{pmatrix}\mu\\ \nu\end{pmatrix}$.
   3.2. *Replace* $\begin{pmatrix}\mu\\ \nu\end{pmatrix}$ *by* $\begin{pmatrix}0 & 1\\ -1 & \lfloor \xi_{\mu^{(i)}}/\xi_{\nu^{(i)}} \rfloor\end{pmatrix}\begin{pmatrix}\mu\\ \nu\end{pmatrix}$.
   3.3. *If* $|\eta_\mu|_i = |\eta_\nu|_i$, *replace* $\begin{pmatrix}\mu\\ \nu\end{pmatrix}$ *by* $\begin{pmatrix}1 & -a\\ 0 & 1\end{pmatrix}\begin{pmatrix}\mu\\ \nu\end{pmatrix}$ *where* $a = \operatorname{sgn}(\eta_{\mu^{(i)}})\operatorname{sgn}(\eta_{\nu^{(i)}}^{-1})$.
4. *While* $|\eta_\mu|_i > 1$, *replace* $\begin{pmatrix}\mu\\ \nu\end{pmatrix}$ *by* $\begin{pmatrix}\lfloor \eta_{\nu^{(i)}}/\eta_{\mu^{(i)}} \rfloor & -1\\ 1 & 0\end{pmatrix}\begin{pmatrix}\mu\\ \nu\end{pmatrix}$.
5. *Replace* μ *by* $\mu - \lfloor \zeta_{\mu^{(i)}} \rfloor/2$ *and* ν *by* $\nu - \lfloor \zeta_{\nu^{(i)}} \rfloor/2$.
6. *Return* (μ, ν).

Here, for $\alpha = \sum_{i=-\infty}^{m} a_i x^i \in k\langle x^{-1}\rangle$, the expression $\lfloor \alpha \rfloor = \sum_{i=0}^{m} a_i x^i$ denotes the polynomial part of α.

7 Approximations

In order to compute expressions such as $\lfloor \xi_{\mu^{(i)}}/\xi_{\nu^{(i)}} \rfloor$ or even just $|\eta_\mu|_i = |\eta_{\mu^{(i)}}|$ in Algorithm 6.2, it is necessary to have a sufficiently good approximation for the basis elements ρ, ω (and their conjugates in the unit rank 2 case). These elements lie in $k\langle x^{-1}\rangle$ and thus have an infinite expansion in x^{-1}, of which we can only carry finitely many terms. We in turn require sufficiently good approximations for the root(s) y_0 (and in the unit rank 2 case, y_1 and y_2 as well) of $f(Y) = Y^3 - AY + B$. In purely cubic fields, this can be accomplished by explicitly extracting a cube root of $-B$ to a sufficient precision. However, if the extension is not purely cubic, i.e. $A \neq 0$, we need to proceed differently; in fact, we essentially use Newton's method.

Analogous to [10] and [6], we define for a nonzero element $\alpha \in k\langle x^{-1}\rangle$ of degree m a *relative approximation of precision* $n \in \mathbb{Z}^{\geq 0}$ *to* α to be a truncated Laurent series $\hat{\alpha}$ with $|1 - \hat{\alpha}/\alpha| < q^{-n}$. If $\alpha = \sum_{i=-\infty}^{m} a_i x^i$, then we can set $\hat{\alpha} = \sum_{i=m-n}^{m} a_i x^i$. For purely cubic fields, the analysis in [10] revealed that precision $n = \deg(\Delta)/2$ for relative approximations to the basis elements ρ and ω was sufficient to guarantee that the reduction algorithm (with ρ and ω replaced by their respective approximations) produces correct results. We suspect that the same is true for arbitrary cubic fields, but a more careful investigation of this question is warranted and is the subject of future research.

Theorem 7.1. *Let* $\alpha \in k\langle x^{-1}\rangle$ *be any root of* $f(Y) = Y^3 - AY + B$ $(A, B \in k[x],$ $AB \neq 0)$. *Set* $l = \max\{0, -\deg(A\alpha^{-2} - 3)\} \in \mathbb{Z}^{\geq 0}$, *and let* α_0 *be a relative approximation of precision* l *to* α. *For* $j \in \mathbb{N}$, *define*

$$\alpha_j = \left\lfloor \frac{(2\alpha_{j-1}^3 - B)x^{r_j}}{3\alpha_{j-1}^2 - A} \right\rfloor x^{-r_j} \text{ with } r_j = \max\{0, 2^j - 1 + l - \deg(\alpha)\} \in \mathbb{Z}^{\geq 0}.$$

Then $\left|1 - \dfrac{\alpha}{\alpha_j}\right| \leq q^{-(2^j + l)}$ *for all* $j \geq 0$.

Proof. The claim holds for $j = 0$. For $j \geq 0$, we have

$$|\alpha - \alpha_{j+1}| = \left| \left(\alpha - \frac{2\alpha_j^3 - B}{3\alpha_j^2 - A}\right) + \left(\frac{2\alpha_j^3 - B}{3\alpha_j^2 - A} - \left\lfloor \frac{(2\alpha_j^3 - B)x^{r_{j+1}}}{3\alpha_j^2 - A} \right\rfloor x^{-r_{j+1}}\right)\right|.$$

The expression in the second set of parentheses has absolute value at most $q^{-r_{j+1}-1} \leq q^{-(2^{j+1}+l)}|\alpha|$. For the term in the first set of parentheses, write

$$\begin{aligned} \alpha - \frac{2\alpha_j^3 - B}{3\alpha_j^2 - A} &= (\alpha - \alpha_j) + \left(\alpha_j - \frac{2\alpha_j^3 - B}{3\alpha_j^2 - A}\right) = \alpha - \alpha_j + \frac{\alpha_j^3 - A\alpha_j + B}{3\alpha_j^2 - A} \\ &= \alpha - \alpha_j + \frac{(\alpha_j^3 - A\alpha_j + B) - (\alpha^3 - A\alpha + B)}{3\alpha_j^2 - A} \\ &= (\alpha - \alpha_j)\left(1 - \frac{\alpha_j^2 + \alpha\alpha_j + \alpha^2 - A}{3\alpha_j^2 - A}\right) = -(\alpha - \alpha_j)^2 \frac{\alpha + 2\alpha_j}{3\alpha_j^2 - A}. \end{aligned}$$

Now $|3\alpha_j^2 - A| = |3(\alpha_j^2 - \alpha^2) + 3\alpha^2 - A|$. By induction hypothesis and assumption, $|\alpha_j^2 - \alpha^2| < q^{-l}|\alpha|^2 \le |3\alpha^2 - A|$, so since $|\alpha + 2\alpha_j| = |\alpha|$, again by induction hypothesis,

$$\left|\alpha - \frac{2\alpha_j^3 - B}{3\alpha_j^2 - A}\right| \le (q^{-(2^j+l)})^2|\alpha|^2 \frac{|\alpha|}{|3\alpha^2 - A|} \le q^{-2^{j+1}-2l}|\alpha|^3 \frac{q^l}{|\alpha|^2} = q^{-(2^{j+1}+l)}|\alpha|.$$

□

Lemma 7.2. *Let $f(Y) = Y^3 - AY + B$ $(A, B \in k[x], B \neq 0)$ and let y_0, y_1, y_2 be the zeros of $f(Y)$ with $y_0 \in k\langle x^{-1}\rangle$ ($y_1, y_2 \in k\langle x^{-1}\rangle$ if K has unit rank 2).*

- *If $|A|^3 > |B|^2$, then $|y_0| = |B|/|A|$, $\mathrm{sgn}(y_0) = \mathrm{sgn}(B)/\mathrm{sgn}(A)$, $|y_1| = |y_2| = |A|^{1/2}$ and $\mathrm{sgn}(y_1) = -\mathrm{sgn}(y_2) = \mathrm{sgn}(A)^{1/2}$.*
- *If $|A|^3 < |B|^2$, then $|y_i| = |B|^{1/3}$ and $\mathrm{sgn}(y_i) = -u^i\mathrm{sgn}(B)^{1/3}$ for $i = 0, 1, 2$, where u is a primitive cube root of unity.*
- *If $|A|^3 = |B|^2$, then $|y_i| = |A|^{1/2}$ and the values of $\mathrm{sgn}(y_i)$ are the roots of the equation $t^3 - \mathrm{sgn}(A)t + \mathrm{sgn}(B) = 0$ for $i = 0, 1, 2$. If $4\,\mathrm{sgn}(A)^3 \neq 27\,\mathrm{sgn}(B)^2$, then these roots are distinct, otherwise, the roots are $-2c, c, c$ where $c = 3\,\mathrm{sgn}(B)/2\,\mathrm{sgn}(A)$, so $\mathrm{sgn}(A) = 3c^2$ and $\mathrm{sgn}(B) = 2c^3$.*

Lemma 7.2 shows that the quantity l in Theorem 7.1 is almost always zero, in which case we can determine $\alpha_0 = \mathrm{sgn}(\alpha)x^{\deg(\alpha)}$ from the lemma. In order to obtain a desired precision n for our root approximation, we then simply compute $\alpha_0, \alpha_1, \ldots, \alpha_m$ where $m = \lceil \log_2(n+1) \rceil$. The only problematic case which requires a better initial approximation α_0 to α happens when $|A|^3 = |B|^2$ and $4\,\mathrm{sgn}(A)^3 = 27\,\mathrm{sgn}(B)^2$. The smaller $|A\alpha^{-2} - 3|$ is, the closer our situation resembles a repeated root scenario (as expected), with two roots y_1, y_2 of $f(Y)$ lying close together (and close to one of the square roots of $A/3$ as well as one of the cube roots of $B/2$). Then $4A^3 \approx 27B^2$, i.e. $|D|$ is small as well (note that $|D| = |A|^2|y_1 - y_2|^2$ in this case).

Note that in order to determine $|\eta_\mu|_i$ in step 4 of Algorithm 6.2, we need to compute $|y' - y''|_i$ by (6.1). If $|y' - y''|_i \ge |y'|_i$, this can be done using Lemma 7.2; otherwise, we have $|y' - y''|_i^2 = |\Delta||(y - y')(y - y'')|_i^{-2}$, and the denominator can again be computed using Lemma 7.2.

We will present an implementation of the ideas presented here as well as numerical results in a future paper.

References

1. S. Amrita, Construction of secure C_{ab} curves using modular curves. *Proc. Fourth Algorithmic Number Theory Symp. ANTS-IV, Lect. Notes Comp. Sci.* **1838**, Springer, Berlin 2000, 113-126.
2. M. L. Bauer, The arithmetic of certain cubic function fields. *Math. Comp.* **73** (2004), 387-413.
3. J. A. Buchmann, A generalization of Voronoi's algorithm I, II. *J. Number Theory* **20** (1985), 177-209.

4. B. N. Delone & D. K. Fadeev, *The Theory of Irrationalities of the Third Degree. Transl. Math. Monographs* **10**, Amer. Math. Soc., Providence, Rhode Island 1964.
5. S. D. Galbraith, S. Paulus & N. P. Smart, Arithmetic on superelliptic curves. *Math. Comp.* **71** (2002), 393–405.
6. Y. Lee, R. Scheidler & C. Yarrish, Computation of the fundamental units and the regulator of a cyclic cubic function field. *Exper. Math.* **12** (2003), 211-225.
7. Y. Lee, *The unit rank classification of a general cubic function field by its discriminant.* Preprint.
8. P. Llorente & E. Nart, Effective determination of the decomposition of the rational primes in a cuvic field. *Proc. Amer. Math. Soc.* **87** (1983), 579-585.
9. M. Rosen, *Number Theory in Function Fields.* Springer, New York 2002.
10. R. Scheidler, Reduction in purely cubic function fields of unit rank one. *Proc. Fourth Algorithmic Number Theory Symp. ANTS-IV, Lect. Notes Comp. Sci.* **1838**, Springer, Berlin 2000, 515-532.
11. R. Scheidler Ideal arithmetic and infrastructure in purely cubic function fields. *J. Thèorie Nombr. Bordeaux* **13** (2001), 609-631.
12. R. Scheidler & A. Stein, Voronoi's algorithm in purely cubic congruence function fields of unit rank 1. *Math. Comp.* **69** (2000), 1245-1266.
13. F. K. Schmidt, Analytische Zahlentheorie in Körpern der Charakteristik p. *Math. Zeitschr.* **33** (1931), 1-32.
14. G. F. Voronoi, *On a Generalization of the Algorithm of Continued Fractions* (in Russian). Doctoral Dissertation, Warsaw 1896.

A Binary Recursive Gcd Algorithm

Damien Stehlé and Paul Zimmermann

LORIA/INRIA Lorraine, 615 rue du jardin botanique, BP 101, F-54602
Villers-lès-Nancy, France
{stehle,zimmerma}@loria.fr

Abstract. The binary algorithm is a variant of the Euclidean algorithm that performs well in practice. We present a quasi-linear time recursive algorithm that computes the greatest common divisor of two integers by simulating a slightly modified version of the binary algorithm. The structure of our algorithm is very close to the one of the well-known Knuth-Schönhage fast gcd algorithm; although it does not improve on its $O(M(n)\log n)$ complexity, the description and the proof of correctness are significantly simpler in our case. This leads to a simplification of the implementation and to better running times.

1 Introduction

Gcd computation is a central task in computer algebra, in particular when computing over rational numbers or over modular integers. The well-known Euclidean algorithm solves this problem in time quadratic in the size n of the inputs. This algorithm has been extensively studied and analyzed over the past decades. We refer to the very complete average complexity analysis of Vallée for a large family of gcd algorithms, see [10]. The first quasi-linear algorithm for the integer gcd was proposed by Knuth in 1970, see [4]: he showed how to calculate the gcd of two n-bit integers in time $O(n\log^5 n\log\log n)$. The complexity of this algorithm was improved by Schönhage [6] to $O(n\log^2 n\log\log n)$. A comprehensive description of the Knuth-Schönhage algorithm can be found in [12]. The correctness of this algorithm is quite hard to establish, essentially because of the technical details around the so-called “fix-up procedure”, and a formal proof is by far out of reach. As an example, several mistakes can be noticed in the proof of [12] and can be found at `http://www.cs.nyu.edu/cs/faculty/yap/book/errata.html`. This “fix-up procedure” is a tedious case analysis and is quite difficult to implement. This usually makes the implementations of this algorithm uninteresting except for very large numbers (of length significantly higher than 10^5 bits).

In this paper, we present a variant of the Knuth-Schönhage algorithm that does not have the “fix-up procedure” drawback. To achieve this, we introduce a new division (GB for Generalized Binary), which can be seen as a natural generalization of the binary division and which has some natural meaning over the 2-adic integers. It does not seem possible to use the binary division itself in a Knuth-Schönhage-like algorithm, because its definition is asymmetric: it eliminates least significant bits but it also considers most significant bits to perform

D.A. Buell (Ed.): ANTS 2004, LNCS 3076, pp. 411–425, 2004.

comparisons. There is no such asymmetry in the GB division. The recursive GB Euclidean algorithm is much simpler to describe and to prove than the Knuth-Schönhage algorithm, while admitting the same asymptotic complexity.

This simplification of the description turns out to be an important advantage in practice: we implemented the algorithm in GNU MP, and it ran between three and four times faster than the implementations of the Knuth-Schönhage algorithm in Magma and Mathematica.

The rest of the paper is organized as follows. In §2 we introduce the GB division and give some of its basic properties. In §3, we describe precisely the new recursive algorithm. We prove its correctness in §4 and analyze its complexity in §5. Some implementation issues are discussed in §6.

Notations: The standard notation $O(.)$ is used. The complexity is measured in elementary operations on bits. Unless specified explicitly, all the logarithms are taken in base 2. If a is a non-zero integer, $\ell(a)$ denotes the length of the binary representation of a, i.e. $\ell(a) = \lfloor \log |a| \rfloor + 1$; $\nu_2(a)$ denotes the 2-adic valuation of a, i.e. the number of consecutive zeroes in the least significant bits of the binary representation of a; by definition, $\nu_2(0) = \infty$. $r := a \text{ cmod } b$ denotes the centered remainder of a modulo b, i.e. $a = r \bmod b$ and $-\frac{b}{2} < r \leq \frac{b}{2}$. We recall that $M(n) = \Theta(n \log n \log \log n)$ is the asymptotic time required to multiply two n-bit integers with Schönhage-Strassen multiplication [7]. We assume the reader is familiar with basic arithmetic operations such as fast multiplication and fast division based on Newton's iteration. We refer to [11] for a complete description of these algorithms.

2 The Generalized Binary Division

In this section we first recall the binary algorithm. Then we define the generalized binary division — GB division for short — and give some basic properties about it. Subsection 2.4 explains how to compute modular inverses from the output of the Euclidean algorithm based on the GB division.

2.1 The Binary Euclidean Algorithm

The binary division is based on the following properties: $\gcd(2a, 2b) = 2\gcd(a, b)$, $\gcd(2a+1, 2b) = \gcd(2a+1, b)$, and $\gcd(2a+1, 2b+1) = \gcd(2b+1, a-b)$. It consists of eliminating the least significant bit at each loop iteration. Fig. 1 is a description of the binary algorithm. The behavior of this algorithm is very well understood (see [1] and the references there). Although it is still quadratic in the size of the inputs, there is a significant gain over the usual Euclidean algorithm, in particular because there is no need to compute any quotient.

2.2 The Generalized Binary Division

In the case of the standard Euclidean division of a by b with $|a| > |b|$, one computes a quotient q such that when qb is added to a, the obtained remainder

Algorithm Binary-Gcd.
Input: $a, b \in \mathbb{Z}$.
Output: $\gcd(a, b)$.

1. If $|b| > |a|$, return Binary-Gcd(b, a).
2. If $b = 0$, return a.
3. If a and b are both even then return $2 \cdot$ Binary-Gcd$(a/2, b/2)$.
4. If a is even and b is odd then return Binary-Gcd$(a/2, b)$.
5. If a is odd and b is even then return Binary-Gcd$(a, b/2)$.
6. Otherwise return Binary-Gcd$((|a| - |b|)/2, b)$.

Fig. 1. The binary Euclidean algorithm.

is smaller than b. Roughly speaking, left shifts of b are subtracted from a as long as possible, that is to say until a has lost its $\ell(a) - \ell(b)$ most significant bits (approximately). The GB division is the dual: in order to GB-divide a by b with $|a|_2 > |b|_2$, where $|a|_2 := 2^{-\nu_2(a)}$ is the 2-adic norm of a, one computes a quotient $\frac{q}{2^k}$ such that when $\frac{q}{2^k}b$ is added to a, the obtained remainder is smaller than b for the 2-adic norm. Roughly speaking, right shifts of b are subtracted from a as long as possible, that is to say until a has lost its $\nu_2(b) - \nu_2(a)$ least significant bits.

Lemma 1 (GB Division). *Let a, b be non-zero integers with $\nu_2(a) < \nu_2(b)$. Then there exists a unique pair of integers (q, r) such that:*

$$r = a + q\frac{b}{2^{\nu_2(b)-\nu_2(a)}}, \tag{1}$$

$$|q| < 2^{\nu_2(b)-\nu_2(a)}, \tag{2}$$

$$\nu_2(r) > \nu_2(b). \tag{3}$$

The integers q and r are called respectively the GB quotient and the GB remainder of (a, b). We define GB(a, b) as the pair (q, r).

Proof. From (1), $q = -\frac{a}{2^{\nu_2(a)}} \cdot (\frac{b}{2^{\nu_2(b)}})^{-1} \bmod 2^{\nu_2(b)-\nu_2(a)+1}$. Since q is odd, the second condition is fulfilled and gives the uniqueness of q. As a consequence, r is uniquely defined by (1). Moreover, since $r = a + q\frac{b}{2^{\nu_2(b)-\nu_2(a)}}$, we have $r = 0 \bmod 2^{\nu_2(b)+1}$, which gives condition (3).

The GB division resembles Hensel's odd division, which was introduced by Hensel around 1900. A description can be found in [8]. For two integers a and b with b odd, it computes q and r in $\mathbb{Z}$, such that: $a = -bq + 2^p r$ and $r < 2b$, where $p = \ell(a) - \ell(b)$ is the difference between the bit lengths of a and b. In other words, Hensel's odd division computes $r := (2^{-p}a) \bmod b$, which may be found efficiently as shown in [5]. Besides, there is also some similarity with the PM algorithm of Brent and Kung [2]. When $\nu_2(a) = 0$ and $\nu_2(b) = 1$, i.e. a is odd and $b = 2b'$ with b' odd, the GB division finds $q = \pm 1$ such that $(a + qb')/2$ is even, which is exactly the PM algorithm. Unlike the binary, Hensel and PM algorithms, the GB division considers only low order bits of a and b: there is no need comparing a and b nor computing their bit lengths. It can be seen as a natural operation when considering a and b as 2-adic integers.

We now give two algorithms to perform the GB division. The first one is the equivalent of the naive division algorithm, and is asymptotically slower than the second one, which is the equivalent of the fast division algorithm. For most of the input pairs (a, b), the Euclidean algorithm based on the GB division performs almost all its divisions on pairs (c, d) for which $\nu_2(d) - \nu_2(c)$ is small. For this reason the first algorithm suffices in practice.

Algorithm Elementary-GB.
Input: Two integers a, b satisfying $\nu_2(a) < \nu_2(b) < \infty$.
Output: $(q, r) = GB(a, b)$.

1. $q := 0$, $r := a$.
2. While $\nu_2(r) \leq \nu_2(b)$ do
3. $\quad q := q - 2^{\nu_2(r)-\nu_2(a)}$,
4. $\quad r := r - 2^{\nu_2(r)-\nu_2(b)} b$.
5. $q := q \text{ cmod } 2^{\nu_2(b)-\nu_2(a)+1}$, $r := q \frac{b}{2^{\nu_2(b)-\nu_2(a)}} + a$.
6. Return (q, r).

Fig. 2. Algorithm **Elementary-GB**.

Lemma 2. *The algorithm* **Elementary-GB** *of Fig. 2 is correct and if the input* (a, b) *satisfies* $\ell(a), \ell(b) \leq n$, *then it finishes in time* $O(n \cdot [\nu_2(b) - \nu_2(a)])$.

It is also possible to compute $GB(a, b)$ in quasi-linear time, in the case of Schönhage-Strassen multiplication, by using Hensel's lifting (which is the p-adic dual of Newton's iteration).

Lemma 3. *The algorithm* **Fast-GB** *of Fig. 3 is correct and with Schönhage-Strassen multiplication, if* a *and* b *satisfy the conditions* $\ell(a), \ell(b) \leq 2n$ *and* $\nu_2(b) - \nu_2(a) \leq n$, *then it finishes in time* $O(M(n))$.

2.3 The GB Euclidean Algorithm

A GB Euclidean algorithm can be derived very naturally from the definition of the GB division, see Fig. 4.

Lemma 4. *The GB Euclidean algorithm of Fig. 4 is correct, and if we use the algorithm* **Elementary-GB** *of Fig. 2, then for any input* (a, b) *satisfying* $\ell(a), \ell(b) \leq n$, *it finishes in time* $O(n^2)$.

Algorithm Fast-GB.
Input: Two integers a, b satisfying $\nu_2(a) < \nu_2(b) < \infty$.
Output: $(q, r) = GB(a, b)$.

1. $A := -\frac{a}{2^{\nu_2(a)}}$, $B := \frac{b}{2^{\nu_2(b)}}$, $n := \nu_2(b) - \nu_2(a) + 1$.
2. $q := 1$.
3. For i from 1 to $\lceil \log n \rceil$ do
4. $\quad q := q + q(1 - Bq) \bmod 2^{2^i}$.
5. $q := Aq \operatorname{cmod} 2^n$.
6. $r := a + \frac{q}{2^{n-1}} b$.
7. Return (q, r).

Fig. 3. Algorithm **Fast-GB**.

Algorithm GB-gcd.
Input: Two integers a, b satisfying $\nu_2(a) < \nu_2(b)$.
Output: The odd part $\frac{g}{2^{\nu_2(g)}}$ of the greatest common divisor g of a and b.

1. If $b = 0$, return $\frac{a}{2^{\nu_2(a)}}$.
2. $(q, r) := GB(a, b)$.
3. Return GB-gcd (b, r).

Fig. 4. The GB Euclidean algorithm.

Proof. Let $r_0 = a, r_1 = b, r_2, \ldots$ be the sequence of remainders that appear in the execution of the algorithm. We first show that this sequence is finite and thus that the algorithm terminates.

For any $k \geq 0$, Eqs. (1) and (2) give $|r_{k+2}| \leq |r_{k+1}| + |r_k|$, so that $|r_k| \leq 2^{n+1}\left(\frac{1+\sqrt{5}}{2}\right)^k$. Moreover, 2^k divides $|r_k|$, which gives $2^k \leq |r_k| \leq 2^{n+1}\left(\frac{1+\sqrt{5}}{2}\right)^k$ and $(1 - \log \frac{1+\sqrt{5}}{2})k \leq n+1$. Therefore there are $O(n)$ remainders in the remainder sequence. Let $t = O(n)$ be the length of the remainder sequence. Suppose that r_t is the last non-zero remainder. From Lemma 2, we know that each of the calls to a GB-division involves a number of bit operations bounded by $O(\log |r_k| \cdot [\nu_2(r_{k+1}) - \nu_2(r_k)]) = O(n \cdot [\nu_2(r_{k+1}) - \nu_2(r_k)])$, so that the overall complexity is bounded by $O(n \cdot \nu_2(r_t))$.

For the correctness, remark that the GB remainder r from a and b satisfies $r = a \bmod b'$ where $b' = \frac{b}{2^{\nu_2(b)}}$ is the odd part of b, thus $\gcd(a, b') = \gcd(r, b')$.

Remark. For a practical implementation, one should remove factors of two in Algorithm GB-gcd. If one replaces the return value in Step 1 by a, and in Step 3 by $2^{\nu_2(a)}\text{GB-gcd}(\frac{b}{2^{\nu_2(b)}}, \frac{r}{2^{\nu_2(b)}})$, the algorithm directly computes $g = \gcd(a, b)$.

A better bound on $|r_k|$ and thus on t is proved in §5.1. Nonetheless the present bound is sufficient to guarantee the quasi-linear time complexity of the recursive algorithm. The improved bound only decreases the multiplying constant of the asymptotic complexity.

For $n \geq 1$ and $q \in [-2^n + 1, 2^n - 1]$ we define the matrix $[q]_n = \begin{pmatrix} 0 & 2^n \\ 2^n & q \end{pmatrix}$. Let r_0, r_1 be two non-zero integers with $0 = \nu_2(r_0) < \nu_2(r_1)$, and $r_0, r_1, r_2, \ldots$ be their GB remainder sequence, and $q_1, q_2, \ldots$ be the corresponding quotients: $r_{i+1} = r_{i-1} + q_i \frac{r_i}{2^{\nu_2(r_i) - \nu_2(r_{i-1})}}$ for any $i \geq 1$. Then the following relation holds for any $i \geq 1$:

$$\begin{pmatrix} r_i \\ r_{i+1} \end{pmatrix} = \frac{1}{2^{\nu_2(r_i)}} [q_i]_{n_i} \cdots [q_1]_{n_1} \cdot \begin{pmatrix} r_0 \\ r_1 \end{pmatrix},$$

where $n_j = \nu_2(r_j) - \nu_2(r_{j-1}) \geq 1$ for any $j \geq 1$.

In what follows, we use implicitly the following simple fact several times.

Lemma 5. *Let r_0, r_1 be two non-zero integers with $0 = \nu_2(r_0) < \nu_2(r_1)$, and $r_0, r_1, r_2, \ldots$ be their GB remainder sequence. Let $d \geq 0$. Then there exists a unique $i \geq 0$ such that $\nu_2(r_i) \leq d < \nu_2(r_{i+1})$.*

2.4 Computing Modular Inverses

This subsection is independent of the remainder of the paper but is justified by the fact that computing modular inverses is a standard application of the Euclidean algorithm.

Let a, b be two non-zero integers with $0 = \nu_2(a) < \nu_2(b)$ and $\ell(a), \ell(b) \leq n$. Suppose that we want to compute the inverse of b modulo a, by using an extended version of the Euclidean algorithm based on the GB division. The execution of the extended GB Euclidean algorithm gives two integers A and B such that $Aa + Bb = 2^\alpha g$, where $\alpha = O(n)$ and $g = \gcd(a, b)$. From such a relation, it is easy to check that $g = 1$. Suppose now that the inverse B' of b modulo a does exist. From the relation $Aa + Bb = 2^\alpha$, we know that:

$$B' = \frac{B}{2^\alpha} \bmod a.$$

Therefore, in order to obtain B', it is sufficient to compute the inverse of 2^α modulo a. By using Hensel's lifting (like in the algorithm **Fast-GB** of Fig. 3), we obtain the inverse of a modulo 2^α. This gives x and y satisfying: $xa + y2^\alpha = 1$. Clearly y is the inverse of 2^α modulo a.

Since multiplication, Hensel's lifting and division on numbers of size $O(n)$ can be performed in time $O(M(n))$ (see [11]), given two n-bit integers a and b, the additional cost to compute the inverse of b modulo a given the output of an extended Euclidean algorithm based on the GB division is $O(M(n))$.

3 The Recursive Algorithm

We now describe the recursive algorithm based on the GB division. This description closely resembles the one of [12]. It uses two routines: the algorithm **Half-GB-gcd** and the algorithm **Fast-GB-gcd**. Given two non-zero integers r_0 and r_1 with $0 = \nu_2(r_0) < \nu_2(r_1)$, the algorithm **Half-GB-gcd** outputs the GB remainders r_i and r_{i+1} of the GB remainder sequence of (r_0, r_1) that satisfy $\nu_2(r_i) \leq \ell(r_0)/2 < \nu_2(r_{i+1})$. It also outputs the corresponding matrix $2^{-(n_1+\ldots+n_i)}[q_i]_{n_i} \cdots [q_1]_{n_1}$. Then we describe the algorithm **Fast-GB-gcd**, which, given two integers a and b, outputs the gcd of a and b by making successive calls to the algorithm **Half-GB-gcd**.

The algorithm **Half-GB-gcd** works as follows: a quarter of the least significant bits of a and b are eliminated by doing a recursive call on the low $\ell(a)/2$ of the bits of a and b. The crucial point is that the GB quotients computed for the truncated numbers are exactly the same as the first GB quotients of a and b. Therefore, by multiplying a and b by the matrix obtained recursively one gets two remainders (a', b') of the GB remainder sequence of (a, b). A single step of the GB Euclidean algorithm is performed on (a', b'), which gives a new remainder pair (b', r). Then there is a second recursive call on approximately $\ell(a)/2$ of the least significant bits of (b', r). The size of the inputs of this second recursive call is similar to the one of the first recursive call. Finally, the corresponding remainders (c, d) of the GB remainder sequence of (a, b) are computed using the returned matrix R_2, and the output matrix R is calculated from R_1, R_2 and the GB quotient of (a', b'). Fig. 5 illustrates the execution of this algorithm.

Note that in the description of the algorithm **Half-GB-gcd** in Fig. 6, a routine GB' is used. This is a simple modification of the GB division: given a and b as input with $0 = \nu_2(a) < \nu_2(b)$, it outputs their GB quotient q, and $\frac{r}{2^{\nu_2(b)}}$ if r is their GB remainder. The algorithm **Fast-GB-gcd** uses several times the algorithm **Half-GB-gcd** to decrease the lengths of the remainders quickly.

The main advantage over the other quasi-linear time algorithms for the integer gcd is that if a matrix R is returned by a recursive call of the algorithm **Half-GB-gcd**, then it contains only "correct quotients". There is no need to go back in the GB remainder sequence in order to make the quotients correct, and thus no need to store the sequence of quotients. The underlying reason is that the remainders are shortened by the least significant bits, and since the carries go in the direction of the most significant bits, these two phenomena do not interfere. For that reason, the algorithm is as simple as the Knuth-Schönhage algorithm in the case of polynomials.

4 Correctness of the Recursive Algorithm

In this section, we show that the algorithm **Fast-GB-gcd** of Fig. 7 is correct. We first give some results about the GB division, and then we show the correctness of the algorithm **Half-GB-gcd** which clearly implies the correctness of the algorithm **Fast-GB-gcd**.

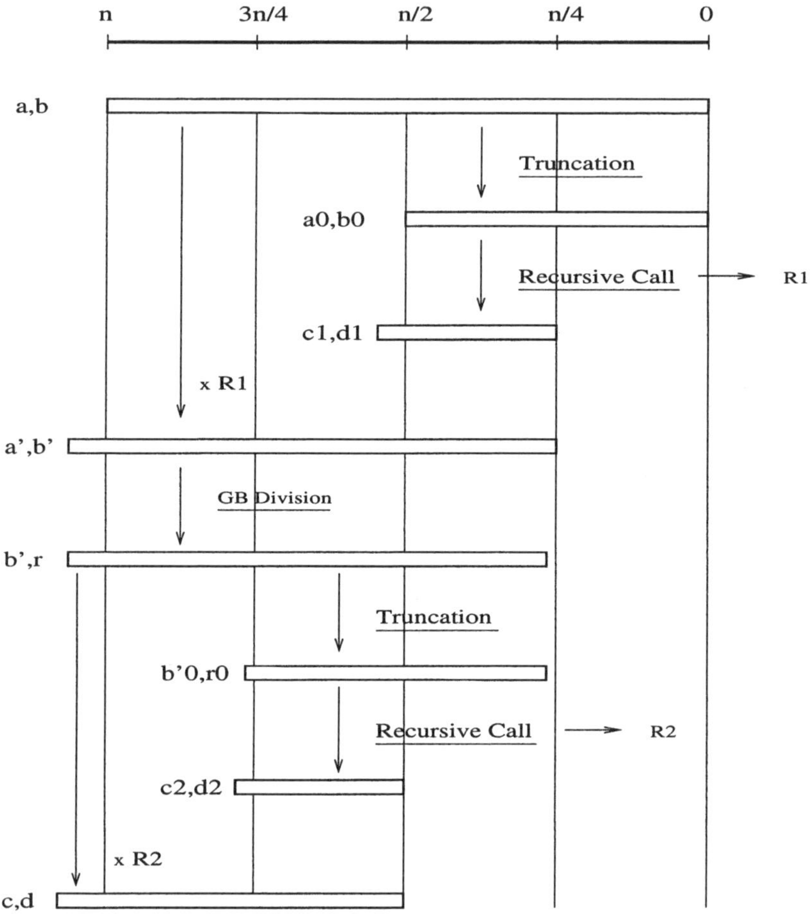

Fig. 5. The recursive structure of the algorithm **Half-GB-gcd**.

4.1 Some Properties of the GB Division

The properties described below are very similar to the ones of the standard Euclidean division that make the Knuth-Schönhage algorithm possible. The first result states that the $\nu_2(b) - \nu_2(a) + 1$ last non-zero bits of a and of b suffice to compute the GB quotient of two integers a and b.

Lemma 6. *Let a, b, a' and b' be such that $a' = a \bmod 2^l$ and $b' = b \bmod 2^l$ with $l \geq 2\nu_2(b) + 1$. Assume that $0 = \nu_2(a) < \nu_2(b)$. Let $(q, r) = GB(a, b)$ and $(q', r') = GB(a', b')$. Then $q = q'$ and $r = r' \bmod 2^{l-\nu_2(b)}$.*

Proof. By definition, $q := -a\left(\frac{b}{2^{\nu_2(b)}}\right)^{-1} \operatorname{cmod} 2^{\nu_2(b)+1}$. Therefore, since $l \geq 2\nu_2(b) + 1$, $\frac{b}{2^{\nu_2(b)}} = \frac{b'}{2^{\nu_2(b')}} \bmod 2^{\nu_2(b)+1}$, we have $a = a' \bmod 2^{\nu_2(b)+1}$ and $q = q'$. Moreover, $r = a + q\frac{b}{2^{\nu_2(b)}}$ and $r' = a' + q\frac{b'}{2^{\nu_2(b')}}$. Consequently $r = r' \bmod 2^{l-\nu_2(b)}$.

Algorithm Half-GB-gcd.
Input: a, b satisfying $0 = \nu_2(a) < \nu_2(b)$.
Output: An integer j, an integer matrix R and two integers c and d with $0 = \nu_2(c) < \nu_2(d)$, such that $\begin{pmatrix} c \\ d \end{pmatrix} = 2^{-2j} R \cdot \begin{pmatrix} a \\ b \end{pmatrix}$, and $c^* = 2^j c$, $d^* = 2^j d$ are the two consecutive remainders of the GB remainder sequence of (a, b) that satisfy $\nu_2(c^*) \leq \ell(a)/2 < \nu_2(d^*)$.

1. $k := \lfloor \ell(a)/2 \rfloor$.
2. If $\nu_2(b) > k$, then return $0, \begin{pmatrix} 1 & 0 \\ 0 & 1 \end{pmatrix}, a, b$.
3. $k_1 := \lfloor k/2 \rfloor$.
4. $a := a_1 2^{2k_1+1} + a_0, \ b := b_1 2^{2k_1+1} + b_0$ with $0 \leq a_0, b_0 < 2^{2k_1+1}$.
5. $j_1, R_1, c_1, d_1 :=$ **Half-GB-gcd**(a_0, b_0).
6. $\begin{pmatrix} a' \\ b' \end{pmatrix} := 2^{2k_1+1-2j_1} R_1 \cdot \begin{pmatrix} a_1 \\ b_1 \end{pmatrix} + \begin{pmatrix} c_1 \\ d_1 \end{pmatrix}, \quad j_0 := \nu_2(b')$.
7. If $j_0 + j_1 > k$, then return j_1, R_1, a', b'.
8. $(q, r) := GB'(a', b')$.
9. $k_2 := k - (j_0 + j_1)$.
10. $\frac{b'}{2^{j_0}} := b'_1 2^{2k_2+1} + b'_0, \ r := r_1 2^{2k_2+1} + r_0$ with $0 \leq b'_0, r_0 < 2^{2k_2+1}$.
11. $j_2, R_2, c_2, d_2 :=$ **Half-GB-gcd**(b'_0, r_0).
12. $\begin{pmatrix} c \\ d \end{pmatrix} := 2^{2k_2+1-2j_2} R_2 \cdot \begin{pmatrix} b'_1 \\ r_1 \end{pmatrix} + \begin{pmatrix} c_2 \\ d_2 \end{pmatrix}$.
13. Return $j_1 + j_0 + j_2, \ R_2 \cdot [q]_{j_0} \cdot R_1, \ c, \ d$.

Fig. 6. Algorithm **Half-GB-gcd**.

This result can be seen as a continuity statement: two pairs of 2-adic integers (a, b) and (a', b') which are sufficiently close for the 2-adic norm (i.e. some least significant bits of a and a', and some of b and b' are equal) have the same quotient and similar remainders (the closer the pairs, the closer the remainders). The second lemma extends this result to the GB continued fraction expansions: if (a, b) and (a', b') are sufficiently close, their first GB quotients are identical. We obtain this result by applying the first one several times.

Lemma 7. *Let a, b, a' and b' such that $a' = a \bmod 2^{2k+1}$ and $b' = b \bmod 2^{2k+1}$, with $k \geq 0$. Suppose that $0 = \nu_2(a) < \nu_2(b)$. Let $r_0 = a, r_1 = b, r_2, \ldots$ be the GB remainder sequence of (a, b), and let $q_1, q_2, \ldots$ be the corresponding GB quotients: $r_{j+1} = r_{j-1} + q_j \frac{r_j}{2^{n_j}}$, with $n_j = \nu_2(r_j) - \nu_2(r_{j-1})$. Let $r'_0 = a', r'_1 = b', r'_2, \ldots$ be the GB remainder sequence of (a', b'), and let $q'_1, q'_2, \ldots$ be the corresponding GB quotients: $r'_{j+1} = r'_{j-1} + q'_j \frac{r'_j}{2^{n'_j}}$, with $n'_j = \nu_2(r'_j) - \nu_2(r'_{j-1})$.*

Algorithm Fast-GB-gcd.
Input: a, b satisfying $0 = \nu_2(a) < \nu_2(b)$.
Output: $g = \gcd(a, b)$.

1. $j, R, a', b' :=$ **Half-GB-gcd**(a, b).
2. If $b' = 0$, return a'.
3. $(q, r) := GB'(a', b')$.
4. Return **Fast-GB-gcd**(b', r').

Fig. 7. Algorithm **Fast-GB-gcd**.

Then, if r_{i+1} is the first remainder such that $\nu_2(r_{i+1}) > k$, we have $q_j = q'_j$ and $r_{j+1} = r'_{j+1}$ mod $2^{2k+1-\nu_2(r_j)}$, for any $j \leq i$.

Proof. We prove this result by induction on $j \geq 0$. This is true for $j = 0$, because $a' = a \bmod 2^{2k+1}$ and $b' = b \bmod 2^{2k+1}$. Suppose now that $1 \leq j \leq i$. We use Lemma 6 with $\frac{r_{j-1}}{2^{\nu_2(r_{j-1})}}, \frac{r_j}{2^{\nu_2(r_{j-1})}}, \frac{r'_{j-1}}{2^{\nu_2(r_{j-1})}}, \frac{r'_j}{2^{\nu_2(r_{j-1})}}$ and $l = 2k+1-2\nu_2(r_{j-1})$. By induction, modulo $2^{2k+1-\nu_2(r_{j-1})}$, $r_{j-1} = r'_{j-1}$ and $r_j = r'_j$. Since $j \leq i$, we have, by definition of i, $2k+1-2\nu_2(r_{j-1}) \geq 2(\nu_2(r_j) - \nu_2(r_{j-1}))+1$, and consequently we can apply Lemma 6. Thus $q_j = q'_j$ and $r_{j+1} = r'_{j+1} \bmod 2^{2k+1-\nu_2(r_j)}$.

Practically, this lemma says that k bits can be gained as regard to the initial pair (a, b) by using only $2k$ bits of a and $2k$ bits of b. This is the advantage of using the GB division instead of the standard division: in the case of the standard Euclidean division, this lemma is only "almost true", because some of the last quotients before gaining k bits can differ, and have to be repaired.

4.2 Correctness of the Half-GB-gcd Algorithm

To show the correctness of the algorithm **Fast-GB-gcd**, it suffices to show the correctness of the algorithm **Half-GB-gcd**, which is established in the following theorem. (Since each call to **Fast-GB-gcd** which does not return in Step 2 performs at least one GB division, the termination is ensured.)

Theorem 1. *The algorithm* **Half-GB-gcd** *of Fig. 6 is correct.*

Proof. We prove the correctness of the algorithm by induction on the size of the inputs. If $\ell(a) = 1$, then the algorithm finishes at Step 2 because $\nu_2(b) \geq 1$. Suppose now that $k \geq 2$ and that $\nu_2(b) \leq k$.

Since $2\lfloor \frac{k}{2} \rfloor + 1 < \ell(a)$, Step 5 is a recursive call (its inputs satisfy the input conditions). By induction j_1, R_1, c_1 and d_1 satisfy $\begin{pmatrix} c_1 \\ d_1 \end{pmatrix} = 2^{-2j_1} R_1 \cdot \begin{pmatrix} a_0 \\ b_0 \end{pmatrix}$, and $2^{j_1} c_1$ and $2^{j_1} d_1$ are the consecutive remainders r'_{i_1} and r'_{i_1+1} of the GB remainder sequence of $r'_0 = a_0$ and $r'_1 = b_0$ that satisfy $\nu_2(r'_{i_1}) \leq k_1 < \nu_2(r'_{i_1+1})$. From

Lemma 7, we know that $2^{-j_1} R_1 \cdot \begin{pmatrix} a \\ b \end{pmatrix}$ are two consecutive remainders $2^{j_1} a' = r_{i_1}$ and $2^{j_1} b' = r_{i_1+1}$ of the GB remainder sequence of $r_0 = a$ and $r_1 = b$, and they satisfy $r_{i_1} = r'_{i_1}$ and $r_{i_1+1} = r'_{i_1+1}$ modulo 2^{k_1+1}. From these last equalities, we have that $\nu_2(r_{i_1}) = \nu_2(r'_{i_1}) \leq k_1 < \nu_2(r_{i_1+1}) \leq \nu_2(r'_{i_1+1})$. Thus, if the execution of the algorithm stops at Step 7, the output is correct.

Otherwise, r_{i_1+2} is computed at Step 8. At Step 9 we compute $k_2 = k - \nu_2(r_{i_1+1})$. Step 7 ensures that $k_2 \geq 0$. Since $\nu_2(r_{i_1+1}) > \lfloor k/2 \rfloor$, we have $k_2 \leq \lceil k/2 \rceil - 1$. Therefore Step 11 is a recursive call (and the inputs satisfy the input conditions). By induction, j_2, S_2, c_2 and d_2 satisfy: $\begin{pmatrix} c_2 \\ d_2 \end{pmatrix} = 2^{-2j_2} R_2 \cdot \begin{pmatrix} b'_0 \\ r_0 \end{pmatrix}$, and $2^{j_2} c_2$ and $2^{j_2} d_2$ are the consecutive remainders r'_{i_2} and r'_{i_2+1} of the GB remainder sequence of (b_0, r'_0). Moreover, $\nu_2(r'_{i_2}) \leq k_2 < \nu_2(r'_{i_2+1})$. From Lemma 7, we know that $2^{-j_2} S_2 \cdot \begin{pmatrix} 2^{j_1} b' \\ 2^{j_1} r' \end{pmatrix}$ are two consecutive remainders r_i and r_{i+1} (with $i = i_1 + i_2 + 1$) of the GB remainder sequence of (a, b), that $\frac{r_i}{2^{j_1+\nu_2(b')}} = 2^{j_2} c_2 \bmod 2^{k_2+2}$ and that $\frac{r_{i+1}}{2^{j_1+\nu_2(b')}} = 2^{j_2} d_2 \bmod 2^{k_2+1}$. Therefore the following sequence of inequalities is valid: $\nu_2(r_i) = j_1 + j_2 + \nu_2(b') \leq k < \nu_2(r_{i+1})$. This ends the proof of the theorem.

5 Analysis of the Algorithms

In this section, we first study the GB Euclidean algorithm. In particular we give a worst-case bound regarding the length of the GB remainder sequence. Then we bound the complexity of the recursive algorithm in the case of the use of Schönhage-Strassen multiplication, and we give some intuition about the average complexity.

5.1 Length of the GB Remainder Sequence

In this subsection, we bound the size of the matrix $\frac{1}{2^{\nu_2(r_i)}} [q_i]_{n_i} \cdots [q_1]_{n_1}$, where the q_j's are the GB quotients of a truncated GB remainder sequence $r_0, \ldots, r_{i+1}$ with $\nu_2(r_i) \leq \nu_2(r_0) + d < \nu_2(r_{i+1})$ for some $d \geq 0$. This will make possible the analysis of the lengths and the number of the GB remainders. As mentioned in §2, this subsection is not necessary to prove the quasi-linear time complexity of the algorithm.

Theorem 2. *Let $d \geq 1$. Let r_0, r_1 with $0 = \nu_2(r_0) < \nu_2(r_1)$, and $r_0, r_1, \ldots, r_{i+1}$ their first GB remainders, where i is such that $\nu_2(r_i) \leq d < \nu_2(r_{i+1})$. We consider the matrix $\frac{1}{2^{\nu_2(r_i)}} [q_i]_{n_i} \cdots [q_1]_{n_1}$, where the q_j's are the GB quotients and $n_j = \nu_2(r_j) - \nu_2(r_{j-1})$ for any $1 \leq j \leq i$. Let M be the maximum of the absolute values of the four entries of this matrix. Then we have:*

- *If $d = 0$ or 1, $M = 1$,*
- *If $d = 3$, $M \leq 11/8$,*

– *If* $d = 5$, $M \leq 67/32$,
– *If* $d = 2, 4$ *or* $d \geq 6$, $M \leq \frac{2}{\sqrt{17}}\left(\left(\frac{1+\sqrt{17}}{4}\right)^{d+1} - \left(\frac{1-\sqrt{17}}{4}\right)^{d+1}\right)$.

Moreover, all these bounds are reached, in particular, the last one is reached when $n_j = 1$ *and* $q_j = 1$ *for any* $1 \leq j \leq i$.

The proof resembles the worst case analysis of the Gaussian algorithm in [9].

Proof. We introduce a partial order on the 2×2 matrices: $A < B$ if and only if for any coordinate $[i, j]$, $A[i, j] \leq B[i, j]$. First, the proof can be restricted to the case where the q_j's are non-negative integers, because we have the inequality $|[q_i]_{n_i} \cdots [q_1]_{n_1}| \leq [|q_i|]_{n_i} \cdots [|q_1|]_{n_1}$. This can be easily showed by induction on i by using the properties: $|A \cdot B| \leq |A| \cdot |B|$, and if the entries of A, A', B, B' are non-negative, $A \leq A'$ and $B \leq B'$ implies $A \cdot B \leq A' \cdot B'$.

Consequently we are looking for the maximal elements for the partial order $>$ in the set: $\{\Pi_{n_1+\ldots+n_i \leq d}\ [q_i]_{n_i} \cdots [q_1]_{n_1}\ /\ \forall 1 \leq j \leq i,\ 0 < q_j \leq 2^{n_j} - 1 \text{ and } n_j \geq 1\}$. We can restrict the analysis to the case where $n_1 + \ldots + n_i = d$ and all the q_j's are maximal, which gives the set:

$$\{\Pi_{n_1+\ldots+n_i=d}\ [2^{n_i} - 1]_{n_i} \cdots [2^{n_1} - 1]_{n_1}\}.$$

Remark now that $[2^n - 1]_n \leq [1]_1^n$ for any $n \geq 3$. Therefore, it is sufficient to consider the case where the n_j's are in $\{1, 2\}$. Moreover, for any integer $j \geq 0$, $[3]_2 \cdot [1]_1^j \cdot [3]_2 \leq [1]_1^{j+4}$, and we also have the inequalities $[3]_2^2 \leq [1]_1^4$, $[1]_1^5 \cdot [3]_2 \leq [1]^7$, $[3]_2 \cdot [1]_1^5 \leq [1]^7$ and $[1]_1^2 \cdot [3]_2 \cdot [1]_1^2 \leq [1]_1^6$.

From these relations, we easily obtain the maximal elements:

– For $d = 1$, $[1]_1$.
– For $d = 2$, $[1]_1^2$ and $[3]_2$.
– For $d = 3$, $[1]_1^3$, $[3]_2 \cdot [1]_1$ and $[1]_1 \cdot [3]_2$.
– For $d = 4$, $[1]_1^4$, $[3]_2 \cdot [1]_1^2$, $[1]_1 \cdot [3]_2 \cdot [1]_1$ and $[1]_1^2 \cdot [3]_2$.
– For $d = 5$, $[1]_1^5$, $[3]_2 \cdot [1]_1^3$, $[1]_1^2 \cdot [3]_2 \cdot [1]_1$, $[1]_1 \cdot [3]_2 \cdot [1]_1^2$ and $[1]_1^3 \cdot [3]_2$.
– For $d = 6$, $[1]_1^6$, $[3]_2 \cdot [1]_1^4$, $[1]_1^3 \cdot [3]_2 \cdot [1]_1$, $[1]_1 \cdot [3]_2 \cdot [1]_1^3$ and $[1]_1^4 \cdot [3]_2$.
– For $d = 7$, $[1]_1^7$, $[1]_1 \cdot [3]_2 \cdot [1]_1^4$ and $[1]_1^4 \cdot [3]_2 \cdot [1]_1$.
– For $d \geq 8$, $[1]_1^d$.

The end of the proof is obvious once we note that $2^{-d}[1]_1^d = \begin{pmatrix} u_{d-1} & u_d \\ u_d & u_{d+1} \end{pmatrix}$, where $u_0 = 0$, $u_1 = 1$ and $u_i = u_{i-2} + \frac{1}{2}u_{i-1}$.

From this result on the quotient matrices, we can easily deduce the following on the size of the remainders and on the length of the remainder sequence.

Theorem 3. *Let* r_0, r_1 *be two non-zero integers with* $0 = \nu_2(r_0) < \nu_2(r_1)$, *and* $r_0, r_1, \ldots, r_{t+1}$ *their complete GB remainder sequence, i.e. with* $r_{t+1} = 0$. *Assume that* $9 \leq j \leq t$. *Then:*

$$2^{\nu_2(r_j)} \leq |r_j| \leq \frac{2}{\sqrt{17}}\left[|r_0|\left(\left(\frac{1+\sqrt{17}}{4}\right)^{\nu_2(r_j)-1} - \left(\frac{1-\sqrt{17}}{4}\right)^{\nu_2(r_j)-1}\right)\right.$$

$$+ |r_1| \left(\left(\frac{1+\sqrt{17}}{4} \right)^{\nu_2(r_j)} - \left(\frac{1-\sqrt{17}}{4} \right)^{\nu_2(r_j)} \right) \Bigg] .$$

The upper bound is reached by $\begin{pmatrix} r_0 \\ r_1 \end{pmatrix} = [1]_1^{-t} \cdot \begin{pmatrix} 2^t \\ 0 \end{pmatrix}$.

Moreover, if $\ell(r_0), \ell(r_1) \leq n$*, then we have:* $t \leq n / \log(\frac{\sqrt{17}-1}{2})$.

Proof. (Sketch) For the inequations concerning $|r_j|$, use the maximal element $[1]_1^d$ from the proof of Theorem 2, with $d = \nu_2(r_i)$, where $i = j - 1$. Remark then that the upper bound grows far slower with $\nu_2(r_j)$ than the lower bound: this fact gives an upper bound on $\nu_2(r_t)$ and therefore on t.

As a comparison, we recall that the worst case for the standard Euclidean division corresponds to the Fibonacci sequence, with $d \leq n / \log(\frac{1+\sqrt{5}}{2}) + o(n)$. Remark that $1/\log(\frac{1+\sqrt{5}}{2}) \approx 1.440$ and $1/\log(\frac{\sqrt{17}-1}{2}) \approx 1.555$.

5.2 Complexity Bound for the Recursive Algorithm

In what follows, $H(n)$ and $G(n)$ respectively denote the maximum of the number of bit operations performed by the algorithms **Half-GB-gcd** and **Fast-GB-gcd**, given as inputs two integers of lengths at most n.

Lemma 8. *Let* $c = \frac{1}{2} \log \frac{1+\sqrt{17}}{2} \approx 0.679$*. The following two relations hold:*

- $G(n) = H(n) + G(\lceil cn \rceil) + O(n)$,
- $H(n) = 2H(\lfloor \frac{n}{2} \rfloor + 1) + O(M(n))$.

Proof. The first relation is an obvious consequence of Theorem 2. We now prove the second relation. The costs of Steps 1, 2, 3, 4, 7, 9 and 10 are negligible. Steps 5 and 11 are recursive calls and the cost of each one is bounded by $H(\lfloor \frac{n}{2} \rfloor + 1)$. Steps 6, 12 and 13 consist of multiplications of integers of size $O(n)$. Finally, Step 8 is a single GB division, and we proved in Lemma 3 that it can be performed in time $O(M(n))$.

From this result and the fact that $c < 1$, one easily obtains the following theorem:

Theorem 4. *The algorithm* **Fast-GB-gcd** *of Fig. 7 runs in quasi-linear time. More precisely,* $G(n) = O(M(n) \log n)$.

The constants that one can derive from the previous proofs are rather large, and not very significant in practice. In fact, for randomly chosen n-bit integers, the quotients of the GB remainder sequence are $O(1)$, and therefore Step 8 of the algorithm **Half-GB-gcd** has a negligible cost. Moreover, the worst-case analysis on the size of the coefficients of the returned matrices gives a worst-case bound $O\left(\left(\frac{1+\sqrt{17}}{2}\right)^n\right)$, which happens to be $O(2^n)$ in practice. With these two heuristics, the "practical cost" of the algorithm **Half-GB-gcd** satisfies the same recurrence than the Knuth-Schönhage algorithm:

$$H(n) \approx 2H(\frac{n}{2}) + kM(n),$$

where the constant k depends from the implementation [11].

6 Implementation Issues

We have implemented the algorithms described in this paper in GNU MP [3]. In this section, we first give some "tricks" that we implemented to improve the efficiency of the algorithm, and then we give some benchmarks.

6.1 Some Savings

First of all, note that some multiplications can be saved easily from the fact that when the algorithm **Fast-GB-gcd** calls the algorithm **Half-GB-gcd**, the returned matrix is not used. Therefore, for such "top-level" calls to the algorithm **Half-GB-gcd**, there is no need to compute the product $R_2 \cdot [q]_{j_0} \cdot R_1$.

Note also that for interesting sizes of inputs (our implementation of the recursive algorithm is faster than usual Euclidean algorithms for several thousands of bits), we are in the domains of Karatsuba and Toom-Cook multiplications, and below the domain of FFT-based multiplication. This leads to some improvements. For example, the algorithm **Fast-GB-gcd** should use calls to the algorithm **Half-GB-gcd** in order to gain γn bits instead of $\frac{n}{2}$, with a constant $\gamma \neq \frac{1}{2}$ that has to be optimized.

Below a certain threshold in the size of the inputs (namely several hundreds of bits), a naive quadratic algorithm that has the requirements of algorithm **Half-GB-gcd** is used. Moreover, each time the algorithm has to compute a GB quotient, it computes several of them in order to obtain a 2×2 matrix with entries of length as close to the size of machine words as possible. This is done by considering only the two least significant machine words of the remainders (which gives a correct result, because of Lemma 6).

6.2 Comparison to Other Implementations of Subquadratic Gcd Algorithms

We compared our implementation in GNU MP — using the ordinary integer interface `mpz` — with those of Magma V2.10-12 and Mathematica 5.0, which both provide a subquadratic integer gcd. This comparison was performed on `laurent3.medicis.polytechnique.fr`, an Athlon MP 2200+. Our implementation wins over the quadratic gcd of GNU MP up from about 2500 words of 32 bits, i.e. about 24000 digits. We used as test numbers both the worst case of the classical subquadratic gcd, i.e. consecutive Fibonacci numbers F_n and F_{n-1}, and the worst case of the binary variant, i.e. G_n and $2G_{n-1}$, where $G_0 = 0$, $G_1 = 1$, $G_n = -G_{n-1} + 4G_{n-2}$, which gives all binary quotients equal to 1. Our experiments show that our implementation in GNU MP of the binary recursive gcd is 3 to 4 times faster than the implementations of the classical recursive gcd in Magma or Mathematica. This ratio does not vary much with the inputs. For example, ratios for F_n and G_n are quite similar.

type, n	Magma V2.10-12	Mathematica 5.0	Fast-GB-gcd (GNU MP)
F_n, 10^6	2.89	2.11	0.70
F_n, $2 \cdot 10^6$	7.74	5.46	1.91
F_n, $5 \cdot 10^6$	23.3	17.53	6.74
F_n, 10^7	59.1	43.59	17.34
G_n, $5 \cdot 10^5$	2.78	2.06	0.71
G_n, 10^6	7.99	5.30	1.94

Fig. 8. Timings in seconds of the gcd routines of Magma, Mathematica and our implementation in GNU MP.

Acknowledgements. The first author thanks Brigitte Vallée for introducing him to fast gcd algorithms and for very helpful discussions on that topic. We thank Jean Vuillemin for the pointer to Hensel's odd division, the Medicis center for allowing the comparison to Magma and Mathematica, and Richard Brent for his comments, which helped to improve both the contents and the form of the paper.

References

1. R. P. Brent. Twenty years' analysis of the binary Euclidean algorithm. In A. W. Roscoe J. Davies and J. Woodcock, editors, *Millenial Perspectives in Computer Science: Proceedings of the 1999 Oxford-Microsoft Symposium in honour of Professor Sir Antony Hoare*, pages 41–53, Palgrave, New York, 2000.
2. R. P. Brent and H. T. Kung. A systolic VLSI array for integer GCD computation. In K. Hwang, editor, *Proceedings of the 7th Symposium on Computer Arithmetic (ARITH-7)*. IEEE CS Press, 1985.
3. T. Granlund. *GNU MP: The GNU Multiple Precision Arithmetic Library*, 4.1.2 edition, 2002. `http://www.swox.se/gmp/\#DOC`.
4. D. Knuth. The analysis of algorithms. In *Actes du Congrès International des Mathématiciens de 1970*, volume 3, pages 269–274, Paris, 1971. Gauthiers-Villars.
5. P. L. Montgomery. Modular multiplication without trial division. *Math. Comp.*, 44(170):519–521, 1985.
6. A. Schönhage. Schnelle Berechnung von Kettenbruchentwicklungen. *Acta Informatica*, 1:139–144, 1971.
7. A. Schönhage and V. Strassen. Schnelle Multiplikation grosser Zahlen. *Computing*, 7:281–292, 1971.
8. M. Shand and J. E. Vuillemin. Fast implementations of RSA cryptography. In E. E. Swartzlander, M. J. Irwin, and J. Jullien, editors, *Proceedings of the 11th IEEE Symposium on Computer Arithmetic (ARITH-11)*, pages 252–259. IEEE Computer Society Press, Los Alamitos, CA, 1993.
9. B. Vallée. Gauss' algorithm revisited. *Journal of Algorithms*, 12:556–572, 1991.
10. B. Vallée. Dynamical analysis of a class of Euclidean algorithms. *Th. Computer Science*, 297(1-3):447–486, 2003.
11. J. von zur Gathen and J. Gerhard. *Modern Computer Algebra.* Cambridge University Press, 2nd edition, 2003.
12. C. K. Yap. *Fundamental Problems in Algorithmic Algebra.* Oxford University Press, 2000.

Lagrange Resolvents Constructed from Stark Units

Brett A. Tangedal

College of Charleston, Charleston SC, 29424, USA
tangedalb@cofc.edu

Abstract. Classical Gauss sums are Lagrange resolvents formed from the Gaussian periods lying in a cyclic extension K over $\mathbb{Q}$ of prime conductor. Elliptic Gauss sums and elliptic resolvents (which are particular instances of Lagrange resolvents) play an important role in the theory of abelian extensions of imaginary quadratic fields. Motivated by the close relationship between the Stark units and Gaussian periods in a cyclic extension $K \subset \mathbb{Q}(\zeta_p)$ and the analogies between Stark units over totally real fields and elliptic units over imaginary quadratic fields, we consider for the first time Lagrange resolvents constructed from Stark units over totally real fields and study the differences and similarities they share with classical Gauss sums.

1 Introduction

In 1801, Gauss introduced what are now called "Gaussian periods" in the seventh section of his *Disquisitiones Arithmeticae.* Lagrange had already introduced his "résolvantes" in 1770 but didn't coin this particular name for them until 1808. Lagrange gave an exposition of Gauss's 7th section in 1808 and formed "classical Gauss sums" from Gaussian periods as a special instance of his resolvents. Gauss apparently preferred his periods to the corresponding resolvents except in the quadratic case (see [We]). During the 19th century, "elliptic Gauss sums" and "elliptic resolvents" were defined by replacing the Gaussian periods by division values of elliptic functions. Elliptic Gauss sums were introduced by Eisenstein to prove certain reciprocity laws (see [Le], p. 299). Fairly recently, the elliptic resolvents introduced by Abel and Hermite have been employed in the study of Galois module structure questions involving abelian extensions of imaginary quadratic fields (see [CT], p. 80; see also [Sc], p. 288, for another appearance of elliptic resolvents).

Our goal in the present paper is to define and study an analogous resolvent construction over a given totally real number field F by replacing the Gaussian periods by special algebraic integers known as "Stark units" that are conjectured to lie in abelian extensions of F. This study is motivated by the relationship and analogies that exist between Stark units, Gaussian periods, and elliptic units.

In the remainder of this section, we establish some conventions and basic properties of Lagrange resolvents used throughout the paper. In Section 2, we

D.A. Buell (Ed.): ANTS 2004, LNCS 3076, pp. 426–441, 2004.

recall some famous results concerning classical Gauss sums which serve as a source of comparison with the Lagrange resolvents constructed from Stark units in Sections 2 and 3. The relationship between the Stark units and Gaussian periods in a real cyclic extension $K \subset \mathbb{Q}(\zeta_p)$ is studied at the end of Section 2. Throughout Section 2, the base field F is $\mathbb{Q}$ but in Section 3 we allow F to be any totally real number field and study Lagrange resolvents constructed from Stark units in this more general setting. The computations we carried out over real quadratic base fields are discussed in Section 3.

If K is an algebraic number field, we write $\mathcal{O}_K$ for the ring of algebraic integers in K. The letter m will always denote a fixed positive integer ≥ 2. The symbol ζ_m will denote a given primitive mth root of unity in the algebraic closure $\mathbb{Q}^c$ of $\mathbb{Q}$. The field diagram we will consider throughout has the following form,

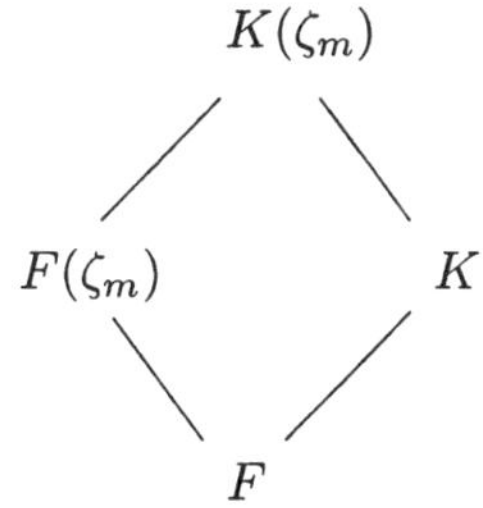

where we assume all of the following:

(a) F is a number field,
(b) $[F(\zeta_m) : F] = \phi(m)$,
(c) K/F is a cyclic Galois extension with $[K : F] = m$,
(d) $F(\zeta_m) \cap K = F$.

Let $G := \mathrm{Gal}(K/F) = \langle \tau \rangle$. The extension $F(\zeta_m)/F$ is relative abelian and $H := \mathrm{Gal}(F(\zeta_m)/F) \cong (\mathbb{Z}/m\mathbb{Z})^\times$ by assumption (b). As usual, the $\phi(m)$ elements of H are denoted by σ_t, where $1 \leq t < m$, $(t, m) = 1$, and $\sigma_t(\zeta_m) = \zeta_m^t$. By assumption (c), the extension $K(\zeta_m)/F$ is relative abelian. By assumption (d), we have an isomorphism $\psi : \mathrm{Gal}(K(\zeta_m)/F) \to H \times G$, with ψ defined as follows: Given $\rho \in \mathrm{Gal}(K(\zeta_m)/F)$,

$$\psi(\rho) = (\rho|_{F(\zeta_m)}, \rho|_K) \in H \times G.$$

From now on, by abuse of notation, we define

$$\tau := \psi^{-1}(\sigma_1, \tau),$$
$$\sigma_t := \psi^{-1}(\sigma_t, \tau^0), \quad 1 \leq t < m, \ (t, m) = 1,$$

as elements in $\mathrm{Gal}(K(\zeta_m)/F)$. Note that $G \cong \mathrm{Gal}(K(\zeta_m)/F(\zeta_m)) = \langle \tau \rangle$.

Given an element $\theta \in K$ and a character $\chi \in \widehat{G}$, we define the corresponding Lagrange resolvent (θ, χ) by

$$(\theta, \chi) := \sum_{i=0}^{m-1} \chi(\tau^i)\tau^i(\theta).$$

(Note: Some authors use $\bar{\chi}(\tau^i)$ instead of $\chi(\tau^i)$.) An element $\theta \in K$ such that the discriminant $d_{K/F}(\theta, \tau(\theta), \ldots, \tau^{m-1}(\theta))$ is non-zero is called a "normal basis element for K/F". In the statement of the following properties of Lagrange resolvents, we assume that $\theta \in \mathcal{O}_K$ is a normal basis element for K/F (the normal basis theorem guarantees the existence of such a θ, see [Na], p. 171).

(1) $(\theta, \chi) \in \mathcal{O}_{K(\zeta_m)}$.
(2) $(\theta, \chi)^m \in \mathcal{O}_{F(\zeta_m)}$.
(3) If $\alpha \in K$ is expressed in the form $\alpha = \sum_{j=0}^{m-1} a_j \tau^j(\theta)$, where $a_j \in F$ for $0 \leq j \leq m-1$, then $(\alpha, \chi) = A \cdot (\theta, \chi)$, where $A = \sum_{j=0}^{m-1} a_j \bar{\chi}(\tau^j)$ lies in $F(\zeta_m)$.
(4) $d_{K/F}(\theta, \tau(\theta), \ldots, \tau^{m-1}(\theta)) = \prod_{\chi \in \widehat{G}} (\theta, \chi)^2$ and so $(\theta, \chi) \neq 0$ for all $\chi \in \widehat{G}$.

The proof of (1) is immediate since $\theta \in \mathcal{O}_K$ and $\chi(\tau^i) \in \mathcal{O}_{F(\zeta_m)}$. Property (4) follows from the evaluation of the determinant of the circulant matrix whose first row is $(\theta \;\; \tau(\theta) \;\cdots\; \tau^{m-1}(\theta))$ ([Zh], p. 107). Property (3) follows easily from

Lemma 1. *If $a, b \in F$ and $\theta, \theta_1, \theta_2 \in K$, then*
(i) $(a\theta_1 + b\theta_2, \chi) = a(\theta_1, \chi) + b(\theta_2, \chi)$,
(ii) $\tau^j((\theta, \chi)) = (\tau^j(\theta), \chi) = \bar{\chi}(\tau^j)(\theta, \chi)$ *for* $0 \leq j \leq m-1$.

Proof. (i) Follows directly from the definition.
(ii) With j fixed, $\tau^j((\theta, \chi)) = \tau^j(\sum_{i=0}^{m-1} \chi(\tau^i)\tau^i(\theta)) = \sum_{i=0}^{m-1} \chi(\tau^i)\tau^i(\tau^j(\theta))$
$= (\tau^j(\theta), \chi) = \bar{\chi}(\tau^j) \sum_{i=0}^{m-1} \chi(\tau^i\tau^j)\tau^i\tau^j(\theta) = \bar{\chi}(\tau^j)(\theta, \chi)$.

To prove property (2), note below that the relative norm lies in $F(\zeta_m)$ and B is a unit in $\mathcal{O}_{F(\zeta_m)}$:
$\mathrm{N}_{K(\zeta_m)/F(\zeta_m)}((\theta, \chi)) = \prod_{j=0}^{m-1} \tau^j((\theta, \chi)) = B \cdot (\theta, \chi)^m$, where $B = \prod_{j=0}^{m-1} \bar{\chi}(\tau^j)$ by (ii). For future use, we note that $\mathrm{N}_{K(\zeta_m)/F(\zeta_m)}((\theta, \chi))$ and $(\theta, \chi)^m$ generate the same principal ideal in $\mathcal{O}_{F(\zeta_m)}$.

2 Classical Gauss Sums

Let $p \in \mathbb{N}$ be a prime, fix a primitive root $r > 0$ modulo p, and let $\mathbb{F}_p$ denote the finite field with p elements. Assume that $p \equiv 1 \pmod{m}$ and let χ be a fixed multiplicative character of order m on $\mathbb{F}_p^\times$. Note that χ is a Dirichlet character of conductor p and $\chi(r) = \zeta_m$, a primitive mth root of unity, since χ is of order m. The corresponding classical Gauss sum is defined by

$$G(\chi) := \sum_{t=1}^{p-1} \chi(t)\zeta_p^t = \sum_{i=0}^{p-2} \chi(r^i)\zeta_p^{r^i} = \sum_{i=0}^{m-1} \zeta_m^i \Big(\sum_{j=0}^{n-1} \zeta_p^{r^{i+mj}} \Big) \quad \text{where} \quad p-1 = mn.$$

The inner sums $\eta_i = \sum_{j=0}^{n-1} \zeta_p^{r^{i+mj}}$, $i = 0, \ldots, m-1$, are known as "Gaussian periods". They are the m roots of a monic irreducible polynomial $f_m(x) \in \mathbb{Z}[x]$, often referred to as an mth period polynomial. These Gaussian periods lie in and generate the unique number field $K \subseteq \mathbb{Q}(\zeta_p)$ such that $[K:\mathbb{Q}] = m$. Note that $K/\mathbb{Q}$ is a cyclic Galois extension with $\mathrm{Gal}(K/\mathbb{Q}) = \langle \tau \rangle$, where $\tau \in \mathrm{Gal}(K/\mathbb{Q})$ is the restriction of the automorphism in $\mathrm{Gal}(\mathbb{Q}(\zeta_p)/\mathbb{Q})$ sending $\zeta_p \mapsto \zeta_p^r$. We set $\chi(\tau) = \chi(r)$, noting that τ is the image of the principal ideal $(r) \subset \mathbb{Z}$ under the global Artin map. We have $\eta_i = \tau^i(\eta_0)$ for $0 \le i \le m-1$ and therefore

$$G(\chi) = \sum_{i=0}^{m-1} \chi(\tau^i)\tau^i(\eta_0) := (\eta_0, \chi) \quad \text{is a Lagrange resolvent.}$$

The Gaussian period η_0 is a normal basis element for $K/\mathbb{Q}$, and even better, we have $\mathcal{O}_K = \mathbb{Z}[\eta_0, \eta_1, \ldots, \eta_{m-1}]$, which means that the η's form a "normal integral basis" for $\mathcal{O}_K$ over $\mathbb{Z}$. For future use, we note that $\eta_0 = \mathrm{Tr}_{\mathbb{Q}(\zeta_p)/K}(\zeta_p)$.

The field K belongs to the group of Dirichlet characters

$$\mathsf{X} = \{\chi_0, \chi, \ldots, \chi^{m-1}\}, \quad \chi_0 \text{ is the trivial character,}$$

according to the terminology in Chapter 3 of [Wa]. By the conductor-discriminant formula, the discriminant of the field K is given by $d_K = p^{m-1}$ since each non- trivial power of χ has conductor p. The field diagram appropriate to the present section is exactly the diagram in Section 1, with $F = \mathbb{Q}$ and $K = \mathbb{Q}(\eta_0)$. We have $\mathbb{Q}(\zeta_m) \cap K = \mathbb{Q}$ since p splits completely in $\mathbb{Q}(\zeta_m)/\mathbb{Q}$ and p is totally (and tamely) ramified in $K/\mathbb{Q}$. Since the η's form a normal integral basis, we have

$$d_{K/\mathbb{Q}}(\eta_0, \tau(\eta_0), \ldots, \tau^{m-1}(\eta_0)) = d_K = p^{m-1}, \tag{1}$$

and so the element $(\eta_0, \chi)^m \in \mathcal{O}_{\mathbb{Q}(\zeta_m)}$ has only primes above p appearing in its factorization by property (4) of Lagrange resolvents in Section 1. Stickelberger's classic theorem makes this precise:

$$((\eta_0, \chi)^m) = \prod_{\substack{0 < t < m \\ (t,m)=1}} (\sigma_t^{-1}(P_1))^t, \tag{2}$$

where P_1 is the unique prime ideal above p in $\mathbb{Q}(\zeta_m)$ such that $r^{(p-1)/m} \equiv \zeta_m^{-1} \pmod{P_1}$ (recall that $\chi(\tau) = \chi(r) = \zeta_m$). A more refined result known as "Stickelberger's congruence" may be formulated in $K(\zeta_m)$. Let $\mathcal{P}_1$ be the unique prime ideal above P_1 in $K(\zeta_m)$, and let $\pi = \mathrm{N}_{\mathbb{Q}(\zeta_p)/K}(1-\zeta_p)$. Then π is a uniformizer in $K(\zeta_m)$ with respect to $\mathcal{P}_1$, and

$$(\eta_0, \chi^k) \equiv (p-1-kn)!\,(-1)^{k+kn}\pi^k \pmod{\mathcal{P}_1^{k+1}} \tag{3}$$

for $k = 1, \ldots, m-1$ (recall that $n = (p-1)/m$). We note that (2) is directly derivable from (3).

In order to make a comparison between the Gaussian periods and Stark units in K, we need to make two specializations that were not necessary above. First, we assume that the field K is totally real which is equivalent to assuming that χ is an even character (or that $n = (p-1)/m$ is even). Second, we need to fix an embedding of K into $\mathbb{R}$ and this is conveniently done by fixing $\zeta_p = e^{2\pi i/p}$ from now on. The Stark unit ϵ_0 for $K/\mathbb{Q}$ is equal to $\mathrm{N}_{\mathbb{Q}(\zeta_p)/K}(1-\zeta_p)$ (see Section 3). We note that $\mathrm{N}_{K/\mathbb{Q}}(\epsilon_0) = p$ (we used $\pi = \epsilon_0$ as the uniformizer above in equation (3)). The relationship between ϵ_0 and η_0 may be found as follows. Let $g(x) = \sum_{j=0}^{p-1}(-1)^j \binom{p}{j+1} x^j$, noting that $g(1-\zeta_p) = 0$. When we factor $g(x)$ over $K[x]$ (it is best to define K using the period polynomial $f_m(x)$), the polynomial

$$x^n + (\eta_0 - n)x^{n-1} + \cdots + \epsilon_0 \in \mathcal{O}_K[x]$$

appears as one of the m factors. These m factors are conjugate under the action of $\mathrm{Gal}(K/\mathbb{Q})$, which allows us to easily obtain the expression $\epsilon_0 = \sum_{i=0}^{m-1} a_i \tau^i(\eta_0)$, where $a_0, \ldots, a_{m-1} \in \mathbb{Z}$.

Example: Let $p = 41$, $m = 4$, $n = 10$, $r = 6$, and $\chi(r) = i$. Then $\epsilon_0 = -44\eta_0 - 23\tau(\eta_0) - 27\tau^2(\eta_0) - 29\tau^3(\eta_0)$. By property (3) of Lagrange resolvents in Section 1, $(\epsilon_0, \chi) = (-17-6i)(\eta_0, \chi)$. Therefore

$$(\epsilon_0, \chi)^4 = \mathfrak{p}_5^8 \mathfrak{p}_{13}^4 \cdot P_1 \bar{P_1}^3$$

and since $(-17-6i, 41) = 1$, we still have the classic Stickelberger factorization above $p = 41$.

Question: Given the above set-up, we know that $(\epsilon_0, \chi) = A \cdot (\eta_0, \chi)$ for some $A \in \mathbb{Z}[\zeta_m]$. Is $(A, p) = 1$?

We have tested this question separately for $m = 2, 3, 4$, and 6, where we allow p to range over every prime such that $1 < p < 5000$ and $p \equiv 1 \pmod{2m}$, and the answer has been found to be "yes" in each case. No clear pattern has emerged yet for the primes dividing A, but A is often divisible by many different primes. For example, given $m = 6$, $p = 3457$, $r = 7$, and $\chi(r) = \rho = e^{2\pi i/6}$, we found that the corresponding algebraic integer $A \in \mathbb{Z}[\rho]$ was divisible by prime ideals above 2, 3, 13, 31, 7069, 31147, 253681, 4757746957, and 817117800695906121725325722880339268134410849. When the answer to this question is yes, then ϵ_0 is a "good substitute" for η_0 since you still have the classic Stickelberger factorization over p with $(\epsilon_0, \chi)^m$ replacing $(\eta_0, \chi)^m$. This question is of interest since a Stark unit ϵ_0 is conjectured to exist when $F \neq \mathbb{Q}$ but there doesn't seem to be a known analogue for η_0 when $F \neq \mathbb{Q}$.

3 Stark Units

The first goal in this section is to find an appropriate generalization of the set-up in Section 2, where we now allow $F \neq \mathbb{Q}$, and replace the Gaussian periods by Stark units. The following assumptions will be made throughout the remainder of the paper:

(i) F is totally real with $[F : \mathbb{Q}] = t \geq 1$, so there are t distinct real embeddings of F into $\mathbb{R}$. Let $F^{(i)}$, for $1 \leq i \leq t$, denote the image of F inside $\mathbb{R}$ under the ith embedding. The embedding for $i = 1$ plays a distinguished role in the following and we remark that any one of the t embeddings may be chosen at the beginning as this "first embedding". Let $\mathfrak{p}_\infty^{(i)}$ denote the infinite prime corresponding to the ith embedding and let $\alpha^{(i)} \in \mathbb{R}$ denote the image of $\alpha \in F$ under this embedding.

(ii) $[F(\zeta_m) : F] = \phi(m)$.

(iii) A prime $p \in \mathbb{N}$ is fixed that splits completely in $F(\zeta_m)$, so $p \equiv 1 \pmod{m}$.

(iv) We fix a prime ideal $\mathfrak{p} \subset \mathcal{O}_F$ lying over p. We have $\mathrm{N}\mathfrak{p} = p$, and $\mathfrak{p}$ splits completely in $F(\zeta_m)$ by (iii). We also fix a primitive root $r > 0$ modulo p which is therefore also a primitive root modulo $\mathfrak{p}$.

(v) Let $J := G(\mathfrak{p}\mathfrak{p}_\infty^{(1)} \cdots \mathfrak{p}_\infty^{(t)})$ denote the narrow ray class group modulo $\mathfrak{p}$. If χ is a character defined on this group, then the conductor $\mathfrak{f}(\chi)$ of χ has a "finite" component $\mathfrak{f}_\chi$ and an "infinite" component $\mathfrak{f}_{\chi,\infty}$. Both of these components are completely determined by the behavior of χ on the principal ideals in $\mathcal{O}_F$ generated by elements relatively prime to $\mathfrak{p}$. The finite component will be either (1) or $\mathfrak{p}$; it is $\mathfrak{p}$ if and only if there exists a totally positive element $\alpha \in \mathcal{O}_F$ such that $(\alpha, \mathfrak{p}) = 1$ and $\chi((\alpha)) \neq 1$. The infinite prime $\mathfrak{p}_\infty^{(i)}$ will appear in $\mathfrak{f}_{\chi,\infty}$ if and only if $\chi((\beta)) \neq 1$ for $\beta \in \mathcal{O}_F$ satisfying $\beta \equiv 1 \pmod{\mathfrak{p}}$, $\beta^{(j)} > 0$ when $j \neq i$, and $\beta^{(i)} < 0$. We assume there is a character $\chi \in \widehat{J}$ having the following properties:

  (a) The conductor of χ has the form $\mathfrak{f}(\chi) = \mathfrak{f}_\chi \mathfrak{f}_{\chi,\infty} = \mathfrak{p}\mathfrak{p}_\infty^{(2)} \cdots \mathfrak{p}_\infty^{(t)}$.

  (b) The order of χ is m.

  (c) We have $\mathfrak{f}_{\chi^j} = \mathfrak{p}$ for $1 \leq j \leq m-1$. This condition implies that $\chi((r))$ is a primitive mth root of unity, denoted by ζ_m from now on, where (r) is the principal ideal in $\mathcal{O}_F$ generated by the primitive root r fixed in (iv).

If $F = \mathbb{Q}$, these assumptions bring us back directly to the set-up at the end of Section 2 (a Dirichlet character χ of conductor p has conductor $\mathfrak{f}(\chi) = (p)$ or $\mathfrak{f}(\chi) = (p)p_\infty^{(1)}$ when considered as a ray class group character, depending upon whether χ is even, or odd, respectively). Our assumption that F be totally real is not completely necessary (the case where F is a complex quadratic field, for example, is also interesting), but Stark's conjecture has a particularly striking form over a totally real field so we keep to this assumption. Assumptions (ii), (iii), (iv), (v)(b), and (v)(c) are all made in order to be consistent with the set-up in Section 2. The special shape of $\mathfrak{f}(\chi)$ in part (v)(a) does, however, require some explanation and for this we recall some basic facts from class field theory as well as the conjectured form of the corresponding Stark unit.

Let $\mathsf{X} = \{\chi_0, \chi, \ldots, \chi^{m-1}\}$ be the group of ray class group characters generated by a character χ satisfying conditions (v)(a)–(c). From class field theory, we know there exists a uniquely defined extension field K over F belonging to the group X of ray class group characters and having the following properties:

- K/F is a cyclic Galois extension with $|\mathrm{Gal}(K/F)| = |\mathsf{X}| = m$. If $\tau \in \mathrm{Gal}(K/F)$ is the automorphism corresponding to the ideal (r) in (v)(c) under the global Artin map, then $\mathrm{Gal}(K/F) = \langle\tau\rangle$. We set $\chi(\tau) = \chi((r)) = \zeta_m$.

K is $\mathfrak{p}$, this ramification being total and tame. Let $\mathfrak{P} \subset \mathcal{O}_K$ be the unique prime above $\mathfrak{p}$ in K. Since $\mathfrak{p}$ splits completely in $F(\zeta_m)$, as noted in (iv), we see that $F(\zeta_m) \cap K = F$, and so conditions (a)–(d) in Section 1 are satisfied.

- The infinite prime $\mathfrak{p}_\infty^{(1)}$ splits completely in K which means that every embedding of K into $\mathbb{C}$ extending the embedding $F \hookrightarrow F^{(1)} \subset \mathbb{R}$ is a real embedding.
- Each infinite prime in the set $\{\mathfrak{p}_\infty^{(2)}, \ldots, \mathfrak{p}_\infty^{(t)}\}$ (this set is empty if $F = \mathbb{Q}$) ramifies in K, i.e., every embedding of K into $\mathbb{C}$ extending an embedding $F \hookrightarrow F^{(j)} \subset \mathbb{R}$, $2 \le j \le t$, is non-real. This implies that $2 \mid m$ if $F \ne \mathbb{Q}$ (m being odd is still interesting if $F = \mathbb{Q}$).
- A prime ideal $\mathfrak{q} \subset \mathcal{O}_F$ splits completely in K/F if and only if $\chi(\mathfrak{q}) = 1$. This exact characterization of completely split primes uniquely determines the extension K/F by a theorem of Bauer (see [Ja], Cor. 5.5, p. 136).

Let $S = \{\mathfrak{p}_\infty^{(1)}, \ldots, \mathfrak{p}_\infty^{(t)}, \mathfrak{p}\}$, and define for each $\psi \in \mathsf{X}$ the corresponding L-function:

$$L_S(s,\psi) = \prod_{\substack{\mathfrak{q} \notin S \\ \mathfrak{q} \text{ prime}}} \left(1 - \frac{\psi(\mathfrak{q})}{\mathrm{N}\mathfrak{q}^s}\right)^{-1}.$$

These infinite products converge only for $\Re(s) > 1$, but there exists a meromorphic continuation of $L_S(s,\psi)$ to all of $\mathbb{C}$ (the extended function is still denoted by $L_S(s,\psi)$), and each of these extended functions is analytic at $s = 0$. If $F = \mathbb{Q}$, each function $L_S(s,\psi)$ here has exactly a first order zero at $s = 0$. If $F \ne \mathbb{Q}$, the order of the zero at $s = 0$ of each $L_S(s,\psi)$, denoted by $r(\psi)$, is given by the following formula because of the conditions that χ satisfies in (v)(a) and (v)(c) (see [DT]):

$$r(\chi^j) = 1 \quad \text{for} \quad j \quad \text{odd}; \tag{4}$$

$$r(\chi^j) = t \ge 2 \quad \text{for} \quad j \quad \text{even}. \tag{5}$$

The most direct way to express Stark's conjecture is via partial ζ-functions. These are defined here by

$$\zeta_S(s,\tau^k) = \frac{1}{m} \sum_{\psi \in \mathsf{X}} \bar{\psi}((r)^k) L_S(s,\psi) \quad \text{for} \quad 0 \le k \le m-1,$$

where again, $\tau \in \mathrm{Gal}(K/F)$ is the automorphism corresponding to the ideal (r) under the global Artin map. Note that each of these partial ζ-functions has at least a first order zero at $s = 0$ and that $\zeta_S'(0,\tau^k) \in \mathbb{R}$ for all k. Indeed, by our special choice of a character χ satisfying the conditions in (v)(a)–(c), it can be shown that $\zeta_S'(0,\tau^k) \ne 0$ in all cases. Since we are directly concerned with first derivatives of partial ζ-functions at $s = 0$ in Stark's conjecture, the assumptions we make regarding χ now have an explanation.

Let K be the extension field of F belonging to the group X as described above and fix an embedding $K \hookrightarrow K^{(1)} \subset \mathbb{R}$ extending the distinguished embedding $F \hookrightarrow F^{(1)} \subset \mathbb{R}$.

Stark's Conjecture ([St2], [Ta]): There exists a unique algebraic integer $\epsilon_0 \in \mathcal{O}_K$ (called the "Stark unit for K/F") such that

$$(\tau^k(\epsilon_0))^{(1)} = e^{-2\zeta_S'(0,\tau^k)} \quad \text{for} \quad k = 0, \dots, m-1. \tag{6}$$

It is also conjectured that if $\epsilon_0 \in \mathcal{O}_K$ satisfies (6), then $K(\sqrt{\epsilon_0})/F$ is an abelian extension. The algebraic integer ϵ_0 we defined in Section 2 when $F = \mathbb{Q}$ satisfies this conjecture and indeed Stark's conjecture has been proved in full over the base field $\mathbb{Q}$ [St2]. However, if F is a totally real field with $[F : \mathbb{Q}] > 1$, then only a few special cases of the conjecture have been proved (see, for example, [DST] and [Sh]). The Stark unit ϵ_0 in Section 2 was not an actual unit (recall that $\mathrm{N}_{K/\mathbb{Q}}(\epsilon_0) = p$), but if F is a totally real field other than $\mathbb{Q}$ then the Stark unit is conjectured to always be an element of $\mathcal{O}_K^\times$ and therefore a "true" unit.

We assume throughout the remainder of the paper that $F \neq \mathbb{Q}$. Recall that $2 \mid [K : F]$ in this setting. Let M denote the unique intermediate field $F \subseteq M \subset K$ such that $[K : M] = 2$ and let $\rho = \tau^{m/2}$, which is the non-trivial automorphism in $\mathrm{Gal}(K/M)$. We have $\zeta_S'(0,\rho) = -\zeta_S'(0,\tau^0)$ by (4), (5), and also since $\chi^j(\rho) = -1$ for j odd. This identity among partial ζ-function values has two important consequences for the conjectured Stark unit:

$$\rho(\epsilon_0) = \frac{1}{\epsilon_0}\,; \tag{7}$$

and

$$|(\tau^k(\epsilon_0))^{(j)}| = 1 \quad \text{for} \quad k = 0, \dots, m-1, \tag{8}$$

for any embedding $K \hookrightarrow K^{(j)} \subset \mathbb{C}$ not extending $F \hookrightarrow F^{(1)} \subset \mathbb{R}$, since $(\rho(\alpha))^{(j)} = \overline{\alpha^{(j)}}$ for all $\alpha \in K$ for such an embedding. If $\epsilon_0 \in \mathcal{O}_K^\times$ satisfies (6), then by Theorem 1 of [St1] we have $K = F(\epsilon_0)$ and even $K = \mathbb{Q}(\epsilon_0)$ since $L_S'(0,\chi^j) \neq 0$ for j odd, as noted in (4). Given (6), (8), and the values $\zeta_S'(0,\tau^k)$ to sufficient accuracy, we may compute a unique candidate for the minimal polynomial

$$f_{\epsilon_0}(x) = \prod_{k=0}^{m-1} (x - \tau^k(\epsilon_0)) \in \mathcal{O}_F[x]$$

of ϵ_0 over F after a finite search which is often surprisingly efficient (see [Co], p. 309). Once $f_{\epsilon_0}(x)$ has been computed, it can be proved independently that a root of $f_{\epsilon_0}(x)$ generates the unique class field K over F belonging to the group X. This explicit construction of the class field K arises directly from the values at $s = 0$ of the meromorphic functions $\zeta_S'(s,\tau^k)$, and these values are computable using only data from the base field F. This gives a beautiful supplement to the classical existence proofs of class field theory and demonstrates why Stark's conjecture gives at least a partial response to Hilbert's twelfth problem.

We now let the Stark unit ϵ_0 play the role that the Gaussian period η_0 played in Section 2 and study how much from the relations in equations (1), (2), and (3) of Section 2 remains true with respect to the Stark unit. Equation (1) is already quite special and it is not guaranteed at all that there exists a normal integral basis for $\mathcal{O}_K$ over $\mathcal{O}_F$. The following proposition shows that there does exist a normal $\mathfrak{p}$-integral basis for a field extension K/F satisfying our assumptions. Let $v_\mathfrak{p}(\ )$ denote the exact power that a prime $\mathfrak{p}$ divides a given element.

Proposition 1. *There exists an element $\alpha \in \mathcal{O}_K$ such that*

$$v_\mathfrak{p}(d_{K/F}(\alpha, \tau(\alpha), \ldots, \tau^{m-1}(\alpha))) = m - 1.$$

Proof. We recall that the ramification of $\mathfrak{P}$ over $\mathfrak{p}$ is tame, which is crucial for the application of Noether's theorem below. Let $K_\mathfrak{P}$ and $F_\mathfrak{p}$ denote the corresponding completions with valuation rings $\mathcal{O}_\mathfrak{P}$ and $\mathcal{O}_\mathfrak{p}$, respectively. For ease of notation, the symbols $\mathfrak{P}$ and $\mathfrak{p}$ are also used to denote the maximal ideals in $\mathcal{O}_\mathfrak{P}$ and $\mathcal{O}_\mathfrak{p}$, respectively. Note that $\mathrm{Gal}(K_\mathfrak{P}/F_\mathfrak{p}) \cong \mathrm{Gal}(K/F)$ and let $\tau' \in \mathrm{Gal}(K_\mathfrak{P}/F_\mathfrak{p})$ be the unique automorphism that restricts to $\tau \in \mathrm{Gal}(K/F)$. By Noether's theorem [Ch], we know there exists an element $\theta \in \mathcal{O}_\mathfrak{P}$ such that $v_\mathfrak{p}(d_{K_\mathfrak{P}/F_\mathfrak{p}}(\theta, \tau'(\theta), \ldots, \tau'^{m-1}(\theta))) = m - 1$. By the strong approximation theorem ([Ha], p. 379), there exists an $\alpha \in \mathcal{O}_K$ such that $\alpha \equiv \theta \pmod{\mathfrak{P}^{m^2}}$. Set $\alpha_j = \tau^j(\alpha)$ and $\theta_j = \tau'^j(\theta)$ for $j = 0, \ldots, m-1$. We have

$$\alpha_i\alpha_j \equiv \theta_i\theta_j \pmod{\mathfrak{P}^{m^2}} \quad \text{and} \quad \mathrm{Tr}_{K/F}(\alpha_i\alpha_j) \equiv \mathrm{Tr}_{K_\mathfrak{P}/F_\mathfrak{p}}(\theta_i\theta_j) \pmod{\mathfrak{p}^m}$$

for $i = 0, \ldots, m-1$; $j = 0, \ldots, m-1$. Therefore

$$d_{K/F}(\alpha, \tau(\alpha), \ldots, \tau^{m-1}(\alpha)) \equiv d_{K_\mathfrak{P}/F_\mathfrak{p}}(\theta, \tau'(\theta), \ldots, \tau'^{m-1}(\theta)) \pmod{\mathfrak{p}^m},$$

from which the proposition follows immediately.

For a Stark unit ϵ_0, we have in general $v_\mathfrak{p}(d_{K/F}(\epsilon_0, \tau(\epsilon_0), \ldots, \tau^{m-1}(\epsilon_0))) \geq m - 1$, with equality usually holding. We will return to this point later in this section. In regard to equation (2) in Section 2, we computed the prime factorization of the ideal $((\epsilon_0, \chi)^m)$ in $\mathcal{O}_{F(\zeta_m)}$ for hundreds of examples. We will see later that it is possible to have $v_\mathfrak{p}(d_{K/F}(\epsilon_0, \tau(\epsilon_0), \ldots, \tau^{m-1}(\epsilon_0))) > m - 1$ and still have the factorization of $((\epsilon_0, \chi)^m)$ above $\mathfrak{p}$ take the classic Stickelberger form. Before we state the results of our computations, we first consider what can be said about the general shape of the prime factorization of (α, χ) in $\mathcal{O}_{K(\zeta_m)}$, when $\alpha \in \mathcal{O}_K$ is an arbitrary normal basis element for K/F. In direct analogy to the definitions given in Section 2, let P_1 be the unique prime ideal in $\mathcal{O}_{F(\zeta_m)}$ above $\mathfrak{p} \subset \mathcal{O}_F$ such that $r^{(p-1)/m} \equiv \zeta_m^{-1} \pmod{P_1}$, let $\mathcal{P}_1$ be the unique prime ideal above P_1 in $K(\zeta_m)$, and let $\pi \in K(\zeta_m)$ be a uniformizer with respect to $\mathcal{P}_1$. Recall that $G := \mathrm{Gal}(K(\zeta_m)/F(\zeta_m)) = \langle\tau\rangle$ and note that the congruence class of $\tau(\pi)/\pi$ modulo $\mathcal{P}_1$ is independent of the choice of π since the inertia group of the prime ideal $\mathcal{P}_1$ over $F(\zeta_m)$ is equal to G.

Lemma 2. *We have*

$$\frac{\tau(\pi)}{\pi} \equiv r^{(p-1)/m} \pmod{\mathcal{P}_1}.$$

Proof. The congruence $(r/\mathcal{P}_1)_m \equiv r^{(p-1)/m} \pmod{\mathcal{P}_1}$ holds for the mth power residue symbol by definition. Let F_{P_1} denote the completion of $F(\zeta_m)$ with respect to P_1 and let $K_{\mathcal{P}_1}$ denote the completion of $K(\zeta_m)$ with respect to $\mathcal{P}_1$. The extension $K_{\mathcal{P}_1}/F_{P_1}$ is tame and totally ramified of relative degree m. If $\tau' \in \mathrm{Gal}(K_{\mathcal{P}_1}/F_{P_1})$ is the unique automorphism restricting to $\tau \in G$, we just need to prove $\tau'(\Pi)/\Pi \equiv r^{(p-1)/m} \pmod{\mathcal{P}_1}$ for any $\Pi \in K_{\mathcal{P}_1}$ such that $v_{\mathcal{P}_1}(\Pi) = 1$. By theorem 5.11 in [Na], there exists an element $\Pi \in K_{\mathcal{P}_1}$ with $v_{\mathcal{P}_1}(\Pi) = 1$ such that $\psi := \Pi^m \in F_{P_1}$ satisfies $v_{P_1}(\psi) = 1$. The restriction of the local Artin symbol $(r^{-1}, K_{\mathcal{P}_1}/F_{P_1}) \in \mathrm{Gal}(K_{\mathcal{P}_1}/F_{P_1})$ to G is equal to the global Artin symbol $((r), K(\zeta_m)/F(\zeta_m)) = \tau$ and therefore $\tau'(\Pi)/\Pi = (r^{-1}, K_{\mathcal{P}_1}/F_{P_1})(\Pi)/\Pi$. The latter expression is equal to the Hilbert symbol $(\frac{r^{-1},\psi}{P_1})$. By the basic properties of the Hilbert symbol ([Ne], p. 334), we have

$$\left(\frac{r^{-1},\psi}{P_1}\right) = \left(\frac{r,\psi}{P_1}\right)^{-1} = \left(\frac{\psi,r}{P_1}\right),$$

and the last expression is equal to $(r/P_1)_m$ ([Ne], p. 415), completing the proof.

Theorem 1. *Given an arbitrary normal basis element $\alpha \in \mathcal{O}_K$ for K/F, the factorization of the ideal $((\alpha,\chi))$ in $\mathcal{O}_{K(\zeta_m)}$ is given by*

$$((\alpha,\chi)) = (I) \prod_{\substack{0<t<m \\ (t,m)=1}} (\sigma_t^{-1}(\mathcal{P}_1))^{c_t},$$

where (I) is the lift to $\mathcal{O}_{K(\zeta_m)}$ of an integral ideal $I \subseteq \mathcal{O}_{F(\zeta_m)}$ relatively prime to $\mathfrak{p}$, and $c_t \in \mathbb{N}$ satisfies $c_t \equiv t \pmod{m}$ for all t as above.

Proof. Fix k in the range $1 \le k \le m-1$. By a calculation similar to the proof of Lemma 1(ii), $\tau((\alpha,\chi^k)) = \zeta_m^{-k}(\alpha,\chi^k)$. Therefore

$$\frac{\tau((\alpha,\chi^k))}{(\alpha,\chi^k)} = \zeta_m^{-k} \equiv (r^{(p-1)/m})^k \equiv \left(\frac{\tau(\pi)}{\pi}\right)^k \pmod{\mathcal{P}_1},$$

where the first congruence holds by the choice of $\mathcal{P}_1$, and the second congruence holds by Lemma 2. By Discussion (4.4) in [Br], we note that the congruence $\tau((\alpha,\chi^k))/(\alpha,\chi^k) \equiv (\tau(\pi)/\pi)^k \pmod{\mathcal{P}_1}$ implies that $v_{\mathcal{P}_1}((\alpha,\chi^k)) \equiv k \pmod{m}$. For a fixed t, we have $\sigma_t((\alpha,\chi)) = (\alpha,\chi^t)$, and so

$$c_t := v_{\sigma_t^{-1}(\mathcal{P}_1)}((\alpha,\chi)) \equiv t \pmod{m}.$$

We now prove the statement about (I). Given a prime ideal $Q \subset \mathcal{O}_{F(\zeta_m)}$ that does not lie over $\mathfrak{p}$, we know that Q is unramified in $K(\zeta_m)$, say $(Q) = \mathcal{Q}_1 \cdots \mathcal{Q}_g$. By Lemma 1(ii), τ fixes the ideal $((\alpha,\chi)) \subset \mathcal{O}_{K(\zeta_m)}$. By the transitive action of G on $\{\mathcal{Q}_1,\ldots,\mathcal{Q}_g\}$, $v_{\mathcal{Q}_1}((\alpha,\chi)) = \cdots v_{\mathcal{Q}_g}((\alpha,\chi))$, which completes the proof.

Corollary 1. *In $\mathcal{O}_{F(\zeta_m)}$, we have the ideal factorization*

$$((\alpha,\chi)^m) = I^m \prod_{\substack{0<t<m\\(t,m)=1}} (\sigma_t^{-1}(P_1))^{c_t},$$

with α, I, and the same exponents c_t as in Theorem 1.

Proof. Take the relative norm of the factorization in Theorem 1, recalling from Section 1 that $\mathrm{N}_{K(\zeta_m)/F(\zeta_m)}((\alpha,\chi))$ and $(\alpha,\chi)^m$ generate the same principal ideal in $\mathcal{O}_{F(\zeta_m)}$.

Note: From Corollary 1 and the congruence $c_t \equiv t \pmod{m}$ for each t, we see that the class of the product $\prod_{\substack{0<t<m\\(t,m)=1}} (\sigma_t^{-1}(P_1))^t$ in the ideal class group of $F(\zeta_m)$ is an mth power.

Corollary 2. *If $\alpha \in \mathcal{O}_K$ is such that $v_{\mathfrak{p}}(d_{K/F}(\alpha,\tau(\alpha),\ldots,\tau^{m-1}(\alpha))) = m-1$, as in Proposition 1, then $c_t = t$ for each t, i.e., the factorization of $((\alpha,\chi)^m)$ above $\mathfrak{p}$ takes the classic Stickelberger form.*

Proof. Set $\alpha_i = \tau^i(\alpha)$ for $0 \le i \le m-1$, and write $\alpha_i \equiv \sum_{j=0}^{m-1} b_{ij}\pi^j \pmod{\mathcal{P}_1^m}$, where $b_{ij} \in \{0,\ldots,p-1\}$. Let $\mathbf{B} = (b_{ij})$, $0 \le i,j \le m-1$, be the corresponding matrix. We also assume here that $\pi \in \mathcal{O}_K$ is a uniformizer with respect to $\mathfrak{P}$ and note that π is a uniformizer in $K(\zeta_m)$ with respect to any prime ideal lying over $\mathfrak{P}$. Given this choice of π, we have $v_{\mathfrak{p}}(d_{K/F}(1,\pi,\ldots,\pi^{m-1})) = v_{\mathfrak{p}}(d(K/F)) = m-1$, and since $v_{\mathfrak{p}}(d_{K/F}(\alpha_0,\alpha_1,\ldots,\alpha_{m-1})) = m-1$ by assumption, it follows that $p \nmid \det(\mathbf{B})$.

Let $\mathbf{V} = (s^{ij})$ for $0 \le i,j \le m-1$, where $s = (r^{(p-1)/m})^{m-1}$. Then

$$\begin{pmatrix} (\alpha,\chi^0) \\ (\alpha,\chi^1) \\ \cdot \\ \cdot \\ \cdot \\ (\alpha,\chi^{m-1}) \end{pmatrix} \equiv \mathbf{V} \begin{pmatrix} \alpha_0 \\ \alpha_1 \\ \cdot \\ \cdot \\ \cdot \\ \alpha_{m-1} \end{pmatrix} \equiv \mathbf{VB} \begin{pmatrix} 1 \\ \pi \\ \cdot \\ \cdot \\ \cdot \\ \pi^{m-1} \end{pmatrix} \pmod{\mathcal{P}_1^m},$$

where there are m congruences to be read along the rows. The matrix $\mathbf{V}$ occurs in the theory of Fourier transforms over $\mathbb{F}_p$ [Po] and appears here when we replace the mth roots of unity in the Lagrange resolvents by the appropriate powers of the primitive root r modulo $\mathcal{P}_1^m$. Let $\mathbf{W} = \mathbf{VB}$, and note that the matrix $\mathbf{W}$ is upper-triangular since $v_{\mathcal{P}_1}((\alpha,\chi^k)) \ge k$ for $0 \le k \le m-1$, as we saw in the proof of Theorem 1. Since $p \nmid \det(\mathbf{W})$, all of the diagonal elements of $\mathbf{W}$ are non-zero and so $v_{\mathcal{P}_1}((\alpha,\chi^k)) = k$ for $0 \le k \le m-1$. This implies in turn that $c_t = t$ for each t.

We now turn to Lagrange resolvents constructed from a Stark unit ϵ_0. Recall from equation (7) that $\rho(\epsilon_0) = \frac{1}{\epsilon_0}$ for an element $\rho \in \mathrm{Gal}(K/F)$. The inertia group of the prime ideal $\mathfrak{P}$ over F is equal to $\mathrm{Gal}(K/F)$, and so $\frac{1}{\epsilon_0} \equiv \epsilon_0 \pmod{\mathfrak{P}}$

by definition of the inertia group, or $\epsilon_0^2 \equiv 1 \pmod{\mathfrak{P}}$. From now on, we let a denote the unique choice of -1 or 1 such that $v_{\mathfrak{P}}(\epsilon_0 - a) > 0$. The case where $v_{\mathfrak{P}}(\epsilon_0 - a) = 1$ is particularly nice since we can make a "canonical" choice for π with $\pi := \epsilon_0 - a$. We assume this holds for now, and prove a partial analogue to the Stickelberger congruence (3) in Section 2 (recall that Stickelberger's congruence depends critically upon the choice of the uniformizer). In the discussion below, we set $\epsilon_k = \tau^k(\epsilon_0)$ for $k = 0, \ldots, m-1$.

Theorem 2. *Assume that* $v_{\mathfrak{P}}(\epsilon_0 - a) = 1$ *and set* $\pi = \epsilon_0 - a$ *. Then*
(i) $(\epsilon_0, \chi^0) \equiv ma \pmod{\mathcal{P}_1}$,
(ii) $(\epsilon_0, \chi^1) \equiv m\pi \pmod{\mathcal{P}_1^2}$.

Proof. (i) This is immediate since $\epsilon_k \equiv a \pmod{\mathcal{P}_1}$ for $k = 0, \ldots, m-1$, and $(\epsilon_0, \chi^0) = \epsilon_0 + \epsilon_1 + \cdots + \epsilon_{m-1}$.
(ii) Clearly $\epsilon_0 \equiv a + 1 \cdot \pi \pmod{\mathcal{P}_1^2}$, and by Lemma 2 we have

$$\begin{aligned} \epsilon_1 &\equiv a + r^{(p-1)/m} \cdot \pi \pmod{\mathcal{P}_1^2} \\ \epsilon_2 &\equiv a + r^{2(p-1)/m} \cdot \pi \pmod{\mathcal{P}_1^2} \\ &\cdot \\ &\cdot \\ &\cdot \\ \epsilon_{m-1} &\equiv a + r^{(m-1)(p-1)/m} \cdot \pi \pmod{\mathcal{P}_1^2}. \end{aligned}$$

We have $(\epsilon_0, \chi) = \epsilon_0 + \zeta_m \epsilon_1 + \cdots + \zeta_m^{m-1} \epsilon_{m-1}$ and also $\zeta_m \equiv r^{(m-1)(p-1)/m} \pmod{\mathcal{P}_1^m}$. Therefore

$$\begin{aligned} (\epsilon_0, \chi) &\equiv \Big[1 + r^{(m-1)(p-1)/m} \cdot r^{(p-1)/m} + \cdots \\ &\quad + r^{(p-1)/m} \cdot r^{(m-1)(p-1)/m}\Big] \pi \pmod{\mathcal{P}_1^2} \\ &\equiv m\pi \pmod{\mathcal{P}_1^2}, \end{aligned}$$

which concludes the proof of (ii).

We also observed that $(\epsilon_0, \chi^2) \equiv (\frac{am}{2})\pi^2 \pmod{\mathcal{P}_1^3}$ in the setting of Theorem 2, when $m \geq 4$. This last congruence seems to be directly tied to the equality $\epsilon_{m/2} = 1/\epsilon_0$, but we have as yet no proof. What we observed with $m \geq 4$ in terms of the matrix $\mathbf{W} = (w_{ij})$, $0 \leq i, j \leq m-1$, that was introduced in the proof of Corollary 2 was the following pattern (recall that $\mathbf{W}$ is upper-triangular and $\pi = \epsilon_0 - a$ is the uniformizer):

$$\mathbf{W} = \begin{pmatrix} ma & 0 & 0 & 0 & \cdot & \cdot & \cdot \\ & m & \frac{m}{2} & * & * & * & * \\ & & \frac{am}{2} & \frac{am}{2} & & & \\ & & & * & & & \\ & & & & * & & \\ & & & & & * & \\ & & & & & & * \end{pmatrix},$$

where an asterisk indicates that several different possible values were observed in these positions with respect to certain constraints. The one constraint in column 4, for example, with the two undetermined coefficients there was that the sum going down the column was always congruent to 0 modulo p. Therefore, the w_{13} and w_{33} entries depend directly upon each other. The following table summarizes some systematic computations we carried out when $m = 4$. If $w_{33} \neq 0$, then $((\epsilon_0, \chi)^4)$ has the classic Stickelberger factorization above $\mathfrak{p}$. We computed 90 examples with F real quadratic, discriminant range $5 \leq d_F < 5000$, $\mathrm{N}\mathfrak{p} = 29$, and $m = 4$. In 3 of these examples, $v_{\mathfrak{P}}(\epsilon_0 - a) = 3$. In the remaining 87 examples, we have $v_{\mathfrak{P}}(\epsilon_0 - a) = 1$. The table below covers these 87 examples, where the second row gives the frequency that w_{33} was equal to the value in the first row.

0	1	2	3	4	5	6	7	8	9	10	11	12	13	14	15	16	17	18
5	1	7	0	3	3	0	2	3	5	5	1	1	1	4	5	4	5	4

19	20	21	22	23	24	25	26	27	28
3	3	2	3	3	3	3	3	2	3

Thus w_{33} was non-zero in 82 of the 87 examples in the table. In terms of the exponents appearing in both Theorem 1 and Corollary 1, we have the following possibilities when $m = 4$:

(i) $v_{\mathfrak{P}}(\epsilon_0 - a) = 1$, $w_{33} \neq 0$: $c_1 = 1$, $c_3 = 3$.
(ii) $v_{\mathfrak{P}}(\epsilon_0 - a) = 1$, $w_{33} = 0$: $c_1 = 1$, $c_3 > 3$.
(iii) $v_{\mathfrak{P}}(\epsilon_0 - a) = 3$: $c_1 > 1$, $c_3 = 3$.
(iv) $v_{\mathfrak{P}}(\epsilon_0 - a) > 3$: $c_1 > 1$, $c_3 > 3$.

Case (i) is the dominant one and gives the classic Stickelberger factorization above $\mathfrak{p}$. Having $v_{\mathfrak{P}}(\epsilon_0 - a) = 2$ is not possible as we'll see below. Case (iv) was only observed once out of hundreds of different examples.

In order to obtain some understanding for the observed frequencies when $m = 4$, we consider the following model which runs along similar lines to the notation and method used in the proof of Corollary 2. Let the indeterminate z play the role of a fixed uniformizer $\pi \in \mathcal{O}_K$ with respect to $\mathfrak{P}$. The congruence classes modulo $\mathcal{P}_1^4$ form a ring isomorphic to the quotient ring $\mathcal{R} := \mathbb{F}_p[z]/(z^4)$. Let $\gamma_0 = a + b_{01}z + b_{02}z^2 + b_{03}z^3 + (z^4) \in \mathcal{R}$ (we drop the $+(z^4)$ below). We define a map $s : \mathcal{R} \to \mathcal{R}$ as follows: s is trivial on $\mathbb{F}_p$ and $s : z \mapsto r^{(p-1)/m}z + e_2z^2 + e_3z^3$. This map s models the action of τ on π recorded in Lemma 2. The elements $\gamma_1, \gamma_2, \gamma_3 \in \mathcal{R}$ defined below are now uniquely determined:

$$\begin{aligned} \gamma_1 &= s(\gamma_0) = a + b_{11}z + b_{12}z^2 + b_{13}z^3 \\ \gamma_2 &= s(\gamma_1) = a + b_{21}z + b_{22}z^2 + b_{23}z^3 \\ \gamma_3 &= s(\gamma_2) = a + b_{31}z + b_{32}z^2 + b_{33}z^3. \end{aligned}$$

Let $\mathbf{B} = (b_{ij})$, $0 \leq i, j \leq 3$, as before, where $b_{i0} = a$ for $0 \leq i \leq 3$. We now compute the matrix $\mathbf{W} = \mathbf{VB}$ ($\mathbf{V}$ is the fixed matrix defined in the proof of Corollary 2) as we vary the 5 parameters $b_{01}, b_{02}, b_{03}, e_2$, and e_3 among the values $\{0, 1, \ldots, p-1\}$. The special form of the map $s : \mathcal{R} \to \mathcal{R}$ guarantees that we have $s(\gamma_3) = \gamma_0$ and that $\mathbf{W}$ is upper-triangular for any combination of values for the 5 parameters above. In order to have $\gamma_0, \gamma_1, \gamma_2, \gamma_3$ model the behavior of $\epsilon_0, \epsilon_1, \epsilon_2, \epsilon_3$, we count how many times a given possibility occurs under the constraint that we must have $\gamma_0\gamma_2 = \gamma_1\gamma_3 = 1 \in \mathcal{R}$ (there does not seem to be any other constraint on the ϵ's) as we let the 5 parameters vary through all possible values. The following counts were obtained with $a = 1$ (the same counts hold for $a = -1$):

$(i)'$ $b_{01} \neq 0$, $w_{33} \neq 0$: count $= (p-1)^2p^2$.
$(ii)'$ $b_{01} \neq 0$, $w_{33} = 0$: count $= (p-1)p^2$.
$(*)$ $b_{01} = 0$, $b_{02} \neq 0$: count $= 0$.
$(iii)'$ $b_{01} = b_{02} = 0$, $b_{03} \neq 0$: count $= (p-1)p^2$.
$(iv)'$ $b_{01} = b_{02} = b_{03} = 0$: count $= p^2$.

Cases $(i)'$, $(ii)'$, $(iii)'$, and $(iv)'$ correspond to the cases (i), (ii), (iii), (iv) defined above. This model shows a good agreement to the results of the computations discussed above. The fact that we obtain a count of 0 in case $(*)$ shows why $v_{\mathfrak{P}}(\epsilon_0 - a) = 2$ is not possible. We also noted with these counts that in the case where $b_{01} \neq 0$ we find that w_{33} takes each value in the set $\{0, 1, \ldots, p-1\}$ equally often.

We noted earlier that the valuation $v_{\mathfrak{p}}(d_{K/F}(\epsilon_0, \ldots, \epsilon_{m-1}))$ is occasionally greater than $m-1$. If this valuation is equal to $m-1$, then the factorization of $((\epsilon_0, \chi)^m)$ above $\mathfrak{p}$ takes the classic Stickelberger form as noted in Corollary 2. The condition for having the classic Stickelberger factorization above $\mathfrak{p}$ is actually weaker, namely, we have this special factorization if and only if $v_{\mathfrak{P}}(\epsilon_0 - a) = 1$ and $w_{tt} \neq 0$ for all $0 < t < m$ with $(t, m) = 1$ when using $\pi = \epsilon_0 - a$ as the uniformizer with respect to $\mathfrak{P}$. For example, if $m = 6$ and $\pi = \epsilon_0 - a$ is a uniformizer with respect to $\mathfrak{P}$, then we can have $w_{44} = 0$ and $w_{55} \neq 0$ which implies that the above valuation is greater than 5 but we still have the classic Stickelberger factorization above $\mathfrak{p}$ (we have $w_{11} \neq 0$ by Theorem 2(ii)). The table below summarizes some systematic computations we carried out when $m = 6$. We computed 61 examples with F real quadratic, discriminant range $5 \leq d_F \leq 5900$, $\mathrm{N}\mathfrak{p} = 19$, and $m = 6$. In 4 of these examples, $v_{\mathfrak{P}}(\epsilon_0 - a) = 3$. In the remaining 57 examples, we have $v_{\mathfrak{P}}(\epsilon_0 - a) = 1$. The table below covers these 57 examples, where the second row gives the frequency that w_{55} was equal to the value in the first row. Note that w_{55} was non-zero in 54 of the 57 examples in the table. All of our computations were carried out with the package PARI/GP [GP] and the method used for computing the first derivatives of the partial zeta-functions at $s = 0$ is based upon the Lavrik-Friedman formula (see [DT]).

0	1	2	3	4	5	6	7	8	9	10	11	12	13	14	15	16	17	18
3	4	2	3	0	5	5	6	2	4	3	4	2	2	2	4	2	2	2

We applied a chi-square test to the data in each table under the null hypothesis that the distribution into the different congruence classes was uniform. In both cases, for the value $\alpha = 0.05$, the null hypothesis was not rejected (we were far from the critical value in both cases). In view of the fact that there are zeros in some of the cells, it might be more appropriate to apply the exact permutation test than the chi-square test but since the margin was quite safe in both cases we believe the same conclusion would have been reached.

Acknowledgements. I'd like to thank Cornelius Greither for his help with the proof of Lemma 2. Thanks also to Werner Bley for making me aware of the truth of Proposition 1 and for several helpful conversations we had while I was enjoying the kind hospitality of the Number Theory group at the University of Augsburg. Finally, I would also like to thank the referees for their helpful comments which enabled me to make several improvements to the original version of this paper.

References

[Br] Jan Brinkhuis, *Gauss sums and their prime factorization*, L'Enseignement Math. **36** (1990), 39–51.

[CT] Philippe Cassou-Noguès and Martin J. Taylor, *Elliptic Functions and Rings of Integers*, Birkhäuser, Boston, 1987.

[Ch] Robin J. Chapman, *A simple proof of Noether's theorem*, Glasgow Math. J. **38** (1996), 49–51.

[Co] Henri Cohen, *Advanced Topics in Computational Number Theory*, Springer-Verlag, New York, 2000.

[DST] David S. Dummit, Jonathan W. Sands, and Brett A. Tangedal, *Stark's conjecture in multi-quadratic extensions, revisited*, J. Théor. Nombres Bordeaux **15** (2003), 83–97.

[DT] David S. Dummit and Brett A. Tangedal, *Computing the lead term of an abelian L-function*, Lecture Notes in Computer Science **1423**, Springer, Berlin, 1998, 400–411.

[GP] C. Batut, K. Belabas, D. Bernardi, H. Cohen, and M. Olivier, *User's guide to PARI/GP version 2.1.3*, 2000.

[Ha] Helmut Hasse, *Number Theory*, Springer-Verlag, New York, 1980.

[Ja] Gerald J. Janusz, *Algebraic Number Fields*, Academic Press, New York, 1973.

[Le] Franz Lemmermeyer, *Reciprocity Laws*, Springer-Verlag, Berlin, 2000.

[Na] Władysław Narkiewicz, *Elementary and Analytic Theory of Algebraic Numbers*, 2nd Ed., Springer-Verlag, New York, 1990.

[Ne] Jürgen Neukirch, *Algebraic Number Theory*, Springer-Verlag, New York, 1999.

[Po] J. M. Pollard, *The fast Fourier transform in a finite field*, Math. Comp. **25** (1971), 365–374.

[Sc] Reinhard Schertz, *Galoismodulstruktur und Elliptische Funktionen*, J. Number Theory **39** (1991), 285–326.

[Sh] Takuro Shintani, *On certain ray class invariants of real quadratic fields*, J. Math. Soc. Japan **30** (1978), no. 1, 139–167.

[St1] Harold M. Stark, *L-functions at $s = 1$. III. Totally real fields and Hilbert's twelfth problem*, Advances in Math. **22** (1976), 64–84.

[St2] ———, *L-functions at $s = 1$. IV. First derivatives at $s = 0$*, Advances in Math. **35** (1980), 197–235.

[Ta] John Tate, *Les conjectures de Stark sur les fonctions L d'Artin en s* $= 0$, Notes d'un cours à Orsay rédigées par Dominique Bernardi et Norbert Schappacher, Birkhäuser, Boston, 1984.

[Wa] Lawrence C. Washington, *Introduction to Cyclotomic Fields*, 2nd Ed., Springer-Verlag, New York, 1997.

[We] André Weil, *La cyclotomie jadis et naguère*, Enseign. Math. **20** (1974), 247–263.

[Zh] Fuzhen Zhang, *Matrix Theory*, Springer-Verlag, New York, 1999.

Cryptanalysis of a Divisor Class Group Based Public-Key Cryptosystem

Aaram Yun, Jaeheon Kim, and Dong Hoon Lee

National Security Research Institute

Abstract. We prove that the public-key cryptosystem proposed by H. Kim and S. Moon, in "New Public-Key Cryptosystem Using Divisor Class Groups", ACISP 2001 [1] is not secure by solving discrete logarithm problem for the divisor class groups used in the proposal.

1 Introduction

In [1], H. Kim and S. Moon proposed a public-key cryptosystem based on ideal arithmetic in the divisor class group of an affine normal subring of $\mathbb{F}[X, Y]$ for some field $\mathbb{F}$. The security of the system is based on the difficulty of the discrete logarithm problem in the divisor class group. There were many proposals of cryptosystems based on divisor class groups of imaginary quadratic orders of quadratic fields, but as far as we know, that was the only proposal based on divisor class groups of polynomial rings.

The authors of the proposal suggest that the discrete logarithm problem for the divisor class group is much more difficult than that of the class group of an order of a quadratic number field. But in this paper we show that actually, there is a 'standard' generator $[P]$ for the divisor class group, so that each element $[I]$ can be written as $[P]^k$ for some k, and moreover the representation for the element $[I]$ contains the information about the exponent k rather transparently.

2 Overview of the Scheme

For mathematical backgrounds and complete treatments, see the original proposal [1]. The goal of this section is mostly to fix notations.

2.1 General Definitions

Throughout this article, $\mathbb{F}$ will denote a fixed base field.

Let R be an integral domain, and K its quotient field. Then K is an R-module. If J is an R-submodule of K, and if $xJ \subseteq R$ for some nonzero $x \in R$, then we say that J is a *fractional ideal.*

When I, J are two fractional ideals, we can multiply and divide them as:

$$IJ = \{\sum_{\text{finite}} \alpha\beta \mid \alpha \in I, \beta \in J\}$$

D.A. Buell (Ed.): ANTS 2004, LNCS 3076, pp. 442–450, 2004.

$$I : J = \{x \in K \mid xJ \subseteq I\}$$

Especially we denote by I^{-1} the ideal quotient $R : I$, and we define the 'divisorial closure' $\overline{I} = (I^{-1})^{-1}$. We say that a fractional ideal I is *divisorial* if $I = \overline{I}$. Note that $\overline{I}$ is equal to the intersection of all principal fractional ideals containing I.

We denote by $\mathcal{D}(R)$ the set of divisorial ideals. It becomes a commutative monoid when we define the product $I * J$ to be $\overline{I \cdot J}$, where $I \cdot J$ is the (fractional) ideal product.

In general $\mathcal{D}(R)$ is *not* a group, and it becomes a group if and only if R is completely integrally closed. It will be the case for the rings of type $R = R_{n,j}$ which we will introduce shortly, so then we can define our divisor class group as follows:

We define the divisor class group $Cl(R)$ to be the quotient $\mathcal{D}(R)/\mathcal{P}(R)$, where $\mathcal{P}(R)$ is the subgroup of the principal fractional ideals. We say that an element of $Cl(R)$ is a *divisor class*, and when J is a divisorial ideal, then we denote its divisor class by $[J]$.

2.2 Cryptosystem Proposed by Kim and Moon

Kim and Moon proposed [1] a public-key cryptosytem based on divisor class groups of certain polynomial rings. Its security is based on the difficulty of the discrete logarithm problem of the divisor class group, and the Gröbner basis computation is used as the basic tool for ideal computation.

For motivation of actual candidates for such polynomial rings, they refer to a theorem proved by D. F. Anderson in [2]:

Theorem 2.1 (Theorem 2.5 of [2]). *Let R be an affine integrally closed subring of $T = \mathbb{F}[X, Y]$ generated by monomials, where $\mathbb{F}$ is a field. Moreover suppose that T is integral over R. Then R is isomorphic either to T itself or to*

$$R_{n,j} := \mathbb{F}[X^n, XY^j, X^2Y^{\overline{2j}}, \dots, X^{n-1}Y^{\overline{(n-1)j}}, Y^n],$$

where $0 < j < n$ and $\gcd(j, n) = 1$. Here $\overline{m}$ denotes the smallest representative in $\mathbb{N}$ of the congruence class of m modulo n.

Since $R_{n,j}$ is integrally closed Noetherian domain, it is a Krull domain by Mori-Nagata Theorem ([4]). So it is completely integrally closed and we may talk about its divisor class group $Cl(R_{n,j})$. D. F. Anderson also gives the divisor class group computation for such rings:

Theorem 2.2 (Theorem 4.4 and its proof of [2]). *Let's use the same notations as in the Theorem 2.1.*

Then $Cl(R_{n,j})$ is isomorphic to $\mathbb{Z}/n\mathbb{Z}$.

Moreover if we define the prime ideal

$$P = \langle X^n, XY^j, X^2Y^{\overline{2j}}, \dots, X^{n-1}Y^{\overline{(n-1)j}} \rangle$$

then P is a generator of the divisor class group $Cl(R_{n,j})$.

Since the general $R_{n,j}$ is rather unwieldy for the Gröbner basis computation, Kim and Moon use $R_{n,1} = \mathbb{F}[X^n, XY, Y^n]$ as the ring of choice. Their cryptosystem can be defined simply as ElGamal and Diffie-Hellman scheme in the divisor class group $Cl(R_{n,1})$.

In this paper we prove most results for the general case $R_{n,j}$, and for the special case $j = 1$ we provide the computation of the complete structure of the cyclic group $Cl(R_{n,1})$ in Proposition 3.4 of Section 3.

For the rest of the article, we will maintain the notations given in this section. So j will be a fixed integer satisfying $0 < j < n$, $\gcd(j,n) = 1$, etc. When n and j are clear, often we'll abbreviate $R_{n,j}$ simply as R. In that case K will denote its quotient field.

3 Preparation

First, we characterize polynomials of $\mathbb{F}[X,Y]$ which are elements of $R_{n,j}$ as follows:

Proposition 3.1. *For any $f(X,Y) \in \mathbb{F}[X,Y]$, $f(X,Y) \in R_{n,j}$ if and only if every monomial X^aY^b occuring in $f(X,Y)$ satisfies $b \equiv aj \pmod{n}$.*

Proof. Recall that

$$R_{n,j} := \mathbb{F}[X^n, XY^j, X^2Y^{\overline{2j}}, \dots, X^{n-1}Y^{\overline{(n-1)j}}, Y^n].$$

Each of the monomials X^n, XY^j, $X^2Y^{\overline{2j}}$, ..., $X^{n-1}Y^{\overline{(n-1)j}}$ is of the form in the statement of the proposition. Therefore their products are again monomials of required form. Then any polynomial of $R_{n,j}$ will be a linear combination of monomials of required form.

Now we prove the other direction. We need only to prove that if X^aY^b satisfies the relation $b \equiv aj \pmod{n}$, then $X^aY^b \in R_{n,j}$.

First, let's divide a by n, and denote the quotient and the remainder by q and r, respectively. Then $a = qn + r$. So,

$$X^aY^b = (X^n)^q \cdot X^rY^b.$$

But $X^n \in R_{n,j}$. So we need only to show that $X^rY^b \in R_{n,j}$.

Therefore without loss of generality we may assume from the start that $0 \leq a < n$.

By definition, $\overline{aj} \equiv aj \equiv b \pmod{n}$ and $b \geq \overline{aj}$. Hence $b - \overline{aj} = ni$ for some $i \geq 0$. Then,

$$X^aY^b = X^aY^{\overline{aj}}Y^{b-\overline{aj}} = X^aY^{\overline{aj}} \cdot (Y^n)^i \in R_{n,j}.$$

□

Next, we need to characterize those polynomials of $\mathbb{F}[X,Y]$ which are elements of K, the quotient field of $R = R_{n,j}$.

Proposition 3.2. *We have*

$$K \cap \mathbb{F}[X,Y] = R.$$

Proof. Computational proof is possible, but since $R_{n,j}$ is integrally closed and $\mathbb{F}[X,Y]$ is integral over $R_{n,j}$, any element of $K \cap \mathbb{F}[X,Y]$ should be in R. □

Recall that from the Theorem 2.2,

$$P = \langle X^n, XY^j, X^2Y^{\overline{2j}}, \ldots, X^{n-1}Y^{\overline{(n-1)j}} \rangle.$$

Lemma 3.1. *We have*

$$X^n \in \overline{P^k}.$$

Proof. Since $\overline{P^k}$ is the intersection of all principal fractional ideals containing P^k, in order to show that $X^n \in \overline{P^k}$, we need to prove that whenever $x \in K$ satisfies $x \cdot P^k \subseteq R$, we have $x \cdot X^n \in R$.

Suppose that $x \in K$ is such an element. Since $X^kY^{kj} \in P^k$, we have $x \cdot X^kY^{kj} \in R$. So we can write x as f/X^kY^{kj} for some $f \in R$.

And since $X^{nk} \in P^k$,

$$x \cdot X^{nk} = \frac{f}{X^kY^{kj}} X^{nk} \in R.$$

Therefore $X^kY^{kj} \mid f \cdot X^{nk}$ in the polynomial ring $\mathbb{F}[X,Y]$. Then $Y^{kj} \mid f \cdot X^{(n-1)k}$. Since Y^{kj} and $X^{(n-1)k}$ are mutually prime, we get $Y^{kj} \mid f$.

Now we have to show that $x \cdot X^n \in R$. But,

$$x \cdot X^n = \frac{f}{X^kY^{kj}} \cdot X^n = \frac{f}{Y^{kj}} \cdot X^{n-k}$$

Since f/Y^{kj} and X^{n-k} are in $\mathbb{F}[X,Y]$, $x \cdot X^n$ is in $K \cap \mathbb{F}[X,Y] = R$. Therefore $X^n \in \overline{P^k}$. □

Throughout this article, when $g_1, \ldots, g_r$ are elements of $R_{n,j} \subseteq \mathbb{F}[X,Y]$, $\gcd(g_1, \ldots, g_r)$ is the greatest common divisor of $g_1, \ldots, g_r$ in the polynomial ring $\mathbb{F}[X,Y]$. If S is an arbitrary subset of $R_{n,j}$ (or more generally of $\mathbb{F}[X,Y]$), then $\gcd(S)$ is defined as the greateast common divisor of elements of S. Note that if I is an ideal of $R_{n,j}$ generated by $g_1, \ldots, g_r \in R_{n,j}$, then $\gcd(I) = \gcd(g_1, \ldots, g_r)$. Since $R_{n,j}$ is an affine $\mathbb{F}$-algebra, every ideal I of $R_{n,j}$ is finitely generated, so $\gcd(I)$ is well defined.

Proposition 3.3. *We have*

$$\gcd(\overline{P^k}) = X^k.$$

Proof. Since X^n and X^kY^{kj} are in $\overline{P^k}$, $\gcd(\overline{P^k})$ divides $\gcd(X^n, X^kY^{kj}) = X^k$. So we need only to prove that X^k divides any element of $\overline{P^k}$.

First, we claim that

$$\frac{Y^n}{X^kY^{\overline{kj}}} \cdot P^k \subseteq R.$$

Any element of P^k is of form $F_1F_2 \cdots F_k$, where each F_i is one of X^n, XY^j, $X^2Y^{\overline{2j}}, \ldots, X^{n-1}Y^{\overline{(n-1)j}}$. Now, in case any of F_i is X^n, then clearly

$$\frac{Y^n}{X^kY^{\overline{kj}}} \cdot F_1F_2 \cdots F_k \in K \cap \mathbb{F}[X,Y] = R.$$

In case none of F_i is X^n, then each F_i has at least one factor X, therefore $X^k \mid F_1F_2 \cdots F_k$. Therefore again the above containment is true.

Now suppose that f is any element of $\overline{P^k}$. Then since $Y^n/(X^kY^{\overline{kj}}) \cdot P^k \subseteq R$, we have

$$\frac{Y^n}{X^kY^{\overline{kj}}} \cdot f \in R.$$

Hence $X^k \mid Y^{n-\overline{kj}} \cdot f$. From this it follows that $X^k \mid f$. □

The above proposition gives much information about the ideal $\overline{P^k}$. Since the divisor class $[P]$ is a generator of the divisor class group $Cl(R_{n,j})$, we will use this to analyze the cryptosystem proposed by Kim and Moon in the next section.

For the special case $j = 1$, we can say even further and actually it is possible to explicitly compute $\overline{P^k}$, thereby determining the complete structure of the divisor class group $Cl(R_{n,1})$ of order n:

Proposition 3.4. *Let $j = 1$. Then $R = R_{n,1} = \mathbb{F}[X^n, XY, Y^n]$. Again let P be the prime ideal $\langle X^n, XY \rangle$. Then*

$$R : P^k = \left\langle \frac{(XY)^{n-k}}{X^n}, 1 \right\rangle \quad \textit{and} \quad \overline{P^k} = \langle X^n, (XY)^k \rangle$$

Proof. Suppose that $f/g \in R : P^k$. Then $f/g \cdot P^k \subseteq R$. Since $(XY)^k \in P^k$, $f/g \cdot (XY)^k = h$ for some $h \in \mathbb{F}[X^n, XY, Y^n]$. Then $f/g = h/(XY)^k$. Therefore we may assume that $g = (XY)^k$.

Also, since $(X^n)^k \in P^k$, $f/(XY)^k \cdot X^{nk} = h_1$ for some $h_1 \in R$. Simplifying, we have

$$f \cdot X^{(n-1)k} = h_1 \cdot Y^k$$

Therefore $Y^k \mid f$. But f is in $\mathbb{F}[X^n, XY, Y^n]$. Each monomial occuring in f is then a product of X^n, XY, and Y^n's. Collect all monomials of f which contain Y^n, and we denote it by $f_1 \cdot Y^n$. This part is divisible by Y^k. Now, since f should be divisible by Y^k, rest of the monomials should contain at least k factors of XY. Therefore if we collect all of them, this can be written as $f_2 \cdot (XY)^k$. Therefore

$$f = f_1 \cdot Y^n + f_2 \cdot (XY)^k \quad \text{for some } f_1,\, f_2 \in R.$$

Then, f/g is equal to

$$\begin{aligned}\frac{f}{g} &= \frac{f_1 \cdot Y^n + f_2 \cdot (XY)^k}{(XY)^k} \\ &= f_1 \cdot \frac{Y^n}{(XY)^k} + f_2 \cdot 1 \\ &= f_1 \cdot \frac{X^n Y^n}{X^n (XY)^k} + f_2 \cdot 1 \\ &= f_1 \cdot \frac{(XY)^{n-k}}{X^n} + f_2 \cdot 1\end{aligned}$$

Therefore $R : P^k \subseteq \langle (XY)^{n-k}/X^n, 1\rangle$. And it is trivial to check that both $(XY)^{n-k}/X^n$ and 1 are in $R : P^k$. So $R : P^k = \langle (XY)^{n-k}/X^n, 1\rangle$.

Now let's prove that the inverse of $\langle (XY)^{n-k}/X^n, 1\rangle$ is $\langle X^n, (XY)^k\rangle$. Suppose that f/g is in the inverse of $\langle (XY)^{n-k}/X^n, 1\rangle$. Then by definition $f/g \cdot (XY)^{n-k}/X^n \in R$ and $f/g \cdot 1 \in R$. Therefore again without loss of generality we may let $g = 1$. Then $f \cdot (XY)^{n-k}/X^n = h$ for some $h \in R = \mathbb{F}[X^n, XY, Y^n]$. Simplifying, $f \cdot Y^{n-k} = h \cdot X^k$. So $X^k \mid f$. As before, collect all terms occuring in f which contain at least one X^n and write $f_1 \cdot X^n$. Then for the rest of the terms, in order that $X^k \mid f$, they should contain at least k factors of XY. So we may write that

$$f = f_1 \cdot X^n + f_2 \cdot (XY)^k \quad \text{for some } f_1,\, f_2 \in R.$$

Hence $f \in \langle X^n, (XY)^k\rangle$. Again it is trivial to verify that both X^n and $(XY)^k$ are in the inverse of $\langle (XY)^{n-k}/X^n, 1\rangle$, proving that $\overline{P^k} = \langle X^n, (XY)^k\rangle$. □

Remark The above computation is consistent with the proposition that $[P]$ generates the divisor class group $Cl(R_{n,1})$ of order n: note that

$$\begin{aligned}[P^k]^{-1} &= \left[\left\langle \frac{(XY)^{n-k}}{X^n}, 1\right\rangle\right] \\ &= [\langle (XY)^{n-k}, X^n\rangle] \\ &= [P^{n-k}]\end{aligned}$$

4 Cryptanalysis of the Scheme

In their proposal [1], Kim and Moon choose the ring $R_{n,1}$ for their divisor class group construction. Divisor classes are represented by reduced ideals of $R_{n,1}$, and various ideal computations (product, quotient, 'divisorial closure', etc.) are done by the Gröbner basis method.

Considered as an abstract group, there is no problem in applying public-key primitives based on discrete logarithm problem to the divisor class group $Cl(R_{n,j})$ for general j, but Kim and Moon restricts their choice to $j = 1$ in order to use the isomorphism $R_{n,1} = \mathbb{F}[X^n, XY, Y^n] \cong \mathbb{F}[x, y, z]/\langle z^n - xy \rangle$. They use this isomorphism to define and use the concept of reduced ideals which are unique representatives of divisor classes.

In this section we will solve the discrete logarithm problem for $Cl(R_{n,j})$.

Suppose that we are given a divisorial ideal I. Without loss of generality we may assume that I is in fact in $R_{n,j}$, that is, an ordinary ideal. We would like to compute k satisfying $[I] = [P]^k$ $(k = 0, \ldots, n-1)$.

Lemma 4.1. *For any ordinary divisorial ideal I, there exists a $f \in R_{n,j}$ and $k \in \{0, \ldots, n-1\}$ so that*

$$I = \frac{f}{X^n}\overline{P^k}.$$

Proof. Since $[P]$ generates the divisor class group $Cl(R_{n,j})$, there exists f, $g \in R_{n,j}$ and some $k \in \{0, \ldots, n-1\}$ so that

$$I = \frac{f}{g}\overline{P^k}.$$

So we need to show that in fact it suffices to take $g = X^n$.

Let $d = \gcd(f, g)$. Note that $d \in \mathbb{F}[X, Y]$ and in general $d \notin R_{n,j}$. Then there are two mutually prime polynomials f', $g' \in \mathbb{F}[X, Y]$ so that $f = d \cdot f'$, $g = d \cdot g'$.

Since $X^n \in \overline{P^k}$, we have

$$\frac{f}{g}X^n = \frac{f'}{g'}X^n \in I \subseteq R \subseteq \mathbb{F}[X, Y].$$

Therefore $g' \mid f' \cdot X^n$. Since f' and g' are mutually prime, $g' \mid X^n$. Therefore $g' = X^r$ for some $r \leq n$.

Then

$$\frac{f}{g} = \frac{f'}{X^r} = \frac{f' \cdot X^{n-r}}{X^n}.$$

Let f'' be the polynomial $f' \cdot X^{n-r}$. Then $f/g = f''/X^n$. But then $f'' = f/g \cdot X^n$, which is in the quotient field K. Since it is also a polynomial, it follows that $f'' \in R_{n,j}$. Therefore

$$I = \frac{f''}{X^n}\overline{P^k}$$

is the equation that we were searching for. □

Using the above lemma and the fact that $\gcd(\overline{P^k}) = k$, we may solve the discrete logarithm problem for $Cl(R_{n,j})$ as follows:

Given an ordinary divisorial ideal I, we have, by the above lemma, $I = f/X^n \cdot \overline{P^k}$ for some $f \in R_{n,j}$ and k. Then,

$$X^n \cdot I = f \cdot \overline{P^k}.$$

Taking the greatest common divisors of the both sides, we get

$$X^n \cdot \gcd(I) = f \cdot \gcd(\overline{P^k}) = f \cdot X^k.$$

Then $f = X^{n-k} \cdot \gcd(I)$.

We have to determine the unknown k. But f is in $R_{n,j}$, and there is a criterion for polynomial membership in $R_{n,j}$ (Proposition 3.1).

Take any monomial X^aY^b occuring in the polynomial $\gcd(I) \in \mathbb{F}[X,Y]$. Then $X^{n-k+a}Y^b$ occurs as a monomial in f. Since $f \in R_{n,j}$, it follows that $b \equiv (n-k+a)j \equiv (a-k)j \pmod{n}$. Therefore we have

$$k \equiv a - bj^{-1} \pmod{n}$$

where j^{-1} is taken modulo n.

5 Some Examples

Example 1. Consider the following prime ideal

$$P' = \langle XY^j, X^2Y^{\overline{2j}}, X^3Y^{\overline{3j}}, \dots, X^{n-1}Y^{\overline{(n-1)j}}, Y^n \rangle.$$

By the symmetry of X and Y, P' also generates $Cl(R_{n,j})$. What is the exponent k such that $[P'] = [P]^k$? Clearly, $\gcd(P') = Y$. Therefore in order that $X^{n-k} \cdot Y \in R_{n,j}$, $1 \equiv (n-k)j \pmod{n}$, that is $k \equiv n - j^{-1}$, so $k = n - j^{-1}$.

Example 2. Consider the following ideal

$$I = \langle\, X^n + X^{n-1}Y^{(n-1)j}, XY^j + Y^{j+(n-1)j}, X^2Y^{\overline{2j}} + XY^{\overline{2j}+(n-1)j}, \dots, \\ X^{n-1}Y^{\overline{(n-1)j}} + X^{n-2}Y^{\overline{(n-1)j}+(n-1)j} \rangle.$$

In fact I is simply

$$\frac{X^n + X^{n-1}Y^{(n-1)j}}{X^n} P$$

but let's follow our methodology to see that if $k = 1$ indeed comes out.

Since $X^n + X^{n-1}Y^{(n-1)j} = X^{n-1}(X + Y^{(n-1)j})$, $XY^j + Y^{j+(n-1)j} = Y^j(X + Y^{(n-1)j}), \dots, X^{n-1}Y^{\overline{(n-1)j}} + X^{n-2}Y^{\overline{(n-1)j}+(n-1)j} = X^{n-2}Y^{\overline{(n-1)j}}(X + Y^{(n-1)j})$, we have

$$\gcd(I) = X + Y^{(n-1)j}.$$

Therefore in order that $X^{n-k} \cdot \gcd(I)$ be in $R_{n,j}$, the monomials X^{n-k+1} and $X^{n-k}Y^{(n-1)j}$ should satisfy the relation $b \equiv aj$ when a and b are exponents of X and Y, respectively. The unique solution in this case is $k = 1$. Therefore $[I] = [P]$.

6 Conclusions

In this paper we have shown that in case of the divisor class group $Cl(R_{n,j})$, the representation of elements of the group reveals the exponent with respect a generator $[P]$ transparently, thereby proving that the public key cryptosystem proposed by Kim and Moon is insecure.

In general, we suspect that the use of the Gröbner basis and large polynomial rings in construction of a cryptosystem is at best problematic, because the efficiency of the ideal computation is only loosely related to the security parameter of the divisor class group, thus making efficiency and security assertions about the system very difficult.

References

1. H. Kim, S. Moon, New Public-Key Cryptosystem Using Divisor Class Groups, in *V. Varadharajan and Y. Mu (Eds.): ACISP 2001*, LNCS 2119, pp. 74–83, 2001.
2. D. F. Anderson, Subrings of $k[X, Y]$ generated by monomials. Canad. J. Math., Vol. 30, pp. 215–224, 1978.
3. D. Cox, J. Little, D. O'Shea, Ideals, Varieties, and Algorithms: An Introduction to Computational Algebraic Geometry and Commutative Algebra, Springer-Verlag, New York, 1997.
4. M. Nagata, Local Rings, Interscience, New York, 1962.

Author Index

Lecture Notes in Computer Science

For information about Vols. 1–2972

please contact your bookseller or Springer-Verlag

Vol. 3083: W. Emmerich, A.L. Wolf (Eds.), Component Deployment. X, 249 pages. 2004.

Vol. 3076: D. Buell (Ed.), Algorithmic Number Theory. XI, 451 pages. 2004.

Vol. 3074: B. Kuijpers, P. Revesz (Eds.), Constraint Databases and Applications. XII, 181 pages. 2004.

Vol. 3065: A. Lomuscio, D. Nute (Eds.), Deontic Logic. X, 275 pages. 2004. (Subseries LNAI).

Vol. 3064: D. Bienstock, G. Nemhauser (Eds.), Integer Programming and Combinatorial Optimization. XI, 445 pages. 2004.

Vol. 3063: A. Llamosí, A. Strohmeier (Eds.), Reliable Software Technologies - Ada-Europe 2004. XIII, 333 pages. 2004.

Vol. 3062: J.L. Pfaltz, M. Nagl, B. Böhlen (Eds.), Applications of Graph Transformations with Industrial Relevance. XV, 500 pages. 2004.

Vol. 3060: A.Y. Tawfik, S.D. Goodwin (Eds.), Advances in Artificial Intelligence. XIII, 582 pages. 2004. (Subseries LNAI).

Vol. 3059: C.C. Ribeiro, S.L. Martins (Eds.), Experimental and Efficient Algorithms. X, 586 pages. 2004.

Vol. 3058: N. Sebe, M.S. Lew, T.S. Huang (Eds.), Computer Vision in Human-Computer Interaction. X, 233 pages. 2004.

Vol. 3056: H. Dai, R. Srikant, C. Zhang (Eds.), Advances in Knowledge Discovery and Data Mining. XIX, 713 pages. 2004. (Subseries LNAI).

Vol. 3054: I. Crnkovic, J.A. Stafford, H.W. Schmidt, K. Wallnau (Eds.), Component-Based Software Engineering. XI, 311 pages. 2004.

Vol. 3053: C. Bussler, J. Davies, D. Fensel, R. Studer (Eds.), The Semantic Web: Research and Applications. XIII, 490 pages. 2004.

Vol. 3052: W. Zimmermann, B. Thalheim (Eds.), Abstract State Machines 2004. Advances in Theory and Practice. XII, 235 pages. 2004.

Vol. 3047: F. Oquendo, B. Warboys, R. Morrison (Eds.), Software Architecture. X, 279 pages. 2004.

Vol. 3046: A. Laganà, M.L. Gavrilova, V. Kumar, Y. Mun, C.K. Tan, O. Gervasi (Eds.), Computational Science and Its Applications - ICCSA 2004. LIII, 1016 pages. 2004.

Vol. 3045: A. Laganà, M.L. Gavrilova, V. Kumar, Y. Mun, C.K. Tan, O. Gervasi (Eds.), Computational Science and Its Applications – ICCSA 2004. LIII, 1040 pages. 2004.

Vol. 3044: A. Laganà, M.L. Gavrilova, V. Kumar, Y. Mun, C.K. Tan, O. Gervasi (Eds.), Computational Science and Its Applications – ICCSA 2004. LIII, 1140 pages. 2004.

Vol. 3043: A. Laganà, M.L. Gavrilova, V. Kumar, Y. Mun, C.K. Tan, O. Gervasi (Eds.), Computational Science and Its Applications – ICCSA 2004. LIII, 1180 pages. 2004.

Vol. 3042: N. Mitrou, K. Kontovasilis, G.N. Rouskas, I. Iliadis, L. Merakos (Eds.), NETWORKING 2004, Networking Technologies, Services, and Protocols; Performance of Computer and Communication Networks; Mobile and Wireless Communications. XXXIII, 1519 pages. 2004.

Vol. 3039: M. Bubak, G.D.v. Albada, P.M. Sloot, J. Dongarra (Eds.), Computational Science - 2004. LXVI, 1271 pages. 2004.

Vol. 3038: M. Bubak, G.D.v. Albada, P.M. Sloot, J. Dongarra (Eds.), Computational Science - ICCS 2004. LXVI, 1311 pages. 2004.

Vol. 3037: M. Bubak, G.D.v. Albada, P.M. Sloot, J. Dongarra (Eds.), Computational Science - ICCS 2004. LXVI, 745 pages. 2004.

Vol. 3036: M. Bubak, G.D.v. Albada, P.M. Sloot, J. Dongarra (Eds.), Computational Science - ICCS 2004. LXVI, 713 pages. 2004.

Vol. 3035: M.A. Wimmer (Ed.), Knowledge Management in Electronic Government. XII, 326 pages. 2004. (Subseries LNAI).

Vol. 3034: J. Favela, E. Menasalvas, E. Chávez (Eds.), Advances in Web Intelligence. XIII, 227 pages. 2004. (Subseries LNAI).

Vol. 3033: M. Li, X.-H. Sun, Q. Deng, J. Ni (Eds.), Grid and Cooperative Computing. XXXVIII, 1076 pages. 2004.

Vol. 3032: M. Li, X.-H. Sun, Q. Deng, J. Ni (Eds.), Grid and Cooperative Computing. XXXVII, 1112 pages. 2004.

Vol. 3031: A. Butz, A. Krüger, P. Olivier (Eds.), Smart Graphics. X, 165 pages. 2004.

Vol. 3030: P. Giorgini, B. Henderson-Sellers, M. Winikoff (Eds.), Agent-Oriented Information Systems. XIV, 207 pages. 2004. (Subseries LNAI).

Vol. 3029: B. Orchard, C. Yang, M. Ali (Eds.), Innovations in Applied Artificial Intelligence. XXI, 1272 pages. 2004.

Vol. 3028: D. Neuenschwander, Probabilistic and Statistical Methods in Cryptology. X, 158 pages. 2004.

Vol. 3027: C. Cachin, J. Camenisch (Eds.), Advances in Cryptology - EUROCRYPT 2004. XI, 628 pages. 2004.

Vol. 3026: C. Ramamoorthy, R. Lee, K.W. Lee (Eds.), Software Engineering Research and Applications. XV, 377 pages. 2004.

Vol. 3025: G.A. Vouros, T. Panayiotopoulos (Eds.), Methods and Applications of Artificial Intelligence. XV, 546 pages. 2004. (Subseries LNAI).

Vol. 3024: T. Pajdla, J. Matas (Eds.), Computer Vision - ECCV 2004. XXVIII, 621 pages. 2004.

Vol. 3023: T. Pajdla, J. Matas (Eds.), Computer Vision - ECCV 2004. XXVIII, 611 pages. 2004.

Vol. 3022: T. Pajdla, J. Matas (Eds.), Computer Vision - ECCV 2004. XXVIII, 621 pages. 2004.

Vol. 3021: T. Pajdla, J. Matas (Eds.), Computer Vision - ECCV 2004. XXVIII, 633 pages. 2004.

Vol. 3019: R. Wyrzykowski, J. Dongarra, M. Paprzycki, J. Wasniewski (Eds.), Parallel Processing and Applied Mathematics. XIX, 1174 pages. 2004.

Vol. 3016: C. Lengauer, D. Batory, C. Consel, M. Odersky (Eds.), Domain-Specific Program Generation. XII, 325 pages. 2004.

Vol. 3015: C. Barakat, I. Pratt (Eds.), Passive and Active Network Measurement. XI, 300 pages. 2004.

Vol. 3014: F. van der Linden (Ed.), Software Product-Family Engineering. IX, 486 pages. 2004.

Vol. 3012: K. Kurumatani, S.-H. Chen, A. Ohuchi (Eds.), Multi-Agnets for Mass User Support. X, 217 pages. 2004. (Subseries LNAI).

Vol. 3011: J.-C. Régin, M. Rueher (Eds.), Integration of AI and OR Techniques in Constraint Programming for Combinatorial Optimization Problems. XI, 415 pages. 2004.

Vol. 3010: K.R. Apt, F. Fages, F. Rossi, P. Szeredi, J. Váncza (Eds.), Recent Advances in Constraints. VIII, 285 pages. 2004. (Subseries LNAI).

Vol. 3009: F. Bomarius, H. Iida (Eds.), Product Focused Software Process Improvement. XIV, 584 pages. 2004.

Vol. 3008: S. Heuel, Uncertain Projective Geometry. XVII, 205 pages. 2004.

Vol. 3007: J.X. Yu, X. Lin, H. Lu, Y. Zhang (Eds.), Advanced Web Technologies and Applications. XXII, 936 pages. 2004.

Vol. 3006: M. Matsui, R. Zuccherato (Eds.), Selected Areas in Cryptography. XI, 361 pages. 2004.

Vol. 3005: G.R. Raidl, S. Cagnoni, J. Branke, D.W. Corne, R. Drechsler, Y. Jin, C.G. Johnson, P. Machado, E. Marchiori, F. Rothlauf, G.D. Smith, G. Squillero (Eds.), Applications of Evolutionary Computing. XVII, 562 pages. 2004.

Vol. 3004: J. Gottlieb, G.R. Raidl (Eds.), Evolutionary Computation in Combinatorial Optimization. X, 241 pages. 2004.

Vol. 3003: M. Keijzer, U.-M. O'Reilly, S.M. Lucas, E. Costa, T. Soule (Eds.), Genetic Programming. XI, 410 pages. 2004.

Vol. 3002: D.L. Hicks (Ed.), Metainformatics. X, 213 pages. 2004.

Vol. 3001: A. Ferscha, F. Mattern (Eds.), Pervasive Computing. XVII, 358 pages. 2004.

Vol. 2999: E.A. Boiten, J. Derrick, G. Smith (Eds.), Integrated Formal Methods. XI, 541 pages. 2004.

Vol. 2998: Y. Kameyama, P.J. Stuckey (Eds.), Functional and Logic Programming. X, 307 pages. 2004.

Vol. 2997: S. McDonald, J. Tait (Eds.), Advances in Information Retrieval. XIII, 427 pages. 2004.

Vol. 2996: V. Diekert, M. Habib (Eds.), STACS 2004. XVI, 658 pages. 2004.

Vol. 2995: C. Jensen, S. Poslad, T. Dimitrakos (Eds.), Trust Management. XIII, 377 pages. 2004.

Vol. 2994: E. Rahm (Ed.), Data Integration in the Life Sciences. X, 221 pages. 2004. (Subseries LNBI).

Vol. 2993: R. Alur, G.J. Pappas (Eds.), Hybrid Systems: Computation and Control. XII, 674 pages. 2004.

Vol. 2992: E. Bertino, S. Christodoulakis, D. Plexousakis, V. Christophides, M. Koubarakis, K. Böhm, E. Ferrari (Eds.), Advances in Database Technology - EDBT 2004. XVIII, 877 pages. 2004.

Vol. 2991: R. Alt, A. Frommer, R.B. Kearfott, W. Luther (Eds.), Numerical Software with Result Verification. X, 315 pages. 2004.

Vol. 2990: J. Leite, A. Omicini, L. Sterling, P. Torroni (Eds.), Declarative Agent Languages and Techniques. XII, 281 pages. 2004. (Subseries LNAI).

Vol. 2989: S. Graf, L. Mounier (Eds.), Model Checking Software. X, 309 pages. 2004.

Vol. 2988: K. Jensen, A. Podelski (Eds.), Tools and Algorithms for the Construction and Analysis of Systems. XIV, 608 pages. 2004.

Vol. 2987: I. Walukiewicz (Ed.), Foundations of Software Science and Computation Structures. XIII, 529 pages. 2004.

Vol. 2986: D. Schmidt (Ed.), Programming Languages and Systems. XII, 417 pages. 2004.

Vol. 2985: E. Duesterwald (Ed.), Compiler Construction. X, 313 pages. 2004.

Vol. 2984: M. Wermelinger, T. Margaria-Steffen (Eds.), Fundamental Approaches to Software Engineering. XII, 389 pages. 2004.

Vol. 2983: S. Istrail, M.S. Waterman, A. Clark (Eds.), Computational Methods for SNPs and Haplotype Inference. IX, 153 pages. 2004. (Subseries LNBI).

Vol. 2982: N. Wakamiya, M. Solarski, J. Sterbenz (Eds.), Active Networks. XI, 308 pages. 2004.

Vol. 2981: C. Müller-Schloer, T. Ungerer, B. Bauer (Eds.), Organic and Pervasive Computing – ARCS 2004. XI, 339 pages. 2004.

Vol. 2980: A. Blackwell, K. Marriott, A. Shimojima (Eds.), Diagrammatic Representation and Inference. XV, 448 pages. 2004. (Subseries LNAI).

Vol. 2979: I. Stoica, Stateless Core: A Scalable Approach for Quality of Service in the Internet. XVI, 219 pages. 2004.

Vol. 2978: R. Groz, R.M. Hierons (Eds.), Testing of Communicating Systems. XII, 225 pages. 2004.

Vol. 2977: G. Di Marzo Serugendo, A. Karageorgos, O.F. Rana, F. Zambonelli (Eds.), Engineering Self-Organising Systems. X, 299 pages. 2004. (Subseries LNAI).

Vol. 2976: M. Farach-Colton (Ed.), LATIN 2004: Theoretical Informatics. XV, 626 pages. 2004.

Vol. 2973: Y. Lee, J. Li, K.-Y. Whang, D. Lee (Eds.), Database Systems for Advanced Applications. XXIV, 925 pages. 2004.